T0178212

Signals and Systems

D. Sundararajan

Signals and Systems

A Practical Approach

Second Edition

D. Sundararajan
Formerly at Concordia University
Montreal, QC, Canada

ISBN 978-3-031-19379-8 ISBN 978-3-031-19377-4 (eBook)
https://doi.org/10.1007/978-3-031-19377-4

This Springer imprint is published by the registered company Springer Nature Switzerland AG
The registered company address is: Gewerbestrasse 11, 6330 Cham, Switzerland

Preface to the Second Edition

A thorough revision of all the chapters has been carried out. Some sections have been significantly expanded. Some sections have been interchanged for a better sequence of presentation. Minor necessary corrections have been made throughout the text.

The new topics included are: (i) Bode plots, (ii) Nyquist diagrams, (iii) down-sampling, (iv) bilinear transformation, (v) diagonalization, (vi) similarity transformation, (vii) controllability, (viii) observability, (ix) complex numbers, and (x) more applications. Expanded sets of MATLAB programs are available at the book's website. Further, simulation programs are introduced to make the learning of mathematical concepts easier.

I am grateful to my editor and his team at Springer for their help and encouragement in completing the second edition of the book. I thank my family for their support during this endeavor.

D. Sundararajan

Preface to the First Edition

The increasing number of applications, requiring a knowledge of the theory of signals and systems, and the rapid developments in digital systems technology and fast numerical algorithms call for a change in the content and approach used in teaching the subject. I believe that a modern signals and systems course should emphasize the practical and computational aspects in presenting the basic theory. This approach of teaching the subject makes the student more effective in subsequent courses. In addition, the student is exposed to practical and computational solutions he will be using in his professional career. This book is my attempt to adapt the theory of signals and systems to use the digital computer efficiently as an analysis tool.

A good knowledge of the fundamentals of the analysis of signals and systems is required to specialize in such areas as signal processing, communication, and control. As most of the practical signals are continuous functions of time and digital systems are mostly used to process them due to several advantages, the study of both continuous and discrete signals and systems is required. The primary objective of writing this book is to present the fundamentals of time-domain and frequency-domain methods of signal and linear time-invariant system analysis from a practical viewpoint. As the discrete signals and systems are more often used in practice and their concepts are relatively easier to understand, for each topic, the discrete version is presented first followed by the corresponding continuous version. Typical applications of the methods of analysis are also provided. Comprehensive coverage of the transform methods, and emphasis on practical methods of analysis and physical interpretation of the concepts are the key features of this book. The well documented software, which is a supplement to this book and available on the Internet, further greatly reduces much of the difficulty in understanding the concepts. Based on this software, a laboratory course can be tailored to suit the individual course requirements.

This book is intended to be a textbook for junior undergraduate level one-semester signals and systems course. This book will also be useful for self-study. Answers to selected exercises, marked $*$, are given at the end of the book. I assume the responsibility for all the errors in this book and in the accompanying supplements, and would very much appreciate receiving readers' suggestions and pointing of any errors (email address: d_sundararajan@yahoo.com). I am grateful to my editor and his team at Wiley for their help and encouragement in completing this project. I thank my family and my friend Dr. A. Pedar for their support during this endeavor.

D. Sundararajan

Contents

Abbreviations

DC	Constant
DFT	Discrete Fourier Transform
DTFT	Discrete-Time Fourier Transform
FT	Fourier Transform
FS	Fourier Series
IDFT	Inverse Discrete Fourier Transform
Im	Imaginary part of a complex number or expression
LTI	Linear Time-Invariant
Re	Real part of a complex number or expression
ROC	Region of Convergence

Chapter 1
Discrete Signals

1.1 Introduction

Signals convey some information about the nature of some physical phenomenon. Based on that, we (the human system) or a machine take some action. Depending on the weather (atmospheric conditions in terms of temperature, wind speed, clouds, and precipitation), we decide to stay indoors or go out. A doctor decides on the state of the health of a person based on the heartbeat, the blood pressure, and the temperature of the body. A control system controls the input power to a heating system based on a preset temperature. A communication system enables an audio or video signal to reach its destination from its source. A system, in response to some input signals, takes some action or produces new signals. The systems have to be designed, and the signals have to be processed. The fundamentals of these aspects constitute the signals and systems course. As such, it is required to further study the applications in all areas of science and engineering, such as control systems, communication systems, signal and image processing systems, etc.

Signals are abundant in the applications of science and engineering. Typical signals are audio, video, biomedical, seismic, radar, vibration, communication, and sonar. In typical applications, we have to process some signals using some systems. While the applications vary from communication to control, the basic analysis and design tools are the same. In signals and systems course, we study these tools, that is, system models differential/difference equation, convolution, transfer function and state-space, and transforms Fourier transform with all its versions, z-transform and Laplace transform, and the associated basic signals. The use of these tools in the analysis of linear time-invariant (LTI) systems with deterministic signals is presented in this book. While most practical systems are nonlinear to some extent, they can be analyzed, with acceptable accuracy, assuming linearity. In addition, the analysis is much easier with this assumption. Good grounding in LTI system analysis is also essential for further study of nonlinear systems and systems with random signals.

D. Sundararajan, *Signals and Systems*,
https://doi.org/10.1007/978-3-031-19377-4_1

For most of the practical systems, the input and output signals are continuous signals and these signals can be processed using continuous systems. However, due to advances in digital systems technology and numerical algorithms, it is advantageous to process continuous signals using digital systems, systems using digital devices, by converting the input signal into a digital signal. Therefore, the study of both continuous and digital systems is required. As the implementations of most practical systems are digital and the concepts are relatively easier to understand, we describe the discrete signals and systems first immediately followed by the corresponding description of continuous signals and systems.

The major problem in signal and system analysis is their representation. Appropriate representation facilitates the analysis. The representation is common for all applications of science and engineering. The necessity for the representation is that, in their naturally occurring form, the amplitude profile of signals and the response of systems are arbitrary. Therefore, it is expedient to explore into more efficient form of representation. The signals occur usually in time-domain form. That is, signals change with respect to time. It is found that the representation of signals in the frequency domain provides efficient signal analysis. Another change that is invariably required is to digitize the signal from the naturally occurring continuous form. Therefore, most of the analysis of the digitized version of signals is carried out in the frequency domain.

The frequency-domain analysis is similar to the use of logarithms to reduce a multiplication operation into a much simpler addition operation. For example, let us try to find the product of 8 and 16. The numbers can be represented in exponential form as 2^3 and 2^4. Then,

$$8 \times 16 = 2^3 2^4 = 2^{3+4} = 2^7 = 128$$

It is assumed that a table is available to find the exponential form of the numbers. Similarly, fast algorithms are available to find the frequency-domain representation of signals. This results in faster processing. For example, the system output can be found faster in the frequency domain. Therefore, the study of signals and systems, which is common to all applications of science and engineering, primarily involves digitization and representation in the frequency domain. It is similar to the use of logarithms. The analysis looks complex due to the details involved, but it is simple in principle. One can become proficient in the indispensable signal and system analysis with sufficient paper-and-pencil and computer programming practice.

The basic problem in the study of systems is how to analyze systems with arbitrary input signals. The solution, in the case of linear time-invariant (LTI) systems, is to decompose the signal in terms of basic signals, such as the impulse or the sinusoid. Then, with the knowledge of the response of a system to these basic signals, the response of the system to any arbitrary signal, that we shall ever encounter in practice, can be obtained. Therefore, the study of the response of systems to the basic signals, along with the methods of decomposition of arbitrary signals in terms of the basic signals and modeling of practical systems, constitutes the core of the study of the analysis of systems to arbitrary input signals.

1.2 Basic Signals

A signal represents some information. Systems carry out some tasks or produce some output signals in response to some input signals. A control system may set the speed of a motor in accordance with an input signal. In a room temperature control system, the power to the heating system is regulated with respect to the room temperature. While signals may be electrical, mechanical, or any other form, they are usually converted to electrical form for processing efficiency. The speech signal is converted from a pressure signal to an electrical signal in the microphone. Signals, in almost all practical systems, have arbitrary amplitude profile. These signals must be represented in terms of simple and well-defined mathematical signals for ease of representation and processing. The response of a system is also represented in terms of these simple signals. Commonly used basic discrete signals are described in Sect. 1.2. In Sect. 1.3, signals are classified according to some properties. Discrete signal operations are presented in Sect. 1.4.

As we have already mentioned, most practical signals have arbitrary amplitude profile. These signals are, for processing efficiency, decomposed in terms of some mathematically well-defined and simple signals. These simple signals, such as the sinusoid with infinite duration, are not practical signals. However, they can be approximated to a desired accuracy. They are used as intermediaries in signal and system analysis. The values of discrete signals are defined only at discrete intervals. The discrete signal is a sequence of numbers with infinite precision. It is represented as $\{x(n)\}$, where n is the independent variable and $x(n)$ is the dependent variable. While $x(n)$ is a single value, although it is incorrect, $x(n)$ is usually used to represent the discrete sequence also.

Time-Domain Representation of Discrete Signals
There are variety of ways of representing a sequence. Consider a sequence

$$\{4.1, 3.1, 2.1, 1.1\}$$

Assuming the value 4.1 corresponds to index 0, it could be written as

$$\{x(0) = 4.1, x(1) = 3.1, x(2) = 2.1, x(3) = 1.1\}$$

or

$$\{x(n), n = 0, 1, 2, 3\} = \{4.1, 3.1, 2.1, 1.1\}$$

or

$$\{\check{4}.1, 3.1, 2.1, 1.1\}$$

or

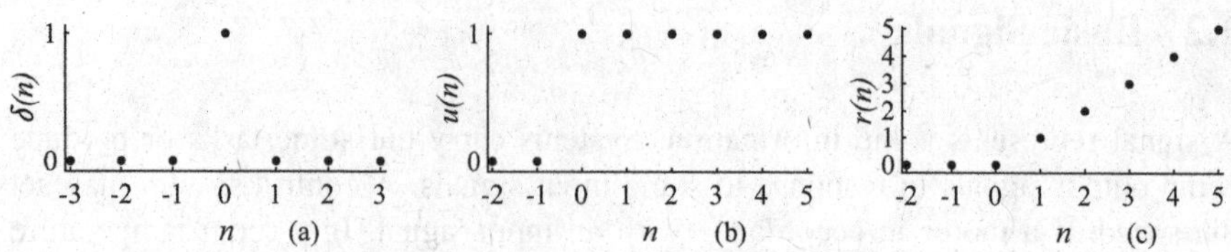

Fig. 1.1 (**a**) The unit-impulse signal, $\delta(n)$; (**b**) the unit-step signal, $u(n)$; (**c**) the unit-ramp signal, $r(n)$

$$x(n) = \begin{cases} -n + 4.1 & \text{for } n = 0, 1, 2, 3 \\ 0 & \text{otherwise} \end{cases}$$

or graphically. The check symbol ˇ indicates that the index of that element is 0 and the samples to the right have positive indices and those to the left have negative indices.

1.2.1 Unit-Impulse Signal

The unit-impulse signal, shown in Fig. 1.1a, is defined as

$$\delta(n) = \begin{cases} 1 & \text{for } n = 0 \\ 0 & \text{for } n \neq 0 \end{cases}$$

The unit-impulse signal is an all-zero sequence except that it has a value of one when its argument is equal to zero. A time-shifted unit-impulse signal $\delta(n - m)$, with argument $(n - m)$, has its only nonzero value at $n = m$. Therefore, $\sum_{n=-\infty}^{\infty} x(n)\delta(n - m) = x(m)$ is called the sampling or sifting property of the impulse. For example,

$$\sum_{n=-\infty}^{\infty} 2^n \delta(n) = 1, \sum_{n=-2}^{0} 2^n \delta(n - 1) = 0, \sum_{n=-2}^{0} 2^n \delta(-n - 1) = 0.5,$$

$$\sum_{n=-2}^{0} 2^n \delta(n + 1) = 0.5, \sum_{n=-\infty}^{\infty} 2^n \delta(n + 2) = 0.25, \sum_{n=-\infty}^{\infty} 2^n \delta(n - 3) = 8$$

In the second summation, the argument $n - 1$ of the impulse never becomes zero within the limits of the summation.

The decomposition of an arbitrary signal in terms of scaled and shifted impulses is a major application of this signal. Consider the product of a signal with a shifted impulse $x(n)\delta(n - m) = x(m)\delta(n - m)$. Summing both sides with respect to m, we get

$$\sum_{m=-\infty}^{\infty} x(n)\delta(n-m) = x(n)\sum_{m=-\infty}^{\infty} \delta(n-m) = x(n) = \sum_{m=-\infty}^{\infty} x(m)\delta(n-m)$$

The general term $x(m)\delta(n-m)$ of the last sum, which is one of the constituent impulses of $x(n)$, is a shifted impulse $\delta(n-m)$ located at $n = m$ with value $x(m)$. The summation operation sums all these impulses to form $x(n)$. Therefore, the signal $x(n)$ is represented by the sum of scaled and shifted impulses with the value of the impulse at any n being $x(n)$. The unit-impulse is the basis function and $x(n)$ is its coefficient. As the value of the sum is nonzero only at $n = m$, the sum is effective only at that point. By varying the value of n, we can sift out all the values of $x(n)$. For example, consider the signal $x(-2) = 2$, $x(0) = 3$, $x(2) = -4$, $x(3) = 1$, and $x(n) = 0$ otherwise. This signal can be expressed, in terms of impulses, as

$$x(n) = 2\delta(n+2) + 3\delta(n) - 4\delta(n-2) + \delta(n-3)$$

With $n = 2$, for instance,

$$x(2) = 2\delta(4) + 3\delta(2) - 4\delta(0) + \delta(-1) = -4$$

1.2.2 Unit-Step Signal

The unit-step signal, shown in Fig. 1.1b, is defined as

$$u(n) = \begin{cases} 1 \text{ for } n \geq 0 \\ 0 \text{ for } n < 0 \end{cases}$$

The unit-step signal is an all-one sequence for positive values of its argument and is an all-zero sequence for negative values of its argument. The causal form of a signal $x(n)$ ($x(n)$ is zero for $n < 0$) is obtained by multiplying it with the unit-step signal as $x(n)u(n)$. For example, $\sin(\frac{2\pi}{6}n)$ has nonzero values in the range $-\infty < n < \infty$, whereas the values of $\sin(\frac{2\pi}{6}n)u(n)$ are zero for $n < 0$ and $\sin(\frac{2\pi}{6}n)$ for $n \geq 0$. A shifted unit-step signal, for example, $u(n-1)$, is $u(n)$ shifted by one sample interval to the right (the first nonzero value occurs at $n = 1$). Using scaled and shifted unit-step signals, any signal, described differently over different intervals, can be specified, for easier mathematical analysis, by a single expression, valid for all n. For example, a pulse with its only nonzero values defined as $x(-1) = 2$, $x(0) = 2$, $x(1) = -3$, and $x(2) = -3$ can be expressed as $x(n) = 2u(n+1) - 5u(n-1) + 3u(n-3)$. Figure 1.2a shows the pulse and Fig. 1.2b shows its constituent scaled and shifted unit-step components.

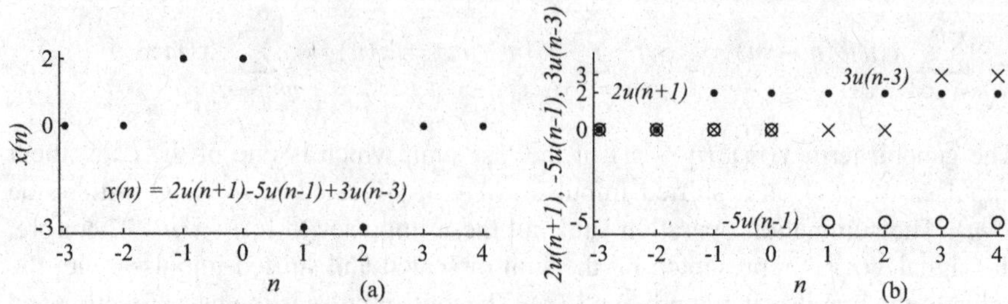

Fig. 1.2 (**a**) A pulse and (**b**) its scaled and shifted unit-step components

1.2.3 Unit-Ramp Signal

Another signal that is often used in the analysis of systems is the unit-ramp signal, shown in Fig. 1.1c. It is defined as

$$r(n) = \begin{cases} n \text{ for } n \geq 0 \\ 0 \text{ for } n < 0 \end{cases}$$

The unit-ramp signal linearly increases for positive values of its argument and is an all-zero sequence for negative values of its argument.

The three signals, the unit-impulse, the unit-step, and the unit-ramp, are closely related. The unit-impulse signal $\delta(n)$ is equal to $u(n)-u(n-1)$. The unit-step signal $u(n)$ is equal to $\sum_{k=0}^{\infty} \delta(n-k)$. The shifted unit-step signal $u(n-1)$ is equal to $r(n) - r(n-1)$. The unit-ramp signal $r(n)$ is equal to

$$r(n) = nu(n) = \sum_{k=0}^{\infty} k\delta(n-k)$$

1.2.4 Sinusoids and Exponentials

The sinusoidal waveform or sinusoid is the well-known trigonometric sine and cosine functions, with arbitrary shift along the horizontal axis. Figure 1.3 shows 32 discrete points on the unit circle, characterized by

$$x^2 + y^2 = 1$$

The unit circle is a circle with its center at the origin and radius 1. For each point on the unit circle, the cosine and sine functions are defined in terms its x and y coordinates as

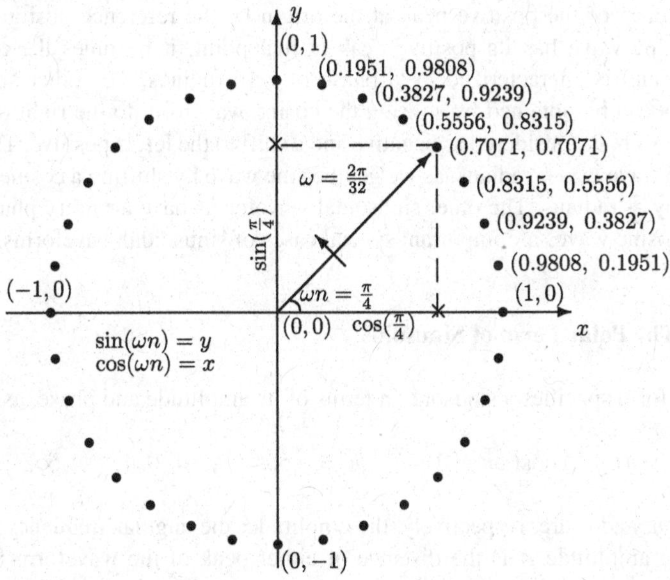

Fig. 1.3 $\cos(\omega n)$ and $\sin(\omega n)$ functions defined on the unit circle, $\omega = \frac{2\pi}{32}$ and $n = -\infty, \ldots, -1, 0, 1, \ldots, \infty$

$$\cos(\omega n) = x, \quad \text{and} \quad \sin(\omega n) = y$$

The coordinates of the points corresponding to angles

$$\omega n = \frac{2\pi}{32}\{0, 1, 2, 3, 4, 5, 6, 7, 8, 16, 24\}$$

are shown in the figure. By using appropriate signs, the coordinates in the other three quadrants can be found. The radian measure of the angle is the arc length. Since the circumference of a circle is 2π, one rotation of the circle is 2π radians. In degree measure, $2\pi = 360$ degrees. Therefore, one radian is $180/\pi$, and it is 57.2958 degrees approximately. Any function defined on a circle is a periodic function of the angle. The function has the same value after any number of rotations on the circle. For example,

$$\cos\left(\frac{\pi}{4}\right) = \cos\left(2\pi + \frac{\pi}{4}\right) = 0.7071$$

The sinusoidal waveforms are oscillatory with peaks occurring at equal distance from the horizontal axis. The waveforms have two zero-crossings in each cycle. As the sinusoidal waveforms of a particular frequency and amplitude have the same shape with the peaks occurring at different instants, we have to define a reference position to distinguish the innumerable number of different sinusoids. Let the occurrence of the positive peak at the origin be the reference position. Then, as the cosine wave has its positive peak at that point, it becomes the reference waveform and is characterized by a phase of zero radians. The other sinusoidal waveforms can be obtained by shifting the cosine waveform to the right or left. A shift to the right is considered as negative and a shift to the left is positive. The phase of the sine wave is $-\frac{\pi}{2}$ radians, as we get the sine wave by shifting a cosine wave to the right by $\frac{\pi}{2}$ radians. The other sinusoidal waveforms have arbitrary phases. The sine and cosine waves are important special cases of sinusoidal waveforms.

1.2.4.1 The Polar Form of Sinusoids

The polar form specifies a sinusoid, in terms of its amplitude and phase, as

$$x(n) = A\cos(\omega n + \theta), \qquad n = -\infty, \ldots, -1, 0, 1, \ldots, \infty,$$

where A, ω, and θ are, respectively, the amplitude, the angular frequency, and the phase. The amplitude A is the distance of either peak of the waveform from the horizontal axis ($A = 2$ for the waveform shown in Fig. 1.4a). A discrete sinusoid has to complete an integral number of cycles (say k, where $k > 0$ is an integer) over an integral number of sample points, called its period (denoted by N, where $N > 0$ is an integer), if it is periodic. Then, as

$$\cos(\omega(n + N) + \theta) = \cos(\omega n + \omega N + \theta) = \cos(\omega n + \theta) = \cos(\omega n + \theta + 2k\pi),$$

$N = \frac{2k\pi}{\omega}$. Note that k is the smallest integer that will make $\frac{2k\pi}{\omega}$ an integer. The cyclic frequency, denoted by f, of a sinusoid is the number of cycles per sample and is equal to the number of cycles the sinusoid makes in a period divided by the period, $f = \frac{k}{N} = \frac{\omega}{2\pi}$ cycles per sample. Therefore, the cyclic frequency of a discrete periodic sinusoid is a rational number. The angular frequency, the number of radians per sample, of a sinusoid is 2π times its cyclic frequency, that is, $\omega = 2\pi f$ radians per sample.

The angular frequency of the sinusoid, shown in Fig. 1.4a, is $\omega = \frac{2\pi}{8}$ radians per sample. The period of the discrete sinusoid is $N = \frac{2k\pi}{\omega} = 8$ samples, with $k = 1$. The cyclic frequency of the sinusoid $\sin\left(\frac{2\sqrt{2}\pi}{16}n + \frac{\pi}{3}\right)$ is $\frac{\sqrt{2}}{16}$. As it is an irrational number, the sinusoid is not periodic. The cyclic frequency of the sinusoid in Fig. 1.4a is $f = \frac{k}{N} = \frac{1}{8}$ cycles per sample. The phase of the sinusoid $2\cos(\frac{2\pi}{8}n + \frac{\pi}{3})$ in Fig. 1.4a is $\theta = \frac{\pi}{3}$ radians. The peak of the waveform does not occur at a sample point. If only the sample values are known, the amplitude and phase can be determined using any two adjacent sample values. As it repeats a pattern over its period, the sinusoid remains the same by a shift of an integral number of its period. A phase-shifted sine wave can be expressed in terms of a phase-shifted cosine wave as $A\sin(\omega n + \theta) = A\cos(\omega n + (\theta - \frac{\pi}{2}))$. The phase of the sinusoid

$$\sin\left(\frac{2\pi}{16}n + \frac{\pi}{3}\right) = \cos\left(\frac{2\pi}{16}n + \left(\frac{\pi}{3} - \frac{\pi}{2}\right)\right) = \cos\left(\frac{2\pi}{16}n - \frac{\pi}{6}\right)$$

is $-\frac{\pi}{6}$ radians. The phase of the sinusoid in Fig. 1.4b is $\frac{5\pi}{6}$ radians. A phase-shifted cosine wave can be expressed in terms of a phase-shifted sine wave as $A\cos(\omega n + \theta) = A\sin(\omega n + (\theta + \frac{\pi}{2}))$.

1.2.4.2 The Rectangular Form of Sinusoids

An arbitrary sinusoid is neither even- nor odd-symmetric. The even and odd components of a sinusoid are, respectively, cosine and sine waveforms. That is, a sinusoid is a linear combination of cosine and sine waveforms of the same frequency as that of the sinusoid. Expressing a sinusoid in terms of its cosine and sine components is called its rectangular form and is given as

$$A\cos(\omega n + \theta) = A\cos(\theta)\cos(\omega n) - A\sin(\theta)\sin(\omega n) = C\cos(\omega n) + D\sin(\omega n),$$

where $C = A\cos\theta$ and $D = -A\sin\theta$. The inverse relation is $A = \sqrt{C^2 + D^2}$ and $\theta = \cos^{-1}(\frac{C}{A}) = \sin^{-1}(\frac{-D}{A})$. For example,

$$2\cos\left(\frac{2\pi}{16}n + \frac{\pi}{3}\right) = \cos\left(\frac{2\pi}{16}n\right) - \sqrt{3}\sin\left(\frac{2\pi}{16}n\right)$$

$$\frac{3}{\sqrt{2}}\cos\left(\frac{2\pi}{16}n\right) + \frac{3}{\sqrt{2}}\sin\left(\frac{2\pi}{16}n\right) = 3\cos\left(\frac{2\pi}{16}n - \frac{\pi}{4}\right)$$

1.2.4.3 The Sum of Sinusoids of the Same Frequency

The sum of sinusoids of arbitrary amplitudes and phases but with the same frequency is also a sinusoid of the same frequency. Let

$$x_1(n) = A_1\cos(\omega n + \theta_1) \qquad \text{and} \qquad x_2(n) = A_2\cos(\omega n + \theta_2)$$

Then,

$$\begin{aligned} x(n) &= x_1(n) + x_2(n) = A_1\cos(\omega n + \theta_1) + A_2\cos(\omega n + \theta_2) \\ &= \cos(\omega n)(A_1\cos(\theta_1) + A_2\cos(\theta_2)) - \sin(\omega n)(A_1\sin(\theta_1) + A_2\sin(\theta_2)) \\ &= A\cos(\omega n + \theta) = \cos(\omega n)(A\cos(\theta)) - \sin(\omega n)(A\sin(\theta)) \end{aligned}$$

Solving for A and θ, we get

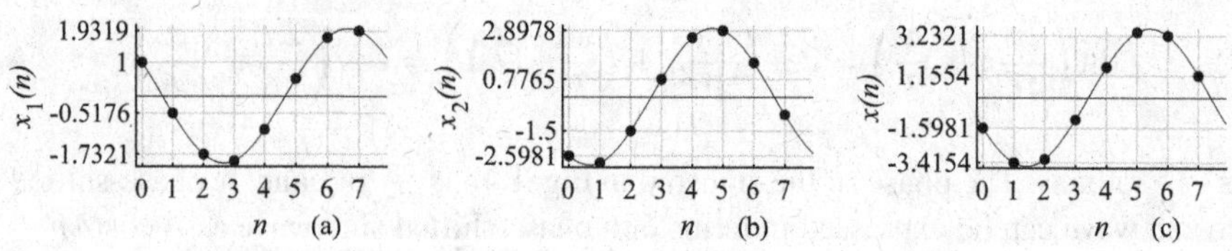

Fig. 1.4 **(a)** The sinusoid $x_1(n) = 2\cos(\frac{2\pi}{8}n + \frac{\pi}{3})$; **(b)** the sinusoid $x_2(n) = 3\cos(\frac{2\pi}{8}n + \frac{5\pi}{6})$; **(c)** the sum of $x_1(n)$ and $x_2(n)$, $x(n) = 3.6056\cos(\frac{2\pi}{8}n + 2.03)$

$$A = \sqrt{A_1^2 + A_2^2 + 2A_1A_2\cos(\theta_1 - \theta_2)}$$

$$\theta = \tan^{-1}\frac{A_1\sin(\theta_1) + A_2\sin(\theta_2)}{A_1\cos(\theta_1) + A_2\cos(\theta_2)}.$$

Any number of sinusoids can be combined into a single sinusoid by repeatedly using the formulas. Note that the formula for the rectangular form of the sinusoid is a special case of the sum of two sinusoids, one sinusoid being the cosine and the other being sine.

Example 1.1 Determine the sum of the two sinusoids $x_1(n) = 2\cos(\frac{2\pi}{8}n + \frac{\pi}{3})$ and $x_2(n) = -3\cos(\frac{2\pi}{8}n - \frac{\pi}{6})$.

Solution As $x_2(n) = -3\cos(\frac{2\pi}{8}n - \frac{\pi}{6}) = 3\cos(\frac{2\pi}{8}n - \frac{\pi}{6} + \pi) = 3\cos(\frac{2\pi}{8}n + \frac{5\pi}{6})$,

$$A_1 = 2, \quad A_2 = 3, \quad \theta_1 = \frac{\pi}{3}, \text{ and } \theta_2 = \frac{5\pi}{6}.$$

Substituting the numerical values in the equations, we get

$$A = \sqrt{2^2 + 3^2 + 2(2)(3)\cos\left(\frac{\pi}{3} - \frac{5\pi}{6}\right)} = \sqrt{13} = 3.6056$$

$$\theta = \tan^{-1}\frac{2\sin\left(\frac{\pi}{3}\right) + 3\sin\left(\frac{5\pi}{6}\right)}{2\cos\left(\frac{\pi}{3}\right) + 3\cos\left(\frac{5\pi}{6}\right)} = 2.03 \text{ radians}$$

The waveforms of the two sinusoids and their sum, $x(n) = 3.6056\cos(\frac{2\pi}{8}n + 2.03)$, are shown, respectively, in Fig. 1.4a, b, and c. ∎

1.2.4.4 Exponentials

A real constant $a \neq 1$ raised to the power of a variable n, $x(n) = a^n$ is the exponential function. We are more familiar with the exponential of the form e^{-2t}

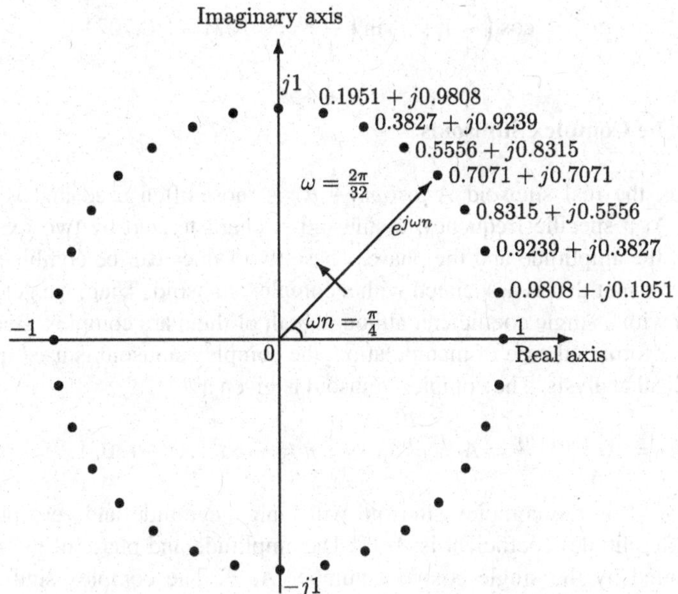

Fig. 1.5 $e^{j\omega n}$, the unit rotating vector, $\omega = \frac{2\pi}{32}$ and $n = -\infty, \dots, -1, 0, 1, \dots, \infty$

with base e, and this form is used in the analysis of continuous signals and systems. The exponential e^{sn} is the same as a^n, where $s = \log_e a$ and $a = e^s$. For example, $e^{-0.2231n} = (0.8)^n$ is a decaying discrete exponential. As both the forms are used in the analysis of discrete signals and systems, it is necessary to get used to both of them.

With base e, the most general form of the continuous exponential is Pe^{st}, where P or s or both may be complex-valued. Let $s = \sigma + j\omega$. Then, $e^{st} = e^{(\sigma+j\omega)t} = e^{\sigma t}e^{j\omega t}$. Exponential $e^{j\omega t} = \cos(\omega t) + j\sin(\omega t)$ is a constant amplitude oscillating signal with the frequency of oscillation in the range $0 \le \omega \le \infty$. When the real part of s is positive ($\sigma > 0$), e^{st} is a growing exponential. When $\sigma < 0$, e^{st} is a decaying exponential. When $\sigma = 0$, e^{st} oscillates with constant amplitude. When $s = 0$, e^{st} is a constant signal.

With base a, the most general form of the discrete exponential is Pa^n, where P or a or both may be complex-valued. Let $a = r\,e^{j\omega}$. Then, $a^n = r^n e^{j\omega n}$. When $|a| = r > 1$, a^n is a growing exponential. When $|a| = r < 1$, a^n is a decaying exponential. When $|a| = r = 1$, a^n is a constant amplitude signal.

Exponential $e^{j\omega n} = \cos(\omega n) + j\sin(\omega n)$, shown in Fig. 1.5 with $\omega = \frac{2\pi}{32}$, is a constant amplitude oscillating signal with the frequency of oscillation in the range $0 \le \omega \le 2\pi$, since $e^{\pm j\omega n} = e^{j(2\pi\pm\omega)n} = e^{j(4\pi\pm\omega)n} = \cdots$. Essentially, it carries the same information about the cosine and sine functions, shown in Fig. 1.3, in a complex form, which is an ordered pair of real functions. For example,

$$e^{j\frac{\pi}{4}} = \cos\left(\frac{\pi}{4}\right) + j\sin\left(\frac{\pi}{4}\right) = 0.7071 + j0.7071$$

1.2.4.5 The Complex Sinusoids

In practice, the real sinusoid $A\cos(\omega n + \theta)$ is most often used and is easy to visualize. At a specific frequency, a sinusoid is characterized by two real-valued quantities, the amplitude and the phase. These two values can be combined into a complex constant that is associated with a complex sinusoid. Then, we get a single waveform with a single coefficient, although both of them are complex. Because of its compact form and ease of manipulation, the complex sinusoid is used in almost all theoretical analysis. The complex sinusoid is given as

$$x(n) = Ae^{j(\omega n+\theta)} = Ae^{j\theta}e^{j\omega n}, \qquad n = -\infty, \ldots, -1, 0, 1, \ldots, \infty$$

The term $e^{j\omega n}$ is the complex sinusoid with unit magnitude and zero phase. Its complex (amplitude) coefficient is $Ae^{j\theta}$. The amplitude and phase of the sinusoid is represented by the single complex number $Ae^{j\theta}$. The complex sinusoid is a functionally equivalent mathematical representation of a real sinusoid. By adding its complex conjugate, $Ae^{-j(\omega n+\theta)}$, with itself and dividing by two, due to Euler's identity, we get

$$x(n) = \frac{A}{2}\left(e^{j(\omega n+\theta)} + e^{-j(\omega n+\theta)}\right) = A\cos(\omega n + \theta)$$

The use of two complex sinusoids to represent a single real sinusoid requires four real quantities instead of two. This redundancy in terms of storage and operations can be taken care of in the implementation of algorithms for processing the exponentials. Figure 1.6b shows the complex sinusoid $e^{j(\frac{2\pi}{16}n)}$ with complex coefficient $1e^{j\frac{\pi}{3}}$.

One cycle of the continuous complex exponential signal, $x(t) = e^{j(\frac{2\pi}{16}t+\frac{\pi}{3})}$, is shown in Fig. 1.6a. We denote a continuous signal, using the independent variable t, as $x(t)$. We call this representation the time-domain representation, although the independent variable is not time for some signals. Using Euler's identity, the signal

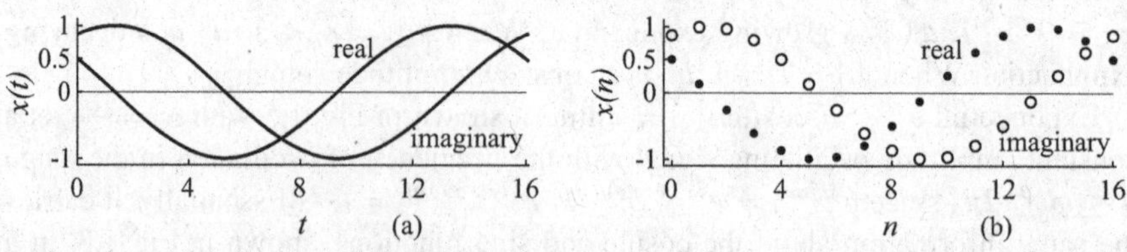

Fig. 1.6 **(a)** The continuous complex exponential signal, $x(t) = e^{j(\frac{2\pi}{16}t+\frac{\pi}{3})}$; **(b)** the discrete complex exponential signal, $x(n) = e^{j(\frac{2\pi}{16}n+\frac{\pi}{3})}$

can be expressed, in terms of cosine and sine signals, as

$$x(t) = e^{j(\frac{2\pi}{16}t+\frac{\pi}{3})} = \cos\left(\frac{2\pi}{16}t + \frac{\pi}{3}\right) + j\sin\left(\frac{2\pi}{16}t + \frac{\pi}{3}\right)$$

The real part of $x(t)$ is the real sinusoid $\cos(\frac{2\pi}{16}t + \frac{\pi}{3})$, and the imaginary part is the real sinusoid $\sin(\frac{2\pi}{16}t + \frac{\pi}{3})$, as any complex signal is an ordered pair of real signals. While practical signals are real-valued with arbitrary amplitude profile, the mathematically well-defined complex exponential is predominantly used in signal and system analysis. One cycle of the discrete complex exponential signal, $x(n) = e^{j(\frac{2\pi}{16}n+\frac{\pi}{3})}$, is shown in Fig. 1.6b. This signal is obtained by sampling the continuous signal $x(t)$, replacing t by nT_s, in Fig. 1.6a with $T_s = 1$ second.

1.2.4.6 Exponentially Varying Amplitude Sinusoids

An exponentially varying amplitude sinusoid, $Ar^n \cos(\omega n + \theta)$, is obtained by multiplying a sinusoidal sequence, $A\cos(\omega n+\theta)$, with a real exponential sequence, r^n. The more familiar constant amplitude sinusoid results when the base of the real exponential, r, is equal to one. If ω is equal to zero, then we get real exponential sequences.

Sinusoid, $x(n) = (0.9)^n \cos(\frac{2\pi}{8}n)$, with exponentially decreasing amplitude is shown in Fig. 1.7a. The amplitude of the sinusoid $\cos(\frac{2\pi}{8}n)$ is constrained by the exponential $(0.9)^n$. When the value of the cosine function is equal to one, the waveform reduces to $(0.9)^n$. Therefore, the graph of the function $(0.9)^n$ is the envelope of the positive peaks of the waveform, as shown in Fig. 1.7a. Similarly, the graph of the function $-(0.9)^n$ is the envelope of the negative peaks of the waveform. Sinusoid, $x(n) = (1.1)^n \cos(\frac{2\pi}{8}n)$, with exponentially increasing amplitude is shown in Fig. 1.7b.

The complex exponential representation of an exponentially varying amplitude sinusoid is given as

$$x(n) = \frac{A}{2}r^n\left(e^{j(\omega n+\theta)} + e^{-j(\omega n+\theta)}\right) = Ar^n\cos(\omega n + \theta)$$

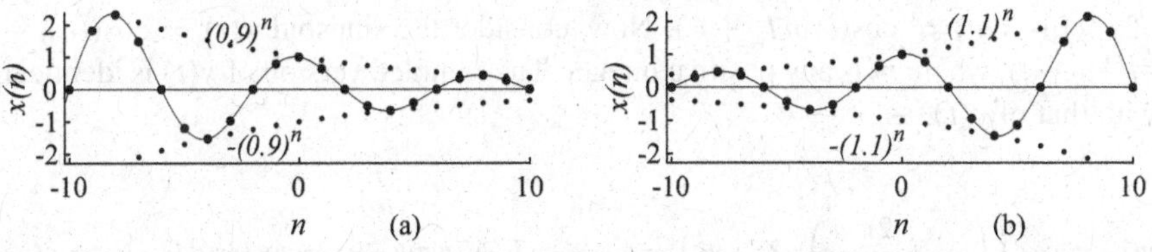

Fig. 1.7 **(a)** Exponentially decreasing amplitude sinusoid, $x(n) = (0.9)^n \cos(\frac{2\pi}{8}n)$; **(b)** exponentially increasing amplitude sinusoid, $x(n) = (1.1)^n \cos(\frac{2\pi}{8}n)$

1.2.4.7 The Sampling Theorem and the Aliasing Effect

As we have already mentioned, most of the practical signals are continuous signals. However, digital signal processing is so advantageous that we prefer to convert the continuous signals into digital form and then process it. This process involves sampling in time and in amplitude of a signal. The sampling in time involves observing the signal only at discrete instants of time. By sampling a signal, we are reducing the number of samples from infinite (of the continuous signal over any finite duration) to finite (of the corresponding discrete signal over the same duration). This reduction in the number of samples restricts the ability to represent rapid time variations of a signal and, consequently, reduces the effective frequency range of discrete signals. Note that high-frequency components of a signal provide its rapid variations. As practical signals have negligible spectral values beyond some finite frequency range, the representation of a continuous signal by a finite set of samples is possible, satisfying a required accuracy. Therefore, we should be able to determine the sampling interval required for a specific signal.

The sampling theorem states that a continuous signal $x(t)$ can be uniquely determined from its sampled version $x(n)$ if the sampling interval T_s is less than $\frac{1}{2f_m}$, where f_m is the cyclic frequency of the highest-frequency component of $x(t)$. That is, a signal bandlimited to f_m Hz can be reconstructed from its samples taken at a sampling frequency f_s greater than $2f_m$ samples/second. This implies that there are more than two samples per cycle of the highest-frequency component. Therefore, a sinusoid, which completes f cycles, has a distinct set of $2f+1$ sample values. A cosine wave, however, can be represented with $2f$ samples. For example, the cyclic frequency of the sinusoid $x(t) = \cos(3(2\pi)t - \frac{\pi}{3})$ is $f = \frac{3(2\pi)}{2\pi} = 3\,\text{Hz}$ and, therefore, $T_s < \frac{1}{2(3)} = \frac{1}{6}$ seconds, and the minimum sampling frequency is $f_s = \frac{1}{T_s} = 2f + 1 = 6 + 1 = 7$ samples per second. In practice, due to nonideal response of physical devices, the sampling frequency used is typically more than twice the theoretical minimum. Given a sampling interval T_s, the cyclic frequency f_m of the highest-frequency component of $x(t)$, for the unambiguous representation of its sampled version, must be less than $\frac{1}{2T_s}$. The corresponding angular frequency ω_m is equal to $2\pi f_m < \frac{\pi}{T_s}$ radians per second. Therefore, the frequency range of the frequency components of the signal $x(t)$, for the unambiguous representation of its sampled version, must be $0 \leq \omega < \frac{\pi}{T_s}$.

To find out why the frequency range is limited, due to sampling of a signal, consider the sinusoid $x(t) = \cos(\omega_0 t + \theta)$ with $0 \leq \omega_0 < \frac{\pi}{T_s}$. The sampled version of $x(t)$ is $x(n) = \cos(\omega_0 n T_s + \theta)$. Now, consider the sinusoid $y(t) = \cos((\omega_0 + \frac{2\pi m}{T_s})t + \theta)$, where m is any positive integer. The sampled version of $y(t)$ is identical with that of $x(t)$, as

$$y(n) = \cos\left(\left(\omega_0 + \frac{2\pi m}{T_s}\right)nT_s + \theta\right) = \cos(\omega_0 n T_s + 2\pi nm + \theta) = \cos(\omega_0 n T_s + \theta) = x(n)$$

Therefore, the effective frequency range is limited to $\frac{2\pi}{T_s}$.

Now, consider the sinusoid $z(t) = \cos((\frac{2\pi m}{T_s} - \omega_0)t - \theta)$, where m is any positive integer. The sampled version of $z(t)$ is identical with that of $x(t)$, as

$$z(n) = \cos\left(\left(\frac{2\pi m}{T_s} - \omega_0\right) nT_s - \theta\right) = \cos(2\pi nm - \omega_0 nT_s - \theta) = \cos(\omega_0 nT_s + \theta) = x(n)$$

We conclude that it is impossible to differentiate between the sampled versions of two continuous sinusoids with the sum or difference of their angular frequencies equal to an integral multiple of $\frac{2\pi}{T_s}$. Therefore, the effective frequency range is further limited to $\frac{\pi}{T_s}$, as given by the sampling theorem. For example, with 512 samples, the uniquely identifiable frequency components are

$$x(n) = \cos\left(\frac{2\pi}{512}kn + \theta\right), \quad k = 0, 1, \ldots, 255$$

The frequency $\frac{\pi}{T_s}$ is called the folding frequency, since higher frequencies are folded back and forth into the frequency range from zero to $\frac{\pi}{T_s}$. The problem is similar to the number of bits available and the possible range of binary numbers. With N bits, unsigned binary numbers in the range 0 to $2^N - 1$ can only be unambiguously represented. Similarly, with N samples, sinusoids with frequencies in the range 0 to $(N/2) - 1$ can only be represented.

Consider the continuous sinusoids $x(t) = \cos(2\pi t + \frac{\pi}{3})$ and $x(t) = \cos(5(2\pi)t + \frac{\pi}{3})$, and their sampled versions, obtained from the corresponding continuous sinusoids by replacing t by $nT_s = n\frac{1}{4}$ with the sampling interval $T_s = \frac{1}{4}$ seconds, $x(n) = \cos(\frac{2\pi}{4}n + \frac{\pi}{3})$ and $x(n) = \cos(5\frac{2\pi}{4}n + \frac{\pi}{3})$, shown in Fig. 1.8. We can easily distinguish one continuous sinusoid from the other, as they are clearly different. However, the set of sample values, shown by dots, of the two discrete sinusoids are the same, and it is impossible to differentiate them. The sample values of both the sinusoids are the same, since

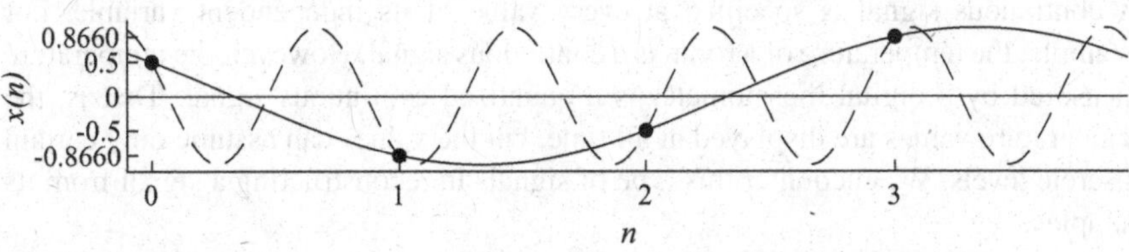

Fig. 1.8 The continuous sinusoids $x(t) = \cos(2\pi t + \frac{\pi}{3})$ and $x(t) = \cos(5(2\pi)t + \frac{\pi}{3})$, and their sampled versions, with the sampling interval $T_s = \frac{1}{4}$ seconds, $x(n) = \cos(\frac{2\pi}{4}n + \frac{\pi}{3})$ and $x(n) = \cos(5\frac{2\pi}{4}n + \frac{\pi}{3}) = \cos(\frac{2\pi}{4}n + \frac{\pi}{3})$

$$\cos\left(5\frac{2\pi}{4}n + \frac{\pi}{3}\right) = \cos\left((4+1)\frac{2\pi}{4}n + \frac{\pi}{3}\right) = \cos\left(\frac{2\pi}{4}n + \frac{\pi}{3}\right)$$

With the sampling interval $T_s = \frac{1}{4}$ seconds, the effective frequency range is limited to $\frac{\pi}{T_s} = 4\pi$. Therefore, the continuous sinusoid $\cos(5(2\pi)t + \frac{\pi}{3})$, with its angular frequency 10π greater than the folding frequency 4π, appears as or impersonates a lower-frequency discrete sinusoid. The impersonation of high-frequency continuous sinusoids as low-frequency discrete sinusoids, due to insufficient number of samples in a cycle (the sampling interval is not short enough), is called the aliasing effect.

As only scaling of the frequency axis is required for any other sampling interval, most of the analysis of discrete signals is carried out assuming that the sampling interval is 1 s. The effective frequency range becomes 0 to π, and it is called half the fundamental range. Low frequencies are those near zero and high frequencies are those near π. The range, 0 to 2π or $-\pi$ to π, is called the fundamental range of frequencies.

1.2.4.8 Frequency-Sampling Theorem

A signal time-limited to T seconds can be reconstructed from its spectral samples taken at frequency intervals not greater than $\Delta f = 1/T$ Hz.

1.3 Classification of Signals

Signals are classified into different types, and the representation and processing of a signal depend on its type.

1.3.1 Continuous, Discrete, and Digital Signals

A continuous signal is specified at every value of its independent variable. For example, the temperature of a room is a continuous signal. However, the temperature measured by a digital thermometer is a quantized continuous signal. That is, the temperature values are displayed at all time, but the values can assume only certain discrete levels. We encounter this type of signals in reconstructing a signal from its samples.

A discrete signal is specified only at discrete values of its independent variable. For example, a signal $x(t)$ is represented only at $t = nT_s$ as $x(nT_s)$, where T_s is the sampling interval and n is an integer. The discrete signal is usually denoted as $x(n)$, suppressing T_s in the argument of $x(nT_s)$. The important advantage of discrete signals is that they can be stored and processed efficiently using digital devices

Table 1.1 Signal classification

Characteristic	Continuous	Sampled
Unquantized	Continuous	Discrete
Quantized	Quantized continuous	Digital

and fast numerical algorithms. As most practical signals are continuous signals, the discrete signal is often obtained by sampling the continuous signal. However, signals such as yearly population of a country and monthly sales of a company are inherently discrete signals. Whether a discrete signal arises inherently or by sampling, it is represented as a sequence of numbers $\{x(n), \ -\infty < n < \infty\}$, where the independent variable n is an integer. Although $x(n)$ represents a single sample, it is also used to denote the sequence instead of $\{x(n)\}$. In this book, we assume that the sampling interval, T_s, is a constant. In sampling a signal, the sampling interval, which depends on the frequency content of the signal, is an important parameter. The sampling interval is required again to convert the discrete signal back to its corresponding continuous form. However, when the signal is in discrete form, most of the processing is independent of the sampling interval. For example, summing of a set of samples of a signal is independent of the sampling interval.

When the sample values of a discrete signal are quantized, it becomes a digital signal. That is, both the dependent and independent variables of a digital signal are in discrete form. This form is actually used to process signals using digital devices, such as a digital computer. Table 1.1 shows the signal classification based on sampling the amplitude and time. Figure 1.9a shows a continuous-time signal. This type of signals, with continuum of values both for its dependent and independent variables, often occurs in practice. However, it is often converted to other types for the convenience of analysis and implementation. When the independent variable only is discrete, it is called a discrete-time signal. Figure 1.9b shows this type of signal with its sample values known only at the interval of 0.1 s. When the dependent variable only is discrete, it is called a quantized continuous-time signal. Figure 1.9c shows this type of signal with its sample values known only at certain discrete levels. When both the dependent and independent variables are discrete, it is called a digital signal. Figure 1.9d shows this type of signal with its sample values known only at certain time points with a predefined discrete levels. Discretization of both the dependent and independent variables becomes necessary in practical signal processing. Of course, sampling introduces some error, which has to be kept sufficiently low by appropriately choosing the sampling intervals.

1.3.2 *Periodic and Aperiodic Signals*

The smallest positive integer $N > 0$ satisfying the condition $x(n + N) = x(n)$, for all n, is the period of the periodic signal $x(n)$. Over the interval $-\infty < n < \infty$, a periodic signal repeats its values in any interval equal to its period, at intervals of

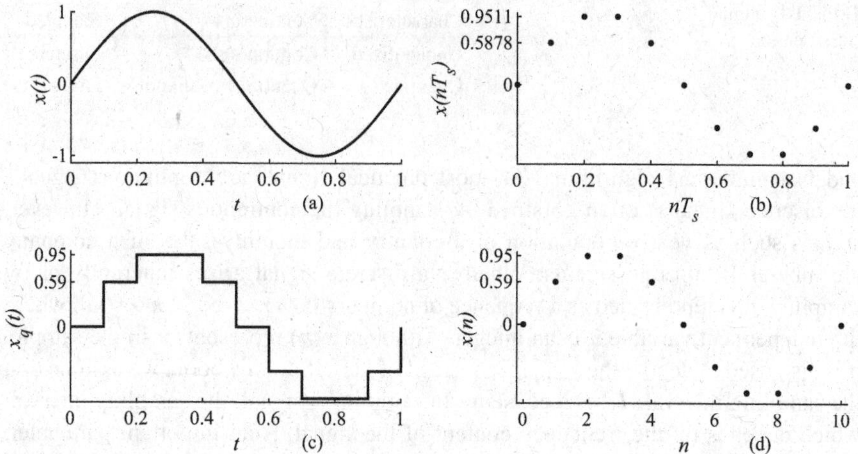

Fig. 1.9 (**a**) Continuous-time signal; (**b**) discrete-time signal; (**c**) quantized continuous-time signal; (**d**) digital signal

its period. Cosine and sine waves, and the complex exponential, shown in Figs. 1.4 and 1.6, are typical examples of a periodic signal. A signal with constant value (DC) is periodic with any period. In Fourier analysis, it is considered as $A\cos(\omega n)$ or $Ae^{j\omega n}$ with the frequency ω equal to zero (period equal to ∞).

If a discrete signal $x(n)$ satisfies the condition

$$x(k \bmod N) = x(n) \quad \text{for all} \quad n,$$

then it is said to be periodic, $k = lN + n$, and l is an integer. This alternate definition emphasizes the cyclic nature of the signal. The smallest $N > 0$ satisfying the constraint is its period. The mod function, $r = k \bmod N$, yields the remainder r of dividing k by N.

$$r = k - \lfloor (k/N) \rfloor N, \quad N \neq 0$$

and r has the same sign as n. The floor function rounds the number to the nearest integer less than or equal to its argument. For example, with $N = 4$, $r = k \bmod 4$ yields

k	−8	−7	−6	−5	−4	−3	−2	−1	0	1	2	3	4	5	6	7
r	0	1	2	3	0	1	2	3	0	1	2	3	0	1	2	3

with a negative k, the corresponding positive value in the range 0 to $N - 1$ is obtained. For example, with $k = -2$,

$$\lfloor (-2/4) \rfloor 4 = \lfloor (-0.5) \rfloor 4 = (-1)4 = -4 \quad \text{and} \quad r = -2 - (-4) = 2$$

with $k = -7$, we get

$$\lfloor(-7/4)\rfloor 4 = \lfloor(-1.75)\rfloor 4 = (-2)4 = -8 \quad \text{and} \quad r = -7 - (-8) = 1$$

$r = k \bmod N$ is periodic with period N since

$$r = k \bmod N = (k + N) \bmod N$$

Let $x(n) = \{\check{2}, -1, 3, 5\}$ a periodic sequence with period 4. The check mark on 2 indicates that its index is zero. That is, $x(0) = 2$. The 17th number in the sequence is -1. The index is obtained as $17 \bmod 4 = 1$ and $x(1) = -1$. $x(-5) = x(3) = 5$.

When the period of a periodic signal approaches infinity, there is no repetition of a pattern indefinitely, and it degenerates into an aperiodic signal. Typical aperiodic signals are shown in Fig. 1.1. It is easier to decompose an arbitrary signal in terms of some periodic signals, such as complex exponentials, and the input-output relationship of LTI systems becomes a multiplication operation for this type of input signals. For these reasons, most of the analysis of practical signals, which are mostly aperiodic having arbitrary amplitude profile, is carried out using periodic basic signals.

1.3.3 Energy and Power Signals

The power or energy of a signal is also as important as its amplitude in its characterization. This measure involves the amplitude and the duration of the signal. Devices, such as amplifiers, transmitters, and motors, are specified by their output power. In signal processing systems, the desired signal is usually mixed up with certain amount of noise. The quality of these systems is indicated by the signal to noise power ratio. Note that noise signals, which are typically of random type, are usually characterized by their average power. In the most common signal approximation method, the Fourier analysis, the goodness of the approximation improves as more and more frequency components are used to represent a signal. The quality of the approximation is measured in terms of the square error, which is an indicator of the difference between the energy or power of a signal and that of its approximate version.

The instantaneous power dissipated in a resistor of one ohm is $x^2(t)$, where $x(t)$ may be the voltage across it or the current through it. By integrating the power over the interval the power is applied, we get the energy dissipated. Similarly, the sum of the squared magnitude of the values of a discrete signal $x(n)$ is an indicator of its energy and is given as

$$E = \sum_{n=-\infty}^{\infty} |x(n)|^2$$

The use of the magnitude $|x(n)|$ makes the expression applicable to complex signals as well. Due to the squaring operation, the energy of a signal $2x(n)$, with double the amplitude, is four times as that of $x(n)$. Aperiodic signals with finite energy are called energy signals. The energy of $x(n) = 4(0.5)^n, \ n \geq 0$ is

$$E = \sum_{n=0}^{\infty} |4(0.5)^n|^2 = \frac{16}{1 - 0.25} = \frac{64}{3}$$

If the energy of a signal is infinite, then it may be possible to characterize it in terms of its average power. The average power is defined as

$$P = \lim_{N \to \infty} \frac{1}{2N + 1} \sum_{n=-N}^{N} |x(n)|^2$$

For a periodic signal with period N, the average power can be determined as

$$P = \frac{1}{N} \sum_{n=0}^{N-1} |x(n)|^2$$

Signals, periodic or aperiodic, with finite average power are called power signals. Cosine and sine waveforms are typical examples of power signals. The average power of the cosine wave $2\cos(\frac{2\pi}{4}n)$ is

$$P = \frac{1}{4} \sum_{n=0}^{3} |x(n)|^2 = \frac{1}{4}(2^2 + 0^2 + (-2)^2 + 0^2) = 2$$

A signal is an energy signal or a power signal, since the average power of an energy signal is zero while that of a power signal is finite. Signals with infinite average power and infinite energy, such as $x(n) = n, \ 0 \leq n < \infty$, are neither power signals nor energy signals. The measures of signal power and energy are indicators of the signal size, since the actual energy or power depends on the load.

1.3.4 Even- and Odd-Symmetric Signals

Figure 1.10a shows a signal, which is neither even- nor odd-symmetric. The storage and processing requirements of a signal can be reduced by exploiting its symmetry. A signal $x(n)$ is even-symmetric, if $x(-n) = x(n)$ for all n. The signal is symmetrical about the vertical axis at the origin. The cosine waveform, shown in Fig. 1.10b, is the even-symmetric component of the signal in (a). A signal $x(n)$ is odd-symmetric, if $x(-n) = -x(n)$ for all n. The signal is asymmetrical about

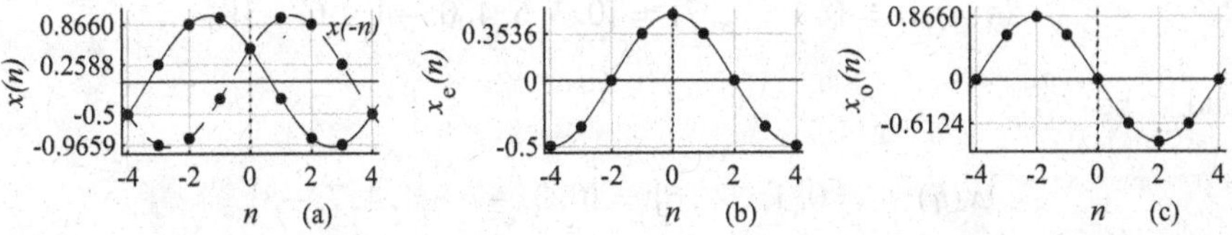

Fig. 1.10 (**a**) The sinusoid $x(n) = \cos(\frac{2\pi}{8}n + \frac{\pi}{3})$ and its time reversed version $x(-n)$; (**b**) its even component $x_e(n) = \frac{1}{2}\cos(\frac{2\pi}{8}n)$; (**c**) its odd component $x_0(n) = -\frac{\sqrt{3}}{2}\sin(\frac{2\pi}{8}n)$

the vertical axis at the origin. For an odd-symmetric signal, $x(0) = 0$. The sine waveform, shown in Fig. 1.10c, is the odd-symmetric component of the signal in (a).

If a N-point signal $x(n)$ satisfies the condition

$$x((-n) \bmod N) = x(n) \quad \text{for all} \quad n,$$

then it is said to be circularly even. That is, its periodic extension is an even-symmetric signal. For example, the sequences

$$\{x(n), n = 0, 1, \ldots, 7\} = \{-7, 1, 7, 5, 3, 5, 7, 1\}$$

and

$$\{x(n), n = 0, 1, \ldots, 6\} = \{-7, 1, 7, 5, 5, 7, 1\}$$

are even. The values at the beginning and at the middle can be arbitrary for a signal with even number of elements, and the other values satisfy $x(n) = x(-n)$, when placed on a circle. Considering the finite extent alone, the condition is

$$x(N - n) = x(n), \quad 1 \leq n \leq N - 1$$

The cosine waveform, with period N,

$$x(n) = \cos\left(\frac{2\pi}{N}n\right)$$

is even.

If a N-point signal $x(n)$ satisfies the condition

$$-x((-n) \bmod N) = x(n) \quad \text{for all} \quad n,$$

then it is said to be circularly odd. That is, its periodic extension is an odd-symmetric signal. For example, the sequences

$$\{x(n), n = 0, 1, \ldots, 7\} = \{0, 3, 6, 4, 0, -4, -6, -3\}$$

and

$$\{x(n), n = 0, 1, \ldots, 6\} = \{0, 3, -7, -4, 4, 7, -3\}$$

are odd. The values at the beginning and at the middle must be zero for a signal with even number of elements, and the other values satisfy $x(n) = -x(-n)$, when placed on a circle. Considering the finite extent alone, the condition is

$$x(N-n) = -x(n), \quad 1 \leq n \leq N-1$$

The sine waveform, with period N,

$$x(n) = \sin\left(\frac{2\pi}{N}n\right)$$

is odd.

The sum $(x(n) + y(n))$ of two odd-symmetric signals, $x(n)$ and $y(n)$, is an odd-symmetric signal, since $x(-n) + y(-n) = -x(n) - y(n) = -(x(n) + y(n))$. For example, the sum of two sine signals is an odd-symmetric signal. The sum $(x(n) + y(n))$ of two even-symmetric signals, $x(n)$ and $y(n)$, is an even-symmetric signal, since $x(-n)+y(-n) = (x(n)+y(n))$. For example, the sum of two cosine signals is an even-symmetric signal. The sum $(x(n) + y(n))$ of an odd-symmetric signal $x(n)$ and an even-symmetric signal $y(n)$ is neither even-symmetric nor odd-symmetric, since $x(-n) + y(-n) = -x(n) + y(n) = -(x(n) - y(n))$. For example, the sum of cosine and sine signals with nonzero amplitudes is neither even-symmetric nor odd-symmetric.

Since $x(n)y(n) = (-x(-n))(-y(-n)) = x(-n)y(-n)$, the product of two odd-symmetric or two even-symmetric signals is an even-symmetric signal. The product $z(n) = x(n)y(n)$ of an odd-symmetric signal $y(n)$ and an even-symmetric signal $x(n)$ is an odd-symmetric signal, since $z(-n) = x(-n)y(-n) = x(n)(-y(n)) = -z(n)$.

An arbitrary signal $x(n)$ can always be decomposed in terms of its even-symmetric and odd-symmetric components, $x_e(n)$ and $x_o(n)$, respectively. That is, $x(n) = x_e(n) + x_o(n)$. Replacing n by $-n$, we get $x(-n) = x_e(-n) + x_o(-n) = x_e(n) - x_o(n)$. Solving for $x_e(n)$ and $x_o(n)$, we get

$$x_e(n) = \frac{x(n) + x(-n)}{2} \quad \text{and} \quad x_o(n) = \frac{x(n) - x(-n)}{2}$$

As the sum of an odd-symmetric signal $x_o(n)$, over symmetric limits, is zero,

$$\sum_{n=-N}^{N} x_o(n) = 0$$

For an even-symmetric signal,

$$\sum_{n=-N}^{N} x(n) = \sum_{n=-N}^{N} x_e(n) = x_e(0) + 2\sum_{n=1}^{N} x_e(n)$$

For example, the even-symmetric component of $x(n) = \cos(\frac{2\pi}{8}n + \frac{\pi}{3})$ is

$$\begin{aligned} x_e(n) &= \frac{x(n) + x(-n)}{2} = \frac{\cos(\frac{2\pi}{8}n + \frac{\pi}{3}) + \cos(\frac{2\pi}{8}(-n) + \frac{\pi}{3})}{2} \\ &= \frac{2\cos(\frac{2\pi}{8}n)\cos(\frac{\pi}{3})}{2} = \frac{\cos(\frac{2\pi}{8}n)}{2} \end{aligned}$$

The odd-symmetric component is

$$\begin{aligned} x_o(n) &= \frac{x(n) - x(-n)}{2} = \frac{\cos(\frac{2\pi}{8}n + \frac{\pi}{3}) - \cos(\frac{2\pi}{8}(-n) + \frac{\pi}{3})}{2} \\ &= \frac{-2\sin(\frac{2\pi}{8}n)\sin(\frac{\pi}{3})}{2} = -\frac{\sqrt{3}}{2}\sin(\frac{2\pi}{8}n) \end{aligned}$$

The sinusoid $x(n)$ and its time reversed version $x(-n)$, its even component, and its odd component are shown, respectively, in Fig. 1.10a, b, and c. As the even and odd components of a sinusoid are, respectively, cosine and sine functions of the same frequency as that of the sinusoid, these results can also be obtained by expanding the expression characterizing the sinusoid.

If a continuous signal is sampled with an adequate sampling rate, the samples uniquely correspond to that signal. Assuming that the sampling rate is adequate, in most of the figures in this book, we have shown the corresponding continuous waveform only for clarity. It should be remembered that a discrete signal is represented only by its sample values.

1.3.5 *Causal and Noncausal Signals*

Most signals, in practice, occur at some finite time instant, usually chosen as $n = 0$, and are considered identically zero before this instant. These signals, with $x(n) = 0$ for $n < 0$, are called causal signals. Signals, with $x(n) \neq 0$ for $n < 0$, are called noncausal signals. Sine and cosine signals, shown in Fig. 1.10, are noncausal signals. Typical causal signals are shown in Fig. 1.1.

1.3.6 Deterministic and Random Signals

Signals such as $x(n) = \sin(\frac{2\pi}{8}n)$, whose values are known for any value of n, are called deterministic signals. Signals such as thermal noise generated in conductors or speech signal, whose future values are not exactly known, are called random signals. Despite the fact that rainfall record is available for several years in the past, the amount of future rainfall at a place cannot be exactly predicted. This type of signals is characterized by a probability model or a statistical model. The study of random signals is important in practice, since all practical signals are random to some extent. However, the analysis of systems is much simpler, mathematically, with deterministic signals. The input-output relationship of a system remains the same whether the input signal is random or deterministic. The time-domain and frequency-domain methods of system analysis is common to both types of signals. The key difference is to find a suitable mathematical model for random signals. In this book, we confine to the study of deterministic signals only.

1.4 Signal Operations

In addition to the arithmetic operations, time shifting, time reversal, and time scaling operations are also commonly used in the analysis of discrete signals. The three operations described in this section are with respect to the independent variable, n.

1.4.1 Time Shifting

By replacing n by $n-N$, where N is an integer, we get the shifted version, $x(n-N)$, of the signal $x(n)$. The value of $x(n)$ at $n = n_0$ occurs at $n = n_0 + N$ in $x(n-N)$. If N is positive (negative), the values of the function are retarded (advanced) by N. Graphically, it amounts to shifting the plot of the function forward (N positive) or backward by N. The exponential signal $x(n) = (0.7)^n u(n)$ is shown in Fig. 1.11 by dots. The signal $x(n-1)$, shown in Fig. 1.11 by crosses, is the signal $x(n)$ shifted by

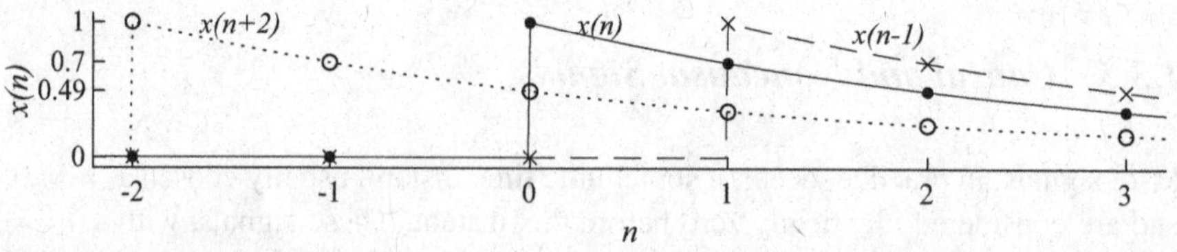

Fig. 1.11 The exponential signal $x(n) = (0.7)^n u(n)$, the right shifted signal, $x(n-1) = (0.7)^{(n-1)} u(n-1)$, and the left shifted signal, $x(n+2) = (0.7)^{(n+2)} u(n+2)$

one sample interval to the right (delayed by one sample interval, as the sample values of $x(n)$ occur one sample interval later). For example, the first nonzero sample value occurs at $n = 1$ as $(0.7)^{1-1}u(1-1) = (0.7)^0u(0) = 1$. That is, the value of the function $x(n)$ at n_0 occurs in the shifted signal one sample interval later at $n_0 + 1$. The signal $x(n+2)$, shown in Fig. 1.11 by unfilled circles, is the signal $x(n)$ shifted by two sample intervals to the left (advanced by two sample intervals, as the sample values of $x(n)$ occur two sample intervals earlier). For example, the first nonzero sample value occurs at $n = -2$ as $(0.7)^{-2+2}u(-2+2) = (0.7)^0u(0) = 1$. That is, the value of the function $x(n)$ at n_0 occurs in the shifted signal two sample intervals earlier at $n_0 - 2$.

1.4.1.1 Circular Shifting

Circular shifting is simply the shifting of the values of a signal placed on a circle. The right circular shift of a N-point signal $x(n)$ by k sample intervals results in

$$x((n-k) \bmod N)$$

Let $x(n) = \{x(0), x(1), x(2), x(3)\} = \{3, 1, 2, 4\}$. Then,

$$x(n-1) = \{x(3), x(0), x(1), x(2)\} = \{4, 3, 1, 2\}$$
$$x(n+1) = \{x(1), x(2), x(3), x(0)\} = \{1, 2, 4, 3\}$$

1.4.2 *Time Reversal*

Forming the mirror image of a signal about the vertical axis at the origin is the time reversal or folding operation. This is achieved by replacing the independent variable n in $x(n)$ by $-n$, and we get $x(-n)$. The value of $x(n)$ at $n = n_0$ occurs at $n = -n_0$ in $x(-n)$. The exponential signal $x(n) = (0.7)^n u(n)$ is shown in Fig. 1.12 by a solid line. The folded signal $x(-n)$ is shown in Fig. 1.12 by crosses. Consider the folded

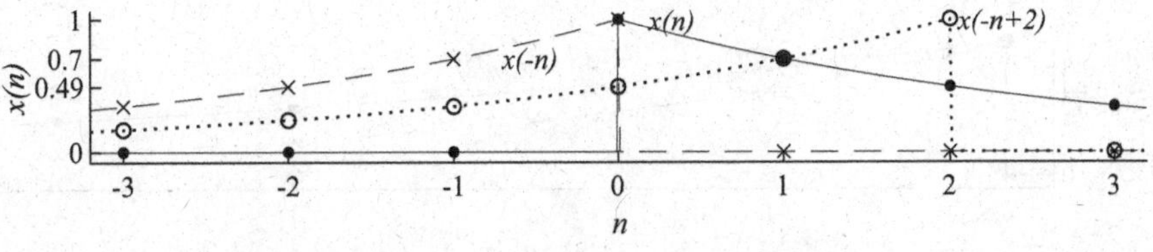

Fig. 1.12 The exponential signal $x(n) = (0.7)^n u(n)$, the folded signal, $x(-n) = (0.7)^{-n}u(-n)$, and the shifted and folded signal, $x(-n+2) = (0.7)^{(-n+2)}u(-n+2)$

and shifted signal $x(-n+2) = x(-(n-2))$, shown in Fig. 1.12 by unfilled circles. This signal can be formed by first folding $x(n)$ to get $x(-n)$ and then shifting it to the right by two sample intervals (n is replaced by $(n-2)$). This signal can also be formed by first shifting $x(n)$ to the left by two sample intervals to get $x(n+2)$ and then folding it about the vertical axis (n is replaced by $-n$). That is, the value of the function $x(n)$ at n_0 occurs in the reversed and shifted signal at $-(n_0-2)$.

1.4.2.1 Circular Time Reversal

Circular time reversal of a N-point signal $x(n)$ is given by

$$\begin{cases} x(0) & \text{for } n = 0 \\ x(N-n) & 1 \le n \le N-1 \end{cases}$$

This is just plotting $x(n)$ in the other direction on a circle. For example, the samples of one cycle of $x(n) = \sin(2\pi n/8)$ are

$$\{0, 0.7071, 1, 0.7071, 0, -0.7071, -1, -0.7071\}$$

and those of $xr(n) = \sin(2\pi(-n)/8) = -\sin(2\pi n/8)$ are

$$\{0, -0.7071, -1, -0.7071, 0, 0.7071, 1, 0.7071\}$$

1.4.3 Time Scaling

Replacing the independent variable n in $x(n)$ by an or $\frac{n}{a}$ results in the time scaled signal $x(an)$ (time compressed version of $x(n)$) or $x(\frac{n}{a})$ (time expanded version of $x(n)$), with $a \neq 0$ being an integer. The value of $x(n)$ at $n = n_0$ occurs at $n = \frac{n_0}{a}$ (for n_0 being an integral multiple of a) in $x(an)$ and at $n = an_0$ in $x(\frac{n}{a})$. Consider the signal $x(n) = (0.8)^n u(n)$, shown in Fig. 1.13 by dots. The time compressed version with $a = 2$, $y(n) = x(2n)$, is shown in Fig. 1.13 by crosses. The values of

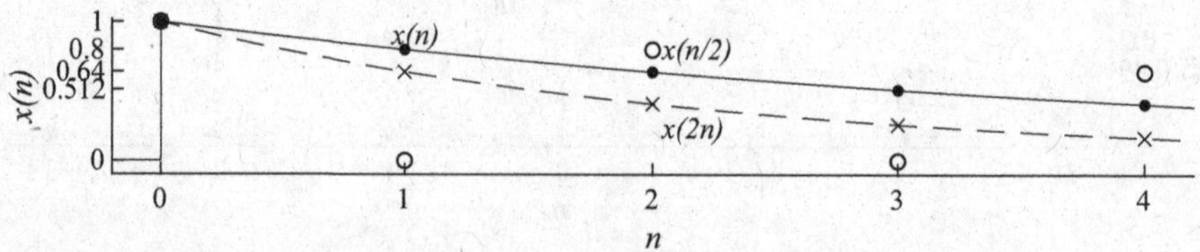

Fig. 1.13 The exponential $x(n) = (0.8)^n u(n)$, $x(2n)$, and $x(\frac{n}{2})$

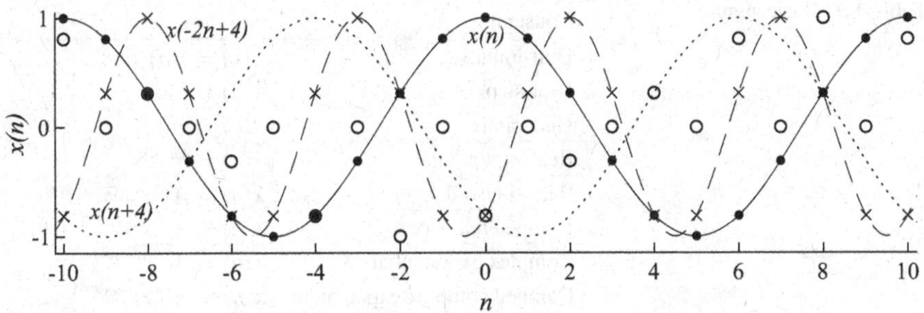

Fig. 1.14 The sinusoid, $x(n) = \cos(\frac{\pi}{5}n)$, $x(-2n+4)$, and $x(-\frac{n}{2}+4)$

the signal $y(n) = x(2n)$ are the even-indexed values of $x(n)$. That is, $y(0) = x(0)$, $y(1) = x(2)$, $y(2) = x(4)$, and so on. The odd-indexed values of $x(n)$ are lost in the time compression operation. In general, $x(an)$ is composed only of every ath sample of $x(n)$.

The time expanded version with $a = 2$, $y(n) = x(\frac{n}{2})$, is shown in Fig. 1.14 by unfilled circles. The values of the time expanded signal are defined from that of $x(n)$ only for the even-indexed values of $y(n)$. That is, $y(0) = x(0)$, $y(2) = x(1)$, $y(4) = x(2)$, and so on. Odd-indexed values of $y(n)$ are assigned the value zero. In general, $y(n) = x(\frac{n}{a})$ is defined only for $n = 0, \pm a, \pm 2a, \pm 3a, \ldots$, and the rest of the values of $y(n)$ are undefined. Interpolation by assigning the value zero is often used in practice. Of course, the undefined values can also be defined using a suitable interpolation formula.

In general, the three operations described on a signal $x(n)$ can be expressed as $y(n) = x(an - b)$ or $y(n) = x(\frac{n}{a} - b)$. The signal $y(n)$ can be generated by replacing n by $(an - b)$ or $(\frac{n}{a} - b)$ in $x(n)$. However, it is instructive to consider it as the result of a sequence of two steps: (i) first shifting the signal $x(n)$ by b to get $x(n - b)$ and then (ii) time scaling (replace n by an or $\frac{n}{a}$) the shifted signal by a to get $y(n) = x(an - b)$ or $y(n) = x(\frac{n}{a} - b)$. Note that, time reversal operation is a special case of the time scaling operation with $a = -1$.

Let $x(n) = \cos(\frac{\pi}{5}n)$, shown in Fig. 1.14 by dots. It is required to find $x(-2n+4)$. The shifted signal is $x(n+4) = \cos(\frac{\pi}{5}(n+4))$, shown by a dotted line. Now scaling this signal by -2 yields the signal $x(-2n + 4) = \cos(\frac{\pi}{5}(-2n + 4))$, shown in Fig. 1.14 by crosses. The value of the function $x(n)$ at an even n_0 occurs in the scaled and shifted signal at $-\frac{(n_0-4)}{2}$. The period of $x(n)$ is 10 and that of the compressed signal is 5.

Let us find $x(-\frac{n}{2}+4)$. Scaling the shifted signal by $-\frac{1}{2}$ yields the signal $x(-\frac{n}{2}+4) = \cos(\frac{\pi}{5}(-\frac{n}{2} + 4))$, shown in Fig. 1.14 by unfilled circles. The value of the function $x(n)$ at n_0 occurs in the scaled and shifted signal at $-2(n_0 - 4)$.

Table 1.2 Basic signals

Constant	$x(n) = c$
Unit-impulse	$x(n) = \delta(n)$
Unit-step	$x(n) = u(n)$
Unit-ramp	$x(n) = nu(n)$
Real exponential	$x(n) = r^n$
Real sinusoid	$x(n) = A\cos(\omega n + \theta)$
Damped real sinusoid	$x(n) = Ar^n \cos(\omega n + \theta)$
Complex exponential	$x(n) = Ae^{j(\omega n+\theta)}$
Damped complex exponential	$x(n) = Ar^n e^{j(\omega n+\theta)}$

1.4.4 Zero Padding

In this operation, a sequence is appended by zero-valued samples. A N-point sequence $x(n)$ is expanded to a M-point, $M > N$, signal $x_z(n), n = 0, 1, \dots, M-1$, defined as

$$x_z(n) = \begin{cases} x(n) & \text{for } n = 0, 1, \dots, N-1 \\ 0 & \text{for } n = N, N+1, \dots, M-1 \end{cases}$$

For example, $x_z(n) = \{4, 1, 3, 0\}$ is the zero-padded version of $x(n) = \{4, 1, 3\}$ to make its length equal to the nearest power of 2.

Table 1.2 shows a list of basic signals.

1.5 Numerical Integration

Integration, which is determining the area under a function, is one of the often used operations in signal and system analysis. Numerical integration is the numerical evaluation of a definite integral

$$I = \int_a^b x(t)dt,$$

where a and b are the limits of integration and $x(t)$ is the function to be integrated. As stated earlier, the amplitude profile of functions encountered in practical applications is usually arbitrary. Then, it becomes a necessity to resort to numerical integration. Numerical integration requires sampling the continuous function with an appropriate sampling interval, T_s.

In numerical integration, the area to be integrated is divided into subintervals, and an approximation function is used to find the area enclosed in each subinterval. One of the approximation functions often used is based on the rectangle. Let the sampling interval be T_s. The interval of integration is subdivided into N equal subintervals of

Table 1.3 Numerical integration

Iteration No.	1	2	3	4	5	6	7	8
$x(n)$	0.5	0.3679	0.1353	0.0498	0.0183	0.0067	0.0025	0.0009
$I(n)$	1	1.7358	2.0064	2.1060	2.1426	2.1561	2.1611	2.1629

length $T_s = (b - a)/N$. Then, the rectangular rule of numerical integration is given by

$$I(n) = I(n-1) + T_s x(n), \quad n = 0, 1, 2, \ldots, N-1$$

with the initial value $I(-1) = 0$.

Let us find the area under the continuous function $x(t) = e^{-0.5t}u(t)$, the real causal exponential function. Analytically,

$$I = \int_{t=0}^{\infty} e^{-0.5t}\,dt = -2e^{-0.5t}|_0^{\infty} = 2$$

Let the sampling interval be $T_s = 2$ s and the record length be 16 s. Then, the first eight samples of the signal are shown in the second row of Table 1.3. While the actual value of the first sample is 1, a value of 0.5 is assumed for better accuracy, as there is a discontinuity. The value 0.5 is the average at the discontinuity. Row 2 shows the partial results of numerical integration. The first partial result is $0+0.5 \times 2 = 1$.

The second partial result is $1+0.3679\times2 = 1.7358$ and so on. With this sampling interval, the final result of integration is 2.1629. The length of the record over which the samples are taken and the sampling interval can be set to obtain the integration value with a required accuracy.

1.6 The Organization of this Book

Four topics are covered in this book. The time-domain analysis of signals and systems is presented in Chaps. 1–4. The four versions of the Fourier analysis are described in Chaps. 5–8. The generalized Fourier analysis, the z-transform and the Laplace transform, is presented in Chaps. 9 and 10. State-space analysis is introduced in Chaps. 11 and 12.

The amplitude profile of practical signals is usually arbitrary. It is a necessity to represent these signals in terms of well-defined basic signals in order to carry out efficient signal and system analysis. The impulse and sinusoidal signals are fundamental in signal and system analysis. In Chap. 1, **Discrete Signals**, we present the basic signals, discrete signal classifications, and signal operations. In Chap. 2, **Continuous Signals**, we present the basic signals, continuous signal classifications, and signal operations.

The study of systems involves modeling, analysis, and design. In Chap. 3, **Time-Domain Analysis of Discrete Systems**, we start with the modeling of a system with the difference equation. The classification of systems is presented next. Then, the convolution-summation model is introduced. The zero-input, zero-state, transient, and steady-state responses of a system are derived from its model. System stability is considered in terms of impulse response. The basic components of discrete systems are identified. In Chap. 4, **Time-Domain Analysis of Continuous Systems**, we start with the classification of systems. The modeling of a system with the differential equation is presented next. Then, the convolution-integral model is introduced. The zero-input, zero-state, transient, and steady-state responses of a system are derived from its model. System stability is considered in terms of impulse response. The basic components of continuous systems are identified.

Basically, the analysis of signals and systems is carried out using impulse or sinusoidal signals. The impulse signal is used in the time-domain analysis, which is presented in Chaps. 3 and 4. Sinusoids (more generally complex exponentials) are used as the basic signals in the frequency-domain analysis. As the frequency-domain analysis is, in general, more efficient, it is most often used. Signals occur usually in the time-domain. In order to use the frequency-domain analysis, signals and systems must be represented in the frequency-domain. Transforms are used to obtain the frequency-domain representation of a signal or a system from its time-domain representation. All the essential transforms required in signal and system analysis use the same family of basis signals, a set of complex exponential signals. However, each transform is more advantageous to analyze certain type of signals and to carry out certain type of system operations, since the basis signals consists of a finite or infinite set of complex exponential signals with different characteristics, continuous or discrete, and the exponent being complex or pure imaginary. The transforms that use the complex exponential with a pure imaginary exponent comes under the heading Fourier analysis. The other transforms use exponentials with complex exponents as their basis signals.

There are four versions of the Fourier analysis, each primarily applicable to a different type of signals such as continuous or discrete and periodic or aperiodic. The discrete Fourier transform (DFT) is the only one in which both the time- and frequency-domain representations are in finite and discrete form. Therefore, it can approximate other versions of Fourier analysis through efficient numerical algorithms. In addition, the physical interpretation of the DFT is much easier. The basis signals of this transform is a finite set of harmonically related discrete exponentials with pure imaginary exponent. In Chap. 5, **The Discrete Fourier Transform**, the DFT, its properties, and some of its applications are presented.

The Fourier analysis of a continuous periodic signal, which is a generalization of the DFT, is called the Fourier series (FS). The FS uses an infinite set of harmonically related continuous exponentials with pure imaginary exponent as the basis signals. This transform is useful in the frequency-domain analysis and design of periodic signals and systems with continuous periodic signals. In Chap. 6, **Fourier Series**, the FS, its properties, and some of its applications are presented.

The Fourier analysis of a discrete aperiodic signal, which is also a generalization of the DFT, is called the discrete-time Fourier transform (DTFT). The DTFT uses a continuum of discrete exponentials, with pure imaginary exponent, over a finite frequency range as the basis signals. This transform is useful in the frequency-domain analysis and design of discrete signals and systems. In Chap. 7, **The Discrete-Time Fourier Transform**, the DTFT, its properties, and some of its applications are presented.

The Fourier analysis of a continuous aperiodic signal, which can be considered as a generalization of the FS or the DTFT, is called the Fourier transform (FT). The FT uses a continuum of continuous exponentials, with pure imaginary exponent, over an infinite frequency range as the basis signals. This transform is useful in the frequency-domain analysis and design of continuous signals and systems. In addition, as the most general version of the Fourier analysis, it can represent all types of signals and is very useful to analyze a system with different types of signals, such as continuous or discrete and periodic or aperiodic. In Chap. 8, **The Fourier Transform**, the FT, its properties, and some of its applications are presented.

Generalization of the Fourier analysis for discrete signals results in the z-transform. This transform uses a continuum of discrete exponentials, with complex exponent, over a finite frequency range of oscillation as the basis signals. With a much larger set of basis signals, this transform is required for the design and transient and stability analysis of discrete systems. In Chap. 9, **The z-Transform**, the z-transform is derived from the DTFT, and its properties and some of its applications are presented. The procedures for obtaining the forward and inverse z-transforms are described.

Generalization of the Fourier analysis for continuous signals results in the Laplace transform. This transform uses a continuum of continuous exponentials, with complex exponent, over an infinite frequency range of oscillation as the basis signals. With a much larger set of basis signals, this transform is required for the design and transient and stability analysis of continuous systems. In Chap. 10, **The Laplace Transform**, the Laplace transform is derived from the FT, and its properties and some of its applications are presented. The procedures for obtaining the forward and inverse Laplace transforms are described.

In Chap. 11, **State-Space Analysis of Discrete Systems**, the state-space analysis of discrete systems is presented. This type of analysis is more general in that it includes the internal description of a system in contrast to the input-output description of other types of analysis. In addition, this method is easier to extend to system analysis with multiple inputs and outputs and nonlinear and time-varying system analysis. In Chap. 12, **State-Space Analysis of Continuous Systems**, the state-space analysis of continuous systems is presented.

In **Appendix A**, the complex number system is reviewed. In **Appendix B**, transform pairs and properties are listed. In **Appendix C**, useful mathematical formulas are given.

1.7 Summary

- In this chapter, basic discrete signals, signal classifications, and signal operations have been presented.
- Signals are used for communication of information about some behavior or nature of some physical phenomenon, such as audio, pressure, and temperature.
- In mathematical form, it is a function of one or more independent variables.
- Signals usually need some processing for their effective use.
- Continuous-time signal is defined at each and every instant of time over the period of its occurrence.
- The values of discrete signals are available only at discrete intervals.
- The values of quantized continuous-time signals are quantized to certain levels.
- Both the dependent and independent variables of digital signals are sampled values.
- A periodic signal repeats its values over any one period indefinitely. A signal that is not periodic is aperiodic.
- As practical signals have arbitrary amplitude profile, these signals are usually decomposed and processed in terms of basic signals, such as the sinusoid or the impulse.
- The impulse is the basis signal in the time domain, while the sinusoid is the basis signal in the frequency domain.
- The sinusoid is an everlasting periodic signal, while the strength of an impulse is concentrated at a single point.
- The sinusoid is a linear combination of the well-known trigonometric functions, sine and cosine.
- While sinusoidal signals are produced by physical devices, a mathematically equivalent form, the complex exponential, is found to be more efficient and compact in signal analysis.
- The sinusoidal signals dominate the linear signal and system analysis, as it brings out the salient characteristics of signals and provides fast processing of operations.
- Storage and processing requirements of a signal depend on its type.
- In addition to the arithmetic operations, time shifting, time reversal, and time scaling operations are also commonly used in the analysis of signals.

Exercises

1.1 Is $x(n)$ an energy signal, a power signal, or neither? If it is an energy signal, find its energy. If it is a power signal, find its average power.

1.1.1 $x(0) = 2, x(-1) = 2, x(-2) = -2, x(-3) = -2$, and $x(n) = 0$ otherwise.

* 1.1.2 $x(n) = 2(0.8)^n u(n)$.

1.1.3 $x(n) = 2^n$.
1.1.4 $x(n) = Ce^{j(\frac{6\pi n}{8})}$.
1.1.5 $x(n) = 3\cos(\frac{\pi n}{2} + \frac{\pi}{4})$.
1.1.6 $x(n) = u(n)$.
1.1.7 $x(n) = 2$.
1.1.8 $x(n) = \frac{2}{n}u(n-1)$.
1.1.9 $x(n) = n$.

1.2 Is $x(n)$ even-symmetric, odd-symmetric, or neither? List the values of $x(n)$ for $n = -3, -2, -1, 0, 1, 2, 3$.

1.2.1 $x(n) = 2\sin(\frac{\pi}{5}n - \frac{\pi}{3})$.
1.2.2 $x(n) = \sin(\frac{\pi}{5}n)$.
1.2.3 $x(n) = 2\cos(\frac{\pi}{5}n)$.
1.2.4 $x(n) = 3$.
1.2.5 $x(n) = n$.
1.2.6 $x(n) = \frac{2\sin(\frac{\pi}{3}n)}{n}$.
1.2.7 $x(n) = \frac{2\sin^2(\frac{\pi}{3}n)}{n}$.
1.2.8 $x(0) = 0$ and $x(n) = \frac{(-1)^n}{n}$ otherwise.
1.2.9 $x(n) = 3\delta(n)$

1.3 Find the even and odd components of the signal. List the values of the signal and its components for $n = -3, -2, -1, 0, 1, 2, 3$. Verify that the values of the components add up to the values of the signal. Verify that the sum of the values of the even component and that of the signal are equal.

1.3.1 $x(0) = 1, x(1) = 1, x(2) = -1, x(3) = -1$, and $x(n) = 0$ otherwise.
1.3.2 $x(n) = 3\cos(\frac{\pi}{5}n + \frac{\pi}{6})$
* 1.3.3 $x(n) = (0.4)^n u(n)$
1.3.4 $x(n) = u(n+1)$
1.3.5 $x(n) = e^{-j(\frac{\pi}{3}n)}$
1.3.6 $x(n) = n\,u(n)$

1.4 Evaluate the summation.

1.4.1 $\sum_{n=0}^{\infty} \delta(n)(0.5)^n u(n)$.
* 1.4.2 $\sum_{n=0}^{\infty} \delta(n+1)(0.5)^n$.
1.4.3 $\sum_{n=0}^{\infty} \delta(n-2)(0.5)^n u(n)$.
1.4.4 $\sum_{n=-\infty}^{\infty} \delta(n+1)(0.5)^n$.

1.5 Express the signal in terms of scaled and shifted impulses.

1.5.1 $x(0) = 2, x(1) = 3, x(2) = -1, x(3) = -4$, and $x(n) = 0$ otherwise.
1.5.2 $x(0) = 5, x(-1) = 3, x(2) = -7, x(-3) = -4$, and $x(n) = 0$ otherwise.

1.6 If the waveform is periodic, what is its period?

1.6.1 $x(n) = 4\cos(0.7\pi n)$.
1.6.2 $x(n) = 2\cos(\sqrt{2}n)$.
1.6.3 $x(n) = 43 + 2\cos(\frac{2\pi}{7}n)$.
1.6.4 $x(n) = 2\cos(\frac{\pi}{5\sqrt{2}}n)$.
* 1.6.5 $x(n) = 4\cos(\frac{4\pi}{9}n)$.

1.7 Find the rectangular form of the sinusoid. List the sample values of one cycle, starting from $n = 0$, of the sinusoid.

1.7.1 $x(n) = -2\sin(\frac{\pi}{6}n - \frac{\pi}{3})$.
1.7.2 $x(n) = -2\cos(\frac{\pi}{6}n - \frac{\pi}{4})$.
1.7.3 $x(n) = \cos(\frac{\pi}{6}n)$.
1.7.4 $x(n) = 3\sin(\frac{\pi}{6}n + \frac{\pi}{3})$.
1.7.5 $x(n) = -\sin(\frac{\pi}{6}n)$.
* 1.7.6 $x(n) = 4\cos(\frac{\pi}{6}n - \frac{\pi}{6})$.

1.8 Find the polar form of the sinusoid. List the sample values of one cycle, starting from $n = 0$, of the sinusoid.

1.8.1 $x(n) = -2\sin(\frac{\pi}{6}n)$.
1.8.2 $x(n) = -2\cos(\frac{\pi}{6}n) - 2\sin(\frac{\pi}{6}n)$.
* 1.8.3 $x(n) = 3\cos(\frac{\pi}{6}n) + \sqrt{3}\sin(\frac{\pi}{6}n)$.
1.8.4 $x(n) = -3\cos(\frac{\pi}{6}n)$.
1.8.5 $x(n) = \sqrt{3}\cos(\frac{\pi}{6}n) - \sin(\frac{\pi}{6}n)$.

1.9 Given $x_1(n) = A_1e^{j(\omega n+\theta_1)}$ and $x_2(n) = A_2e^{j(\omega n+\theta_2)}$, derive expressions for A and θ of the complex sinusoid $x(n) = x_1(n) + x_2(n) = Ae^{j(\omega n+\theta)}$ in terms of those of $x_1(n)$ and $x_2(n)$.

1.10 Given the complex sinusoids $x_1(n) = A_1e^{j(\omega n+\theta_1)}$ and $x_2(n) = A_2e^{j(\omega n+\theta_2)}$, find the complex sinusoid $x(n) = x_1(n) + x_2(n) = Ae^{j(\omega n+\theta)}$, using the formulas derived in Exercise 1.9. Find the sample values of one cycle, starting from $n = 0$, of the complex sinusoids $x_1(n)$ and $x_2(n)$ and verify that the sample values of $x_1(n) + x_2(n)$ are the same as those of $x(n)$.

1.10.1 $x_1(n) = -2e^{j(\frac{\pi}{3}n+\frac{\pi}{3})}$, $x_2(n) = 3e^{j(\frac{\pi}{3}n-\frac{\pi}{6})}$.
1.10.2 $x_1(n) = 3e^{-j(\frac{\pi}{3}n+\frac{\pi}{3})}$, $x_2(n) = 2e^{-j(\frac{\pi}{3}n-\frac{\pi}{3})}$.
1.10.3 $x_1(n) = 2e^{j(\frac{\pi}{3}n)}$, $x_2(n) = 3e^{j(\frac{\pi}{3}n)}$.
1.10.4 $x_1(n) = e^{j(\frac{\pi}{3}n-\frac{\pi}{2})}$, $x_2(n) = e^{j(\frac{\pi}{3}n)}$.
* 1.10.5 $x_1(n) = 2e^{j(\frac{\pi}{3}n+\frac{\pi}{6})}$, $x_2(n) = 4e^{j(\frac{\pi}{3}n+\frac{\pi}{4})}$.

1.11 Find the corresponding exponential of the form a^n. List the values of the exponential for $n = 0, 1, 2, 3, 4, 5$.

1.11.1 $x(n) = e^{0.6931n}$.
1.11.2 $x(n) = e^{n}$.
* 1.11.3 $x(n) = e^{-0.6931n}$.
1.11.4 $x(n) = e^{-0.3567n}$.

1.12 Give the sample values of the exponentially varying amplitude sinusoid for $n = -2, -1, 0, 1, 2, 3, 4$.

1.12.1 $x(n) = (0.8)^n \sin(\frac{2\pi}{6}n - \frac{\pi}{6})$.
1.12.2 $x(n) = (-0.6)^n \cos(\frac{2\pi}{6}n + \frac{\pi}{3})$.
1.12.3 $x(n) = (1.1)^n \sin(\frac{2\pi}{6}n - \frac{\pi}{4})$.
1.12.4 $x(n) = (-1.2)^n \cos(\frac{2\pi}{6}n + \frac{\pi}{6})$.
1.12.5 $x(n) = (0.7)^n \cos(\pi n)$.

1.13 Find the next three higher-frequency sinusoids with the same set of sample values as that of $x(n)$.

1.13.1 $x(n) = 2\cos(2\frac{2\pi}{9}n + \frac{\pi}{6})$.
1.13.2 $x(n) = 4\sin(3\frac{2\pi}{7}n - \frac{\pi}{3})$.
1.13.3 $x(n) = \cos(4\frac{2\pi}{9}n - \frac{\pi}{6})$.
* 1.13.4 $x(n) = 3\sin(3\frac{2\pi}{8}n - \frac{\pi}{3})$.
1.13.5 $x(n) = 3\cos(\pi n)$.
1.13.6 $x(n) = 5\cos(0n)$.

1.14 Find the minimum sampling rate required to represent the continuous signal unambiguously.

1.14.1 $x(t) = 3\cos(10\pi t)$.
1.14.2 $x(t) = 3\cos(10\pi t + \frac{\pi}{3})$.
* 1.14.3 $x(t) = 2\sin(10\pi t)$.
1.14.4 $x(t) = 2\sin(10\pi t - \frac{\pi}{6})$.

1.15 The sinusoid $x(n)$ and the value k are specified. Express the sinusoid $x(n+k)$ in polar form. List the sample values of one cycle, starting from $n = 0$, of the sinusoids $x(n)$ and $x(n+k)$.

1.15.1 $x(n) = 2\cos(\frac{2\pi}{6}n - \frac{\pi}{3})$, $k = 2$.
1.15.2 $x(n) = -3\sin(\frac{2\pi}{6}n + \frac{\pi}{6})$, $k = -1$.
1.15.3 $x(n) = \cos(\frac{2\pi}{6}n - \frac{\pi}{6})$, $k = 3$.
1.15.4 $x(n) = -\sin(\frac{2\pi}{6}n + \frac{\pi}{3})$, $k = 6$.
* 1.15.5 $x(n) = \cos(\frac{2\pi}{6}n + \frac{\pi}{2})$, $k = -7$.
1.15.6 $x(n) = \sin(\frac{2\pi}{6}n + \frac{2\pi}{3})$, $k = 15$.

1.16 The sinusoid $x(n)$ and the value k are specified. Express the sinusoid $x(-n+k)$ in polar form. List the sample values of one cycle, starting from $n = 0$, of the sinusoids $x(n)$ and $x(-n+k)$.

1.16.1 $x(n) = \sin(\frac{2\pi}{6}n - \frac{\pi}{3})$, $k = 0$.
1.16.2 $x(n) = \sin(\frac{2\pi}{6}n + \frac{\pi}{3})$, $k = -2$.
* 1.16.3 $x(n) = \cos(\frac{2\pi}{6}n - \frac{\pi}{6})$, $k = 1$.
1.16.4 $x(n) = \sin(\frac{2\pi}{6}n + \frac{\pi}{2})$, $k = -3$.
1.16.5 $x(n) = \cos(\frac{2\pi}{6}n - \frac{\pi}{2})$, $k = 6$.
1.16.6 $x(n) = \sin(\frac{2\pi}{6}n + \frac{\pi}{6})$, $k = 7$.
1.16.7 $x(n) = \cos(\frac{2\pi}{6}n + \frac{\pi}{3})$, $k = 14$.

1.17 The sinusoid $x(n)$ and the values k and a are specified. List the sample values of one cycle, starting from $n = 0$, of the sinusoid $x(n)$ and $x(an+k)$. Assume interpolation using zero-valued samples, if necessary.

1.17.1 $x(n) = -\sin(\frac{2\pi}{6}n + \frac{\pi}{3})$, $a = -2$, $k = 0$.
1.17.2 $x(n) = 2\cos(\frac{2\pi}{6}n - \frac{\pi}{6})$, $a = \frac{1}{2}$, $k = -2$.
* 1.17.3 $x(n) = \sin(\frac{2\pi}{6}n + \frac{\pi}{6})$, $a = -1$, $k = 1$.
1.17.4 $x(n) = 3\cos(\frac{2\pi}{6}n + \frac{\pi}{3})$, $a = \frac{1}{3}$, $k = 6$.
1.17.5 $x(n) = \sin(\frac{2\pi}{6}n - \frac{\pi}{2})$, $a = -3$, $k = 7$.
1.17.6 $x(n) = \cos(\frac{2\pi}{6}n - \frac{\pi}{6})$, $a = -1$, $k = 15$.

1.18 The waveform $x(n)$ and the values k and a are specified. List the sample values with indices $n = -3, -2, -1, 0, 1, 2, 3$ of the waveforms $x(n)$ and $x(an+k)$. Assume interpolation using zero-valued samples, if necessary.

1.18.1 $x(0) = 2, x(1) = 3, x(2) = -4, x(3) = 1$, and $x(n) = 0$ otherwise. $a = -2$, $k = 2$.
1.18.2 $x(0) = 2, x(1) = 3, x(2) = -4, x(3) = 1$, and $x(n) = 0$ otherwise. $a = -\frac{1}{2}$, $k = 1$.
1.18.3 $x(n) = (0.8)^n$. $a = -3$, $k = -1$.
1.18.4 $x(n) = (0.8)^n$. $a = \frac{1}{3}$, $k = 2$.
1.18.5 $x(n) = (1.1)^n$. $a = 2$, $k = 2$.
1.18.6 $x(n) = (1.1)^n$. $a = -\frac{1}{2}$, $k = 1$.
1.18.7 $x(n) = -2\sin(\frac{2\pi}{6}n + \frac{\pi}{6})u(n)$. $a = \frac{1}{2}$, $k = 3$.
* 1.18.8 $x(n) = -2\sin(\frac{2\pi}{6}n + \frac{\pi}{6})u(n)$. $a = -2$, $k = 2$.
1.18.9 $x(n) = (0.7)^n\cos(\frac{2\pi}{6}n - \frac{\pi}{3})u(n)$. $a = \frac{1}{3}$, $k = 3$.
1.18.10 $x(n) = (0.7)^n\cos(\frac{2\pi}{6}n - \frac{\pi}{3})u(n)$. $a = -2$, $k = 2$.

Chapter 2
Continuous Signals

While the analysis of continuous signals is essentially the same as that of the discrete signals, there are differences due to the continuous nature. For example, the summation operation on a discrete signal corresponds to the integration operation on a continuous signal, the difference operation corresponds to the derivative, and the continuous impulse signal is defined in terms of area in contrast to the discrete impulse signal defined by its amplitude. A continuous signal can be thought of as a discrete signal in which the sampling interval is allowed to approach zero. Both the continuous and discrete form of signals are important. The discrete signals are amenable for numerical analysis and, therefore, mostly used in practice. Most of the naturally occurring signals are, however, of continuous nature, and more formulas are available for their analysis. In this chapter, basic signals, signal classifications, and signal operations of continuous signals are described in Sects. 2.1, 2.2, and 2.3, respectively.

2.1 Basic Signals

While the input signal to a system, in practice, is arbitrary, some mathematically well-defined and simple signals are used for testing systems and decomposition of the arbitrary signals for analysis. These signals, for example, the sinusoid with infinite duration and the impulse with infinite bandwidth, are mathematical idealizations and are not practical signals. However, they are convenient in the analysis of signals and systems as intermediaries. In practice, they can be approximated to a desired accuracy.

D. Sundararajan, *Signals and Systems*,
https://doi.org/10.1007/978-3-031-19377-4_2

2.1.1 The Unit-Step Signal

A system is usually turned on by closing a switch. While practical switches have finite switching time, in theoretical analysis, zero switching time is assumed for convenience. This implies that the input signal is applied instantaneously. A function representing such a signal does not exist in the normal function theory, since the derivative of a function, at a discontinuity, is not defined. As this type of function is required frequently in the analysis of systems, we define such a function and its derivative and denote them by special symbols. The unit-step function $u(t)$, shown in Fig. 2.1a, is defined as

$$u(t) = \begin{cases} 1 & \text{for } t > 0 \\ 0 & \text{for } t < 0 \\ \text{undefined} & \text{for } t = 0 \end{cases}$$

The unit-step signal has a value of one for $t > 0$ and has a value of zero for $t < 0$. The value $u(0)$, if required, can be assigned values such as 0, $\frac{1}{2}$, or 1 to suit a specific application. For example, the value $\frac{1}{2}$ is assigned in Fourier analysis.

The causal form of a signal $x(t)$, $x(t)$ is zero for $t < 0$ is obtained by multiplying it with the unit-step signal as $x(t)u(t)$. For example, $\cos(\frac{2\pi}{6}t)$ has nonzero values in the range $-\infty < t < \infty$, whereas the values of $\cos(\frac{2\pi}{6}t)u(t)$ are zero for $t < 0$ and $\cos(\frac{2\pi}{6}t)$ for $t > 0$. A time-shifted unit-step signal, for example $u(t-2)$, is $u(t)$ shifted by two units to the right (changes from 0 to 1 at $t = 2$). Using scaled and shifted unit-step signals, a discontinuous signal, described differently over different intervals, can be specified, for easier mathematical analysis, by a single expression, valid for all t. For example, a signal that is identical to the first half period, beginning at $t = 0$, of the sine wave $\sin(t)$ and is zero otherwise can be expressed as

$$x(t) = \sin(t)\,(u(t) - u(t-\pi)) = \sin(t)\,u(t) + \sin(t-\pi)\,u(t-\pi)$$

The first expression can be interpreted as the sine wave multiplied by a pulse of unit height over the interval $0 < t < \pi$. The second expression can be interpreted as the sum of the causal form of the sine wave and its right shifted version by π (a half

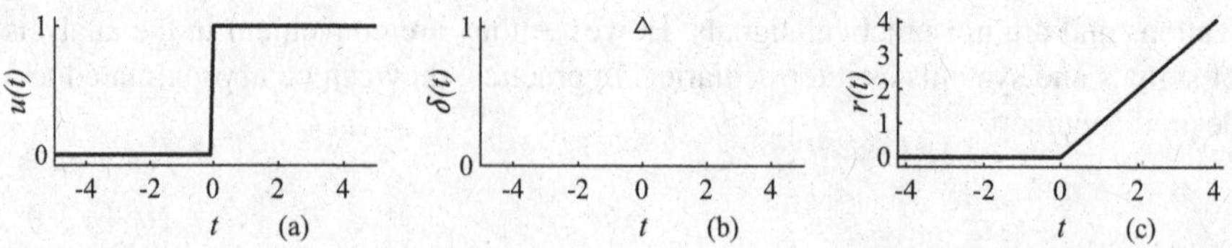

Fig. 2.1 (**a**) The unit-step signal, $u(t)$; (**b**) the unit-impulse signal, $\delta(t)$; (**c**) the unit-ramp signal, $r(t)$

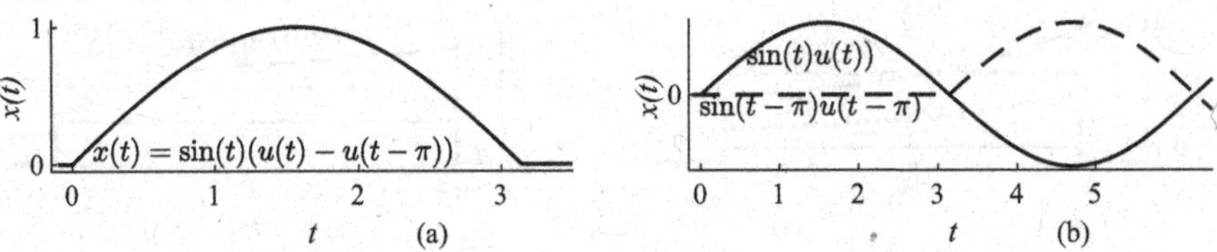

Fig. 2.2 (**a**) $\sin(t)\,(u(t) - u(t-\pi))$; (**b**) $\sin(t)\,u(t)$ and $\sin(t-\pi)\,u(t-\pi)$

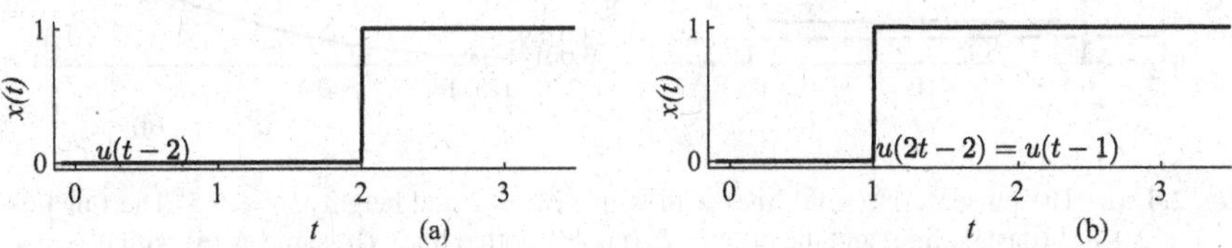

Fig. 2.3 (**a**) The delayed unit-step signal, $u(t-2)$; (**b**) the delayed and compressed unit-step signal, $u(2t-2) = u(t-1)$

period). Figure 2.2a and b show, respectively, $x(t) = \sin(t)\,(u(t) - u(t-\pi))$ and its two components.

The time scaled and shifted unit-step function $u(\pm at - t_0)$ is the same as $u(\pm t - \frac{t_0}{a})$, where $a \neq 0$ is a positive number. Figure 2.3a and b show, respectively, $u(t-2)$ and $u(2t-2) = u(t-1)$.

2.1.2 The Unit-Impulse Signal

Consider a narrow unit area rectangular pulse, $\delta_q(t)$, of width $2a$ and height $\frac{1}{2a}$ centered at $t = 0$ and the function $x(t) = 2 + e^{-t}$. The integral of their product, which is the local average of $x(t)$, is

$$\int_{-\infty}^{\infty} x(t)\delta_q(t)\,dt = \frac{1}{2a}\int_{-a}^{a}(2+e^{-t})\,dt = 2 + \frac{e^{a}-e^{-a}}{2a}$$

The limiting value of the integral, as $a \to 0$, is

$$\lim_{a\to 0}\left(2 + \frac{e^{a}-e^{-a}}{2a}\right) = 2 + \lim_{a\to 0}\left(\frac{e^{a}+e^{-a}}{2}\right) = 3 = x(0)$$

In evaluating the limit, we used the L'Hôpital's rule. For $a = \{0.1, 0.4, 1\}$, the integral evaluates to, respectively,

$$\{3.0017, \quad 3.0269, \quad 3.1752\}$$

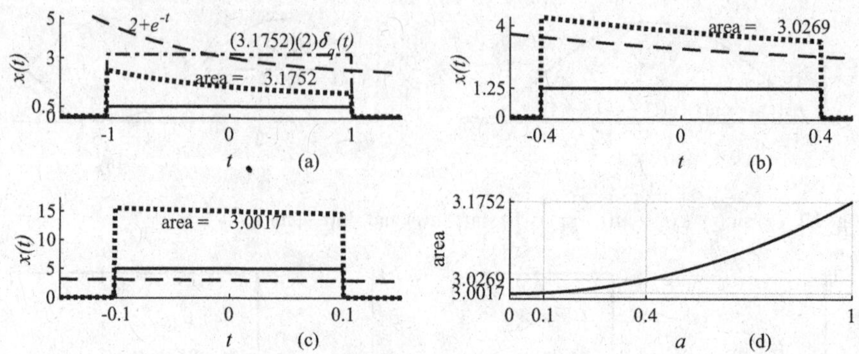

Fig. 2.4 (**a**) The pulse $\delta_q(t)$ (solid line) with width $2a = 2$ and height $\frac{1}{2a} = 0.5$. The function $x(t) = 2+e^{-t}$ (dashed line) and the product $\delta_q(t)x(t)$ (dotted line); (**b**) same as (**a**) with $a = 0.4$; (**c**) same as (**a**) with $a = 0.1$; (**d**) the area enclosed by the product $\delta_q(t)x(t)$ for various values of a

As long as a is not equal to zero, the pulse is clearly defined by its width and height. The integral is an integral in the conventional sense. As $a \to 0$, the rectangular pulse, $\delta_q(t)$, degenerates into an impulse $\delta(t)$ and it is characterized only by its unit area at $t = 0$. Then, the integral becomes a definition

$$\lim_{a\to 0} \int_{-\infty}^{\infty} x(t)\delta_q(t)\,dt = \int_{-\infty}^{\infty} x(t)\delta(t)\,dt = x(0)\int_{0^-}^{0^+} \delta(t)\,dt = x(0)$$

The pulse, $\delta_q(t)$, and the signal $x(t) = 2+e^{-t}$ are shown in Fig. 2.4a with $a = 1$. Their product $x(t)\delta_q(t)$ is shown by the dotted line. The integral of the product is 3.1752 with four-digit precision. The signal $x(t) = 2 + e^{-t}$ is approximated by the rectangle, shown in dash-dot line, defined as $3.1752(2)\delta_q(t)$ in the interval $-1 < t < 1$. That is,

$$x(t) = 2 + e^{-t} \approx 3.1752(2)\delta_q(t), \quad -1 < t < 1$$

Obviously, the representation of the function by the pulse becomes better with a shorter one.

Figure 2.4b and c show the functions with $a = 0.4$, and $a = 0.1$, respectively. As the pulse width a is reduced, the variation in the amplitude of the function $x(t) = 2+e^{-t}$ is also reduced and the integral of the product $\delta_q(t)x(t)$ (the local average of $x(t)$) approaches the value $x(0)$, as shown in Fig. 2.4d. The reason for associating the impulse in deriving the value of $x(0)$, rather than replacing t by 0 in $x(t)$, is to express $x(t)$ in terms of shifted and scaled impulses, as we shall see later.

The continuous unit-impulse signal $\delta(t)$, located at $t = 0$, is defined, in terms of an integral, as

$$\int_{-\infty}^{\infty} x(t)\delta(t)\,dt = x(0)$$

assuming that $x(t)$ is continuous at $t = 0$ (so that the value $x(0)$ is unique). The value of the function $x(t)$ at $t = 0$ has been sifted out or sampled by the defining operation. The impulse function is called a generalized function, since it is defined by the result of its operation (integration) on an ordinary function, rather than by its amplitude profile. A time-shifted unit-impulse signal $\delta(t-\tau)$, located at $t = \tau$, sifts out the value $x(\tau)$,

$$\int_{-\infty}^{\infty} x(t)\delta(t-\tau)\,dt = x(\tau),$$

assuming that $x(t)$ is continuous at $t = \tau$. As the amplitude profile of the impulse is undefined, the unit-impulse is characterized by its unit area concentrated at $t = 0$ (in general, whenever its argument becomes zero), called the strength of the impulse. The unit-impulse is represented by a small triangle (pointing upward for a positive impulse and pointing downward for a negative impulse), as shown in Fig. 2.1b. The power or energy of the impulse signal is undefined.

The area enclosed by a function over some finite duration is easy to visualize. For example, the distribution of mass along a line is defined by its density $\rho(x)$, and the mass between $x = 0$ and $x = 1$ is given by $\int_0^1 \rho(x)dx$. *However, the symbol $\delta(t)$ stands for a function, whose shape and amplitude is such that its integral at the point $t = 0$ is unity.* This is the limiting case of the density $\rho(x)$, when unit mass is concentrated at a single point $x = 0$. It is difficult to visualize such a function. But, it is easy to visualize a function of arbitrarily brief but nonzero duration. For example, the impulse can be considered, for practical purposes, as a sufficiently narrow rectangular pulse of unit area. The width of the pulse Δt should be so short that the variation of any ordinary function $x(t)$, appearing in an expression involving an impulse, is negligible in Δt seconds. Therefore, to understand any operation involving the impulse, we start with a brief pulse, perform the operation, and take the limiting form as the width of the pulse approaches zero. As only its area is specified, it is possible to start with many functions of brief duration and apply the limiting process. The only condition is that its area must be unity throughout the limiting process. Some other functions, besides the rectangular pulse, those degenerate into the unit-impulse signal in the limit, are shown in Fig. 2.5. For practical purposes, any of these functions with a sufficiently short duration is adequate. The point is that practical devices can produce a pulse of finite width only, whereas, in theory, we use zero-width pulses for the sake of mathematical convenience.

The product of an ordinary function $x(t)$, which is continuous at $t = \tau$, and $\delta(t-\tau)$ is given as $x(t)\delta(t-\tau) = x(\tau)\delta(t-\tau)$, since the impulse has unit area concentrated at $t = \tau$ and the value of $x(t)$ at that point is $x(\tau)$. That is, the product of an ordinary function with the unit-impulse is an impulse with its area or strength equal to the value of the function at the location of the impulse. As the impulse is defined by an integral, any expression involving an impulse has to be eventually

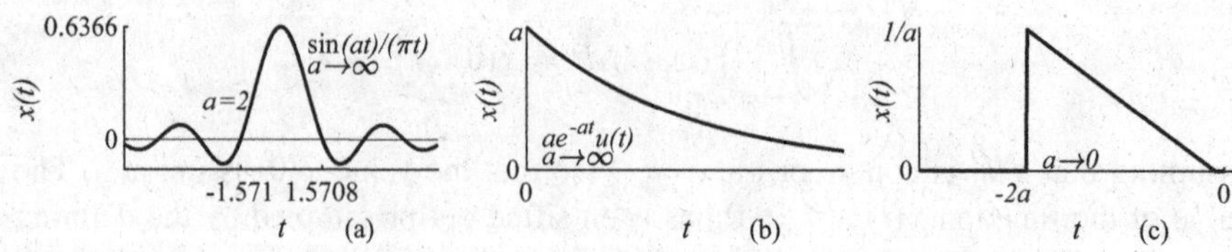

Fig. 2.5 Some functions those degenerate into unit-impulse signal, $\delta(t)$, in the limit. (**a**) $\delta(t) = \lim_{a\to\infty} \frac{\sin(at)}{\pi t}$; (**b**) $\delta(t) = \lim_{a\to\infty} ae^{-at}u(t)$; (**c**) $\delta(t) = \lim_{a\to 0} \frac{t}{2a^2}(u(t) - u(t+2a))$

integrated to have a numerical value. An expression like the product given above implies that the integral of the two sides are equal. For example,

$$\int_{-\infty}^{\infty} e^t\delta(t)dt = 1, \quad \int_{-\infty}^{\infty} e^t\delta(t+2)dt = e^{-2}, \quad \int_{-2}^{2} e^t\delta(-t)dt = 1,$$

$$\int_{2}^{4} e^t\delta(t)dt = 0, \quad \int_{0^-}^{0^+} e^t\delta(t)dt = 1, \quad \int_{-\infty}^{\infty} e^{t-3}\delta(t-3)dt = 1$$

In the fourth integral, the argument t of the impulse never becomes zero within the limits of the integral.

2.1.2.1 The Impulse Representation of Signals

A major application of the impulse is to decompose an arbitrary signal $x(t)$ into scaled and shifted impulses, so that the representation and analysis of $x(t)$ becomes easier. In the well-known rectangular rule of numerical integration, an arbitrary signal $x(t)$ is approximated by a series of rectangles. Each rectangle is of fixed width, say a, and height equal to a known value of $x(t)$ in that interval. The area of the rectangle is an approximation to that of $x(t)$ in that interval. The sum of areas of all such rectangles is an approximation of the *area* enclosed by the signal.

We can as well *represent* $x(t)$ approximately, in each interval of width a, by the area of the corresponding rectangle located at $t = t_0$ multiplied by a unit area rectangular pulse, $\delta_q(t-t_0)$ of width a and height $\frac{1}{a}$, since the amplitude of the pulse $x(t_0)a\delta_q(t-t_0)$ is $x(t_0)$. For example, $x(t)$ can be represented by $(3.1752)(2)\delta_q(t)$ in Fig. 2.4a, shown in dashdot line. The sum of a succession of all such rectangles is an approximation to $x(t)$. As the width a is made shorter, the approximation becomes better. For example, $x(t)$ is represented by $(3.0269)(0.8)\delta_q(t)$ in Fig. 2.4b and $(3.0017)(0.2)\delta_q(t)$ in Fig. 2.4c. Eventually, as $a \to 0$, the pulse degenerates into impulse, and the representation becomes exact.

Consider the product of a signal with a shifted impulse $x(t)\delta(t-\tau) = x(\tau)\delta(t-\tau)$. Integrating both sides with respect to τ, we get

$$\int_{-\infty}^{\infty} x(t)\delta(t-\tau)d\tau = x(t)\int_{-\infty}^{\infty} \delta(t-\tau)d\tau = x(t) = \int_{-\infty}^{\infty} x(\tau)\delta(t-\tau)d\tau$$

The integrand $x(\tau)\delta(t-\tau)d\tau$, which is one of the constituent impulses of $x(t)$, is a shifted impulse $\delta(t-\tau)$ located at $t=\tau$ with strength $x(\tau)d\tau$. The integration operation, with respect to τ, sums all these impulses to form $x(t)$. It should be emphasized that the integral, in this instance, represents a sum of continuum of impulses (not of evaluating any area). Therefore, the signal $x(t)$ is represented by the sum of scaled and shifted continuum of impulses with the strength of the impulse at any t being $x(t)dt$. The unit-impulse is the basis function and $x(t)dt$ is its coefficient. As the area enclosed by the integrand is nonzero only at the point $t=\tau$, the integral is effective only at that point. By varying the value of t, we can sift out all the values of $x(t)$.

Let a quasi-impulse, $\delta_q(t)$, be defined by a rectangular pulse with its base of width a, from $t=0$ to $t=a$, and height $\frac{1}{a}$. Assume that the signal, $x(t) = e^{-1.2t}(u(t)-u(t-1.5))$, is approximated by rectangles with width a and height equal to the value of $x(t)$ at the beginning of the corresponding rectangle. Figure 2.6 shows the approximation of $x(t)$ by rectangular pulses of width $a=0.5$. We break up $x(t)$ so that it is expressed as a sum of sections of width $a=0.5$.

$$\begin{aligned} x(t) &= x0(t) + x1(t) + x2(t) \\ &= e^{-1.2t}(u(t)-u(t-0.5)) \\ &\quad + e^{-1.2t}(u(t-0.5)-u(t-1)) \\ &\quad + e^{-1.2t}(u(t-1)-u(t-1.5)) \end{aligned}$$

By replacing each section by a function that is constant with a value equal to that of $x(t)$ at the beginning of the section, we get

$$\begin{aligned} x(t) &\approx xa(t) + xb(t) + xc(t) \\ &= e^{-1.2(0)(0.5)}(u(t)-u(t-0.5)) \\ &\quad + e^{-1.2(1)(0.5)}(u(t-0.5)-u(t-1)) \\ &\quad + e^{-1.2(2)(0.5)}(u(t-1)-u(t-1.5)) \end{aligned}$$

By multiplying and dividing by $a=0.5$, we get

$$\begin{aligned} x(t) &\approx xa(t) + xb(t) + xc(t) \\ &= e^{-1.2(0)(0.5)}\left(\frac{u(t)-u(t-0.5)}{0.5}\right)(0.5) \\ &\quad + e^{-1.2(1)(0.5)}\left(\frac{u(t-0.5)-u(t-1)}{0.5}\right)(0.5) \end{aligned}$$

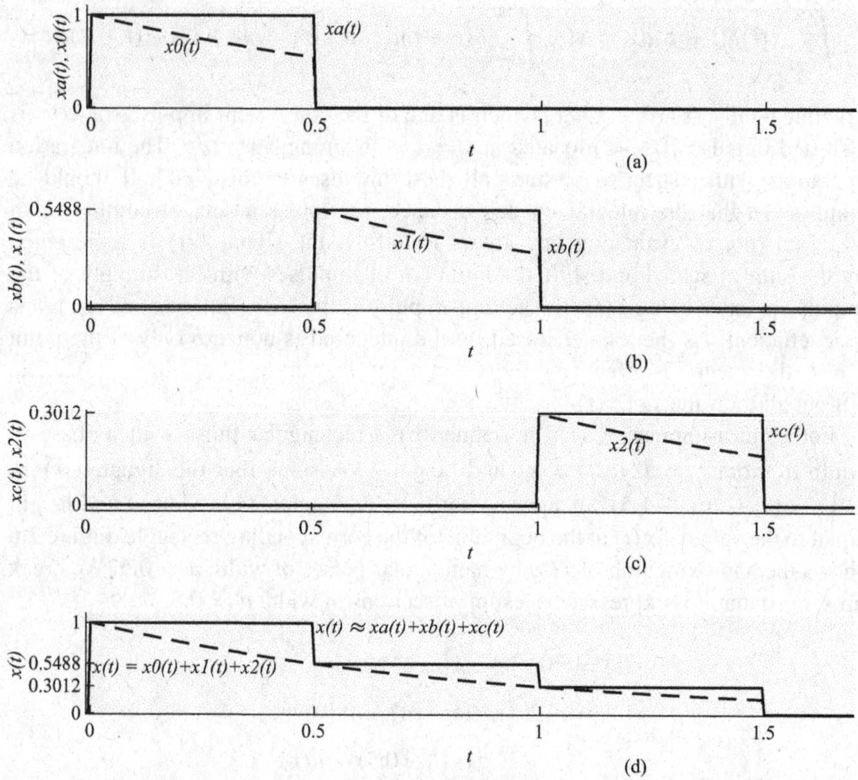

Fig. 2.6 The approximation of a signal by a sum of scaled and shifted rectangular pulses: (**a**) $xa(t) = \delta_q(t)(0.5)$, $x0(t) = e^{-1.2t}(u(t) - u(t-0.5))$; (**b**) $xb(t) = e^{-0.6}\delta_q(t-0.5)(0.5)$, $x1(t) = e^{-1.2t}(u(t-0.5) - u(t-1))$; (**c**) $xc(t) = e^{-1.2}\delta_q(t-1)(0.5)$, $x2(t) = e^{-1.2t}(u(t-1) - u(t-1.5))$; (**d**) $x(t) = x0(t) + x1(t) + x2(t) = e^{-1.2t}(u(t) - u(t-1.5)) \approx \delta_q(t)(0.5) + e^{-0.6}\delta_q(t-0.5)(0.5) + e^{-1.2}\delta_q(t-1)(0.5)$

$$+ e^{-1.2(2)(0.5)}\left(\frac{u(t-1) - u(t-1.5)}{0.5}\right)(0.5)$$

$$\begin{aligned} x(t) &\approx xa(t) + xb(t) + xc(t) \\ &= \delta_q(t)(0.5) + e^{-0.6}\delta_q(t-0.5)(0.5) + e^{-1.2}\delta_q(t-1)(0.5) \\ &= \sum_{n=0}^{2} e^{-1.2(n)(0.5)}\delta_q(t-(n)(0.5))(0.5) \end{aligned}$$

In general, we approximate an arbitrary $x(t)$ as

$$x(t) \approx \sum_{n=-\infty}^{\infty} x((n)(a))\delta_q(t-(n)(a))(a),$$

which reverts to the exact representation of $x(t)$

$$x(t) = \int_{-\infty}^{\infty} x(\tau)\delta(t-\tau)d\tau,$$

as $a \to 0$ (a is replaced by the differential $d\tau$ and $(n)(a)$ becomes the continuous variable τ).

2.1.2.2 The Unit-Impulse as the Derivative of the Unit-Step

A function, which is the derivative of the unit-step function, must have its integral equal to zero for $t < 0$ and one for $t > 0$. Therefore, such a function must be defined to have unit area at $t = 0$ and zero area elsewhere. Figure 2.7a shows the quasi-impulse $\delta_q(t)$ with width 1 and height $\frac{1}{1} = 1$ (solid line), and Fig. 2.7b shows its integral $u_q(t)$ (solid line), which is an approximation to the unit-step function. As the width of $\delta_q(t)$ is reduced and its height correspondingly increased, as shown in Fig. 2.7a (dashed line with width 0.5 and dotted line with width 0.2), $\delta_q(t)$ resembles more like an impulse, and the corresponding integrals, shown in Fig. 2.7b (dashed and dotted lines), become better approximations to the unit-step function. At any stage in the limiting process, $u_q(t)$ remains the integral of $\delta_q(t)$, and $\delta_q(t)$ remains the derivative (except at the corners) of $u_q(t)$ and is defined to be so even in the limit (for the sake of mathematical convenience) as the width of $\delta_q(t)$ tends to zero. $\delta_q(t)$ and $u_q(t)$ become, respectively, the unit-impulse and unit-step functions in the limit and

$$\frac{du(t)}{dt} = \delta(t) \quad \text{and} \quad \int_{-\infty}^{t} \delta(\tau)d\tau = u(t)$$

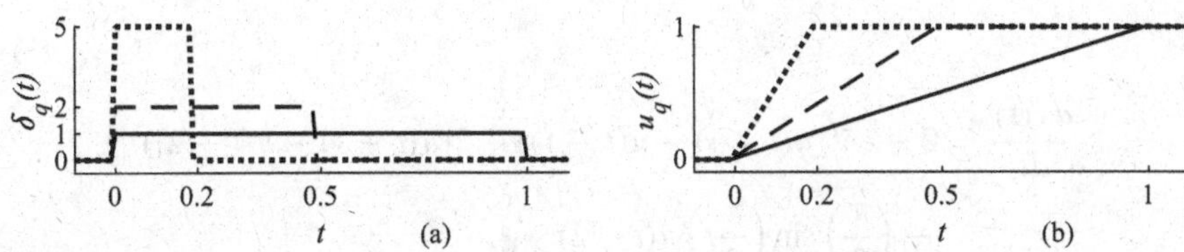

Fig. 2.7 **(a)** The quasi-impulse $\delta_q(t)$ with width 1 and height $\frac{1}{1} = 1$ (solid line), and with width 0.5 and height $\frac{1}{0.5} = 2$ (dashed line), and with width 0.2 and height $\frac{1}{0.2} = 5$ (dotted line), and **(b)** their integrals $u_q(t)$, approaching the unit-step function as the width of the quasi-impulse tends to zero

For example, the voltage across a capacitor is proportional to the integral of the current through it. Therefore, a unit-impulse current passing through a capacitor of one Farad produces a unit-step voltage across the capacitor.

A signal $x(t)$, with step discontinuities, for example, at $t = t_1$ of height $(x(t_1^+) - x(t_1^-))$ and at $t = t_2$ of height $(x(t_2^+) - x(t_2^-))$, can be expressed as

$$x(t) = x_c(t) + (x(t_1^+) - x(t_1^-))u(t - t_1) + (x(t_2^+) - x(t_2^-))u(t - t_2),$$

where $x_c(t)$ is $x(t)$ with the discontinuities removed and $x(t_1^+)$ and $x(t_1^-)$ are, respectively, the right- and left-hand limits of $x(t)$ at $t = t_1$. The derivative of $x(t)$ is given by the generalized function theory as

$$\frac{dx(t)}{dt} = \frac{dx_c(t)}{dt} + \left(x\left(t_1^+\right) - x\left(t_1^-\right)\right)\delta(t - t_1) + \left(x\left(t_2^+\right) - x\left(t_2^-\right)\right)\delta(t - t_2),$$

where $\frac{dx_c(t)}{dt}$ is the ordinary derivative of $x_c(t)$ at all t except at $t = t_1$ and $t = t_2$. Note that $\frac{dx_c(t)}{dt}$ may have step discontinuities. In the expression for $\frac{dx(t)}{dt}$, the impulse terms serve as *indicators* of step discontinuities in its integral, that is, $x(t)$. Therefore, the use of impulses in this manner prevents the loss of step discontinuities in the integration operation, and we get back $x(t)$ exactly by integrating its derivative. That is,

$$x(t) = x(t_0) + \int_{t_0}^{t} \frac{dx(t)}{dt} dt$$

For example, the derivative of the signal $x(t)$, shown in Fig. 2.8a along with $x_c(t)$ in dashed line,

$$\begin{aligned} x(t) &= u(-t - 1) + e^{-t}(u(t + 1) - u(t - 1)) + 2t(u(t - 1) - u(t - 2)) \\ &\quad + \cos\left(\frac{\pi}{2}t\right) u(t - 2) \\ &= x_c(t) + 1.7183u(t + 1) + 1.6321u(t - 1) - 5u(t - 2), \end{aligned}$$

is

$$\begin{aligned} \frac{dx(t)}{dt} &= 0 - e^{-t}(u(t + 1) - u(t - 1)) + 2(u(t - 1) - u(t - 2)) \\ &\quad - \left(\frac{\pi}{2}\right) \sin\left(\frac{\pi}{2}t\right) u(t - 2) \\ &\quad + 1.7183\delta(t + 1) + 1.6321\delta(t - 1) - 5\delta(t - 2), \end{aligned}$$

shown in Fig. 2.8b. Note that $1/e \approx 0.3679$.

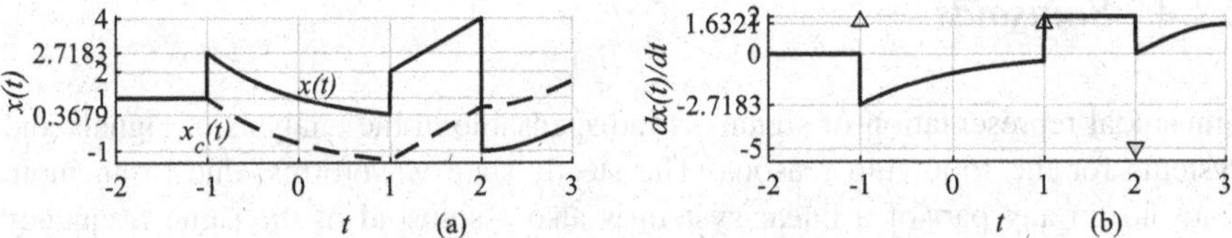

Fig. 2.8 (a) Signal $x(t)$ with step discontinuities and (b) its derivative

2.1.2.3 The Scaling Property of the Impulse

The area enclosed by a time-scaled pulse $x(at)$ and that of its time-reversed version $x(-at)$ is equal to the area enclosed by $x(t)$ divided by $|a|$. Therefore, the scaling property of the impulse is given as

$$\delta(at) = \frac{1}{|a|}\delta(t), \ a \neq 0$$

With $a = -1$, $\delta(-t) = \delta(t)$ implying that the impulse is an even-symmetric signal. For example,

$$\delta(2t+1) = \delta\left(2\left(t+\frac{1}{2}\right)\right) = \frac{1}{2}\delta\left(t+\frac{1}{2}\right) \quad \text{and} \quad \delta\left(\frac{1}{2}t-1\right) = \delta\left(\frac{1}{2}(t-2)\right) = 2\delta(t-2)$$

2.1.3 *The Unit-Ramp Signal*

The unit-ramp signal, shown in Fig. 2.1c, is defined as

$$r(t) = \begin{cases} t \text{ for } t \geq 0 \\ 0 \text{ for } t < 0 \end{cases}$$

The unit-ramp signal linearly increases, with unit slope, for positive values of its argument, and its value is zero for negative values of its argument.

The unit-impulse, unit-step, and unit-ramp signals are closely related. The unit-impulse signal $\delta(t)$ is equal to the derivative of the unit-step signal, $\frac{du(t)}{dt}$, according to the generalized function theory. The unit-step signal $u(t)$ is equal to $\int_{-\infty}^{t} \delta(\tau)d\tau$. The unit-step signal $u(t)$ is equal to $\frac{dr(t)}{dt}$, except at $t = 0$, where no unique derivative exists. The unit-ramp signal $r(t)$ is equal to

$$r(t) = tu(t) = \int_{-\infty}^{t} u(\tau)d\tau$$

2.1.4 Sinusoids

Sinusoidal representation of signals is indispensable in the analysis of signals and systems for the following reasons. The steady-state waveforms, due to an input sinusoid, in any part of a linear system is also a sinusoid of the same frequency as that of the input differing only in its amplitude and phase. In addition, the sum of any number of sinusoids of the same frequency is also a sinusoid of the same frequency. The frequency of a sinusoid remains the same in its derivative form also. Therefore, system models, such as differential equation and convolution, reduce to algebraic equations for a sinusoidal input for linear systems. Further, due to the orthogonal property, an arbitrary signal can be decomposed into a set of sinusoids easily. In addition, this decomposition can be implemented faster, in practice, using fast numerical algorithms resulting in finding the system output faster than other methods. Physical systems also, such as a combination of an inductor and a capacitor, produce an output of sinusoidal nature.

2.1.4.1 The Polar Form of Sinusoids

The polar form specifies a sinusoid, in terms of its amplitude and phase, as

$$x(t) = A\cos(\omega t + \theta), \qquad -\infty < t < \infty$$

where A, ω, and θ are, respectively, the amplitude, the angular frequency, and the phase. The amplitude A is the distance of either peak of the waveform from the horizontal axis. Let the period of the sinusoid be T seconds. Then, as

$$\cos(\omega(t+T)+\theta) = \cos(\omega t + \omega T + \theta) = \cos(\omega t + \theta) = \cos(\omega t + \theta + 2\pi),$$

$T = \frac{2\pi}{\omega}$. The cyclic frequency, denoted by f, of a sinusoid is the number of cycles per second and is equal to the reciprocal of the period, $f = \frac{1}{T} = \frac{\omega}{2\pi}$ cycles per second (Hz). Frequecy is the number of time the signal repeats itself in one unit of time. The angular frequency, the number of radians per second, of a sinusoid is 2π times its cyclic frequency, that is, $\omega = 2\pi f$ radians per second. For example, consider the sinusoid $3\cos(\frac{\pi}{8}t + \frac{\pi}{3})$, with $A = 3$. The angular frequency is $\omega = \frac{\pi}{8}$ radians per second. The period is $T = \frac{2\pi}{\frac{\pi}{8}} = 16$ seconds. The cyclic frequency is $f = \frac{1}{T} = \frac{1}{16}$ Hz. The phase is $\theta = \frac{\pi}{3}$ radians. The phase can also be expressed in terms of seconds, as $\cos(\omega t + \theta) = \cos(\omega(t + \frac{\theta}{\omega}))$. The phase of $\frac{\pi}{3}$ radians corresponds to $\frac{8}{3}$ seconds. As it repeats a pattern over its period, the sinusoid remains the same by a shift of an integral number of its period. A phase-shifted sine wave can be expressed as a phase-shifted cosine wave, $A\sin(\omega t + \theta) = A\cos(\omega t + (\theta - \frac{\pi}{2}))$. The phase of the sinusoid

$$\sin\left(\frac{2\pi}{16}t + \frac{\pi}{3}\right) = \cos\left(\frac{2\pi}{16}t + \left(\frac{\pi}{3} - \frac{\pi}{2}\right)\right) = \cos\left(\frac{2\pi}{16}t - \frac{\pi}{6}\right)$$

is $-\frac{\pi}{6}$ radians. A phase-shifted cosine wave can be expressed as a phase-shifted sine wave, $A\cos(\omega t + \theta) = A\sin(\omega t + (\theta + \frac{\pi}{2}))$.

2.1.4.2 The Rectangular Form of Sinusoids

An arbitrary sinusoid is neither even- nor odd-symmetric. The even component of a sinusoid is the cosine waveform, and the odd component is the sine waveform. That is, a sinusoid is a linear combination of cosine and sine waveforms of the same frequency as that of the sinusoid. Expressing a sinusoid in terms of its cosine and sine components is called its rectangular form and is given as

$$A\cos(\omega t + \theta) = A\cos(\theta)\cos(\omega t) - A\sin(\theta)\sin(\omega t) = C\cos(\omega t) + D\sin(\omega t),$$

where $C = A\cos\theta$ and $D = -A\sin\theta$. The inverse relation is $A = \sqrt{C^2 + D^2}$ and $\theta = \cos^{-1}(\frac{C}{A}) = \sin^{-1}(\frac{-D}{A})$. For example,

$$\cos\left(\frac{2\pi}{16}t + \frac{\pi}{3}\right) = \frac{1}{2}\cos\left(\frac{2\pi}{16}t\right) - \frac{\sqrt{3}}{2}\sin\left(\frac{2\pi}{16}t\right)$$

$$\frac{3}{\sqrt{2}}\cos\left(\frac{2\pi}{16}t\right) + \frac{3}{\sqrt{2}}\sin\left(\frac{2\pi}{16}t\right) = 3\cos\left(\frac{2\pi}{16}t - \frac{\pi}{4}\right)$$

Figure 2.9a shows the continuous cosine waveform $\cos(\frac{2\pi}{16}t)$. Figure 2.9b shows the continuous sine waveform $\sin(\frac{2\pi}{12}t)$. Cosine and sine waves are important special cases of a sinusoid. The peak in (a) occurs at $t = 0$, and its phase is defined as zero degrees. As $\sin(\frac{2\pi}{12}t) = \cos(\frac{2\pi}{12}t - \frac{\pi}{2})$, the peak occurs at $t = 3$ (the peak delayed by one-quarter of a cycle). Figure 2.9c shows the continuous cosine waveform $\cos(\frac{2\pi}{16}t - \frac{\pi}{3})$. With $t = 8/3$, the argument of the waveform becomes zero, and its first peak closest to $t = 0$ occurs. Therefore, its phase shift is $-\pi/3$ radians. Figure 2.9d shows the continuous sine waveform

$$\sin\left(\frac{2\pi}{12}t + \frac{\pi}{6}\right) = \cos\left(\frac{2\pi}{12}t + \frac{\pi}{6} - \frac{\pi}{2}\right) = \cos\left(\frac{2\pi}{12}t - \frac{\pi}{3}\right)$$

With $t = 2$, the argument of the sine waveform becomes $\frac{\pi}{2}$ and that of its equivalent cosine waveform becomes zero. Therefore, its phase shift is $-\pi/3$ radians.

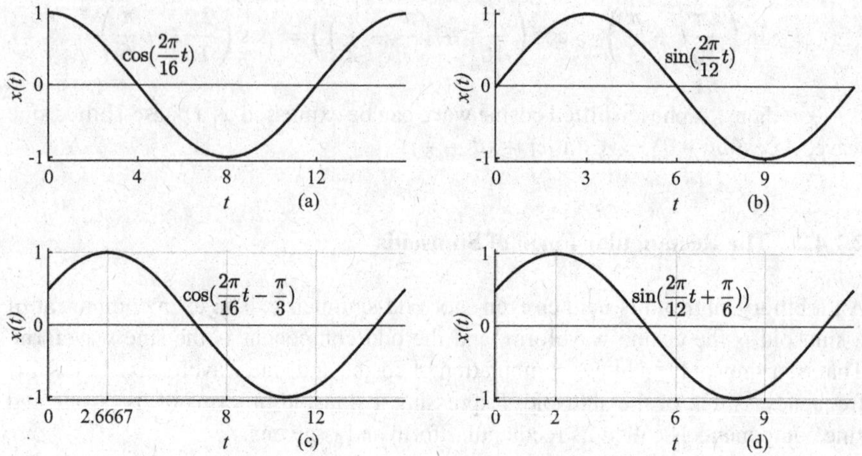

Fig. 2.9 **(a)** $\cos(\frac{2\pi}{16}t)$; **(b)** $\sin(\frac{2\pi}{12}t)$; **(c)** $\cos(\frac{2\pi}{16}t - \frac{\pi}{3})$; **(d)** $\sin(\frac{2\pi}{12}t + \frac{\pi}{6})$

2.1.4.3 The Sum of Sinusoids of the Same Frequency

The sum of sinusoids of arbitrary amplitudes and phases but with the same frequency is also a sinusoid of the same frequency. Let

$$x_1(t) = A_1 \cos(\omega t + \theta_1) \qquad \text{and} \qquad x_2(t) = A_2 \cos(\omega t + \theta_2)$$

Then, $x(t) = x_1(t) + x_2(t) = A \cos(\omega t + \theta)$, where

$$A = \sqrt{A_1^2 + A_2^2 + 2A_1 A_2 \cos(\theta_1 - \theta_2)}$$
$$\theta = \tan^{-1} \frac{A_1 \sin(\theta_1) + A_2 \sin(\theta_2)}{A_1 \cos(\theta_1) + A_2 \cos(\theta_2)}$$

Any number of sinusoids can be combined into a single sinusoid by repeatedly using the formulas. Note that the formula for the rectangular form of the sinusoid is a special case of the sum of two sinusoids, one sinusoid being the cosine and the other being sine.

2.1.4.4 The Complex Sinusoids

The complex sinusoid is given as

$$x(t) = Ae^{j(\omega t+\theta)} = Ae^{j\theta}e^{j\omega t}, \qquad -\infty < t < \infty$$

The term $e^{j\omega t}$ is the complex sinusoid with unit magnitude and zero phase. Its complex (amplitude) coefficient is $Ae^{j\theta}$. The amplitude and phase of the sinusoid is represented by the single complex number $Ae^{j\theta}$. By adding its complex conjugate, $Ae^{-j(\omega t+\theta)}$, with itself and dividing by two, due to Euler's identity, we get

$$x(t) = \frac{A}{2}\left(e^{j(\omega t+\theta)} + e^{-j(\omega t+\theta)}\right) = A\cos(\omega t + \theta)$$

The use of two complex sinusoids to represent a single real sinusoid requires four real quantities instead of two. This redundancy in terms of storage and operations can be reduced by about a factor of 2 in practical implementation of the algorithms.

2.1.4.5 Exponentially Varying Amplitude Sinusoids

An exponentially varying amplitude sinusoid, $Ae^{at}\cos(\omega t + \theta)$, is obtained by multiplying a sinusoid, $A\cos(\omega t+\theta)$, with a real exponential, e^{at}. The more familiar constant amplitude sinusoid results when $a = 0$. If ω is equal to zero, then we get a real exponential. Sinusoids, $x(t) = e^{-0.1t}\cos(\frac{2\pi}{8}t)$ and $x(t) = e^{0.1t}\cos(\frac{2\pi}{8}t)$, with exponentially varying amplitudes are shown, respectively, in Figure 2.10a and b.

The complex exponential representation of a exponentially varying amplitude sinusoid is given as

$$x(t) = \frac{A}{2}e^{at}\left(e^{j(\omega t+\theta)} + e^{-j(\omega t+\theta)}\right) = Ae^{at}\cos(\omega t + \theta)$$

Figure 2.11a and b shows, respectively, exponentially varying amplitude real and complex exponentials, $x(t) = e^{-0.1t}$ and $x(t) = e^{(-0.1+j(3\frac{2\pi}{16}))t}$. The **time constant**, which is the inverse of the coefficient associated with the independent variable t, is 10. The peak value is 1 at $t = 0$ in 2.11(a). At $t = 10$ (one time constant), its value is $1/e \approx 0.37$. At $t = 20$ (two time constants), its value is $(1/e)^2 \approx 0.135$ and so on.

A major advantage of the sinusoidal or complex exponential input functions is that the particular solution of linear differential equations with constant coefficients can be obtained by solving algebraic equations. Consider the circuit with a

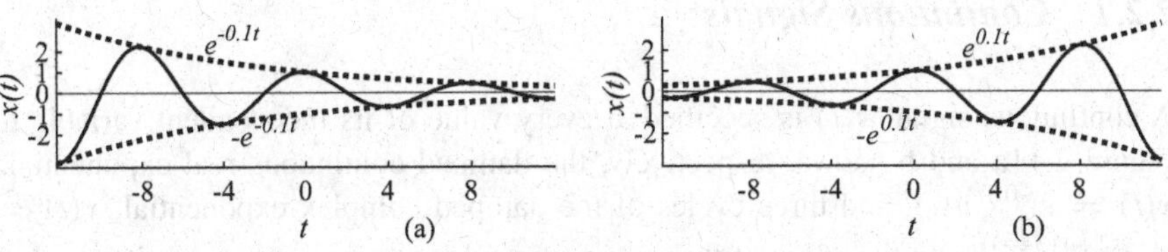

Fig. 2.10 **(a)** Exponentially decreasing amplitude cosine wave, $x(t) = e^{-0.1t}\cos(\frac{2\pi}{8}t)$; **(b)** exponentially increasing amplitude cosine wave, $x(t) = e^{0.1t}\cos(\frac{2\pi}{8}t)$

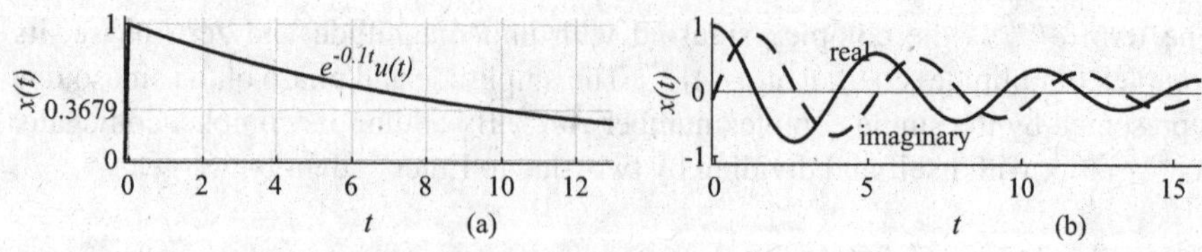

Fig. 2.11 (**a**) The damped continuous real exponential, $x(t) = e^{-0.1t}u(t)$; (**b**) the damped continuous complex exponential, $x(t) = e^{(-0.1+j(3\frac{2\pi}{16}))t}$

resistance R in series with an inductance L. The differential equation characterizing the circuit, with the input $Ee^{j\omega t}$, is

$$L\frac{di(t)}{dt} + Ri(t) = Ee^{j\omega t}$$

The task is to find the current $i(t)$ in the circuit. A function of the form $i(t) = Ie^{j\omega t}$ satisfies the differential equation. Substituting $Ie^{j\omega t}$ for $i(t)$, we get

$$(j\omega L + R)Ie^{j\omega t} = Ee^{j\omega t}$$

Solving for I, we get

$$I = \frac{E}{R + j\omega L} = \frac{E}{Z},$$

where $Z = R + j\omega L$ is called the impedance of the circuit.

2.2 Classification of Signals

Signals are classified into different types, and the representation and analysis of a signal depend on its type.

2.2.1 *Continuous Signals*

A continuous signal $x(t)$ is specified at every value of its independent variable t. Figure 2.11a and b shows, respectively, the damped continuous real exponential, $x(t) = e^{-0.1t}u(t)$, and three cycles of the damped complex exponential, $x(t) = e^{(-0.1+j(3\frac{2\pi}{16}))t}$. As the value of the exponential is decreasing with time, it is called a damped or decaying exponential, characterized by the negative constant, -0.1, in its exponent. An exponential e^{at}, where a is a positive constant, is an example of a growing exponential, as its value is increasing with time. We denote a continuous signal, using the independent variable t, as $x(t)$. We call this representation the time-domain representation, although the independent variable is not time for some signals. While most signals, in practical applications, are real-valued, complex-valued signals are often used in analysis. A complex-valued or complex signal is an ordered pair of real signals. The damped complex exponential signal, shown in Fig. 2.11b, can be expressed, using Euler's identity, in terms of damped cosine and sine signals as

$$x(t) = e^{(-0.1+j(3\frac{2\pi}{16}))t} = e^{(-0.1t)} \cos\left(3\frac{2\pi}{16}t\right) + je^{(-0.1t)} \sin\left(3\frac{2\pi}{16}t\right)$$

The real and imaginary parts of $x(t)$ are, respectively, $e^{(-0.1t)}\cos(3\frac{2\pi}{16}t)$ (shown by the solid line in Fig. 2.11b) and $e^{(-0.1t)}\sin(3\frac{2\pi}{16}t)$ (shown by the dashed line in Fig. 2.11b).

2.2.2 *Periodic and Aperiodic Signals*

The smallest positive number $T > 0$ satisfying the condition $x(t+T) = x(t)$, for all t, is the fundamental period of the continuous periodic signal $x(t)$. The reciprocal of the fundamental period is the fundamental cyclic frequency, $f = \frac{1}{T}$ Hz (cycles per second). The fundamental angular frequency is $\omega = 2\pi f = \frac{2\pi}{T}$ radians per second. Over the interval $-\infty < t < \infty$, a periodic signal repeats its values over any interval equal to its period, at intervals of its period. Cosine and sine waves are typical examples of a periodic signal. A signal with constant value (DC) is periodic with any period. In Fourier analysis, it is considered as $A\cos(\omega t)$ or $Ae^{j\omega t}$ with the frequency ω equal to zero (period equal to ∞). When the period of a periodic signal approaches infinity, it degenerates into an aperiodic signal. The exponential signal, shown in Fig. 2.10a, is an aperiodic signal.

It is easier to decompose an arbitrary signal in terms of some periodic signals, such as complex exponentials, and the input-output relationship of a LTI system becomes a multiplication operation for this type of input signals. For these reasons, most of the analysis of practical signals, which are mostly aperiodic having arbitrary amplitude profile, is carried out using periodic basic signals.

2.2.3 *Energy and Power Signals*

The energy of a signal $x(t)$ is expressed as the integral of the squared magnitude of its values as

$$E = \int_{-\infty}^{\infty} |x(t)|^2 dt$$

Aperiodic signals with finite energy are called energy signals. The energy of $x(t) = 3e^{-t},\ t \geq 0$ is

$$E = \int_{0}^{\infty} |3e^{-t}|^2 dt = \frac{9}{2}$$

If the energy of a signal is infinite, then it may be possible to characterize it in terms of its average power. The average power is defined as

$$P = \lim_{T\to\infty} \frac{1}{T} \int_{-\frac{T}{2}}^{\frac{T}{2}} |x(t)|^2 dt$$

For periodic signals, the average power can be computed over one period as

$$P = \frac{1}{T} \int_{-\frac{T}{2}}^{\frac{T}{2}} |x(t)|^2 dt,$$

where T is the period. Signals, periodic or aperiodic, with finite average power are called power signals. Cosine and sine waveforms are typical examples of power signals. The average power of the cosine wave $3\cos(\frac{\pi}{8}t)$ is

$$P = \frac{1}{16} \int_{-8}^{8} |3\cos\left(\frac{\pi}{8}t\right)|^2 dt = \frac{9}{32} \int_{-8}^{8} \left(1 + \cos\left(2\frac{\pi}{8}t\right)\right) dt = \frac{9}{2}$$

A signal is an energy signal or a power signal, since the average power of an energy signal is zero while that of a power signal is finite. Signals with infinite power and infinite energy, such as $x(t) = t,\ t \geq 0$, are neither power signals nor energy signals. The measures of signal power and energy are indicators of the signal size, since the actual energy or power depends on the load.

2.2.4 *Even- and Odd-Symmetric Signals*

The analysis of a signal can be simplified by exploiting its symmetry. A signal $x(t)$ is even-symmetric, if $x(-t) = x(t)$ for all t. The signal is symmetrical about the vertical axis at the origin. The cosine waveform is an example of an even-symmetric signal. A signal $x(t)$ is odd-symmetric, if $x(-t) = -x(t)$ for all t. The signal is asymmetrical about the vertical axis at the origin. For an odd-symmetric signal, $x(0) = 0$. The sine waveform is an example of an odd-symmetric signal.

The sum $(x(t) + y(t))$ of two odd-symmetric signals, $x(t)$ and $y(t)$, is an odd-symmetric signal, since $x(-t) + y(-t) = -x(t) - y(t) = -(x(t) + y(t))$. For example, the sum of two sine signals is an odd-symmetric signal. The sum $(x(t) + y(t))$ of two even-symmetric signals, $x(t)$ and $y(t)$, is an even-symmetric signal, since $x(-t)+y(-t) = (x(t)+y(t))$. For example, the sum of two cosine signals is an even-symmetric signal. The sum $(x(t) + y(t))$ of an odd-symmetric signal $x(t)$ and an even-symmetric signal $y(t)$ is neither even-symmetric nor odd-symmetric, since $x(-t) + y(-t) = -x(t) + y(t) = -(x(t) - y(t))$. For example, the sum of cosine and sine signals with nonzero amplitudes is neither even-symmetric nor odd-symmetric.

Since $x(t)y(t) = (-x(-t))(-y(-t)) = x(-t)y(-t)$, the product of two odd-symmetric or two even-symmetric signals is an even-symmetric signal. The product $z(t) = x(t)y(t)$ of an odd-symmetric signal $y(t)$ and an even-symmetric signal $x(t)$ is an odd-symmetric signal, since $z(-t) = x(-t)y(-t) = x(t)(-y(t)) = -z(t)$.

An arbitrary signal $x(t)$ can be decomposed in terms of its even-symmetric and odd-symmetric components, $x_e(t)$ and $x_o(t)$, respectively. That is, $x(t) = x_e(t) + x_o(t)$. Replacing t by $-t$, we get $x(-t) = x_e(-t)+x_o(-t) = x_e(t)-x_o(t)$. Solving for $x_e(t)$ and $x_o(t)$, we get

$$x_e(t) = \frac{x(t) + x(-t)}{2} \quad \text{and} \quad x_o(t) = \frac{x(t) - x(-t)}{2}$$

As the integral of an odd-symmetric signal $x_0(t)$, over symmetric limits, is zero,

$$\int_{-t_0}^{t_0} x_0(t)\,dt = 0$$

For an even-symmetric signal

$$\int_{-t_0}^{t_0} x(t)\,dt = \int_{-t_0}^{t_0} x_e(t)\,dt = 2\int_{0}^{t_0} x_e(t)\,dt$$

For example, the even-symmetric component of $x(t) = e^{j(\frac{2\pi}{16}t+\frac{\pi}{3})}$ is

$$x_e(t) = \frac{x(t) + x(-t)}{2} = e^{j(\frac{\pi}{3})}\frac{e^{j(\frac{2\pi}{16}t)} + e^{j(\frac{2\pi}{16}(-t))}}{2} = e^{j(\frac{\pi}{3})}\cos\left(\frac{2\pi}{16}t\right)$$

The odd-symmetric component is

$$x_o(t) = \frac{x(t) - x(-t)}{2} = e^{j(\frac{\pi}{3})}\frac{e^{j(\frac{2\pi}{16}t)} - e^{j(\frac{2\pi}{16}(-t))}}{2} = je^{j(\frac{\pi}{3})}\sin\left(\frac{2\pi}{16}t\right)$$

Note that $x(t) = x_e(t) + x_o(t)$. The complex exponential, its even component, and its odd component are shown, respectively, in Fig. 2.12a, b, and c.

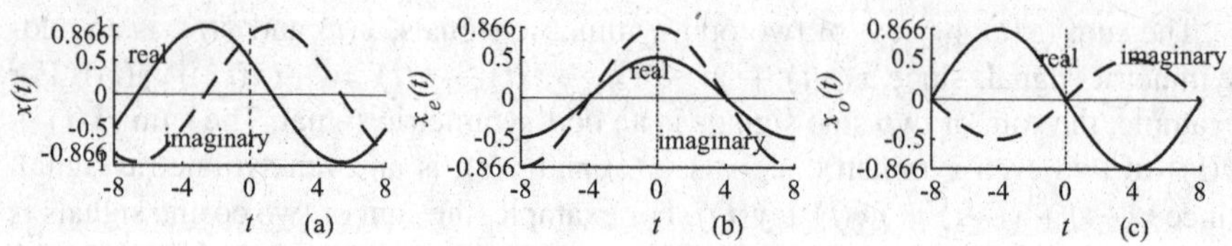

Fig. 2.12 (**a**) The complex exponential $x(t) = e^{j(\frac{2\pi}{16}t+\frac{\pi}{3})}$; (**b**) its even component $x_e(t) = e^{j(\frac{\pi}{3})}\cos(\frac{2\pi}{16}t)$; (**c**) its odd component $x_o(t) = je^{j(\frac{\pi}{3})}\sin(\frac{2\pi}{16}t)$

2.2.5 *Causal and Noncausal Signals*

Most signals, in practice, occur at some finite time instant, usually chosen as $t = 0$, and are considered identically zero before this instant. These signals, with $x(t) = 0$ for $t < 0$, are called causal signals (e.g., the exponential shown in Fig. 2.11a). Signals, with $x(t) \neq 0$ for $t < 0$, are called noncausal signals (e.g., the complex exponential shown in Fig. 2.10b).

2.3 Signal Operations

In addition to the arithmetic operations, time shifting, time reversal, and time scaling operations are also commonly used in the analysis of continuous signals. The three operations described in this section are with respect to the independent variable, t.

2.3.1 *Time Shifting*

A signal $x(t)$ is time shifted by T seconds by replacing t by $t-T$. The value of $x(t)$ at $t = t_0$ occurs at $t = t_0 + T$ in $x(t - T)$. Graphically, it amounts to shifting the plot of the function forward (T positive) or backward by T. The rectangular pulse $x(t) = u(t-1) - u(t-3)$, shown in Fig. 2.13 by a solid line, is a combination of two delayed unit-step signals. The right shifted pulse $x(t-1) = u(t-2) - u(t-4)$, shown in Fig. 2.13 by a dashed line, is $x(t)$ shifted by 1 s to the right (delayed by 1 s, as the values of $x(t)$ occur 1 s late). For example, the first nonzero value occurs at $t = 2$ as $u(2-2) - u(2-4) = 1$. That is, the value of $x(t)$ at t_0 occurs in the shifted pulse 1 s later at $t_0 + 1$. The pulse $x(t + 1.5)$, shown in Fig. 2.13 by a dotted line, is $x(t)$ shifted by 1.5 s to the left (advanced by 1.5 s, as the values of $x(t)$ occur 1.5 s early). For example, the first nonzero value occurs at $t = -0.5$ as $u(-0.5 + 0.5) - u(-0.5 - 1.5) = 1$. That is, the value of $x(t)$ at t_0 occurs in the shifted pulse 1.5 s earlier at $t_0 - 1.5$.

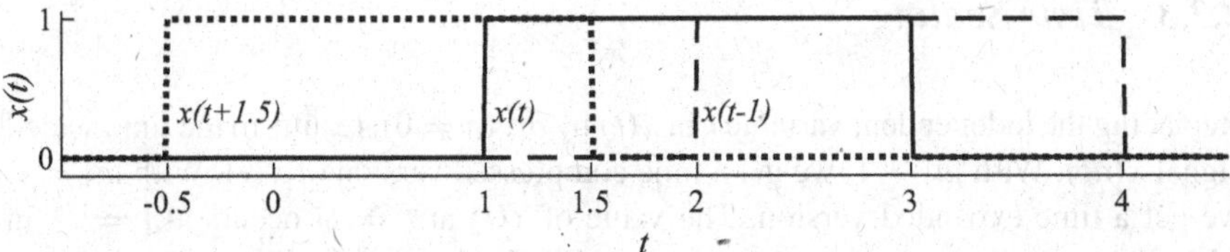

Fig. 2.13 The rectangular pulse, $x(t) = u(t-1) - u(t-3)$; the right shifted pulse, $x(t-1)$; and the left shifted pulse, $x(t+1.5)$

2.3.2 *Time Reversal*

Replacing the independent variable t in $x(t)$ by $-t$ results in the time reversed or folded signal $x(-t)$. The value of $x(t)$ at $t = t_0$ occurs at $t = -t_0$ in $x(-t)$. A signal and its time reversed version are mirror images of each other. The signal $x(t) = r(t+1) - r(t-1)$, shown in Fig. 2.14 by a solid line, is a combination of two shifted unit-ramp signals. Consider the folded and shifted signal $x(-t-1) = x(-(t+1)) = r(-t) - r(-t-2)$, shown in Fig. 2.14 by a dashed line. This signal can be formed by first folding $x(t)$ to get $x(-t)$ and then shifting it to the left by 1 s (t is replaced by $(t+1)$). This signal can also be formed by first shifting $x(t)$ to the right by 1 s to get $x(t-1)$ and then folding it about the vertical axis at the origin (t is replaced by $-t$). The value of the signal $x(t)$ at t_0 occurs in the folded and shifted signal at $-t_0 - 1$. Consider the folded and shifted signal $x(-t+3) = x(-(t-3)) = r(-t+4) - r(-t+2)$, shown in Fig. 2.14 by a dotted line. This signal can be formed by first folding $x(t)$ to get $x(-t)$ and then shifting it to the right by 3 s (t is replaced by $(t-3)$). This signal can also be formed by first shifting $x(t)$ to the left by 3 s to get $x(t+3)$ and then folding it about the vertical axis at the origin (t is replaced by $-t$). The value of $x(t)$ at t_0 occurs in the folded and shifted signal at $-t_0 + 3$.

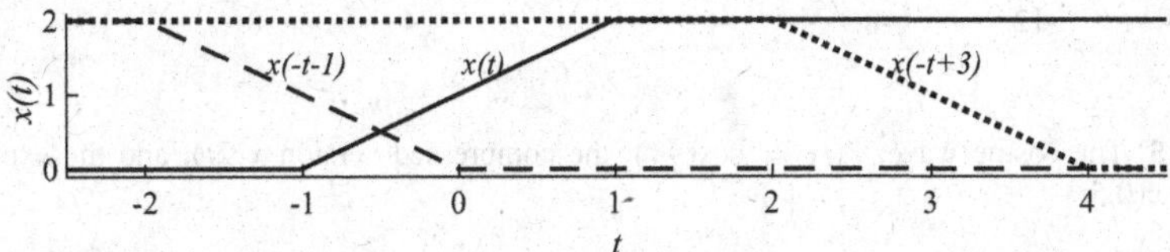

Fig. 2.14 The signal, $x(t) = r(t+1) - r(t-1)$; the shifted and folded signal, $x(-t-1)$; and the shifted and folded signal, $x(-t+3)$

2.3.3 Time Scaling

Replacing the independent variable t in $x(t)$ by at, $(a \neq 0)$, results in the time scaled signal $x(at)$. With $|a| > 1$, we get a time compressed version of $x(t)$. With $|a| < 1$, we get a time expanded version. The value of $x(t)$ at $t = t_0$ occurs at $t = \frac{t_0}{a}$ in $x(at)$. The signal $x(t) = \cos(\frac{\pi}{8}t)$, shown in Fig. 2.15 by a solid line, completes two cycles during 32 s. The time compressed version with $a = 2$, $x(2t) = \cos(\frac{\pi}{8}(2t))$, shown in Fig. 2.15 by a dashed line, completes four cycles during 32 s. The value of the signal $x(t)$ at t occurs at $\frac{t}{2}$ in $x(2t)$. For example, the negative peak at $t = 8$ in $x(t)$ occurs at $t = 4$ in $x(2t)$. The time expanded version with $a = 0.5$, $x(0.5t) = \cos(\frac{\pi}{8}(0.5t))$, shown in Fig. 2.15 by a dotted line, completes one cycle during 32 s. The value of the signal in $x(t)$ at t occurs at $\frac{t}{0.5}$ in $x(0.5t)$. For example, the negative peak at $t = 8$ in $x(t)$ occurs at $t = 16$ in $x(0.5t)$.

In general, the three operations described on a signal $x(t)$ can be expressed as $y(t) = x(at - b)$. The signal $y(t)$ can be generated by replacing t by $(at - b)$. However, it is instructive to consider it as the result of a sequence of two steps: (i) first shifting the signal $x(t)$ by b to get $x(t - b)$ and then (ii) time scaling (replace t by at) the shifted signal by a to get $x(at - b)$. An alternate sequence of two steps is to (i) first time scale the signal $x(t)$ by a to get $x(at)$ and then (ii) shift (replace t by $t - \frac{b}{a}$) the time scaled signal by $\frac{b}{a}$ to get $x(a(t - \frac{b}{a})) = x(at - b)$. Note that time reversal operation is a part of the time scaling operation with a negative.

Let $x(t) = \cos(\frac{\pi}{8}t + \frac{\pi}{4})$, shown in Fig. 2.16 by a solid line. It is required to find $x(-2t+4)$. The shifted signal is $x(t+4) = \cos(\frac{\pi}{8}(t+4)+\frac{\pi}{4})$, shown by a thin line. Now scaling this signal by -2 yields the signal $x(-2t+4) = \cos(\frac{\pi}{8}(-2t+4)+\frac{\pi}{4}) = \cos(\frac{2\pi}{8}t - \frac{3\pi}{4})$, shown in Fig. 2.16 by a dashed line. The value of the signal $x(t)$ at

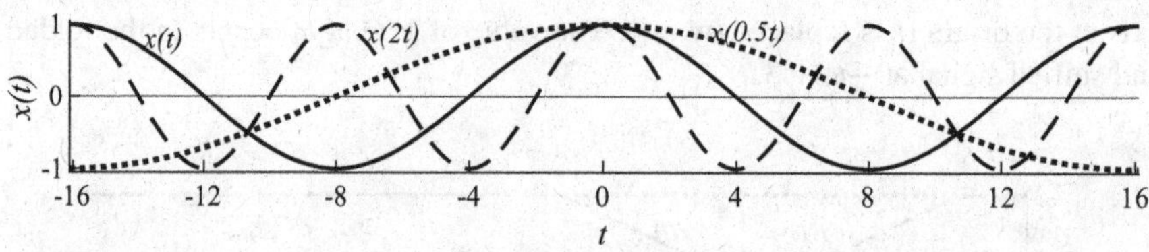

Fig. 2.15 The cosine wave, $x(t) = \cos(\frac{\pi}{8}t)$, the compressed version $x(2t)$, and the expanded version $x(0.5t)$

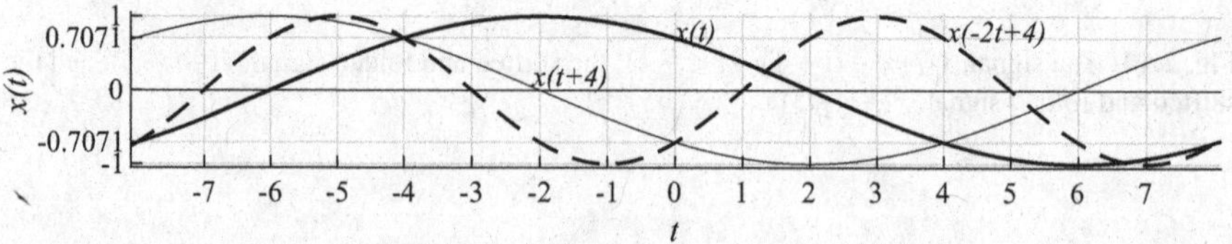

Fig. 2.16 The sinusoid $x(t) = \cos(\frac{\pi}{8}t + \frac{\pi}{4})$ and $x(-2t + 4)$

Table 2.1 Basic signals

Constant	$x(t) = c$
Unit-impulse	$x(t) = \delta(t)$
Unit-step	$x(t) = u(t)$
Unit-ramp	$x(t) = tu(t)$
Real exponential	$x(t) = e^{at}$
Real sinusoid	$x(t) = A\cos(\omega t + \theta)$
Damped real sinusoid	$x(t) = Ae^{at}\cos(\omega t + \theta)$
Complex exponential	$x(t) = Ae^{j(\omega t+\theta)}$
Damped complex exponential	$x(t) = Ae^{at}e^{j(\omega t+\theta)}$

t_0 occurs at $\frac{-t_0+4}{2}$ in $x(-2t+4)$. We could have done the time scaling operation by -2 first to obtain $x(-2t) = \cos(\frac{\pi}{8}(-2t) + \frac{\pi}{4})$. Shifting this signal by $\frac{4}{-2} = -2$ (replace t by $t-2$), we get $x(-2t+4) = \cos(\frac{\pi}{8}(-2t+4) + \frac{\pi}{4})$. The period of $x(t)$ is 16 and that of its compressed version is 8.

Table 2.1 shows a list of basic signals.

2.4 Summary

- In this chapter, continuous signal classifications, basic signals, and signal operations have been presented.
- The representation and analysis of a signal depends on its type.
- Signals have to decompose in terms of some well-defined basic signals, such as the impulse and sinusoid, for compact representation and easier processing. Systems can be characterized by their responses to the basic signals, impulse, unit-step, ramp, and sinusoids.
- The unit-step signal has a value of one for positive values of its argument t, and its value is zero otherwise.
- The impulse function is called a generalized function, since it is defined by the result of its operation (integration) on an ordinary function.
- The unit-ramp signal linearly increases, with unit slope, for positive values of its argument and its value is zero for negative values of its argument.
- A general sinusoidal waveform is a linear combination of trigonometric sine and cosine waveforms or shifted sine and cosine functions.
- While the sinusoidal waveform is generated by practical systems, its mathematically equivalent form, called the complex sinusoid,

$$v(t) = Ve^{j(\omega t+\theta)} = Ve^{j\theta}e^{j\omega t}, \qquad -\infty < t < \infty$$

is found to be indispensable for analysis due to its compact form and ease of manipulation of the exponential function.

- Another commonly encountered signal in signal and systems is the real causal exponential signal.
- In addition to the arithmetic operations, time shifting, time reversal, and time scaling operations are also commonly used in the analysis of continuous signals.

Exercises

2.1 Is $x(t)$ an energy signal, a power signal, or neither? If it is an energy signal, find its energy. If it is a power signal, find its average power.

2.1.1 $x(t) = 3, -1 < t < 1$ and $x(t) = 0$ otherwise.
2.1.2 $x(t) = 2t,\ 0 < t < 1$ and $x(t) = 0$ otherwise.
2.1.3 $x(t) = 4e^{-0.2t}u(t)$.
2.1.4 $x(t) = e^t$.
2.1.5 $x(t) = Ce^{j(\frac{2\pi t}{T})}$.
2.1.6 $x(t) = 2\cos(\frac{\pi t}{4} + \frac{\pi}{3})$.
2.1.7 $x(t) = u(t)$.
2.1.8 $x(t) = t$.
* 2.1.9 $x(t) = 2\frac{1}{t}u(t-1)$.
2.1.10 $x(t) = 3e^{j(\frac{2\pi t}{6})}$.
2.1.11 $x(t) = 3$.
2.1.12 $x(t) = 3\sin(\frac{\pi t}{4} + \frac{\pi}{3})$.

2.2 Is $x(t)$ even-symmetric, odd-symmetric, or neither? List the values of $x(t)$ at $t = -3, -2, -1, 0, 1, 2, 3$.

2.2.1 $x(t) = 3\cos(\frac{2\pi}{6}t + \frac{\pi}{6})$.
2.2.2 $x(t) = 2\sin(\frac{2\pi}{6}t - \frac{\pi}{3})$.
2.2.3 $x(t) = 4\cos(\frac{2\pi}{6}t)$.
2.2.4 $x(t) = 5$.
2.2.5 $x(t) = -2\sin(\frac{2\pi}{6}t)$.
2.2.6 $x(t) = t$.
2.2.7 $x(t) = \frac{\sin(\frac{\pi}{3}t)}{t}$.
2.2.8 $x(t) = \frac{\sin^2(\frac{\pi}{3}t)}{t}$.
2.2.9 $x(t) = e^{-t}$.

2.3 Find the even and odd components of $x(t)$. Verify that the integral of the odd component is zero. Verify that the integral of the even component and that of $x(t)$ are equal.

2.3.1 $x(t) = 2, -1 < t < 1$ and $x(t) = 0$ otherwise.
2.3.2 $x(t) = 3, -1 < t < 2$ and $x(t) = 0$ otherwise.
2.3.3 $x(t) = 2t, -1 < t < 1$ and $x(t) = 0$ otherwise.

* 2.3.4 $x(t) = 3t,\ 0 < t < 1$ and $x(t) = 0$ otherwise.
2.3.5 $x(t) = 2\cos(\frac{2\pi}{6}t - \frac{\pi}{3})$.
2.3.6 $x(t) = e^{-0.2t}u(t)$.
2.3.7 $x(t) = u(t)$.
2.3.8 $x(t) = e^{j(\frac{2\pi}{6}t)}$.
2.3.9 $x(t) = t\,u(t)$.
2.3.10 $x(t) = \sin(\frac{2\pi}{6}t)$.

2.4 Evaluate the integral.

2.4.1 $\int_{-\infty}^{0} u(3t + 1)dt$.
2.4.2 $\int_{-\infty}^{-2} u(\frac{1}{3}t + 2)dt$.
2.4.3 $\int_{-\infty}^{5} u(\frac{1}{2}t - 4)dt$.
* 2.4.4 $\int_{-11}^{\infty} u(-\frac{1}{2}t - 4)dt$.

2.5 Assume that the impulse is approximated by a rectangular pulse, centered at $t = 0$, of width $2a$ and height $\frac{1}{2a}$. Using this quasi-impulse, the signal $x(t)$ is sampled. What are the sample values of $x(t)$ at $t = 0$ with $a = 1$, $a = 0.1$, $a = 0.01$, $a = 0.001$, and $a = 0$.

2.5.1 $x(t) = 4e^{-3t}$.
2.5.2 $x(t) = 2\cos(t)$.
* 2.5.3 $x(t) = 3\sin(t - \frac{\pi}{6})$.
2.5.4 $x(t) = \cos(t + \frac{\pi}{3})$.
2.5.5 $x(t) = \sin(t + \frac{\pi}{4})$.

2.6 Evaluate the integral.

2.6.1 $\int_{-1}^{\infty} \delta(t)e^{t}dt$.
2.6.2 $\int_{0}^{\infty} \delta(t + 1)e^{t}dt$.
* 2.6.3 $\int_{0}^{\infty} \delta(t - 2)e^{t}dt$.
2.6.4 $\int_{-\infty}^{\infty} \delta(t + 1)e^{t}dt$.
2.6.5 $\int_{-\infty}^{\infty} \delta(t + 1)e^{t}u(t)dt$.
2.6.6 $\int_{1}^{5} \delta(t + 1)e^{t}dt$.
2.6.7 $\int_{-4}^{-1} \delta(t + 2)e^{t}dt$.
2.6.8 $\int_{0}^{\infty} \delta(t - 2)e^{(t-2)}dt$.

2.7 A quasi-impulse, $\delta_q(t)$, is defined by a rectangular pulse with its base of width a, from $t = 0$ to $t = a$, and height $\frac{1}{a}$. Assume that the signal $x(t)$ is approximated by a series of rectangles with the height of each rectangle equal to the value of $x(t)$ at the beginning of the corresponding rectangle and width a. Express the signal $x(t)$ in terms of the quasi-impulse with $a = 1$ and $a = 0.5$.

2.7.1 $x(t) = e^{t},\ 0 \le t \le 5$ and $x(t) = 0$ otherwise.

* 2.7.2 $x(t) = \cos(\frac{\pi}{6}t)$, $0 \leq t \leq 4$ and $x(t) = 0$ otherwise.
2.7.3 $x(t) = (t + 3)$, $0 \leq t \leq 3$ and $x(t) = 0$ otherwise.

2.8 Find the derivative of the signal.

2.8.1 $\cos(\pi t)u(t)$.
2.8.2 $\sin(\pi t)u(t)$.
* 2.8.3 $2e^{-3t}u(t)$.

2.9 Evaluate the integral.

2.9.1 $\int_{-\infty}^{\infty} \delta(3t + 1)dt$.
2.9.2 $\int_{-\infty}^{\infty} \delta(\frac{1}{3}t + 2)dt$.
2.9.3 $\int_{-\infty}^{3} \delta(\frac{1}{2}t - 2)dt$.
* 2.9.4 $\int_{-\infty}^{4} \delta(-\frac{1}{3}t + 2)dt$.
2.9.5 $\int_{-\infty}^{4} \delta(-\frac{1}{3}t - 2)dt$.

2.10 Find the rectangular form of the sinusoid. Find the value of $t > 0$ where the first positive peak of the sinusoid occurs. Find the values of t at which the next two consecutive peaks, both negative and positive, occur.

2.10.1 $x(t) = -3\cos(\frac{2\pi}{8}t - \frac{\pi}{3})$.
2.10.2 $x(t) = 2\sin(\frac{2\pi}{6}t + \frac{\pi}{6})$.
* 2.10.3 $x(t) = -5\sin(2\pi t + \frac{\pi}{4})$.
2.10.4 $x(t) = 2\cos(2\pi t + \frac{\pi}{3})$.
2.10.5 $x(t) = 4\cos(\frac{2\pi}{5}t - \frac{13\pi}{6})$.

2.11 Find the polar form of the sinusoid. Find the values of $t > 0$ of the first three zeros of the sinusoid.

2.11.1 $x(t) = -\sqrt{3}\cos(\frac{2\pi}{6}t) - \sin(\frac{2\pi}{6}t)$.
2.11.2 $x(t) = \sqrt{2}\cos(\frac{2\pi}{6}t) - \sqrt{2}\sin(\frac{2\pi}{6}t)$.
2.11.3 $x(t) = -2\cos(\frac{2\pi}{6}t) + 2\sqrt{3}\sin(\frac{2\pi}{6}t)$.
* 2.11.4 $x(t) = \cos(\frac{2\pi}{6}t) + \sin(\frac{2\pi}{6}t)$.
2.11.5 $x(t) = 3\cos(\frac{2\pi}{6}t) - \sqrt{3}\sin(\frac{2\pi}{6}t)$.
2.11.6 $x(t) = -2\sin(\frac{2\pi}{6}t)$.

2.12 Given the sinusoids $x_1(t) = A_1\cos(\omega t + \theta_1)$ and $x_2(t) = A_2\cos(\omega t + \theta_2)$, find the sinusoid $x(t) = x_1(t) - x_2(t) = A\cos(\omega t + \theta)$. First add a phase of π or $-\pi$ to the sinusoid $x_2(t)$ and then use the summation formulas given in the book. Find the sample values of the sinusoids $x_1(t)$ and $x_2(t)$ at $t = 0, 1, 2$ and verify that the sample values of $x_1(t) - x_2(t)$ are the same as those of $x(t)$.

2.12.1 $x_1(t) = -2\cos(\frac{2\pi}{6}t - \frac{\pi}{3})$, $x_2(t) = 3\sin(\frac{2\pi}{6}t + \frac{\pi}{3})$.
2.12.2 $x_1(t) = \sin(\frac{2\pi}{6}t + \frac{\pi}{4})$, $x_2(t) = \cos(\frac{2\pi}{6}t + \frac{5\pi}{6})$.

* 2.12.3 $x_1(t) = 3\cos(\frac{2\pi}{6}t + \frac{\pi}{3}), x_2(t) = 4\cos(\frac{2\pi}{6}t + \frac{\pi}{4})$.
2.12.4 $x_1(t) = 2\cos(\frac{2\pi}{6}t + \frac{\pi}{6}), x_2(t) = 5\cos(\frac{2\pi}{6}t + \frac{\pi}{3})$.

2.13 Give the sample values of the exponentially varying amplitude sinusoid for $t = -1, 0, 1$.

2.13.1 $x(t) = e^{-t}\sin(\frac{2\pi}{6}t + \frac{\pi}{6})$.
2.13.2 $x(t) = e^{2t}\cos(\frac{2\pi}{6}t - \frac{\pi}{3})$.
2.13.3 $x(t) = e^{-2t}\cos(\pi t)$.
2.13.4 $x(t) = e^{2t}\sin(\frac{2\pi}{6}t)$.

2.14 The sinusoid $x(t)$ and the value k are specified. Find the value of $t > 0$ where the first positive peak of the sinusoid $x(t)$ occur. From the sinusoid $x(t + k)$, verify that its first positive peak, after $t > 0$, occurs as expected from the value of k.

2.14.1 $x(t) = 2\cos(\frac{2\pi}{6}t - \frac{\pi}{3}), k = 2$.
* 2.14.2 $x(t) = \sin(\frac{2\pi}{6}t + \frac{\pi}{6}), k = -1$.
2.14.3 $x(t) = \sin(\frac{2\pi}{6}t - \frac{\pi}{4}), k = 15$.
2.14.4 $x(t) = \cos(\frac{2\pi}{6}t + \frac{5\pi}{6}), k = 12$.
2.14.5 $x(t) = \sin(\frac{2\pi}{6}t), k = 1$.

2.15 The sinusoid $x(t)$ and the value k are specified. Find the value of $t > 0$ where the first positive peaks of the sinusoids $x(t)$ and $x(-t + k)$ occur.

2.15.1 $x(t) = 3\sin(\frac{2\pi}{6}t + \frac{\pi}{6}), k = -1$.
* 2.15.2 $x(t) = 2\cos(\frac{2\pi}{6}t - \frac{\pi}{4}), k = 2$.
2.15.3 $x(t) = \sin(\frac{2\pi}{6}t - \frac{\pi}{3}), k = -3$.
2.15.4 $x(t) = \sin(\frac{2\pi}{6}t + \frac{\pi}{3}), k = -12$.
2.15.5 $x(t) = \cos(\frac{2\pi}{6}t + \frac{\pi}{6}), k = 4$.

2.16 The sinusoid $x(t)$ and the values of a and k are specified. Find the value of $t > 0$ where the first positive peaks of the sinusoids $x(t)$ and $x(at + k)$ occur.

2.16.1 $x(t) = \cos(\frac{2\pi}{8}t + \frac{\pi}{3}), a = 2. k = 1$.
2.16.2 $x(t) = \sin(\frac{2\pi}{8}t + \frac{\pi}{6}), a = -\frac{1}{3}. k = -2$.
2.16.3 $x(t) = \cos(\frac{2\pi}{8}t - \frac{\pi}{4}), a = \frac{3}{2}. k = -1$.
* 2.16.4 $x(t) = \sin(\frac{2\pi}{8}t - \frac{\pi}{3}), a = -\frac{2}{3}. k = 2$.
2.16.5 $x(t) = \cos(\frac{2\pi}{8}t), a = 3. k = 1$.

2.17 The waveform $x(t)$ and the values k and a are specified. List the values at $t = -3, -2, -1, 0, 1, 2, 3$ of the waveforms $x(t)$, $x(t + k)$, and $x(at + k)$. Assume that the value of the function is its right-hand limit at any discontinuity.

2.17.1 $x(t) = e^{-0.1t}. a = 2, k = -1$.
2.17.2 $x(t) = e^{-0.2t}. a = \frac{1}{2}, k = 2$.

2.17.3 $x(t) = e^{1.05t}$. $a = \frac{3}{2}, k = 2$.
2.17.4 $x(t) = e^{1.2t}$. $a = \frac{1}{3}, k = -3$.
* 2.17.5 $x(t) = -2\sin(\frac{2\pi}{6}t + \frac{\pi}{3})u(t)$. $a = 2, k = 3$.
2.17.6 $x(t) = -2\sin(\frac{2\pi}{6}t + \frac{\pi}{6})u(t)$. $a = \frac{1}{2}, k = -1$.
2.17.7 $x(t) = e^{-0.3t}\cos(\frac{2\pi}{6}t + \frac{\pi}{4})u(t)$. $a = -2, k = 1$.
2.17.8 $x(t) = e^{-0.4t}\cos(\frac{2\pi}{6}t - \frac{\pi}{6})u(t)$. $a = \frac{1}{3}, k = 3$.

Chapter 3
Time-Domain Analysis of Discrete Systems

A system carries out some task in response to an input signal or produces an output signal that is an altered version of the input signal. For example, when we switch the power on to an electrical motor, it produces mechanical power. A filter produces an output signal in which the various frequency components of the input signal are altered in a predefined way. A system is realized using physical components (hardware realization) or using a computer program (software realization) or a combination of both. In order to analyze a system, a mathematical model of the system has to be derived using the laws governing the behavior of its components and their interconnection. It is usually not possible to develop an accurate model of a system. Therefore, a model, with minimum mathematical complexity, is developed so that it is a sufficiently accurate representation of the actual system. Although systems can have multiple inputs and multiple outputs, we consider single input and single output systems only, for simplicity. For the reason that the frequency-domain methods, described in later chapters, are easier for the analysis of higher-order systems, only first-order systems are considered in this chapter.

The difference equation model of discrete systems is derived in Sect. 3.1. In Sect. 3.2, the various classifications of systems are described. The convolution-summation model of discrete systems is developed in Sect. 3.3. In Sect. 3.4, the stability condition of discrete systems is derived in terms of their impulse response. In Sect. 3.5, the basic components used in the implementation of discrete systems, implementation of a specific system, and the decomposition of higher-order systems are presented.

3.1 Difference Equation Model

The resistor-capacitor (RC) circuit, shown in Fig. 3.1, is a lowpass filter, as the reactance of the capacitor is smaller at higher frequencies and larger at lower

D. Sundararajan, *Signals and Systems*,
https://doi.org/10.1007/978-3-031-19377-4_3

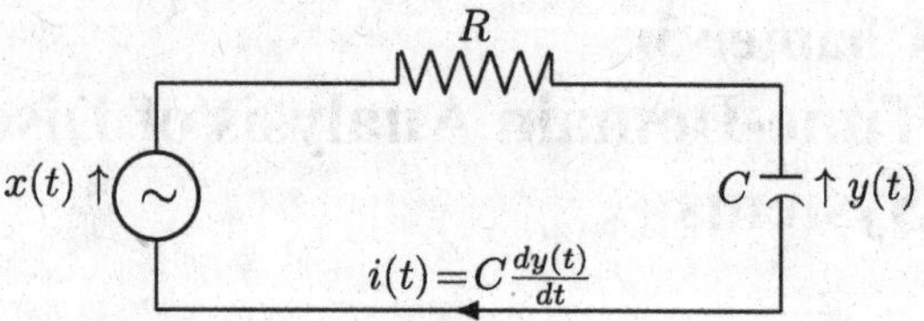

Fig. 3.1 An RC filter circuit

frequencies. Therefore, the output voltage across the capacitor $y(t)$ is a filtered version of the input $x(t)$. The relationship between the current through the capacitor and the voltage across it is $i(t) = C\frac{dy(t)}{dt}$. Then, due to Kirchhoff's voltage law, we get the linear differential equation (an equation that is a linear combination of the derivatives of functions) model of the circuit

$$RC\frac{dy(t)}{dt} + y(t) = x(t),$$

where R is in ohms and C is in farads. This model of the filter circuit can be approximated by a difference equation (an equation that contains differences of functions) by approximating the differential in the differential equation by a difference. One of the ways of this approximation is by replacing the term $\frac{dy(t)}{dt}$ by $\frac{y(nT_s)-y((n-1)T_s)}{T_s}$, where T_s is the sampling interval. The continuous variables $x(t)$ and $y(t)$ become $x(nT_s)$ and $y(nT_s)$, respectively. As usual, the sampling interval T_s in nT_s is suppressed, and we get the difference equation as

$$RC\frac{y(n) - y(n-1)}{T_s} + y(n) = x(n)$$

Let $b_1 = \frac{T_s}{T_s+RC}$ and $a_0 = -\frac{RC}{T_s+RC}$. Then, we get the difference equation characterizing the circuit as

$$y(n) = b_1x(n) - a_0y(n-1) \tag{3.1}$$

Let us assume that the input voltage is applied to the circuit at $n = 0$. Then, the output of the circuit at $n = 0$ is given by

$$y(0) = b_1x(0) - a_0y(0-1)$$

The voltage $y(-1)$ across the capacitor at $n = -1$, called the initial condition of the circuit, is required to find the output. The number of initial conditions required to find the output indicates the number of independent storage devices in the system. This number is also the order of the system. As only one value of initial condition is required, the model of the RC circuit is a first-order difference equation. Given the initial condition and the input, using this model, we can approximate the response of the circuit.

3.1.1 System Response

The response of a linear system is due to two independent causes, the input and the initial condition of the system at the time the input is applied. The response due to the initial condition alone is called the zero-input response, as the input is assumed to be zero. The response due to the input alone is called the zero-state response, as the initial condition or the state of the system is assumed to be zero. The complete response of a linear system is the sum of the zero-input and zero-state responses.

3.1.1.1 Zero-State Response

The difference equation characterizing a system has to be solved to get the system response. One way of solving a difference equation is by iteration. With the given initial condition $y(-1)$ and the inputs $x(0)$ and $x(-1)$, we can find the output $y(0)$ of a first-order difference equation. Then, in the next iteration, using $y(0)$, $x(1)$, and $x(0)$, we can compute $y(1)$. We repeat this process to get the desired number of outputs. Note that this method is suitable for programming in a digital computer. We can also deduce the closed-form solution by looking at the pattern of the expressions of the first few iterations. Let us solve Eq. (3.1) by iteration. Assume that the initial condition is zero, and the input signal is the unit-step, $u(n)$.

$$y(0) = b_1x(0) + (-a_0)y(-1) = b_1$$

$$y(1) = b_1x(1) + (-a_0)y(0) = b_1(1 + (-a_0))$$

$$\vdots$$

$$y(n) = b_1(1 + (-a_0) + (-a_0)^2 + \cdots + (-a_0)^n)$$

$$= b_1\left(\frac{1-(-a_0)^{(n+1)}}{1-(-a_0)}\right), \quad (-a_0) \neq 1, \; n = 0, 1, 2, \ldots$$

3.1.1.2 Zero-Input Response

Assume that the initial condition is $y(-1) = 3$. Since $x(n) = 0$ for all n, Eq. (3.1) reduces to $y(n) = (-a_0)y(n-1), \; y(-1) = 3$. Therefore,

$$y(0) = 3(-a_0), \quad y(1) = 3(-a_0)^2, \cdots, y(n) = 3(-a_0)^{(n+1)}$$

3.1.1.3 Complete Response

The complete response of the system is the sum of the zero-input and zero-state responses.

$$y(n) = \overbrace{b_1\left(\frac{1-(-a_0)^{(n+1)}}{1-(-a_0)}\right)}^{\text{zero-state}} + \overbrace{3(-a_0)^{n+1}}^{\text{zero-input}}, \quad n = 0, 1, 2, \dots$$

$$y(n) = \overbrace{b_1\left(\frac{1}{1-(-a_0)}\right)}^{\text{steady-state}} + \overbrace{b_1\left(\frac{-(-a_0)^{(n+1)}}{1-(-a_0)}\right) + 3(-a_0)^{(n+1)}}^{\text{transient}}$$

3.1.1.4 Transient and Steady-State Responses

The transient response of the system is $b_1\left(\frac{-(-a_0)^{(n+1)}}{1-(-a_0)}\right) + 3(-a_0)^{(n+1)}$. The steady-state response of the system, $b_1\left(\frac{1}{1-(-a_0)}\right)$, is the response of the system after the transient response has decayed. The transient response of a stable system always decays with time. The form of the transient response depends solely on the characteristics of the system while that of the steady-state response solely depends on the input signal.

Figure 3.2 shows the various components of the response of the first-order system governed by the difference equation

$$y(n) = 0.1x(n) + 0.9y(n-1)$$

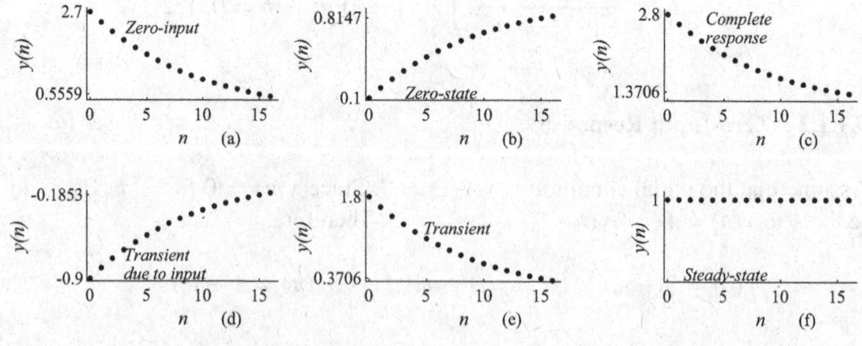

Fig. 3.2 The response of a first-order system for unit-step input signal: (**a**) zero-input response; (**b**) zero-state response; (**c**) complete response; (**d**) transient response due to input; (**e**) transient response; (**f**) steady-state response

with the initial condition $y(-1) = 3$ and the input $x(n) = u(n)$, the unit-step signal. The zero-input response, shown in Fig. 3.2a, is $3(0.9)^{(n+1)}u(n)$. The first ten values are

$$\{2.7000, 2.4300, 2.1870, 1.9683, 1.7715, 1.5943, 1.4349, 1.2914, 1.1623, 1.0460\}$$

The zero-state response, shown in Fig. 3.2b, is $(1 - (0.9)^{(n+1)})u(n)$. The first ten values are

$$\{0.1000, 0.1900, 0.2710, 0.3439, 0.4095, 0.4686, 0.5217, 0.5695, 0.6126, 0.6513\}$$

The sum of the zero-input and zero-state responses is the complete response, shown in Fig. 3.2c, is $3(0.9)^{(n+1)} + 1 - (0.9)^{(n+1)} = (1 + 2(0.9)^{(n+1)})u(n)$. The first ten values are

$$\{2.8000, 2.6200, 2.4580, 2.3122, 2.1810, 2.0629, 1.9566, 1.8609, 1.7748, 1.6974\}$$

The transient response due to input alone, shown in Fig. 3.2d, is $-(0.9)^{(n+1)}u(n)$. The total transient response, shown in Fig. 3.2e, is $3(0.9)^{(n+1)} - (0.9)^{(n+1)} = 2(0.9)^{(n+1)}u(n)$. The steady-state response, shown in Fig. 3.2f, is $u(n)$. The sum of the transient and steady-state responses also forms the complete response.

3.1.1.5 Coding and Simulation

Simulation is the study of the behavior of a system without actually building it. Simulations are easier to achieve by using functional blocks provided by the software. Because simulation diagrams are graphic display of the models, it is easier to read. In writing a coding program, we have to write our own codes, in addition to those built-in in the software. While coding gives a detailed understanding of the system response, it requires more effort. It is better to use both coding and simulation in the analysis of systems. The results by both the methods for the same system can be compared, and any bug in either of the methods can be found and rectified. Further, we obtain two viewpoints from the two methods, which makes our understanding of system analysis better. Simulations allow the analysis of nonlinear systems also. In addition, initial conditions can be specified. It is easier to vary parameters and input of the system and find the variations in the response.

Figure 3.3 shows the block diagram of the simulation model of the first-order difference equation. The block with a triangle symbol is the gain block, where you can set any gain. The square block with z^{-1} written in it is the unit-delay block, which delays its input by one unit of delay. The initial value is also set in the delay block. The sum block with two plus signs finds the sum of its inputs. The step input values are generated and loaded into the simin block, by executing the given input program. The simout variable stores the output values, which can be loaded for external use. The output of the summer unit is

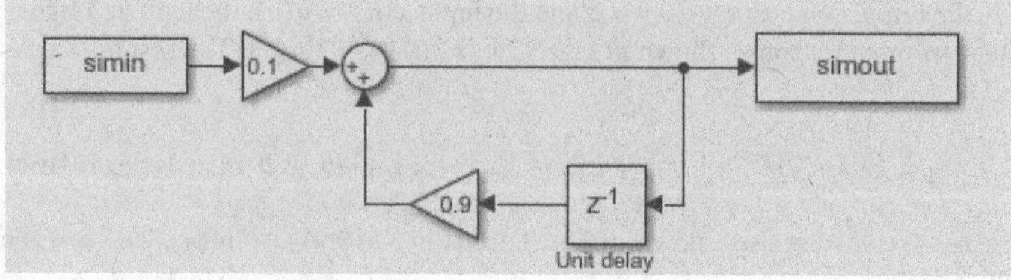

Fig. 3.3 Block diagram of the simulation model of the first-order difference equation

$$0.1x(n) + 0.9y(n-1) = y(n)$$

Running the simulation yields the same values, as given earlier.

3.1.1.6 Zero-Input Response by Solving the Difference Equation

Consider the Nth order difference equation of a causal LTI discrete system relating the output $y(n)$ to the input $x(n)$

$$y(n) + a_{N-1}y(n-1) + a_{N-2}y(n-2) + \cdots + a_0y(n-N) =$$
$$b_Nx(n) + b_{N-1}x(n-1) + \cdots + b_0x(n-N),$$

where N is the order of the system and the coefficients a's and b's are real constants characterizing the system. If the input is zero, the difference equation reduces to

$$y(n) + a_{N-1}y(n-1) + a_{N-2}y(n-2) + \cdots + a_0y(n-N) = 0$$

The solution to this equation gives the zero-input response of the system. This equation is a linear combination of $y(n)$ and its delayed versions equated to zero, for all values of n. Therefore, $y(n)$ and all its delayed versions must be of the same form. Only the exponential function has this property. Therefore, the solution is of the form $C\lambda^n$, where C and λ are to be found. Substituting $y(n) = C\lambda^n$, $y(n-1) = C\lambda^{n-1}$, etc., we get

$$(1 + a_{N-1}\lambda^{-1} + a_{N-2}\lambda^{-2} + \cdots + a_0\lambda^{-N})C\lambda^n = 0$$

Multiplying both sides by λ^N, we get

$$(\lambda^N + a_{N-1}\lambda^{N-1} + a_{N-2}\lambda^{N-2} + \cdots + a_0)C\lambda^n = 0$$

Assuming that the solution $C\lambda^n$ is nontrivial ($C \neq 0$),

$$(\lambda^N + a_{N-1}\lambda^{N-1} + a_{N-2}\lambda^{N-2} + \cdots + a_0) = 0 \tag{3.2}$$

The characteristic polynomial on the left side has N roots, $\lambda_1, \lambda_2, \ldots, \lambda_N$. Therefore, we get N solutions, $C_1\lambda_1^n, C_2\lambda_2^n, \ldots, C_N\lambda_N^n$. As the system is assumed to be linear and the solution has to satisfy N independent initial conditions of the system, the zero-input response of the system is given by

$$y(n) = C_1\lambda_1^n + C_2\lambda_2^n + \cdots + C_N\lambda_N^n,$$

assuming all the roots of the characteristic polynomial are distinct. The constants can be found using the N independent initial conditions of the system. The zero-input response represents a behavior that is characteristic of the system. As the form of the zero-input response of any N-th order system is the same, it is the set of roots of the characteristic polynomial that distinguishes a specific system. Therefore, Eq. (3.2) is called the characteristic equation of the system, and the roots, $\lambda_1, \lambda_2, \ldots, \lambda_N$, are called the characteristic roots or eigenvalues of the system. The corresponding exponentials, $\lambda_1^n, \lambda_2^n, \ldots, \lambda_N^n$, are called the characteristic modes of the system. The characteristic modes of a system are also influential in the determination of the zero-state response.

Example 3.1 Find the zero-input response of the system by solving its difference equation

$$y(n) = 0.1x(n) + 0.9y(n-1)$$

The initial condition is $y(-1) = 3$.

Solution The characteristic equation is $\lambda - 0.9 = 0$. The characteristic value of the system is $\lambda = 0.9$. The characteristic mode of the system is $(0.9)^n$. Therefore, the zero-input response is of the form

$$y(n) = C\,(0.9)^n$$

With $y(-1) = 3$ and letting $n = -1$, we get $C = 2.7$. Therefore, the zero-input response, as shown in Fig. 3.2a, is

$$y(n) = 2.7\,(0.9)^n\,u(n)$$ ∎

3.1.2 Impulse Response

The impulse response, $h(n)$, of a system is its response for a unit-impulse input signal with the initial conditions of the system zero. One way to find the impulse response of a system is by iteration. Another method is to find the zero-input response by solving the characteristic equation.

Example 3.2 Find the closed-form of the impulse response $h(n)$ of the system governed by the difference equation, with input $x(n)$ and output $y(n)$

$$y(n) = 2x(n) + 3x(n-1) + \frac{1}{2}y(n-1)$$

(i) by solving the difference equation and (ii) by iteration. Find the first four values of $h(n)$.

Solution As the system is initially relaxed (initial conditions zero), we get from the difference equation $h(0) = 2$ and $h(1) = 4$ by iteration. As the values of the impulse signal is zero for $n > 0$, the response for $n > 0$ can be considered as zero-input response. The characteristic equation is

$$\left(\lambda - \frac{1}{2}\right) = 0$$

The zero-input response is of the form

$$h(n) = C\left(\frac{1}{2}\right)^n u(n-1)$$

As $u(n-1) = u(n) - \delta(n)$, the response is also given by

$$h(n) = C\left(\frac{1}{2}\right)^n u(n) - C\delta(n),\ n > 0$$

Letting $n = 1$, with $h(1) = 4$, we get $C = 8$. The impulse response is the sum of the response of the system at $n = 0$ and the zero-input response for $n > 0$. Therefore,

$$h(n) = 2\delta(n) + 8\left(\frac{1}{2}\right)^n u(n) - 8\delta(n) = -6\delta(n) + 8\left(\frac{1}{2}\right)^n u(n)$$

By iteration,

$$h(0) = 2$$

$$h(1) = \left(3 + 2\frac{1}{2}\right) = 4$$

$$h(2) = \left(\frac{1}{2}\right)4$$

$$h(3) = \left(\frac{1}{2}\right)^2 4$$

$$\vdots$$

$$h(n) = \left(\frac{1}{2}\right)^{n-1} 4$$

$$h(n) = 2\delta(n) + \left(4\left(\frac{1}{2}\right)^{n-1}\right) u(n-1) = -6\delta(n) + 8\left(\frac{1}{2}\right)^{n} u(n)$$

The first four values of $h(n)$ are

$$\{h(0) = 2, h(1) = 4, h(2) = 2, h(3) = 1\}$$ ∎

In general, the impulse response of a first-order system governed by the difference equation $y(n) + a_0 y(n-1) = b_1 x(n) + b_0 x(n-1)$ is $h(n) = \frac{b_0}{a_0}\delta(n) + (b_1 - \frac{b_0}{a_0})(-a_0)^n u(n)$.

3.1.3 *Characterization of Systems by Their Responses to Impulse and Unit-Step Signals*

We can get information about the system behavior from the impulse and unit-step responses. If the significant values of the impulse response is of longer duration, as shown by filled circles in Fig. 3.4a, then the response of the system is sluggish. The corresponding unit-step response is shown by filled circles in Fig. 3.4b. The time taken for the unit-step response to rise from 10% to 90% of its final value is called the rise time of the system. If the significant values of the impulse response is of shorter duration, as shown by unfilled circles in Fig. 3.4a, then the response of the system is faster, as shown by unfilled circles in Fig. 3.4b. A system with a shorter impulse response has less memory, and it is readily influenced by the recent values of the input signal. Therefore, its response is fast. The faster is the rate of decay of the impulse response, the faster the response approaches its steady-state value.

The unit-step response, at n, is the sum of the first $n + 1$ terms of the impulse response, $y(n) = \sum_{m=0}^{n} h(m)$. As the final value tends to one in Fig. 3.4b and as the unit-step signal, ultimately, acts like a DC signal, the monotonically decreasing impulse response indicates a system that passes low-frequency components of a signal well.

Figure 3.4c shows typical alternating sequence impulse responses. The corresponding unit-step responses are shown in Fig. 3.4d. In these cases also, the system response time is faster with a short duration impulse response. However, note that the final value of the unit-step response approaches a very low value in Fig. 3.4d. This indicates a system that does not pass low-frequency components of a signal well.

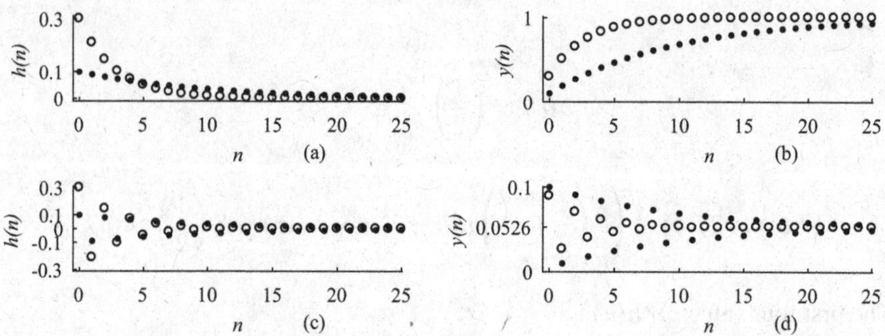

Fig. 3.4 (**a**) Typical monotonically decreasing impulse responses; (**b**) the corresponding unit-step responses; (**c**) typical alternating sequence impulse responses; (**d**) the corresponding unit-step responses

3.2 Classification of Systems

3.2.1 Linear and Nonlinear Systems

Let the response of a system to signal $x_1(n)$ be $y_1(n)$ and the response to $x_2(n)$ be $y_2(n)$. Then, the system is linear if the response to the linear combination $ax_1(n)+bx_2(n)$ is $ay_1(n)+by_2(n)$, where a and b are arbitrary constants. A general proof is required to prove that a system is linear. However, one counterexample is enough to prove that a system is nonlinear. Nonlinear terms, such as $x^2(n)$ or $x(n)y(n-1)$ (terms involving the product of $x(n)$, $y(n)$, and their shifted versions), in the difference equation are an indication that the system is not linear. Any nonzero constant term is also an indication of a nonlinear system. The linearity condition implies that the total response of a linear system is the sum of zero-input and zero-state components. The linearity of a system with respect to zero-input and zero-state responses should be checked individually. In most cases, zero-state linearity implies zero-input linearity.

Example 3.3 Given the difference equation of a system, with input $x(n)$ and output $y(n)$, determine whether the system is linear. Verify the conclusion with the inputs $\{x_1(n), n = 0, 1, 2, 3\} = \{1, 4, 3, 2\}$, $\{x_2(n), n = 0, 1, 2, 3\} = \{2, 3, 4, 1\}$, and $x(n) = 2x_1(n) - 3x_2(n)$ by computing the first four values of the output. Assume that the initial condition $y(-1)$ is zero.

(**a**) $y(n) = x(n) + y(n-1) + 3$
(**b**) $y(n) = x(n) - (2n)y(n-1)$

Solution

(a) As the nonzero term indicates that the system is nonlinear, we try the counterexample method. By iteration, the first four output values of the system to the input signal $x_1(n)$ are

$$y_1(0) = x_1(0) + y_1(0-1) + 3 = 1 + 0 + 3 = 4$$

$$y_1(1) = x_1(1) + y_1(1-1) + 3 = 4 + 4 + 3 = 11$$

$$y_1(2) = x_1(2) + y_1(2-1) + 3 = 3 + 11 + 3 = 17$$

$$y_1(3) = x_1(3) + y_1(3-1) + 3 = 2 + 17 + 3 = 22$$

The output to $x_2(n)$ is $\{y_2(n), n = 0, 1, 2, 3\} = \{5, 11, 18, 22\}$. Now, $y(n) = \{2y_1(n) - 3y_2(n), n = 0, 1, 2, 3\} = \{-7, -11, -20, -22\}$.
The system response to the combined input $\{2x_1(n) - 3x_2(n), n = 0, 1, 2, 3\} = \{-4, -1, -6, 1\}$ is $\{y(n), n = 0, 1, 2, 3\} = \{-1, 1, -2, 2\}$. As this output is different from that computed earlier, the system is nonlinear.

(b) The system output to $x_1(n)$ is $y_1(n) = x_1(n) - (2n)\, y_1(n-1)$. The system output to $x_2(n)$ is $y_2(n) = x_2(n) - (2n)\, y_2(n-1)$. Then,

$$ay_1(n) + by_2(n) = ax_1(n) - (2an)\, y_1(n-1) + bx_2(n) - (2bn)\, y_2(n-1)$$

The system output to $ax_1(n) + bx_2(n)$ is

$$ax_1(n) + bx_2(n) - (2n)(ay_1(n-1) + by_2(n-1))$$

As both the expressions for the output are the same, the system is linear. The output to $x_1(n)$ is $\{y_1(n), n = 0, 1, 2, 3\} = \{1, 2, -5, 32\}$. The output to $x_2(n)$ is $\{y_2(n), n = 0, 1, 2, 3\} = \{2, -1, 8, -47\}$. Now, $y(n) = \{2y_1(n) - 3y_2(n), n = 0, 1, 2, 3\} = \{-4, 7, -34, 205\}$.

The system response to the combined input $\{2x_1(n) - 3x_2(n), n = 0, 1, 2, 3\} = \{-4, -1, -6, 1\}$ is $\{y(n), n = 0, 1, 2, 3\} = \{-4, 7, -34, 205\}$. This output is the same as that computed earlier. ▮

3.2.2 Time-Invariant and Time-Varying Systems

The output of a time-invariant system to the input $x(n-m)$ must be $y(n-m)$ for all m, assuming that the output to the input $x(n)$ is $y(n)$ and the initial conditions are identical. A general proof is required to prove that a system is time-invariant. However, one counterexample is enough to prove that a system is time-variant. Terms, such as $x(2n)$ or $x(-n)$, with a nonzero and nonunity constant associated with the index n in the difference equation indicate a time-variant system. Any coefficient that is an explicit function of n in the difference equation also indicates a time-variant system.

Example 3.4 Given the difference equation of a system, with input $x(n)$ and output $y(n)$, determine whether the system is time-invariant. Verify the conclusion with the inputs $\{x(n), n = 0, 1, 2, 3\} = \{1, 4, 3, 2\}$ and $\{x(n-2), n = 2, 3, 4, 5\} = \{1, 4, 3, 2\}$ by computing the first four values of the output. Assume that the initial condition $y(-1)$ is zero.

(a) $y(n) = nx(n)$
(b) $y(n) = 2x(n)$

Solution

(a) As the coefficient in the difference equation is the independent variable n, we try the counterexample method. The output of the system to $x(n)$ is $\{y(n), n = 0, 1, 2, 3\} = \{0, 4, 6, 6\}$. The output of the system to $x(n-2)$ is $\{y(n), n = 2, 3, 4, 5\} = \{2, 12, 12, 10\}$. As the two outputs are different, the system is time-varying.
(b) The system output to $x(n)$ is $y(n) = 2x(n)$. By replacing n by $(n-2)$, we get $y(n-2) = 2x(n-2)$. The system output to $x(n-2)$ is $2x(n-2)$. As the outputs are the same, the system is time-invariant. The output of the system to $x(n)$ is $\{y(n), n = 0, 1, 2, 3\} = \{2, 8, 6, 4\}$. The output of the system to $x(n-2)$ is $\{y(n), n = 2, 3, 4, 5\} = \{2, 8, 6, 4\}$. ∎

Linear time-invariant (LTI) systems satisfy the linearity and time-invariant properties and are easier to analyze and design. Most practical systems, although not strictly linear and time-invariant, can be considered as LTI systems with acceptable error limits.

3.2.3 *Causal and Noncausal Systems*

Practical systems respond only to present and past input values, but not to future input values. These systems are called causal or nonanticipatory systems. If the present output $y(n)$ depends on the input $x(n+k)$ with $k > 0$, then the system is noncausal. This implies that the impulse response of a causal system $h(n)$ is zero for $n < 0$. Ideal systems, such as ideal filters, are noncausal. However, they are of interest because they set an upper bound for the system response. Practical systems approximate the ideal response while being causal (i.e., physically realizable).

Example 3.5 Given the difference equation of a system, with input $x(n)$ and output $y(n)$, determine whether the system is causal. Find the impulse response.

(a) $y(n) = x(n+2) + 2x(n) - 3x(n-1)$
(b) $y(n) = 2x(n) - x(n-1) + 3x(n-4)$.

Solution

(a) As the output $y(n)$ is a function of the future input sample $x(n+2)$, the system is noncausal. The impulse response of the system is obtained, by substituting $x(n) = \delta(n)$ in the input-output relation, as $y(n) = h(n) = \delta(n+2) + 2\delta(n) - 3\delta(n-1)$. That is, $h(-2) = 1, h(-1) = 0, h(0) = 2$, and $h(1) = -3$.

(b) The system is causal. The impulse response of the system is

$$\{h(0) = 2, h(1) = -1, h(2) = 0, h(3) = 0, h(4) = 3\}$$ ∎

3.2.4 Instantaneous and Dynamic Systems

With regard to system memory, systems are classified as instantaneous or dynamic. A system is instantaneous (no memory) if its output at an instant is a function of the input at that instant only. The system characterized by the difference equation $y(n) = 2x(n)$ is a system with no memory. An example is an electrical circuit consisting of resistors only. Any system with storage elements, such as inductors and capacitors, is a dynamic system, since the output at an instant of such systems is a function of past values of the input also. The discrete model of this type of systems will have terms, such as $x(n-1)$ or $x(n-2)$, that require memory units to implement. If the output depends only on a finite number of past input samples, then it is called a finite memory system. For example, $y(n) = x(n-1) + x(n-2)$ is the difference equation of a system with two memory units. Systems with capacitive or inductive elements are infinite memory systems, since their output is a function of entire past history of the input. Instantaneous systems are a special case of the dynamic systems with zero memory.

3.2.5 Inverse Systems

A system is invertible if its input can be determined from its output. This implies that each input has a unique output. Systems with the input-output relationship such as $y(n) = x^2(n)$ are not invertible. If the impulse response of a system, made up of two systems connected in cascade, is $h(n) = \delta(n)$, then the two systems are the inverses of one another. For example, the inverse of the system with the input-output relationship $y(n) = 2x(n)$ is $x(n) = \frac{1}{2}y(n)$.

3.2.6 Continuous and Discrete Systems

In continuous systems, input, output, and all other signals are of continuous type, and they are processed using devices such as resistors, inductors, and capacitors. In a discrete system, input, output, and all other signals are of discrete type, and they are processed using discrete devices such as a digital computer. While most naturally occurring signals are of continuous type, they are usually analyzed and processed using their discrete approximations, as it is advantageous, by converting the continuous signals to discrete signals by sampling. These types of systems, in which both types of signals appear, are called hybrid systems.

3.3 Convolution-Summation Model

Commonly used models for discrete linear systems are difference equation, convolution summation, transfer function, and state space. All the models are equivalent representations of a system, in the sense that all of them produce the same output for a certain input. Each model is advantageous in bringing out the salient characteristics of the dynamic behavior of a system with some respect. Further, studying different models of a system gives a better understanding of the system.

In the difference equation model of a system, we used some output and input values in formulating the model. In the convolution-summation model, the model is formulated in terms of all the input values applied to the system, assuming that the initial conditions are zero. The input signal is decomposed in terms of scaled and shifted unit-impulses. Therefore, with the knowledge of the response of the system to just the unit-impulse (called the impulse response), we find the response to each of the constituent impulses of an arbitrary input signal and sum the individual responses to find the total response. As the initial conditions are assumed to be zero, the response obtained using this model is the zero-state response.

The study of signals and systems has analogy with what we do in real life. We can easily understand and remember the mathematical process of these models using the analogy. Mathematical concepts such as convolution and Fourier analysis can be presented using real-life processes. Let the problem be finding the amount in our deposit at current time. Let the interest rate be 10% compounded annually. Let the current (index 0) and the last 3 years deposits be

$$\{x(-3) = 300, x(-2) = 400, x(-1) = 200, x(0) = 100\}$$

The balance due to current deposit is obtained by multiplying it with 1. Similarly, the balance due to past deposits are obtained by multiplying them with 1.1, 1,21, and 1.331, respectively. The current balance in our account is

$$100 \times 1 + 200 \times 1.1 + 400 \times 1.21 + 300 \times 1.331 = 1203.3$$

With respect to running time, the computation required is to reverse one of the sequences and find the sum of products. This is the essence of convolution. The contributions of all the impulse components of the input of a system are found and summed to find the total output. Figure 3.5 shows the basics of linear convolution.

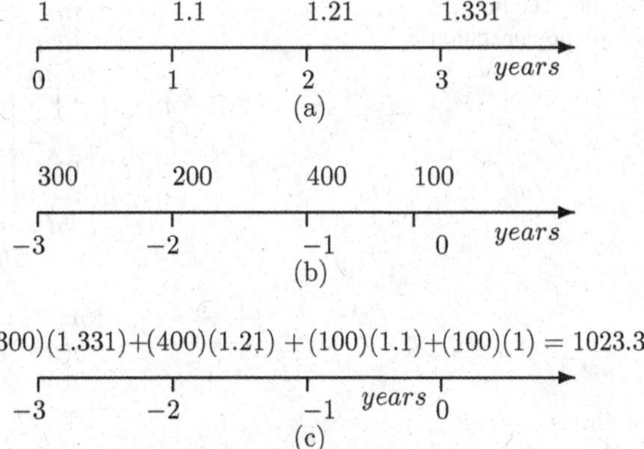

Fig. 3.5 Basics of linear convolution. (**a**) annual interest rate; (**b**) deposits; (**c**) computation of current balance

Alternately, we can compute the total amount using the updated last year deposit and the present deposit. For example, for the present problem, the balance up to last year, called the previous output, is

$$200 \times 1 + 400 \times 1.1 + 300 \times 1.21 = 1003$$

Then, the current balance in our account is

$$1003 \times 1.1 + 100 = 1203.3$$

This is the difference equation model, which uses past outputs, some past inputs, and the present input.

Let us find the convolution of the impulse response $\{h(m), m = 0, 1, 2, 3\} = \{5, 0, 3, 2\}$ and the input $\{x(m), m = 0, 1, 2, 3\} = \{4, 1, 3, 2\}$ shown in Fig. 3.6. The time-reversed impulse response, $\{h(0 - m), m = 3, 2, 1, 0\}$, is $\{2, 3, 0, 5\}$. There is only one nonzero product, $x(0)h(0) = 4 \times 5 = 20$, of $x(m)h(0 - m)$ with $m = 0$, and the convolution output is $y(0) = 20$. The product $x(0)h(0)$ is the response of the system at $n = 0$ to the present input sample $x(0)$. There is no contribution to the output at $n = 0$ due to input samples $x(1)$, $x(2)$, and $x(3)$, since the system is causal. The time-reversed impulse response is shifted to the right by one sample interval to get $h(1-m) = h(-m+1)) = h(-(m-1))$. The convolution output $y(1)$ at $n = 1$ is the sum of products $x(m)h(1 - m)$, $m = 0, 1$. That is, $y(1) = x(0)h(1) + x(1)h(0) = 4 \times 0 + 1 \times 5 = 5$. The product $x(1)h(0)$ is the response of the system at $n = 1$ to the present input sample $x(1)$. The product $x(0)h(1)$ is the response of the system at $n = 1$ to the past input sample $x(0)$. Repeating the process, we find the remaining five output values. While $x(n)$ and $h(n)$ have four elements each, the output sequence $y(n)$ has seven elements. The duration of the convolution of two finite sequences of length N and M is $N + M - 1$ samples, as the overlap of nonzero portions can occur only over that length.

Fig. 3.6 The linear convolution operation

m				0	1	2	3			
$h(m)$				5	0	3	2			
$x(m)$				4	1	3	2			
$h(0-m)$	2	3	0	5						
$h(1-m)$		2	3	0	5					
$h(2-m)$			2	3	0	5				
$h(3-m)$				2	3	0	5			
$h(4-m)$					2	3	0	5		
$h(5-m)$						2	3	0	5	
$h(6-m)$							2	3	0	5
n				0	1	2	3	4	5	6
$y(n)$				20	5	27	21	11	12	4

A more formal development of the convolution operation is as follows. An arbitrary signal can be decomposed, in terms of scaled and shifted impulses, as

$$x(n) = \sum_{m=-\infty}^{\infty} x(m)\delta(n-m)$$

The impulse response $h(n)$ of a LTI system is its response to an impulse $\delta(n)$ with the system initially relaxed (initial conditions zero). Due to the time-invariance property, a delayed impulse $\delta(n-m)$ will produce the response $h(n-m)$. Since a LTI system is linear, a scaled and shifted impulse $x(m)\delta(n-m)$ will produce the response $x(m)h(n-m)$. Therefore, using both the linearity and time-invariance properties, the system response $y(n)$ to an arbitrary signal $x(n)$ can be expressed as

$$y(n) = \sum_{m=-\infty}^{\infty} x(m)h(n-m) = x(n) * h(n)$$

The convolution-summation of the sequences $x(n)$ and $h(n)$ is denoted as $x(n) * h(n)$. For a causal system, as its impulse response $h(n)$ is zero for $n < 0$, the upper limit of the summation is n, instead of ∞, as $h(n-m) = 0$, $m > n$.

$$y(n) = \sum_{m=-\infty}^{\infty} x(m)h(n-m) = \sum_{m=-\infty}^{n} x(m)h(n-m)$$

If the signal $x(n)$ starts at any finite instant $n = n_0$, then the lower limit is equal to n_0. The effective range of the summation is easily determined by observing that if $x(m)$ or $h(n-m)$ or both are zero in a certain range, the product $x(m)h(n-m)$ is zero in that range.

Essentially, convolution operation is finding the sum of products of two sequences, each other's index running in opposite directions. To summarize, the output of a system is found by convolution with the repeated use of four operations (fold, shift, multiply, and add).

1. One of the two sequences to be convolved (say $h(m)$) is time-reversed, that is, folded about the vertical axis at the origin to get $h(-m)$.
2. The time-reversed sequence, $h(-m)$, is shifted by n_0 sample intervals (right shift for positive n_0 and left shift for negative n_0), yielding $h(n_0 - m)$, to find the output at $n = n_0$.
3. The term by term products of the overlapping samples of the two sequences, $x(m)$, and $h(n_0 - m)$, are computed.
4. The sum of all the products is the output sample value at $n = n_0$.

Two finite sequences to be convolved overlap only partly at the beginning and the end of the convolution operation, as can be seen in Fig. 3.6, and less arithmetic is required to find the convolution output in these cases. The convolution expression, requiring minimum arithmetic, for two finite sequences is given as follows. Let $x(n), n = 0, 1, \ldots, N-1$ and $h(n), n = 0, 1, \ldots, M-1$. Then,

$$y(n) = \sum_{m=Max(0,n-M+1)}^{Min(n,N-1)} x(m)h(n-m) = \sum_{m=Max(0,n-N+1)}^{Min(n,M-1)} h(m)x(n-m),$$
$$n = 0, 1, \ldots, N+M-2, \tag{3.3}$$

where *Min* and *Max* stand, respectively, for “minimum of” and “maximum of.” Along with the shift property of convolution presented shortly, this expression can be used to evaluate the convolution of two finite sequences starting from any n.

Discrete Convolution Is Like Polynomial Multiplication

The coefficients of the product polynomial of the product of two polynomials is the convolution sum of the coefficients of the two polynomials getting multiplied. Using the sequences $x(n)$ and $h(n)$ as the coefficients of two polynomials and multiplying, the coefficients of the product polynomial are the same as the convolution output of the two sequences.

$$(4+q+3q^2+2q^3)(5+0q+3q^2+2q^3) = 20+5q+27q^2+21q^3+11q^4+12q^5+4q^6$$

The convolution sum can be checked as follows. The product of the sum of the terms of $x(n)$ and $h(n)$ must equal to the sum of the terms of the sequence $y(n)$. For the example,

$$(4+1+3+2)(5+0+3+2) = 10 \times 10 = 100 = (20+5+27+21+11+12+4)$$

The same test, with the sign of the odd-indexed terms of all the three sequences changed, also holds.

$$(4-1+3-2)(5-0+3-2) = 4 \times 6 = 24 = (20-5+27-21+11-12+4)$$

Table 3.1 shows a list commonly occurring discrete convolution summations.

Table 3.1 Convolution table

$x(n)$	$h(n)$	$x(n) * h(n)$
$x(n)$	$\delta(n-k)$	$x(n-k)$
$a^n u(n)$	$u(n)$	$\frac{a^{n+1}-1}{a-1} u(n)$
$u(n)$	$u(n)$	$(n+1)u(n))$
$a^n u(n)$	$b^n u(n)$	$\frac{a^{n+1}-b^{n+1}}{a-b} u(n), \quad a \neq b$
$a^n u(n)$	$a^n u(n)$	$(n+1)a^n u(n)$
$u(n)$	$nu(n)$	$\frac{n(n+1)}{2} u(n)$

Example 3.6 Find the linear convolution of the sequences $\{x(n), n = 0, 1, 2\} = \{1, 2, 3\}$ and $\{h(n), n = 0, 1\} = \{2, -3\}$.

Solution Using Eq. (3.3), we get

$$y(0) = (1)(2) = 2$$
$$y(1) = (1)(-3) + (2)(2) = 1$$
$$y(2) = (2)(-3) + (3)(2) = 0$$
$$y(3) = (3)(-3) = -9$$

The values of the convolution of $x(n)$ and $h(n)$ are

$$\{y(0) = 2, \quad y(1) = 1, \quad y(2) = 0, \quad y(3) = -9\}$$ ∎

Example 3.7 Find the closed-form expression of the convolution of the sequences $x(n) = (0.6)^n u(n)$ and $h(n) = (0.5)^n u(n)$.

Solution

$$y(n) = \sum_{l=-\infty}^{\infty} x(l)h(n-l) = \sum_{l=0}^{n} (0.6)^l (0.5)^{n-l}, \, n \geq 0$$
$$= (0.5)^n \sum_{l=0}^{n} \left(\frac{0.6}{0.5}\right)^l = (0.5)^n \left(\frac{1 - \left(\frac{0.6}{0.5}\right)^{n+1}}{1 - \left(\frac{0.6}{0.5}\right)} \right)$$
$$= (6(0.6)^n - 5(0.5)^n)u(n)$$

The first four values of the convolution of $x(n)$ and $h(n)$ are

$$\{y(0) = 1, \quad y(1) = 1.1, \quad y(2) = 0.91, \quad y(3) = 0.671\}$$ ∎

3.3.1 Properties of Convolution-Summation

The convolution-summation is commutative, that is, the order of the two sequences to be convolved is immaterial.

$$x(n) * h(n) = h(n) * x(n)$$

The convolution-summation is distributive. That is, the convolution of a sequence with the sum of two sequences is the same as the sum of the individual convolution of the first sequence with the other two sequences.

$$x(n) * (h_1(n) + h_2(n)) = x(n) * h_1(n) + x(n) * h_2(n)$$

The convolution-summation is associative. That is the convolution of a sequence with the convolution of two sequences is the same as the convolution of the convolution of the first two sequences with the third sequence.

$$x(n) * (h_1(n) * h_2(n)) = (x(n) * h_1(n)) * h_2(n)$$

The shift property of convolution is that

$$\text{if} \quad x(n) * h(n) = y(n), \quad \text{then} \quad x(n-l) * h(n-m) = y(n-l-m)$$

The convolution of two shifted sequences is the convolution of the two original sequences shifted by the sum of the shifts of the individual sequences.

Convolution of a sequence $x(n)$ with the unit-impulse leaves the sequence unchanged except for the translation of the origin of the sequence to the location of the impulse.

$$x(n) * \delta(n-k) = \sum_{m=-\infty}^{\infty} \delta(m-k)x(n-m) = x(n-k)$$

Example 3.8 Find the linear convolution of the sequences $\{x(n), n = 0, 1, 2\} = \{3, 2, 4\}$ and $h(n) = \delta(n+3)$.

Solution

$$x(n) * \delta(n+3) = \{x(n+3), n = -3, -2, -1\} = \{3, 2, 4\}$$ ∎

Convolution of $x(n)$ with the unit-step is the running sum of $x(n)$.

$$x(n) * u(n) = \sum_{l=-\infty}^{n} x(l)$$

For example, let $\{x(n), n = 0, 1, 2\} = \{3, 2, 4\}$ and $h(n) = u(n)$.

$$x(n) * h(n) = \{3, 5, 9, 9, 9, 9, \ldots\}$$

3.3.2 The Difference Equation and the Convolution-Summation

The difference equation and the convolution-summation are two different mathematical models of a LTI system producing the same output for the same input. Therefore, these two models are related. Consider the first-order difference equation, with input $x(n)$ and output $y(n)$.

$$y(n) = b_1 x(n) + (-a_0) y(n-1)$$

As the initial conditions are assumed to be zero for the convolution-summation model, $y(-1) = 0$. In order to derive the convolution-summation model, we have to express the past output term in terms of input samples.

$$\begin{aligned}
y(0) &= b_1 x(0) \\
y(1) &= b_1 x(1) + (-a_0) y(0) = b_1 x(1) + (-a_0) b_1 x(0) \\
y(2) &= b_1 x(2) + (-a_0) y(1) = b_1 x(2) + (-a_0) b_1 x(1) + (-a_0)^2 b_1 x(0) \\
&\vdots \\
y(n) &= b_1 x(n) + (-a_0) b_1 x(n-1) + \cdots + (-a_0)^n b_1 x(0)
\end{aligned}$$

Then, the impulse response, with $x(n) = \delta(n)$, is given as

$$h(0) = b_1,\ h(1) = (-a_0) b_1,\ h(2) = (-a_0)^2 b_1,\ \ldots, h(n) = (-a_0)^n b_1$$

The output $y(n)$, using $h(n)$, can be expressed as

$$y(n) = h(0)x(n) + h(1)x(n-1) + \cdots + h(n)x(0) = \sum_{m=0}^{n} h(m)x(n-m),$$

which is the convolution-summation. For any n, $h(0)$ determines the effect of the current input $x(n)$ on the output $y(n)$. In general, $h(m)$ determines the effect of the input $x(n-m)$, applied m iterations before, on the output $y(n)$. A system, whose impulse response is of finite duration, is called a finite impulse response system. A system, whose impulse response is of infinite duration, is called an infinite impulse response system. In the difference equation model of a system, a system is characterized by the coefficients, a's and b's, of its difference equation.

In the convolution-summation model of a system, the system is characterized by its impulse response $h(n)$.

3.3.3 Response to Complex Exponential Input

A complex exponential with frequency ω_0 is given as $x(n) = e^{j\omega_0 n}$, $-\infty < n < \infty$. Assuming a causal and stable system with real-valued impulse response $h(n)$, the output of the system is given by the convolution-summation as

$$y(n) = \sum_{m=0}^{\infty} h(m)e^{j\omega_0(n-m)} = e^{j\omega_0 n}\sum_{m=0}^{\infty} h(m)e^{-j\omega_0 m}$$

As the second summation is independent of n and letting

$$H(e^{j\omega_0}) = \sum_{m=0}^{\infty} h(m)e^{-j\omega_0 m}$$

we get,

$$y(n) = H(e^{j\omega_0})e^{j\omega_0 n} = H(e^{j\omega_0})x(n)$$

$H(e^{j\omega_0})$ is called the frequency response since it is a constant complex scale factor indicating the amount of change in the amplitude and phase of an input complex exponential $e^{j\omega_0 n}$, with frequency ω_0, at the output. Since the impulse response is real-valued for practical systems, the scale factor for an exponential with frequency $-\omega_0$ is $H^*(e^{j\omega_0})$, where the symbol $*$ indicates complex conjugation. The point is that the input-output relationship of a LTI system becomes a multiplication operation rather than the more complex convolution operation. As the complex exponential is the only signal that has this property, it is used predominantly as the basis for signal decomposition. Even if the exponent of the complex exponential input signal has a real part, $x(n) = e^{(\sigma + j\omega_0)n}$, the response of the system is still related to the input by the multiplication operation. A real sinusoidal input $A\cos(\omega_0 n + \theta)$ is also changed at the output by the same amount of amplitude and phase of the complex scale factor $H(e^{j\omega_0})$. That is, $A\cos(\omega_0 n + \theta)$ is changed to $(|H(e^{j\omega_0})|A)\cos(\omega_0 n + (\theta + \angle(H(e^{j\omega_0}))))$. The proof is as follows:

$$A\cos(\omega_0 n + \theta) = 0.5A(e^{j(\omega_0 n+\theta)} + e^{-j(\omega_0 n+\theta)})$$

The response, due to linearity, is

$$\begin{aligned} &0.5A|H(j\omega_0|(e^{j(\omega_0 n+\theta+\angle(H(j\omega_0))} + e^{-j(\omega_0 n+\theta+\angle(H(j\omega_0))})\\ &= A|H(j\omega_0|\cos(\omega_0 n + \theta + \angle(H(j\omega_0))) \end{aligned}$$

There was no transient component in the output expression $y(n)$, since the exponential signal was applied at $n = -\infty$. For finite values of n, any transient component in the output of a stable system must have died out. However, if we apply the exponential at any finite instant, say $n = 0$, there will be a transient component in the response, in addition to the steady-state component $H(e^{j\omega_0})e^{j\omega_0 n}u(n)$.

Example 3.9 Let the input signal to a stable system with impulse response $h(n) = b_1(-a_0)^n u(n)$ be $x(n) = e^{j\omega_0 n}u(n)$. Find the response of the system. Assume that $y(-1) = 0$.

Solution Using the convolution-summation, we get

$$y(n) = \sum_{m=0}^{n} h(m)e^{j\omega_0(n-m)} = b_1 e^{j\omega_0 n} \sum_{m=0}^{n} (-a_0)^m e^{-j\omega_0 m}$$

$$= \left(\frac{b_1}{1-(-a_0)e^{-j\omega_0}}\right)\left(e^{j\omega_0 n} - (-a_0)^{(n+1)} e^{-j\omega_0}\right), \quad n = 0, 1, \ldots$$

The first term, the steady-state component $\left(\frac{b_1}{1-(-a_0)e^{-j\omega_0}}\right) e^{j\omega_0 n}$, is the same as the input complex exponential with a complex scale factor. The second term,

$$\left(\frac{b_1}{1-(-a_0)e^{-j\omega_0}}\right)(-(-a_0)^{(n+1)} e^{-j\omega_0}),$$

is the transient component that will die for sufficiently large values of n. ∎

3.4 System Stability

One of the criteria for the stability of a system is that the system output is bounded if the input is bounded. A sequence $x(n)$ is bounded if $|x(n)| \le M$ for all values of n, where M is a finite positive number. For example, the sequence $x(n) = (0.8)^n u(n)$ is bounded and $x(n) = (1.2)^n u(n)$ is unbounded. As convolution-summation is a sum of products, the sum is bounded if the input signal is bounded and the sum of the magnitude of the terms of the impulse response is finite. Let the sample values of the input signal $x(n)$ are bounded by the positive constant M. From the convolution-summation relation for a causal system with impulse response $h(n)$, we get

$$|y(n)| = |\sum_{m=0}^{\infty} h(m)x(n-m)|$$

$$\leq \sum_{m=0}^{\infty} |h(m)x(n-m)| = \sum_{m=0}^{\infty} |h(m)||x(n-m)|$$

$$|y(n)| \leq \sum_{m=0}^{\infty} |h(m)|M = M \sum_{m=0}^{\infty} |h(m)|$$

Therefore, if $\sum_{m=0}^{\infty} |h(m)|$ is bounded then $|y(n)|$ is bounded. Consequently, a necessary and sufficient stability condition is that the impulse response is absolutely summable:

$$\sum_{m=0}^{\infty} |h(m)| < \infty$$

As we used the convolution-summation to derive the stability condition, the stability condition ensures a bounded zero-state response. The stability of the zero-input response should be separately checked and it is presented in Chap. 9.

Example 3.10 Is the system governed by the difference equation, with input $x(n)$ and output $y(n)$, stable?

(i) $y(n) = 9x(n) + 2y(n-1)$
(ii) $y(n) = 9x(n) + 0.8y(n-1)$

Solution

(i) The impulse response of the system is $h(n) = 9(2)^n u(n)$. As $h(n)$ is not absolutely summable, the system is unstable.
(ii) The impulse response of the system is $h(n) = 9(0.8)^n u(n)$. As $h(n)$ is absolutely summable, the system is stable. ∎

3.5 Realization of Discrete Systems

A discrete system can be realized in software or hardware or as a combination of both. In any case, the three basic components required in the realization of discrete systems are (i) multiplier units, (ii) adder units, and (iii) delay units. A multiplier unit, shown in Fig. 3.7a, produces an output sequence $c\,x(n)$, in which each element is the product of the corresponding element in the input sequence $x(n)$ and the coefficient c. An adder unit, shown in Fig. 3.7b, produces an output sequence $x(n)+y(n)$, in which each element is the sum of the corresponding elements in the input sequences $x(n)$ and $y(n)$. By complementing the subtrahend and then adding it with the minuend, subtraction operation can be realized by an adder unit. A delay unit, shown in Fig. 3.7c, produces an output sequence $x(n-1)$, which is a delayed version of the input sequence $x(n)$ by one sampling interval.

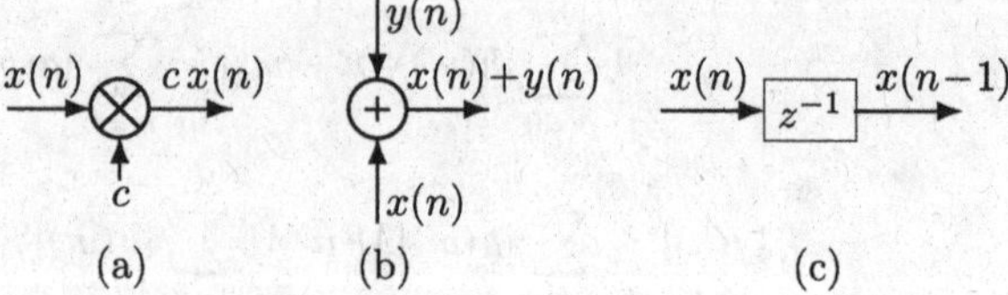

Fig. 3.7 Basic components required in the realization of discrete systems: (**a**) multiplier unit; (**b**) adder unit; (**c**) delay unit

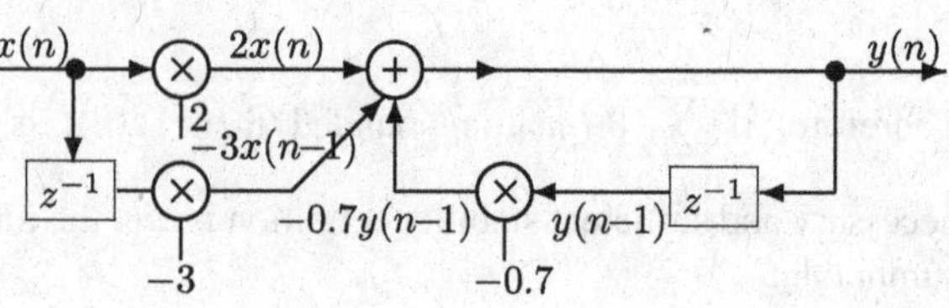

Fig. 3.8 Block diagram of the realization of a discrete system

By using the necessary number of the basic units, any arbitrary difference equation can be realized. Consider the implementation of the difference equation

$$y(n) = 2x(n) - 3x(n-1) - 0.7y(n-1),$$

shown in Fig. 3.8. Let the input be the unit-step:

$$x(0) = 1,\ x(2) = 1,\ x(3) = 1,\ x(4) = 1,\ x(5) = 1, \ldots$$

Let the initial condition be $y(-1) = 0$. We get the delayed output term $y(n-1)$ by passing $y(n)$ through a delay unit. The product term $-0.7y(n-1)$ is obtained by passing $y(n-1)$ through a multiplier unit with coefficient -0.7. The product term $2x(n)$ is obtained by passing $x(n)$ through a multiplier unit with coefficient 2. We get the delayed input term $x(n-1)$ by passing $x(n)$ through a delay unit. The product term $-3x(n-1)$ is obtained by passing $x(n-1)$ through a multiplier unit with coefficient -3. The adder unit combines the partial results to produce the output signal $y(n)$. By iteration, the first few outputs, with $x(-1) = y(-1) = 0$, are

$$y(0) = 2,\ y(1) = -2.4,\ y(2) = 0.68,\ y(3) = -1.476,\ y(4) = 0.0332$$

Figure 3.9 shows the block diagram of the simulation model of the first-order difference equation. The step input values have to be loaded into the simin block by executing the given input program. The system can be implemented using only one delay unit.

3.5.1 Decomposition of Higher-Order Systems

To meet a given specification, a higher-order system is often required. Due to several advantages, a system is usually decomposed into first- and second-order systems

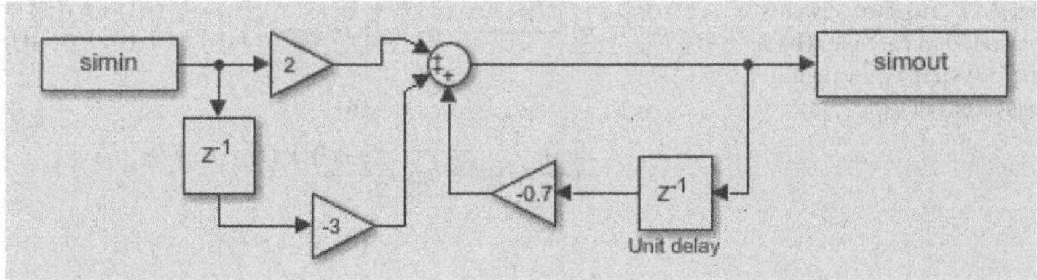

Fig. 3.9 Block diagram of the simulation model of the first-order difference equation

Fig. 3.10 (**a**) Two systems connected in parallel; (**b**) a single system equivalent to the system in (**a**)

$x(n)$ — $h_1(n)$ — $x(n) * h_1(n)$
$h_2(n)$ — $x(n) * h_2(n)$
$x(n) * h_1(n) + x(n) * h_2(n) = x(n) * (h_1(n) + h_2(n))$

(a)

$x(n)$ — $h_1(n) + h_2(n)$ — $x(n) * (h_1(n) + h_2(n))$

(b)

connected in cascade or parallel. Figure 3.10a shows two systems with impulse responses $h_1(n)$ and $h_2(n)$ connected in parallel. The same input is applied to each system, and the total response is the sum of the individual responses. The combined response of the two systems for the input $x(n)$ is $y(n) = x(n)*h_1(n)+x(n)*h_2(n)$. This expression, due to the distributive property of convolution, can be written as $y(n) = x(n) * (h_1(n) + h_2(n))$. That is, the parallel connection of the two systems is equivalent to a single system with impulse response $h(n) = h_1(n) + h_2(n)$, as shown in Fig. 3.10b.

Figure 3.11a shows two systems with impulse responses $h_1(n)$ and $h_2(n)$ connected in cascade. The output of one system is the input to the other. The response of the first system for the input $x(n)$ is $y_1(n) = x(n) * h_1(n)$. The response of the second system for the input $y_1(n) = x(n) * h_1(n)$ is $y(n) = (x(n) * h_1(n)) * h_2(n)$. This expression, due to the associative property of convolution, can be written as $y(n) = x(n) * (h_1(n) * h_2(n))$. That is, the cascade connection of the two systems is equivalent to a single system with impulse response $h(n) = h_1(n) * h_2(n)$, as shown in Fig. 3.11b. Due to the commutative property of convolution, the order of the systems in the cascade connection is immaterial, with respect to the input-output relationship.

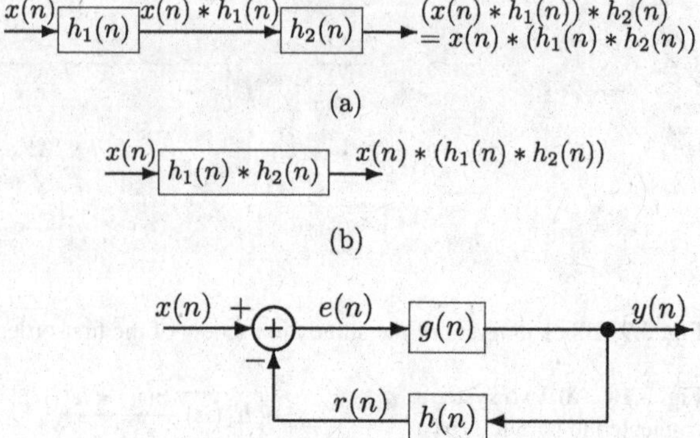

Fig. 3.11 (**a**) Two systems connected in cascade; (**b**) a single system equivalent to the system in (**a**)

Fig. 3.12 Two systems connected in a feedback configuration

3.5.2 *Feedback Systems*

Another configuration of systems, often used in control systems, is the feedback configuration shown in Fig. 3.12. In feedback systems, a fraction of the output signal is fed back and subtracted from the input signal to form the effective input signal. A feedback signal $r(n)$ is produced by a system with impulse response $h(n)$ from the delayed samples of the output signal, $y(n-1)$, $y(n-2)$, etc. That is, $r(n) = \sum_{m=1}^{\infty} h(m)y(n-m)$. This implies that $h(0) = 0$. The error signal $e(n)$ is the difference between the input signal $x(n)$ and the feedback signal $r(n)$, $e(n) = x(n) - r(n)$. This error signal is the input to a system with impulse response $g(n)$, which produces the output signal $y(n)$. That is, $y(n) = \sum_{m=0}^{\infty} g(m)e(n-m)$.

3.6 Summary

- In this chapter, the time-domain analysis of LTI discrete systems has been presented.
- As discrete systems offer several advantages, they are mostly used instead of continuous systems. These systems can be designed to approximate continuous systems with a desired accuracy by selecting a sufficiently short sampling interval.
- The zero-input component of the response of a LTI system is its response due to the initial conditions alone with the input assumed to be zero. The zero-state component of the response of a LTI system is its response due to the input alone with the initial conditions assumed to be zero. The sum of the zero-input and zero-state responses is the complete response of the system.
- Two of the commonly used system models for time-domain analysis are the difference equation and convolution-summation models.

- The convolution-summation model gives the zero-state response of a LTI system. Both the zero-input and zero-state responses can be found using the difference equation model either by solving the difference equation or by iteration.
- The impulse response of a system is its response to the unit-impulse input signal with zero initial conditions.
- The convolution-summation model is based on decomposing the input signal into a set of shifted and scaled impulses. The total response is found by summing the responses to all the constituent impulses of the input signal.
- The complete response of a system can also be considered as the sum of the transient component and the steady-state component. For a stable system, the transient component always decays with time. The steady-state component is the response after the transient response dies down.
- A system is stable if its response is bounded for all bounded input signals. As the convolution-summation is a sum of products of the input and the impulse response, with the input bounded, the impulse response of a stable system must be absolutely summable for the convolution sum to be bounded.
- By interconnecting adder, multiplier, and delay units, any discrete system can be realized. A higher-order system is usually decomposed into a set of first- and second-order systems connected in cascade or parallel. A feedback system is obtained by feeding back some part of the output to the input.

Exercises

3.1 Derive the closed-form expression for the impulse response $h(n)$, by iteration, of the system governed by the difference equation, with input $x(n)$ and output $y(n)$. List the values of the impulse response $h(n)$ at $n = 0, 1, 2, 3, 4, 5$.

3.1.1 $y(n) = x(n) + 2x(n-1) - 3y(n-1)$.
3.1.2 $y(n) = 2x(n) - 3x(n-1) + \frac{1}{2}y(n-1)$.
* 3.1.3 $y(n) = 3x(n) - \frac{1}{3}y(n-1)$.
3.1.4 $y(n) = x(n) - 2x(n-1) + 2y(n-1)$.
3.1.5 $y(n) = 3x(n) - 4x(n-1) + y(n-1)$.

3.2 Find the closed-form expression for the impulse response $h(n)$ of the system by solving its difference equation, with input $x(n)$ and output $y(n)$. List the values of the impulse response $h(n)$ at $n = 0, 1, 2, 3, 4, 5$.

3.2.1 $y(n) = 3x(n) - x(n-1) + 2y(n-1)$.
3.2.2 $2y(n) = x(n) + x(n-1) - y(n-1)$.
3.2.3 $y(n) = 2x(n) + \frac{1}{4}y(n-1)$.
* 3.2.4 $y(n) = 4x(n) + 3x(n-1) - y(n-1)$.
3.2.5 $y(n) = x(n) + x(n-1) - y(n-1)$.

3.3 Is the system governed by the given difference equation, with input $x(n)$ and output $y(n)$, linear? Let $\{x_1(n), n = 0, 1, 2, 3\} = \{1, 2, 3, 2\}$, $\{x_2(n), n = 0, 1, 2, 3\} = \{2, 3, 0, 4\}$ and $x(n) = 2x_1(n) - 3x_2(n)$. Assuming that the initial condition $y(-1)$ is zero, compute the first four output values, and verify the conclusion.

3.3.1 $y(n) = 3x(n) - 2y(n-1) + 1$.
3.3.2 $y(n) = (x(n))^2 + y(n-1)$.
* 3.3.3 $y(n) = x(n) - (n)y(n-1) + 2\cos(\frac{\pi}{2})$.
3.3.4 $y(n) = x(n) + x(n)y(n-1)$.
3.3.5 $y(n) = |x(n)|$.
3.3.6 $y(n) = (n)x(n) + y(n-1) - 3\cos(\pi)$.

3.4 Is the system governed by the given difference equation, with input $x(n)$ and output $y(n)$, time-invariant? Let $\{x(n), n = 0, 1, 2, 3, 4, 5, 6, 7, 8\} = \{2, 1, 3, 3, 4, 2, 5, 1, 3\}$. Assuming that the initial condition is zero, compute the first four output values and verify the conclusion to the input $\{x(n-2), n = 2, 3, 4, 5, 6, 7, 8, 9, 10\} = \{2, 1, 3, 3, 4, 2, 5, 1, 3\}$.

3.4.1 $y(n) = x(2n) + 2y(n-1)$.
3.4.2 $y(n) = 2x(n) - \sin(\frac{\pi}{2}n)y(n-1)$.
* 3.4.3 $y(n) = (x(n))^2 - 2\cos(6\pi n)y(n-1)$.
3.4.4 $y(n) = x(n) + (n)y(n-1)$.
3.4.5 $y(n) = x(8-n)$.

3.5 Find the linear convolution of the sequences $x(n)$ and $h(n)$.

3.5.1 $\{x(n), n = 0, 1, 2\} = \{4, 2, 1\}$ and $\{h(n), n = 0, 1\} = \{-2, -3\}$.
3.5.2 $\{x(n), n = -2, -1, 0\} = \{2, -1, 4\}$ and $\{h(n), n = 3, 4, 5, 6\} = \{2, 1, 4, 3\}$.
* 3.5.3 $\{x(n), n = -3, -2, -1, 0\} = \{2, 2, 1, 4\}$ and $\{h(n), n = 2, 3, 4, 5\} = \{3, 2, 3, 4\}$.

3.6 Find the closed-form expression for the convolution of the sequences $x(n)$ and $h(n)$. List the values of the convolution output at $n = 0, 1, 2, 3, 4, 5$.

3.6.1 $x(n) = u(n-1)$ and $h(n) = u(n-3)$.
3.6.2 $x(n) = (0.5)^n u(n-2)$ and $h(n) = (0.7)^n u(n-1)$.
3.6.3 $x(n) = (0.5)^{n-1} u(n-1)$ and $h(n) = (0.7)^{n-2} u(n-2)$.
3.6.4 $x(n) = (0.6)^n u(n)$ and $h(n) = x(n)$.
* 3.6.5 $x(n) = (0.6)^n u(n-2)$ and $h(n) = u(n-1)$.

3.7 Find the linear convolution of the sequences $x(n)$ and $h(n)$.

3.7.1 $\{x(n), n = 1, 2, 3, 4\} = \{3, 2, 4, 1\}$ and $h(n) = \delta(n)$.
3.7.2 $\{x(n), n = -4, -3, -2\} = \{1, 3, 2\}$ and $h(n) = \delta(n-2)$.
3.7.3 $\{x(n), n = 3, 4, 5\} = \{5, 2, 3\}$ and $h(n) = \delta(n+3)$.

3.7.4 $x(n) = e^{j\frac{2\pi}{6}n}u(n)$ and $h(n) = \delta(n+4)$.
3.7.5 $x(n) = e^{j\frac{2\pi}{6}n}$ and $h(n) = \delta(n-6)$.
3.7.6 $x(n) = \cos(\frac{2\pi}{6}n)$ and $h(n) = \delta(n)$.

3.8 Verify the distributive and associative properties of convolution-summation

$$x(n) * (h_1(n) + h_2(n)) = x(n) * h_1(n) + x(n) * h_2(n)$$

and

$$x(n) * (h_1(n) * h_2(n)) = (x(n) * h_1(n)) * h_2(n)$$

where $\{h_1(n), n = 0, 1, 2, 3\} = \{1, 2, 3, 4\}$, $\{h_2(n), n = 0, 1, 2, 3\} = \{3, 2, 1, 5\}$, and $\{x(n), n = 0, 1, 2, 3\} = \{4, 4, 3, 2\}$.

3.9 Find the steady-state response of the system, with the impulse response

$$h(n) = -\frac{5}{3}\delta(n) + \frac{11}{3}(-0.6)^n u(n),\ n = 0, 1, 2, \dots,$$

to the input $x(n) = 3\sin(\frac{2\pi}{6}n - \frac{\pi}{6})u(n)$. Deduce the response to the input $e^{j\frac{2\pi}{6}n}$.

* **3.10** Find the steady-state response of the system, with the impulse response

$$h(n) = -4\delta(n) + 7(0.5)^n u(n),\ n = 0, 1, 2, \dots,$$

to the input $x(n) = 2\cos(\frac{2\pi}{5}n + \frac{\pi}{4})u(n)$. Deduce the response to the input $e^{j\frac{2\pi}{5}n}$.

3.11 Derive the closed-form expression for the complete response (by finding the zero-state response using the convolution-summation and the zero-input response) of the system governed by the difference equation

$$y(n) = 2x(n) - x(n-1) + \frac{1}{3}y(n-1)$$

with the initial condition $y(-1) = 2$ and the input $x(n) = u(n)$, the unit-step function. List the values of the complete response $y(n)$ at $n = 0, 1, 2, 3, 4, 5$. Deduce the expressions for the transient and steady-state responses of the system.

3.12 Derive the closed-form expression for the complete response (by finding the zero-state response using the convolution-summation and the zero-input response) of the system governed by the difference equation

$$y(n) = x(n) - 2x(n-1) - \frac{1}{2}y(n-1)$$

with the initial condition $y(-1) = -3$ and the input $x(n) = (-1)^n u(n)$. List the values of the complete response $y(n)$ at $n = 0, 1, 2, 3, 4, 5$. Deduce the expressions for the transient and steady-state responses of the system.

* **3.13** Derive the closed-form expression for the complete response (by finding the zero-state response using the convolution-summation and the zero-input response) of the system governed by the difference equation

$$y(n) = 3x(n) - 2x(n-1) + \frac{1}{4}y(n-1)$$

with the initial condition $y(-1) = 1$ and the input $x(n) = nu(n)$. List the values of the complete response $y(n)$ at $n = 0, 1, 2, 3, 4, 5$. Deduce the expressions for the transient and steady-state responses of the system.

3.14 Derive the closed-form expression for the complete response (by finding the zero-state response using the convolution-summation and the zero-input response) of the system governed by the difference equation

$$y(n) = x(n) + 3x(n-1) - \frac{3}{5}y(n-1)$$

with the initial condition $y(-1) = -2$ and the input $x(n) = (\frac{2}{5})^n u(n)$. List the values of the complete response $y(n)$ at $n = 0, 1, 2, 3, 4, 5$. Deduce the expressions for the transient and steady-state responses of the system.

3.15 Derive the closed-form expression for the complete response (by finding the zero-state response using the convolution-summation and the zero-input response) of the system governed by the difference equation

$$y(n) = 2x(n) - 4x(n-1) + \frac{1}{3}y(n-1)$$

with the initial condition $y(-1) = -3$ and the input $x(n) = 2\sin(\frac{2\pi}{6}n + \frac{\pi}{3})u(n)$. List the values of the complete response $y(n)$ at $n = 0, 1, 2, 3, 4, 5$. Deduce the expressions for the transient and steady-state responses of the system.

3.16 The impulse response of a LTI system is given. Is the system stable?

3.16.1 $h(0) = 0,\ h(n) = \frac{(-1)^{n+1}}{n},\ n = 1, 2, \ldots.$

3.16.2 $h(0) = 0,\ h(n) = \frac{1}{n},\ n = 1, 2, \ldots.$

3.16.3 $h(0) = 0,\ h(n) = \frac{1}{n^2},\ n = 1, 2, \ldots.$

3.17 Derive the closed-form expression of the impulse response $h(n)$ of the combined system consisting of systems governed by the given difference equations, with input $x(n)$ and output $y(n)$, if the systems are connected in (i) parallel and (ii) in cascade. List the first four values of the impulse response of the combined system.

3.17.1

$$y_1(n) = 3x_1(n) + 2x_1(n-1) - \frac{1}{3}y_1(n-1) \quad \text{and}$$

$$y_2(n) = 2x_2(n) - 3x_2(n-1) - \frac{1}{4}y_2(n-1)$$

* 3.17.2

$$y_1(n) = x_1(n) - x_1(n-1) + \frac{1}{5}y_1(n-1) \quad \text{and}$$

$$y_2(n) = x_2(n) + 2x_2(n-1) - \frac{3}{5}y_2(n-1)$$

3.17.3

$$y_1(n) = 2x_1(n) + 2x_1(n-1) + \frac{2}{5}y_1(n-1) \quad \text{and}$$

$$y_2(n) = 3x_2(n) - x_2(n-1) + \frac{5}{6}y_2(n-1)$$

Chapter 4
Time-Domain Analysis of Continuous Systems

While discrete systems, in general, are advantageous, we still need the study of continuous systems. Continuous systems offer higher speed of operation. Even if we decide to use a discrete system, as the input and output signals are mostly continuous, we still need continuous systems for the processing of signals before and after the interface between the two types of systems. The design of a discrete system can be made by first designing a continuous system and then using a suitable transformation of that design. As the discrete systems, usually, approximate the continuous systems, comparing with the exact analysis results of the continuous systems with that of the actual performance of the corresponding discrete system gives a measure of the approximation. For these reasons, the study of continuous systems is as much required as that of the discrete systems. In this chapter, we study two time-domain models of LTI continuous systems. We consider only first-order systems in this chapter as frequency-domain methods, described in later chapters, are easier for the analysis of higher-order systems. The analysis procedure remains essentially the same as that of discrete systems except that continuous systems are modeled using differential equation and convolution-integral, as the signals are of continuous type.

In Sect. 4.1, the various classifications of LTI continuous systems are described. In Sects. 4.2 and 4.3, we develop the differential equation and convolution-integral models of a system, respectively. Using these models, in Sect. 4.4, the various components of the system response are derived. The important property of an exponential input signal to remain in the same form at the output of a stable LTI system is demonstrated. In Sect. 4.5, the stability of a system in terms of its impulse response is established. In Sect. 4.6, the basic components used in the implementation of continuous systems are presented, and an implementation of a specific system is given. The decomposition of an higher-order system into a set of lower-order systems is also presented.

D. Sundararajan, *Signals and Systems*,
https://doi.org/10.1007/978-3-031-19377-4_4

4.1 Classification of Systems

4.1.1 Linear and Nonlinear Systems

A system is linear if its response to a linear combination of input signals is the same linear combination of the individual responses to the inputs. Let the response of a system to signal $x_1(t)$ be $y_1(t)$ and the response to $x_2(t)$ be $y_2(t)$. Then, the system is linear if the response to a linear combination, $ax_1(t) + bx_2(t)$, is $ay_1(t) + by_2(t)$, where a and b are arbitrary constants. Nonlinear terms, such as $x^2(t)$ or $x(t)y(t)$ (the products involving $x(t)$, $y(t)$, and their derivatives), in the differential equation are an indication that the system is not linear. Any nonzero constant term is also an indication of a nonlinear system. The linearity condition implies that the total response of a linear system is the sum of zero-input and zero-state components. The zero-input component of the response of a system is its response due to the initial conditions alone with the input assumed to be zero. The zero-state component of the response of a system is its response due to the input alone with the initial conditions assumed to be zero. The linearity of a system with respect to zero-input and zero-state responses should be checked individually. In most cases, zero-state linearity implies zero-input linearity.

Example 4.1 Given the differential equation of a system, with output $y(t)$ and input $x(t)$, determine whether the system is linear. Assume that the initial condition $y(0)$ is zero.

(a) $y(t) = x(t) + t\frac{dy(t)}{dt}$
(b) $y(t) = x(t) + (\frac{dy(t)}{dt})^2$

Solution

(a) Let $y_1(t)$ be the output to $x_1(t)$ and $y_2(t)$ be the output to $x_2(t)$. The system differential equation with $x_1(t)$ is $y_1(t) = x_1(t) + t\frac{dy_1(t)}{dt}$. The system differential equation with $x_2(t)$ is $y_2(t) = x_2(t) + t\frac{dy_2(t)}{dt}$. Then,

$$ay_1(t) + by_2(t) = ax_1(t) + at\frac{dy_1(t)}{dt} + bx_2(t) + bt\frac{dy_2(t)}{dt}$$

$$= ax_1(t) + bx_2(t) + t\frac{d}{dt}(ay_1(t) + by_2(t))$$

The system output to $x(t) = ax_1(t) + bx_2(t)$ is $y(t) = ay_1(t) + by_2(t)$ for a linear system. Substituting in the differential equation, we get

$$ay_1(t) + by_2(t) = ax_1(t) + bx_2(t) + t\frac{d}{dt}(ay_1(t) + by_2(t))$$

As both the differential equations are the same, the system is linear.

(b) The system differential equation with $x_1(t)$ is $y_1(t) = x_1(t) + (\frac{dy_1(t)}{dt})^2$. The system differential equation with $x_2(t)$ is $y_2(t) = x_2(t) + (\frac{dy_2(t)}{dt})^2$. Then,

$$ay_1(t) + by_2(t) = ax_1(t) + a\left(\frac{dy_1(t)}{dt}\right)^2 + bx_2(t) + b\left(\frac{dy_2(t)}{dt}\right)^2$$

The system output to $x(t) = ax_1(t) + bx_2(t)$ is $y(t) = ay_1(t) + by_2(t)$ for a linear system. Substituting in the differential equation, we get

$$ay_1(t) + by_2(t) = ax_1(t) + bx_2(t) + \left(a\frac{dy_1(t)}{dt} + b\frac{dy_2(t)}{dt}\right)^2$$

As the differential equations are different, the system is nonlinear. ∎

4.1.2 *Time-Invariant and Time-Varying Systems*

The output of a time-invariant system to the input $x(t - t_0)$ must be $y(t - t_0)$ for all t_0, assuming that the output to the input $x(t)$ is $y(t)$ and the initial conditions are identical. Terms, such as $x(2t)$ or $x(-t)$, with a nonzero and nonunity constant associated with the argument t in the differential equation indicate a time-variant system. Any coefficient that is an explicit function of t in the differential equation also indicates a time-variant system.

Example 4.2 Given the differential equation of a system, with output $y(t)$ and input $x(t)$, determine whether the system is time-invariant. Assume that the initial condition is zero.

(a) $y(t) = x(t) + t\frac{dy(t)}{dt}$
(b) $y(t) = x(t) + (\frac{dy(t)}{dt})^2$

Solution

(a) By replacing t by $(t - a)$ in the differential equation, we get

$$y(t - a) = x(t - a) + (t - a)\frac{dy(t - a)}{dt}$$

The system output to $x(t-a)$ is $y(t-a)$ for a time-invariant system. Substituting in the differential equation, we get

$$y(t - a) = x(t - a) + t\frac{dy(t - a)}{dt}$$

As the differential equations are different, the system is time-varying.

(b) By replacing t by $(t - a)$ in the differential equation, we get

$$y(t-a) = x(t-a) + \left(\frac{dy(t-a)}{dt}\right)^2$$

The system output to $x(t-a)$ is $y(t-a)$ for a time-invariant system. Substituting in the differential equation, we get

$$y(t-a) = x(t-a) + \left(\frac{dy(t-a)}{dt}\right)^2$$

As both the differential equations are the same, the system is time-invariant. ▮

Linear time-invariant (LTI) systems satisfy the linearity and time-invariant properties and are easier to analyze and design. Most practical systems, although not strictly linear and time-invariant, can be considered as LTI systems with acceptable error limits.

4.1.3 *Causal and Noncausal Systems*

Practical systems respond only to present and past input values, but not to the future input values. These systems are called causal or nonanticipatory systems. This implies that the impulse response of a causal system $h(t)$ is zero for $t < 0$. If the present output $y(t)$ depends on the input $x(t+t_0)$ with $t_0 > 0$, then the system is noncausal. Ideal systems, such as ideal filters, are noncausal. However, they are of interest because they set an upper bound for the system response. Practical systems approximate the ideal response while being causal (i.e., physically realizable).

Example 4.3 Given the differential equation of a system, with output $y(t)$ and input $x(t)$, determine whether the system is causal. Find the impulse response.

(a) $y(t) = x(t+1) + 2x(t) - 3x(t-1)$

(b) $y(t) = 2x(t) - x(t-1) + 3x(t-2)$.

Solution

(a) As the output $y(t)$ is a function of the future input sample $x(t+1)$, the system is noncausal. The impulse response of the system is obtained by substituting $x(t) = \delta(t)$ in the differential equation, as $y(t) = h(t) = \delta(t+1) + 2\delta(t) - 3\delta(t-1)$.

(b) The system is causal. The impulse response of the system is $y(t) = h(t) = 2\delta(t) - \delta(t-1) + 3\delta(t-2)$. ∎

4.1.4 Instantaneous and Dynamic Systems

With regard to system memory, systems are classified as instantaneous or dynamic. A system is instantaneous (no memory) if its output at an instant is a function of the input at that instant only. An example is an electrical circuit consisting of resistors only with input-output relationship such as $v(t) = Ri(t)$. Any system with storage elements, such as inductors and capacitors, is called a dynamic system, since the output at an instant of such systems is a function of past values of the input also. If the output depends only on the input during T seconds of immediate past, then it is called finite memory system. Systems with capacitive or inductive elements are infinite memory systems, since their output is a function of entire past history of the input. Instantaneous systems are a special case of the dynamic systems with zero memory.

4.1.5 Lumped-Parameter and Distributed-Parameter Systems

If the propagation time of a signal through a system is negligible, then that system is called a lumped-parameter system. For example, the current through a resistor in such a system is a function of time only, but not on the dimensions of the resistor. Such systems are modeled using ordinary differential equations. If the dimensions of a component are large compared to the wavelength of the highest frequency of interest, then the signal through that component is a function of time and the dimensions of the component. A system with that type of components is called a distributed-parameter system. Such systems, for example, transmission lines, are modeled using partial differential equations.

4.1.6 Inverse Systems

A system is invertible if its input can be determined from its output. This implies that each input has a unique output. Systems with the input-output relationship such as $y(t) = x^2(t)$ are not invertible. If the impulse response of a system, made up of two systems connected in cascade, is $h(t) = \delta(t)$, then the two systems are the inverses of one another. For example, the inverse of the system with the input-output relationship $y(t) = 4x(t)$ is $x(t) = \frac{1}{4}y(t)$.

4.2 Differential Equation Model

Commonly used models for continuous linear systems are differential equation, convolution-integral, transfer function, and state space. Differential equations are used in one type of time-domain modeling of continuous systems. The input-output relationship of commonly used components of a system, such as inductors and capacitors, is governed by differential equations. Therefore, differential equations naturally arise in modeling systems. The interconnection of several elements leads to a model represented by higher-order differential equations. Consider the Nth order differential equation of a causal LTI continuous system relating the output $y(t)$ to the input $x(t)$

$$\frac{d^N y(t)}{dt^N} + a_{N-1}\frac{d^{N-1}y(t)}{dt^{N-1}} + \cdots + a_1\frac{dy(t)}{dt} + a_0 y(t)$$

$$= b_N\frac{d^N x(t)}{dt^N} + b_{N-1}\frac{d^{N-1}x(t)}{dt^{N-1}} + \cdots + b_1\frac{dx(t)}{dt} + b_0 x(t)$$

where N is the order of the system and the coefficients a's and b's are real constants characterizing the system. If the input is zero, the differential equation reduces to

$$\frac{d^N y(t)}{dt^N} + a_{N-1}\frac{d^{N-1}y(t)}{dt^{N-1}} + \cdots + a_1\frac{dy(t)}{dt} + a_0 y(t) = 0$$

Denoting $\frac{d}{dt} = D$, we get

$$(D^N + a_{N-1}D^{N-1} + \cdots + a_1 D + a_0)y(t) = 0$$

The solution to this equation gives the zero-input response of the system. This equation is a linear combination of $y(t)$ and its N successive derivatives equated to zero, for all values of t. Therefore, $y(t)$ and all its N successive derivatives must be of the same form. Only the exponential function has this property. Therefore, the solution is of the form $Ce^{\lambda t}$, where C and λ are to be found. Substituting $y(t) = Ce^{\lambda t}$, $\frac{dy(t)}{dt} = C\lambda e^{\lambda t}$, etc., we get

$$(\lambda^N + a_{N-1}\lambda^{N-1} + \cdots + a_1\lambda + a_0)Ce^{\lambda t} = 0$$

Assuming that the solution is nontrivial ($C \neq 0$),

$$(\lambda^N + a_{N-1}\lambda^{N-1} + \cdots + a_1\lambda + a_0) = 0 \tag{4.1}$$

The characteristic polynomial on the left side has N roots, $\lambda_1, \lambda_2, \ldots, \lambda_N$. Therefore, we get N solutions, $C_1e^{\lambda_1 t}$, $C_2e^{\lambda_2 t}, \ldots, C_Ne^{\lambda_N t}$. As the system is assumed to be linear and the solution has to satisfy N independent initial conditions of the system, the zero-input response of the system is given by

$$y(t) = C_1 e^{\lambda_1 t} + C_2 e^{\lambda_2 t} + \cdots + C_N e^{\lambda_N t},$$

assuming all the roots are distinct. The constants can be found using the N independent initial conditions of the system. The zero-input response represents a behavior that is characteristic of the system. As the form of the zero-input response of any N-th order system is the same, it is the set of roots of the characteristic polynomial that distinguishes a specific system. Therefore, Eq. (4.1) is called the characteristic equation of the system, and the roots, $\lambda_1, \lambda_2, \ldots, \lambda_N$, are called the characteristic roots of the system. The corresponding exponentials, $e^{\lambda_1 t}$, $e^{\lambda_2 t}, \ldots, e^{\lambda_N t}$, are called the characteristic modes of the system. The characteristic modes of a system are also influential in the determination of the zero-state response.

Example 4.4 Find the complete response of the system characterized by the differential equation

$$\frac{dy(t)}{dt} + 4y(t) = 3x(t)$$

with $x(t) = u(t)$, the unit-step input signal. Assume that $y(0^-) = 2$.

Solution The characteristic equation of the system is $\lambda + 4 = 0$. The zero-input response is of the form Ce^{-4t}. For a unit-step input, the particular response is a constant K. Therefore, the complete response is of the form

$$y(t) = K + Ce^{-4t}$$

Substituting it in the differential equation,

$$-4Ce^{-4t} + 4K + C4e^{-4t} = 3$$

we get $K = 3/4$. Using the initial condition $y(0^-) = 2$,

$$y(t) = \frac{3}{4} + Ce^{-4t}|_{t=0^-} = 2,$$

we get $C = \frac{5}{4}$. Therefore, the total response is

$$y(t) = \left(\frac{3}{4} + \frac{5}{4}e^{-4t}\right)u(t)$$

The first term of the response is the steady-state response, and the second term is the transient. The zero-input and the zero-state components of the response, respectively, are

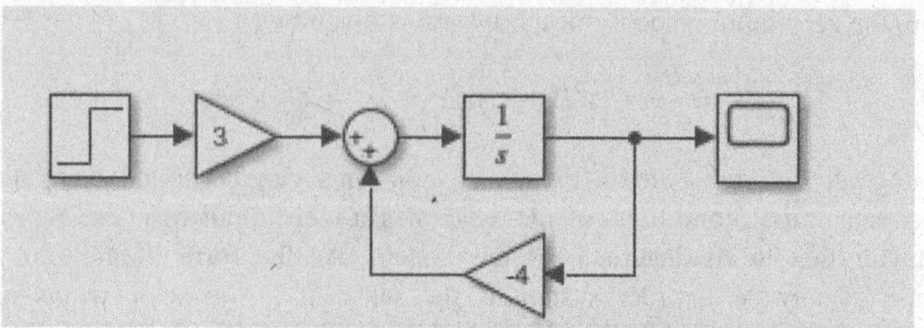

Fig. 4.1 Block diagram of the simulation of the first-order differential equation

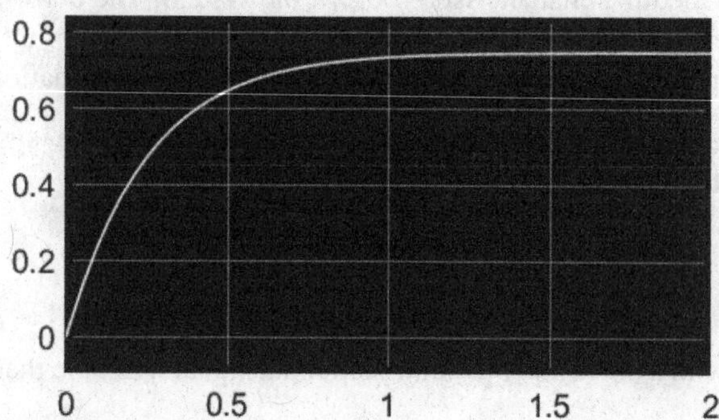

Fig. 4.2 Zero-state response

$$y_{zi}(t) = 2e^{-4t}u(t) \quad \text{and} \quad y_{zs}(t) = \frac{3}{4} - \frac{3}{4}e^{-4t}u(t)$$

These components, of course, add up to the total response. As the impulse is the derivative of the step, the impulse response, $h(t) = 3e^{-4t}u(t)$, can be obtained by differentiating the zero-state step response. ∎

Block diagram of the simulation of the first-order differential equation is shown in Fig. 4.1. The initial condition is set in the integrator block indicated by $\frac{1}{s}$. The input to the integrator block is

$$\frac{dy(t)}{dt} = 3x(t) - 4y(t)$$

The zero-state response is shown in Fig. 4.2. The zero-input response is shown in Fig. 4.3. The total response is shown in Fig. 4.4.

Impulse Response

Block diagram of the simulation of the system to determine the impulse response is shown in Fig. 4.5. The initial condition in the integrator block must be set to zero. Further the pulse width of the pulse generator has to be very short, and the corresponding height must be set such that the area of the pseudo impulse is one. The impulse response, shown in Fig. 4.6, was obtained using a pulse width 0.0001

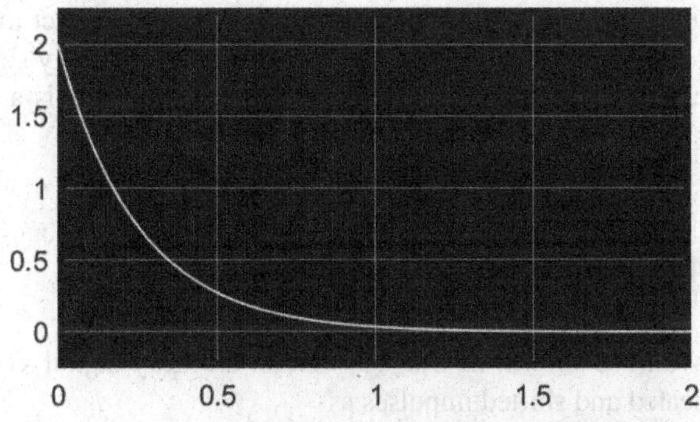

Fig. 4.3 Zero-input response

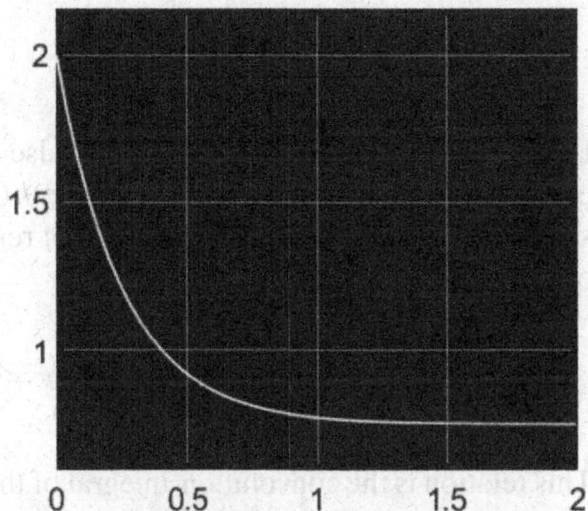

Fig. 4.4 Total response

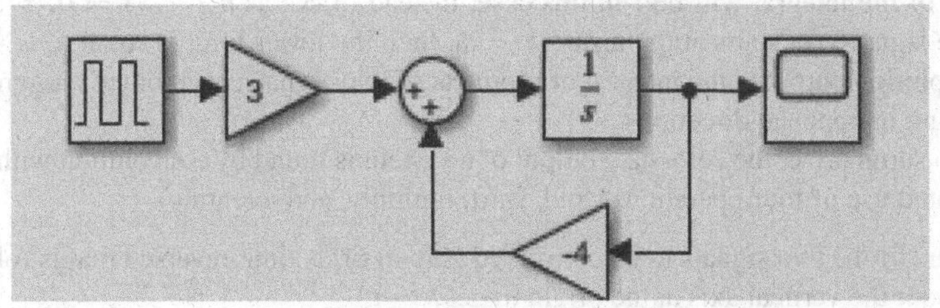

Fig. 4.5 Block diagram of the simulation of the system to determine the impulse response

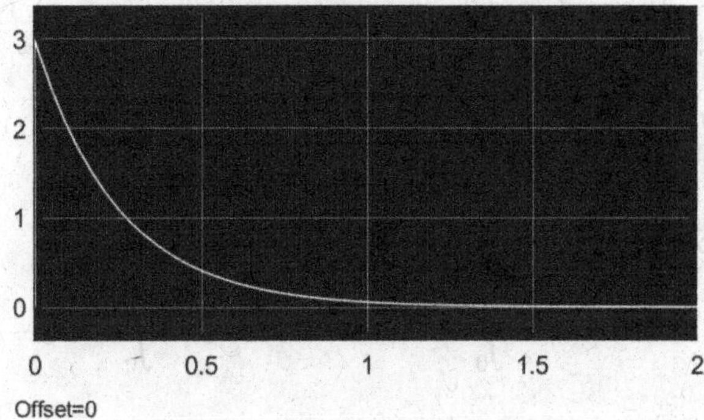

Fig. 4.6 Impulse response

and height 10,000. This simulation confirms the fact that, while the impulse is an ideal signal, it can be approximated by a sufficiently short pulse with unit area for all practical purposes. In all the figures, time versus amplitude of the response is shown.

4.3 Convolution-Integral Model

We have shown, in Chap. 2, that an arbitrary signal $x(t)$ can be decomposed into scaled and shifted impulses as

$$x(t) = \int_{-\infty}^{\infty} x(\tau)\delta(t - \tau)d\tau$$

Let the response of a system to the impulse $\delta(t)$ be $h(t)$. Then, the system response of a LTI system to $x(\tau)\delta(t-\tau)d\tau$ is $x(\tau)h(t-\tau)d\tau$. The total response $y(t)$ of the system to the signal $x(t)$ is the integral of responses of all the constituent continuum of impulse components of $x(t)$:

$$y(t) = \int_{-\infty}^{\infty} x(\tau)h(t - \tau)d\tau = x(t) * h(t)$$

This relation is the convolution-integral of the signals $x(t)$ and $h(t)$ denoted as $x(t)* h(t)$. As the impulse response $h(t)$ of a causal system is zero for $t < 0$, the upper limit of the integral will be t in this case, instead of ∞, as $h(t - \tau) = 0,\ \tau > t$. If the signal $x(t)$ starts at the instant $t = t_0$, then the lower limit is equal to t_0. The convolution output is the integral of products of two signals, each other's argument running in opposite directions.

To summarize, the zero-state output of a system is found by convolution with the repeated use of four operations (fold, shift, multiply, and integrate).

1. One of the two signals to be convolved (say $h(\tau)$) is time-reversed that is folded about the vertical axis at the origin to get $h(-\tau)$.
2. The time-reversed signal, $h(-\tau)$, is shifted by t_0 seconds (right shift for positive t_0 and left shift for negative t_0), yielding $h(t_0 - \tau)$, to find the output at $t = t_0$.
3. The product of the two signals, $x(\tau)$ and $h(t_0 - \tau)$, is found.
4. The integral of the product is the output value at $t = t_0$.

Example 4.5 Find the convolution of the signals $x(t) = e^{-2t}u(t)$ and $h(t) = e^{-3t}u(t)$.

Solution

$$y(t) = \int_{0}^{t} e^{-2\tau}e^{-3(t-\tau)}d\tau = e^{-3t}\int_{0}^{t} e^{\tau}d\tau = (e^{-2t} - e^{-3t})u(t)$$

∎

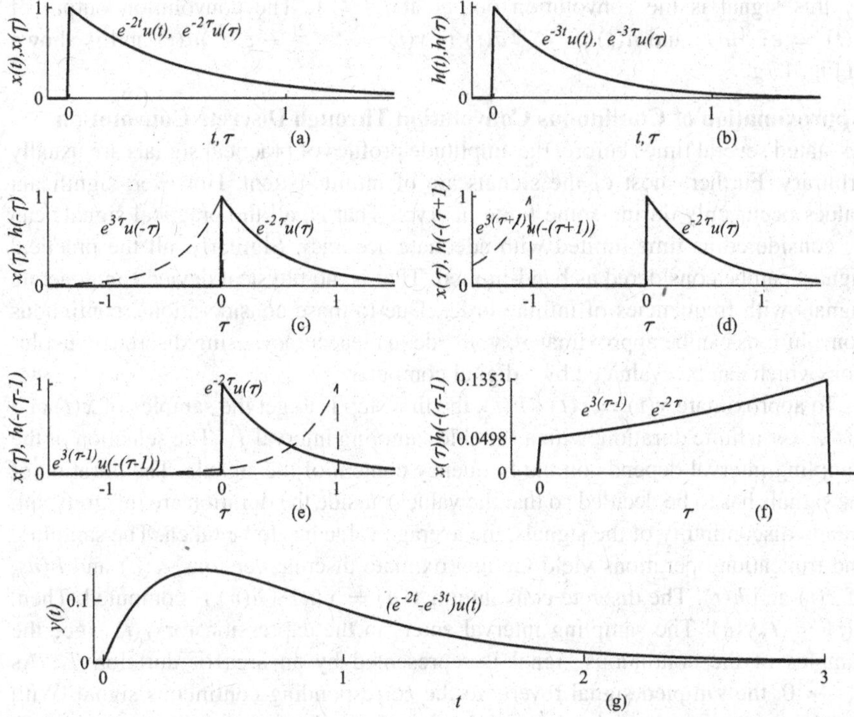

Fig. 4.7 **(a)** $x(t) = e^{-2t}u(t)$, $x(\tau) = e^{-2\tau}u(\tau)$; **(b)** $h(t) = e^{-3t}u(t)$, $h(\tau) = e^{-3\tau}u(\tau)$; **(c)** $x(\tau) = e^{-2\tau}u(\tau)$ and the time-reversed signal $h(-\tau) = e^{3\tau}u(-\tau)$; **(d)** $x(\tau) = e^{-2\tau}u(\tau)$ and the time-reversed and advanced signal $h(-(\tau+1)) = e^{3(\tau+1)}u(-(\tau+1))$; **(e)** $x(\tau) = e^{-2\tau}u(\tau)$ and the time-reversed and delayed signal $h(-(\tau-1)) = e^{3(\tau-1)}u(-(\tau-1))$; **(f)** The product of $x(\tau)$ and $h(-(\tau-1))$, $e^{-2\tau}u(\tau)e^{3(\tau-1)}u(-(\tau-1))$; **(g)** The convolution output of $x(t) = e^{-2t}u(t)$ and $h(t) = e^{-3t}u(t)$, $y(t) = (e^{-2t} - e^{-3t})u(t)$

Figure 4.7a and b shows the two signals to be convolved. These signals and the convolution output, shown in Fig. 4.7g, have the same independent variable t. However, the convolution-integral, for each value of t, is evaluated with respect to the dummy variable τ (a dummy variable exists only during the operation). Therefore, the two signals to be convolved are also shown with respect to τ in Fig. 4.7a and b. Figure 4.7c shows $x(\tau) = e^{-2\tau}u(\tau)$ and the time-reversed signal $h(-\tau) = e^{3\tau}u(-\tau)$. The convolution output at $t = 0$ is zero, since the area enclosed by the signal $e^{-2\tau}u(\tau)e^{3\tau}u(-\tau)$ is zero (there is no overlap of nonzero portions of the signals). Figure 4.7d shows $x(\tau) = e^{-2\tau}u(\tau)$ and the time-reversed and advanced signal $h(-(\tau+1)) = e^{3(\tau+1)}u(-(\tau+1))$. The convolution output at $t = -1$ is zero, since there is no overlap of nonzero portions of the signals. Figure 4.7e shows $x(\tau) = e^{-2\tau}u(\tau)$ and the time-reversed and delayed signal $h(-(\tau-1)) = e^{3(\tau-1)}u(-(\tau-1))$. The nonzero portions of the two signals overlap in the interval from $\tau = 0$ and $\tau = 1$. The product of the signals, $e^{-2\tau}u(\tau)e^{3(\tau-1)}u(-(\tau-1))$, in this interval is shown in Fig. 4.7f. The area enclosed by this signal is the convolution output at $t = 1$. The convolution output of $x(t) = e^{-2t}u(t)$ and $h(t) = e^{-3t}u(t)$ is $y(t) = (e^{-2t} - e^{-3t})u(t)$ and is shown in Fig. 4.7g.

Approximation of Continuous Convolution Through Discrete Convolution
As stated several times before, the amplitude profiles of practical signals are usually arbitrary. Further, most of the signals are of infinite-extent. However, significant values occur only during some finite interval. That is, all the practical signals can be considered as time-limited with adequate accuracy. Similarly, all the practical signals can be considered as band-limited. That is, no physical device can generate signals with frequencies of infinite order. Due to these considerations, continuous convolutions can be approximated, with adequate accuracy, using discrete convolution, which can be evaluated by a digital computer.

To approximate $y(t) = x(t) * h(t)$, the first step is to get the samples of $x(t)$ and $h(t)$, over a finite duration, with a suitable sampling interval T_s. The selection of the sampling interval depends on the frequency content of the signals. The duration of the signals has to be decided so that the values outside the duration are insignificant. At any discontinuity of the signals, the average value has to be taken. The sampling and truncation operations yield the approximate discrete versions, $x(n)$ and $h(n)$, of $x(t)$ and $h(t)$. The discrete convolution $y(n) = x(n) * h(n)$ is computed. Then, $y(t) \approx T_s y(n)$. The sampling interval enters in the expression for $y(t)$ since the samples of the continuous signal is represented by an area of duration T_s. As $T_s \to 0$, the sampled signal reverts to the corresponding continuous signal. With no idea of the effective duration of the signals or the frequency content, simply choose arbitrary values. Compute the convolution. Then, double the duration of the signals and reduce the sampling intervals by a factor of 2 and then recompute the convolution. This is a trial-and-error procedure. A good approximation of the continuous convolution is obtained, when two consecutive iterations yield almost the same output. A program is available at the book's website.

4.3.1 *Properties of Convolution-Integral*

The convolution-integral is commutative, that is, the order of the two signals to be convolved is immaterial.

$$x(t) * h(t) = h(t) * x(t)$$

The convolution-integral is distributive. That is, the convolution of a signal with the sum of two signals is the same as the sum of the individual convolutions of the first signal with the other two signals.

$$x(t) * (h_1(t) + h_2(t)) = x(t) * h_1(t) + x(t) * h_2(t)$$

The convolution-integral is associative. That is, the convolution of a signal with the convolution of two signals is the same as the convolution of the convolution of the first two signals with the third signal.

$$x(t) * (h_1(t) * h_2(t)) = (x(t) * h_1(t)) * h_2(t)$$

The shift property of convolution is that

$$\text{if} \quad x(t) * h(t) = y(t), \quad \text{then} \quad x(t - t_1) * h(t - t_2) = y(t - t_1 - t_2)$$

The convolution of two shifted signals is the convolution of the two original signals shifted by the sum of the shifts of the individual signals.

The duration of the convolution of two finite length signals of duration T_1 and T_2 is $T_1 + T_2$, as the overlap of nonzero portions can occur only over that length.

Convolution of a signal $x(t)$ with the unit-impulse leaves the signal unchanged except for the translation of the origin of the signal to the location of the impulse.

$$x(t) * \delta(t - t_1) = \int_{-\infty}^{\infty} \delta(\tau - t_1)x(t - \tau)d\tau = x(t - t_1)$$

Convolution of $x(t)$ with the unit-step is the running integral of $x(t)$.

$$x(t) * u(t) = \int_{-\infty}^{t} x(\tau)d\tau$$

For example, $u(t) * u(t) = tu(t)$.

4.3.2 *Convolution of a Function with a Narrow Unit Area Pulse*

Looking at the amplitude profile of signals, such as a sinusoid or a unit-step signal or an arbitrary signal, we are able to visualize the function. As the amplitude profile of the continuous unit-impulse signal is undefined, it is presented as an unit area narrow pulse with the width of the pulse approaching zero. It was presented, in Chap. 2, using a narrow rectangular pulse with its width approaching zero and the unit area remaining constant throughout the limit process. As the concept is difficult, we present the unit-impulse from the convolution point of view also.

Let $x(t) = 2 + e^{-t}$ and $h(t) = 5(u(t + 0.1) - u(t - 0.1))$, as shown in Fig. 4.8a and b. Let us find the convolution of $x(t)$ and $h(t)$. Figure 4.9 shows the signal $x(\tau)$ and the delayed by $t = 1$ and time-reversed $h(\tau)$, $h(1 - \tau)$, and their product $x(\tau)h(1 - \tau)$. The area enclosed by $x(\tau)h(1 - \tau)$ is 2.3685, the convolution output at $t = 1$. The value of the function $x(t)$ at $t = 1$ is 2.3679. Both the values are almost the same. By shifting the pulse and convolving at all points of the function

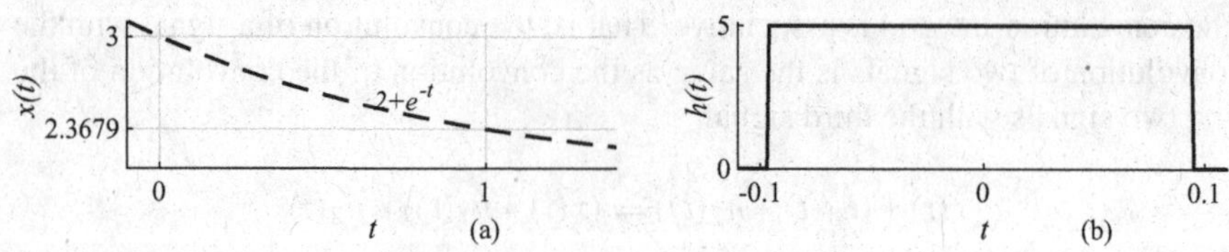

Fig. 4.8 (**a**) $x(t) = 2 + e^{-t}$; (**b**) $h(t) = 5(u(t + 0.1) - u(t - 0.1))$

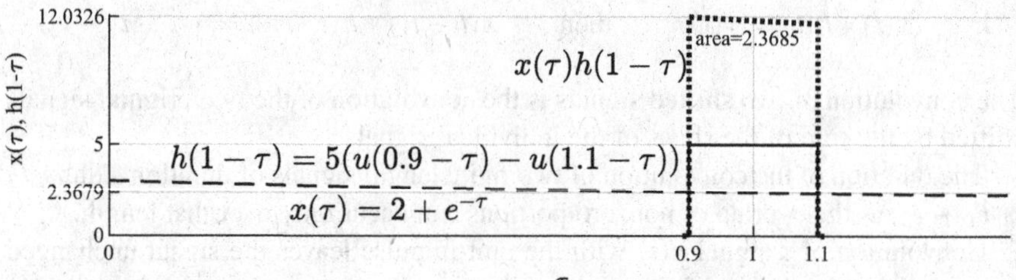

Fig. 4.9 Convolution of $x(\tau) = 2 + e^{-\tau}$ and $h(\tau) = 5(u(\tau + 0.1) - u(\tau - 0.1))$ at $t = 1$

will yield substantially the other function. Reducing the width of the pulse will reduce the error of approximation. It is understood that the variation of the function $x(\tau)$ throughout the width of the pulse is negligible so that the approximation of the function is adequate. Of course, when the width of the pulse approaches zero, the pulse degenerates into an impulse, and the representation becomes exact. Therefore, an important property of the convolution-integral is that convolving any function with the unit-impulse will result in a copy of the function shifted to the location of the impulse. That is,

$$x(t) = \int_{-\infty}^{\infty} x(\tau)\delta(t - \tau)d\tau$$

In experimental work and numerical analysis, finite width pulses of sufficiently short duration should be used.

4.4 System Response

As the amplitude profile of practical signals is usually arbitrary, the output of a system to such signals is found by decomposing the input signals into mathematically well-defined impulse or sinusoidal (in general, exponential) signals. While we are interested in the response of a system to a specific input signal, we use the impulse and the sinusoidal signals as intermediaries. In the convolution-integral model of a system, the impulse signal is used as an intermediary. While these intermediary signals are mathematical idealizations, they can be approximated to a required accuracy for practical purposes. Therefore, it is important to find the response of systems to these signals. In addition, system characteristics, such as rise-time, time-constant, and frequency selectivity, can be obtained from these responses.

4.4.1 *Impulse Response*

The impulse response, $h(t)$, of a system is its response to the unit-impulse input signal with the initial conditions of the system zero.

Example 4.6 Find the closed-form of the impulse response of the system governed by the differential equation, with output $y(t)$ and input $x(t)$:

$$\frac{dy(t)}{dt} + a_0 y(t) = b_1 \frac{dx(t)}{dt} + b_0 x(t)$$

Solution The input signal $x(t) = \delta(t)$ is effective only at the instant $t = 0$ and establishes nonzero initial conditions in the system, by storing energy in system components such as capacitor, at the instant immediately after $t = 0$. Therefore, for $t > 0$, this problem can be considered as finding the zero-input response of the system with the initial condition $y(0^+)$. The symbol $y(0^+)$ indicates the value of $y(t)$ at the instant immediately after $t = 0$ and $y(0^-)$ indicates the value of $y(t)$ at the instant immediately before $t = 0$. Therefore, we have to find the initial condition $y(0^+)$ first and then the response to $\delta(t)$. The response to the input $b_1 \frac{d\delta(t)}{dt} + b_0\delta(t)$ is found using the linearity property of the system. The value $y(0^+)$ is obtained by integrating the differential equation

$$\frac{dy(t)}{dt} + a_0 y(t) = \delta(t)$$

from $t = 0^-$ to $t = 0^+$.

$$\int_{0^-}^{0^+} \frac{dy(t)}{dt} dt + \int_{0^-}^{0^+} a_0 y(t) dt = \int_{0^-}^{0^+} \delta(t) dt$$

The right-hand side is equal to one. The first term on the left-hand side reduces to $y(0^+)$ as $y(0^-) = 0$. Remember that the impulse response is defined as the response of a system to the unit-impulse input with the initial conditions zero. An impulse on the right-hand side implies an impulse on the left-hand side. This impulse must occur in the highest derivative, $\frac{dy(t)}{dt}$, of $y(t)$ since an impulse in $y(t)$ requires the first term to contain the derivative of the impulse and the input does not contain any such function. Therefore, the second term reduces to zero, since the function $y(t)$ is known to be finite (a step function as it the integral of the first term) in the infinitesimal interval of integration. Therefore, the equation reduces to

$y(0^+) = 1$. In general, the integrals of all the lower-order derivative terms of an N-th order differential equation evaluate to zero at $t = 0^+$, and the only nonzero initial condition is $\left.\frac{d^{N-1}y(t)}{dt^{N-1}}\right|_{t=0^+} = 1$.

For a first-order system, the zero-input response is of the form Ce^{-a_0t}. With the initial condition $y(0^+) = 1$, we get the zero-input response as $e^{-a_0t}u(t)$. For the input $b_0\delta(t)$, the response is $b_0e^{-a_0t}u(t)$. For the input $b_1\frac{d\delta(t)}{dt}$, by differentiating $b_1e^{-a_0t}u(t)$, we get the response as $b_1\delta(t) - b_1a_0e^{-a_0t}u(t)$. Note that, for linear systems, if $y(t)$ is the output to $x(t)$, then $\frac{dy(t)}{dt}$ is the output to $\frac{dx(t)}{dt}$. Therefore, the impulse response of the system is

$$h(t) = b_0e^{-a_0t}u(t) + b_1\delta(t) - b_1a_0e^{-a_0t}u(t) = b_1\delta(t) + (b_0 - b_1a_0)e^{-a_0t}u(t)$$

∎

4.4.2 Response to Unit-Step Input

Example 4.7 Find the complete response of the system characterized by the differential equation

$$\frac{dy(t)}{dt} + 4y(t) = 3x(t)$$

with $x(t) = u(t)$, the unit-step input signal. Assume that $y(0^-) = 2$.

Solution

Zero-Input Response
The characteristic equation of the system is $\lambda + 4 = 0$. The zero-input response is of the form Ce^{-4t}. Using the given initial condition, we get $C = 2$ and the zero-input response is $2e^{-4t}u(t)$.

Zero-State Response
The impulse response of the system is $h(t) = 3e^{-4t}u(t)$. Using the convolution-integral, we get the zero-state response as

$$y(t) = \int_0^t u(t-\tau)3e^{-4\tau}d\tau = 3\int_0^t e^{-4\tau}d\tau = \frac{3}{4}(1 - e^{-4t})u(t)$$

As the unit-step signal is the integral of the unit-impulse, the unit-step response is the integral of the of the unit-impulse response. The unit-impulse response $h(t)$ is the derivative of the unit-step response $y(t)$.

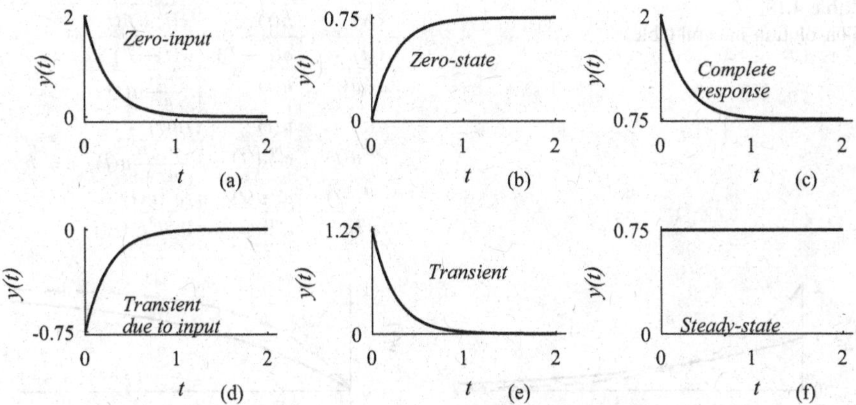

Fig. 4.10 The response of the system for unit-step signal: (**a**) zero-input response; (**b**) zero-state response; (**c**) complete response; (**d**) transient response due to input; (**e**) transient response; (**f**) steady-state response

Complete Response

The complete response of the system is the sum of the zero-input and zero-state responses.

$$y(t) = \overbrace{\frac{3}{4} - \frac{3}{4}e^{-4t}}^{\text{zero-state}} + \overbrace{2e^{-4t}}^{\text{zero-input}}, \ t \geq 0$$

$$= \overbrace{\frac{3}{4}}^{\text{steady-state}} \overbrace{-\frac{3}{4}e^{-4t} + 2e^{-4t}}^{\text{transient}} = \frac{3}{4} + \frac{5}{4}e^{-4t}$$

Transient and Steady-State Responses

The transient response of the system is $\frac{5}{4}e^{-4t}$. The steady-state response of the system, $\frac{3}{4}$, is the response of the system after the transient response dies down. The transient response of a stable system always decays with time. The form of the transient response depends solely on the characteristics of the system while the form of the steady-state response solely depends on the input signal. The various components of the response are shown in Fig. 4.10. The first three responses have been already obtained by simulation. ∎

Table 4.1 shows a list commonly occurring convolution integrals.

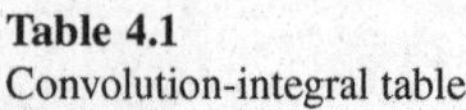

Table 4.1
Convolution-integral table

$x(t)$	$h(t)$	$x(t) * h(t)$
$x(t)$	$\delta(t-T)$	$x(t-T)$
$e^{at}u(t)$	$u(t)$	$\frac{e^{at}-1}{a}u(t)$
$u(t)$	$u(t)$	$tu(t)$
$e^{at}u(t)$	$e^{bt}u(t)$	$\frac{e^{at}-e^{bt}}{a-b}u(t),\ a \neq b$
$e^{at}u(t)$	$e^{at}u(t)$	$te^{at}u(t)$
$te^{at}u(t)$	$e^{at}u(t)$	$0.5t^2e^{at}u(t)$

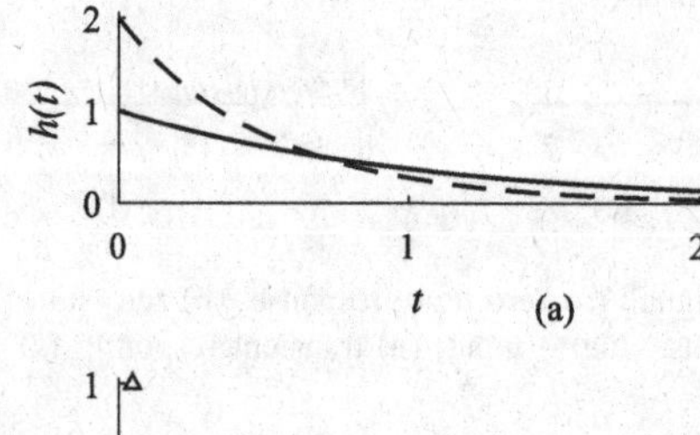

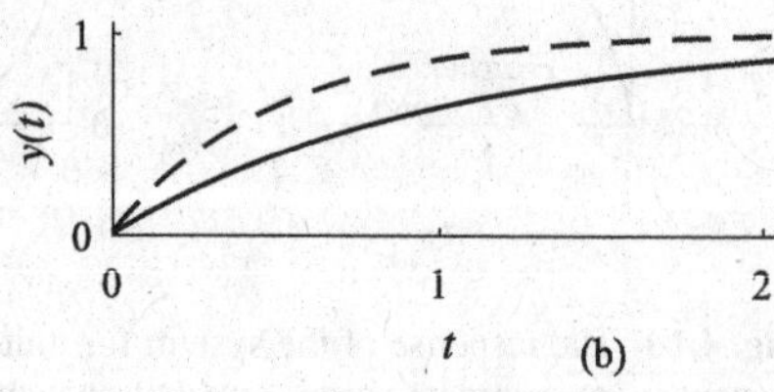

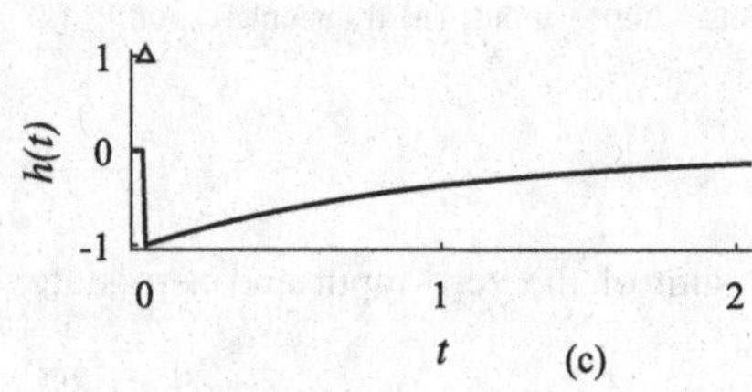

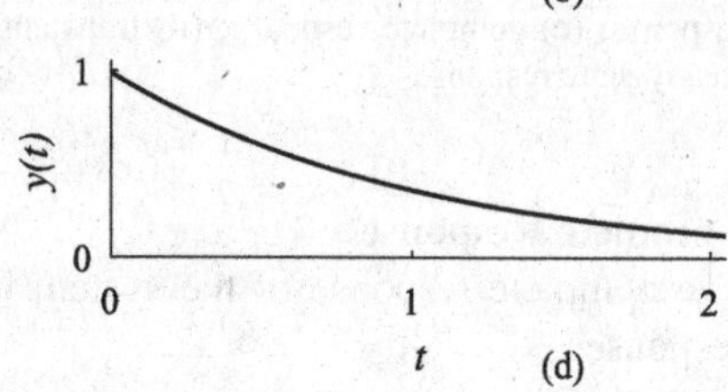

Fig. 4.11 (**a**) Typical monotonically decreasing impulse responses; (**b**) the corresponding unit-step responses; (**c**) an impulse response that is a combination of an impulse and an exponential; (**d**) the corresponding unit-step response

4.4.3 Characterization of Systems by Their Responses to Impulse and Unit-Step Signals

We can get information about the system behavior from the impulse and unit-step responses. If the significant portions of the impulse response is of longer duration, as shown by solid line in Fig. 4.11a, then the response of the system is sluggish. The corresponding unit-step response is shown by solid line in Fig. 4.11b. The time taken for the unit-step response to rise from 10% to 90% of its final value is called the rise time of the system. If the significant portions of the impulse response is of shorter duration, as shown by dashed line in Fig. 4.11a, then the response of the system is faster, as shown by dashed line in Fig. 4.11b. A system with a shorter impulse response has less memory, and it is readily influenced by the recent values of the input signal. Therefore, its response is fast. The faster is the rate of decay of the impulse response, the faster the response approaches its steady-state value.

The unit-step response is the integral of the unit-impulse response, $y(t) = \int_0^t h(\tau)d\tau$. The final value tends to one in Fig. 4.11b, as the unit-step signal, ultimately, acts like a DC signal. The monotonically decreasing impulse response indicates a system that passes low-frequency components of a signal well.

Figure 4.11c shows the impulse response $\delta(t) - e^{-t}u(t)$. The corresponding unit-step response is shown in Fig. 4.11d. Note that the final value of the unit-step response approaches a very low value in Fig. 4.11d. This indicates a system that does not pass low-frequency components of a signal well.

4.4.4 Response to Complex Exponential Input

A complex exponential with frequency ω_0 is given as $x(t) = e^{j\omega_0 t}$, $-\infty < t < \infty$. Assuming a causal and stable system with impulse response $h(t)$, the output is given by the convolution-integral as

$$y(t) = \int_0^{\infty} h(\tau)e^{j\omega_0(t-\tau)}d\tau = e^{j\omega_0 t}\int_0^{\infty} h(\tau)e^{-j\omega_0\tau}d\tau$$

Note that the second integral is independent of t. Let

$$H(j\omega_0) = \int_0^{\infty} h(\tau)e^{-j\omega_0\tau}d\tau$$

Then,

$$y(t) = H(j\omega_0)e^{j\omega_0 t} = H(j\omega_0)x(t)$$

$H(j\omega_0)$ is called the frequency response since it is a constant complex scale factor indicating the amount of change in the amplitude and phase of an input complex exponential $e^{j\omega_0 t}$ with frequency ω_0 at the output. The point is that the input-output relationship of a LTI system becomes a multiplication operation rather than the more complex convolution operation. As the complex exponential is the only signal that has this property, it is used predominantly as the basis for signal decomposition. Even if the exponent of the exponential input signal has a real part, $x(t) = e^{(\sigma+\omega_0)t} = e^{s_0 t}$, the response of the system is still related to the input by the multiplication operation. A real sinusoid input $A\cos(\omega_0 t + \theta)$ is also changed at the output by the same amount of amplitude and phase of the complex scale factor $H(j\omega_0)$. That is, $A\cos(\omega_0 t + \theta)$ is changed to $(|H(j\omega_0)|A)\cos(\omega_0 t + (\theta + \angle(H(j\omega_0)))$.

There was no transient component in the output expression $y(t)$, since the exponential signal was applied at $t = -\infty$. For finite values of t, any transient component in the output of a stable system must have died out. However, if we apply the exponential at any finite instant, say $t = 0$, there will be a transient component, in addition to the steady-state component $H(j\omega_0)e^{j\omega_0 t}u(t)$.

Example 4.8 Let the input signal to a stable system with impulse response $h(t) = e^{-t}u(t)$ be $x(t) = e^{s_0 t}u(t)$. Find the response $y(t)$ of the system. Assume that $y(0^-) = 0$.

Solution Using the convolution-integral, we get

$$y(t) = \int_0^t e^{-\tau} e^{s_0(t-\tau)} d\tau = e^{s_0 t} \int_0^t e^{-\tau(1+s_0)} d\tau$$
$$= \left(\frac{1}{s_0+1}\right)\left(e^{s_0 t} - e^{-t}\right)u(t), \; s_0 \neq -1$$

The steady-state component, $\left(\frac{1}{s_0+1}\right)\left(e^{s_0 t}\right)u(t)$, is the same as the input complex exponential with a complex scale factor. The second term, $\left(\frac{-e^{-t}}{s_0+1}\right)u(t)$, is the transient component that will die for sufficiently large values of t. ∎

4.5 System Stability

One of the criteria for the stability of a system is that the system output is bounded if the input is bounded. A signal $x(t)$ is bounded if $|x(t)| \leq P$ for all values of t, where P is a finite positive number. For example, the signal $x(t) = e^{-0.8t}u(t)$ is bounded and $x(t) = e^{0.8t}u(t)$ is unbounded. As convolution-integral is an integral of products, its value is bounded if the input signal is bounded and the value of the integral of the magnitude of the impulse response is bounded. Let the input signal $x(t)$ be bounded by the positive constant P. From the convolution-integral relation for a causal system with impulse response $h(t)$, we get

$$|y(t)| = \left|\int_0^\infty h(\tau)x(t-\tau)d\tau\right|$$
$$\leq \int_0^\infty |h(\tau)x(t-\tau)d\tau| = \int_0^\infty |h(\tau)||x(t-\tau)|d\tau$$
$$|y(t)| \leq \int_0^\infty |h(\tau)|P d\tau = P\int_0^\infty |h(\tau)|d\tau$$

Therefore, if $\int_0^\infty |h(\tau)|d\tau$ is bounded then $|y(t)|$ is bounded. Consequently, a necessary and sufficient stability condition is that the impulse response is absolutely integrable:

$$\int_0^\infty |h(\tau)|d\tau < \infty$$

As we used the convolution-integral to derive the stability condition, the stability condition ensures a bounded zero-state response. The stability of the zero-input response should be separately checked, and it is presented in Chap. 10.

Example 4.9 Find the condition so that the causal LTI system governed by the differential equation, with output $y(t)$ and input $x(t)$,

$$\frac{dy(t)}{dt} + a_0 y(t) = b_0 x(t)$$

is stable.

Solution As the impulse response of this system, $h(t) = b_0 e^{-a_0 t} u(t)$, is an exponential signal, the condition $a_0 > 0$ ensures that $h(t)$ is absolutely integrable. ∎

4.6 Realization of Continuous Systems

The three basic components required in the realization of continuous systems are (i) multiplier unit, (ii) adder unit, and (iii) integrator unit. A multiplier unit, shown in Fig. 4.12a, produces an output signal $cx(t)$, which is the product of the input signal $x(t)$ with the coefficient c. An adder unit, shown in Fig. 4.12b, produces an output signal $x(t) + y(t)$, which is the sum of the input signals $x(t)$ and $y(t)$. By changing the sign of the subtrahend and then adding it with the minuend, subtraction operation can be realized by an adder unit. An integrator unit, shown in Fig. 4.12c, produces an output $\int_{-\infty}^{t} x(\tau)\,d\tau$ for an input $x(t)$. The output is the integral of the input.

The realization of a continuous system is an interconnection of the basic components. Consider the realization, shown in Fig. 4.13, of a first-order system governed by the differential equation, with output $y(t)$ and input $x(t)$:

$$\frac{dy(t)}{dt} + 3y(t) = 2x(t)$$

A multiplier unit with coefficient -3 and input $y(t)$ produces $-3y(t)$. A multiplier unit with coefficient 2 and input $x(t)$ produces $2x(t)$. The adder unit combines the two partial results to produce the signal $-3y(t) + 2x(t)$, which is equal to $\frac{dy(t)}{dt}$. By passing this signal through an integrator unit, we get $y(t)$.

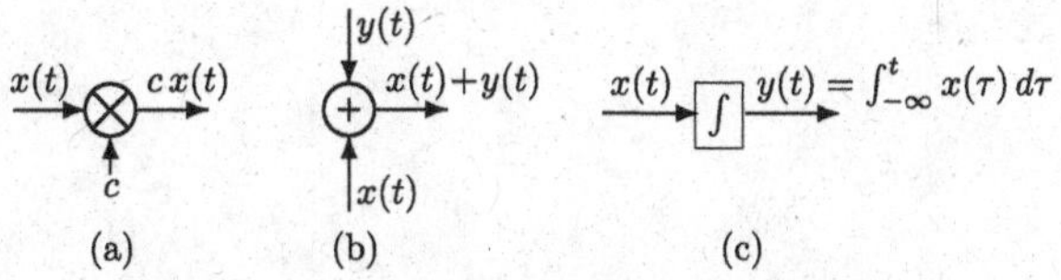

Fig. 4.12 Basic components required in the realization of continuous systems: (**a**) multiplier unit; (**b**) adder unit; (**c**) integrator unit

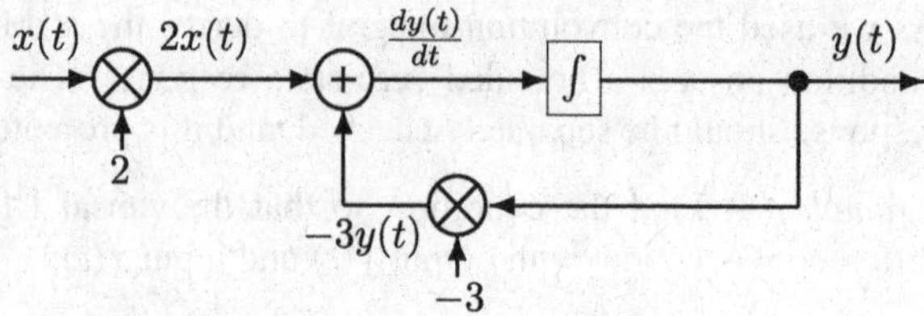

Fig. 4.13 A continuous system realization

Fig. 4.14 (**a**) Two systems connected in parallel: (**b**) a single system equivalent to the system in (**a**)

4.6.1 Decomposition of Higher-Order Systems

To meet a given specification, a higher-order system is often required. Due to several advantages, a system is usually decomposed into first- and second-order systems connected in cascade or parallel. Consider two systems with impulse responses $h_1(t)$ and $h_2(t)$ connected in parallel, shown in Fig. 4.14a. The same input is applied to each system, and the total response is the sum of the individual responses. The combined response of the two systems for the input $x(t)$ is $y(t) = x(t) * h_1(t) + x(t) * h_2(t)$. This expression, due to the distributive property of convolution, can be written as $y(t) = x(t) * (h_1(t) + h_2(t))$. That is, the parallel connection of the two systems is equivalent to a single system with impulse response $h(t) = h_1(t)+h_2(t)$, as shown in Fig. 4.14b.

Consider two systems with impulse responses $h_1(t)$ and $h_2(t)$ connected in cascade, shown in Fig. 4.15a. The output of one system is the input to the other. The response of the first system for the input $x(t)$ is $y_1(t) = x(t)*h_1(t)$. The response of the second system for the input $y_1(t) = x(t) * h_1(t)$ is $y(t) = (x(t) * h_1(t)) * h_2(t)$. This expression, due to the associative property of convolution, can be written as $y(t) = x(t) * (h_1(t) * h_2(t))$. That is, the cascade connection of the two systems is equivalent to a single system with impulse response $h(t) = h_1(t) * h_2(t)$, as shown in Fig. 4.15b. Due to the commutative property of convolution, the order of the systems in the cascade connection is immaterial, with respect to the input-output relationship.

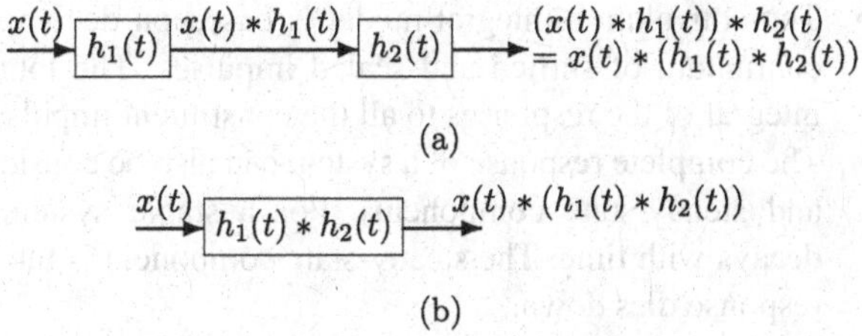

Fig. 4.15 (**a**) Two systems connected in cascade; (**b**) a single system equivalent to the system in (**a**)

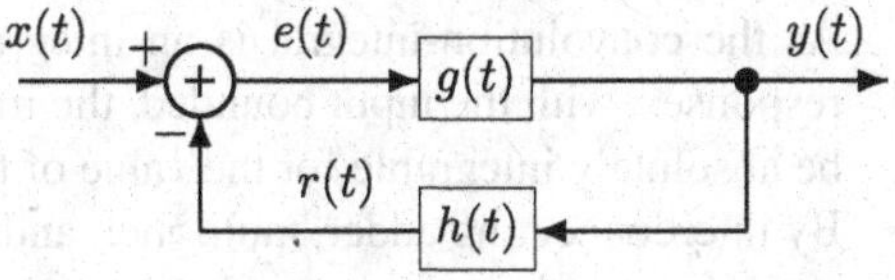

Fig. 4.16 Two systems connected in a feedback configuration

4.6.2 *Feedback Systems*

Another configuration of systems, often used in control systems, is feedback configuration shown in Fig. 4.16. In feedback systems, a fraction of the output signal is fed back and subtracted from the input signal to form the effective input signal. A feedback signal $r(t)$ is produced by a causal system with impulse response $h(t)$ from the the output signal, $y(t)$. That is, $r(t) = \int_0^{\infty} h(\tau)y(t-\tau)d\tau$. The error signal $e(t)$ is difference between the input signal $x(t)$ and the feedback signal $r(t)$, $e(t) = x(t) - r(t)$. This error signal is the input to a causal system with impulse response $g(t)$, which produces the output signal $y(t)$. That is, $y(t) = \int_0^{\infty} g(\tau)e(t-\tau)d\tau$.

4.7 Summary

- In this chapter, the time-domain analysis of LTI continuous systems has been presented.
- The zero-input component of the response of a LTI system is its response due to the initial conditions alone with the input assumed to be zero. The zero-state component of the response of a LTI system is its response due to the input alone with the initial conditions assumed to be zero. The sum of the zero-input and zero-state responses is the complete response of the system.
- Two of the commonly used system models for time-domain analysis are differential equation and convolution-integral models.
- The convolution-integral model gives the zero-state response of a LTI system. Both the zero-input and zero-state responses can be found by solving the differential equation.
- The impulse response of a system is its response to the unit-impulse input signal with the initial conditions zero.

- The convolution-integral model is based on decomposing the input signal into continuum of shifted and scaled impulses. The total response is found by the integral of the responses to all the constituent impulses of the input signal.
- The complete response of a system can also be considered as the sum of transient and steady-state components. For a stable system, the transient component decays with time. The steady-state component is the response after the transient response dies down.
- A system is stable if its response is bounded for all bounded input signals. As the convolution-integral is an integral of the product of input and impulse responses, with the input bounded, the impulse response of a stable system must be absolutely integrable for the value of the convolution-integral to be bounded.
- By interconnecting adder, multiplier, and integrator units, any continuous system can be realized. An higher-order system is usually decomposed into a set of first- and second-order systems connected in cascade or parallel. A feedback system is obtained by feeding back some part of the output to the input.

Exercises

4.1 Is the system governed by the given differential equation, with output $y(t)$ and input $x(t)$, linear?

4.1.1 $\frac{dy(t)}{dt} + 2y(t) + 2 = x(t)$.
4.1.2 $\left(\frac{dy(t)}{dt}\right)^2 + y(t) = \frac{dx(t)}{dt} + x(t)$.
4.1.3 $\frac{dy(t)}{dt} + t\,y(t) = 3\frac{dx(t)}{dt} + 2x(t)$.
4.1.4 $\frac{dy(3t)}{dt} + y(3t) = x(t)$.
4.1.5 $\frac{dy(t)}{dt} + y(t) + \sin(\pi) = x(t)$.
4.1.6 $\frac{dy(t)}{dt} + y(t) + \cos(\pi) = x(t)$.
4.1.7 $\frac{dy(t)}{dt} + y(t) = x(t)\frac{dx(t)}{dt}$.
4.1.8 $\frac{dy(t)}{dt} + e^{y(t)} = x(t)$.
* 4.1.9 $\frac{dy(t)}{dt} = |x(t)|$.

4.2 Is the system governed by the given differential equation, with output $y(t)$ and input $x(t)$, time-invariant?

4.2.1 $\frac{dy(t)}{dt} + y(2t) = x(t)$.
4.2.2 $\frac{dy(t)}{dt} + \cos(\frac{\pi}{2}t)y(t) = x(t)$.
4.2.3 $\frac{dy(t)}{dt} + y(t) = tx(t)$.
* 4.2.4 $y(t) = x(t-5)$.
4.2.5 $\frac{dy(t)}{dt} + ty(t) = x(t)$.
4.2.6 $\frac{dy(t)}{dt} + y(t) = x(t)\frac{dx(t)}{dt}$.

4.2.7 $\frac{dy(t)}{dt} + e^{y(t)} = x(t)$.
4.2.8 $\frac{dy(t)}{dt} + y(-t) = x(t)$.

4.3 Find the closed-form expression for the convolution of the signals $x(t)$ and $h(t)$. List the values of the convolution output at $t = 0, 1, 2, 3, 4, 5$.

4.3.1 $x(t) = 3u(t+1)$ and $h(t) = 2u(t-3)$.
4.3.2 $x(t) = 2e^{-2t}u(t-1)$ and $h(t) = 2e^{-2t}u(t+3)$.
4.3.3 $x(t) = 4u(t-1)$ and $h(t) = 2e^{-2(t+2)}u(t+2)$.
* 4.3.4 $x(t) = (u(t) - u(t-3))$ and $h(t) = (u(t) - u(t-3))$.

4.4 Find the convolution of the signals $x(t)$ and $h(t)$.

4.4.1 $x(t) = e^{j\frac{2\pi}{6}t}u(t)$ and $h(t) = \delta(t+4)$.
4.4.2 $x(t) = e^{j\frac{2\pi}{6}t}$ and $h(t) = \delta(t+12)$.
4.4.3 $x(t) = \cos(\frac{2\pi}{6}t)$ and $h(t) = \delta(t)$.

4.5 Verify the distributive property of convolution integral, $x(t) * (h_1(t) + h_2(t)) = x(t) * h_1(t) + x(t) * h_2(t)$.

4.5.1 $h_1(t) = 2e^{-2t}u(t),\ h_2(t) = 3e^{-2t}u(t),\ x(t) = u(t)$.
4.5.2 $h_1(t) = 3e^{-3t}u(t),\ h_2(t) = 5e^{-3t}u(t),\ x(t) = e^{-t}u(t)$.

4.6 Verify the associative property of convolution integral, $x(t) * (h_1(t) * h_2(t)) = (x(t) * h_1(t)) * h_2(t)$.

4.6.1 $h_1(t) = e^{-2t}u(t),\ h_2(t) = e^{-3t}u(t),\ x(t) = u(t)$.
4.6.2 $h_1(t) = e^{-2t}u(t),\ h_2(t) = e^{-3t}u(t),\ x(t) = e^{-t}u(t)$.

4.7 Find the closed-form expression for the impulse response $h(t)$ of the system characterized by the differential equation, with output $y(t)$ and input $x(t)$. Deduce the closed-form expression for the unit-step response $y(t)$ of the system.

4.7.1 $\frac{dy(t)}{dt} + 2y(t) = -\frac{dx(t)}{dt} + x(t)$.
* 4.7.2 $\frac{dy(t)}{dt} - y(t) = 2\frac{dx(t)}{dt} + 3x(t)$.
4.7.3 $\frac{dy(t)}{dt} + 3y(t) = 2x(t)$.
4.7.4 $\frac{dy(t)}{dt} + 4y(t) = -2\frac{dx(t)}{dt} + x(t)$.
4.7.5 $\frac{dy(t)}{dt} + 2y(t) = 4x(t)$.

4.8 Derive the closed-form expression for the complete response (by finding the zero-state response by convolution and the zero-input response) of the system governed by the differential equation

$$\frac{dy(t)}{dt} + y(t) = 3\frac{dx(t)}{dt} + 2x(t)$$

with the initial condition $y(0^-) = 2$ and the input $x(t) = u(t)$, the unit-step function. Deduce the expressions for the transient and steady-state responses of the system.

4.9 Derive the closed-form expression for the complete response (by finding the zero-state response by convolution and the zero-input response) of the system governed by the differential equation

$$\frac{dy(t)}{dt} - 2y(t) = -3\frac{dx(t)}{dt} + x(t)$$

with the initial condition $y(0^-) = 1$ and the input $x(t) = tu(t)$, the unit-ramp function. Deduce the expressions for the transient and steady-state responses of the system.

4.10 Derive the closed-form expression for the complete response (by finding the zero-state response by convolution and the zero-input response) of the system governed by the differential equation

$$\frac{dy(t)}{dt} + 4y(t) = 2\frac{dx(t)}{dt} - 3x(t)$$

with the initial condition $y(0^-) = 2$ and the input $x(t) = e^{-3t}u(t)$. Deduce the expressions for the transient and steady-state responses of the system.

* **4.11** Derive the closed-form expression for the complete response (by finding the zero-state response by convolution and the zero-input response) of the system governed by the differential equation

$$\frac{dy(t)}{dt} + y(t) = -\frac{dx(t)}{dt} + x(t)$$

with the initial condition $y(0^-) = 3$ and the input $x(t) = 2\cos(t)u(t)$. Deduce the expressions for the transient and steady-state responses of the system.

4.12 Derive the closed-form expression for the complete response (by finding the zero-state response by convolution and the zero-input response) of the system governed by the differential equation

$$\frac{dy(t)}{dt} + 5y(t) = 3\frac{dx(t)}{dt} - x(t)$$

with the initial condition $y(0^-) = -2$ and the input $x(t) = \sin(t)u(t)$. Deduce the expressions for the transient and steady-state responses of the system.

4.13 Find the steady-state response of the system, with the impulse response

$$h(t) = 3\delta(t) - 2e^{-2t}u(t),$$

to the input $x(t) = 3\cos(\frac{2\pi}{8}t + \frac{\pi}{3})u(t)$. Deduce the response for the input $e^{j\frac{2\pi}{8}t}$.

* **4.14** Find the steady-state response of the system, with the impulse response

$$h(t) = 2\delta(t) - 4e^{-t}u(t),$$

to the input $x(t) = 2\sin(\frac{2\pi}{6}t - \frac{\pi}{6})u(t)$. Deduce the response for the input $e^{j\frac{2\pi}{6}t}$.

4.15 The impulse response of a LTI system is given. Use the bounded input bounded output test to find whether the system is stable?

4.15.1 $h(t) = e^{-2t}u(t)$.
4.15.2 $h(t) = u(t)$.
4.15.3 $h(t) = \frac{\sin(2t)}{t}u(t)$.
4.15.4 $h(t) = \left(\frac{\sin(2t)}{t}\right)^2 u(t)$.
4.15.5 $h(t) = -e^{3t}u(t)$.

4.16 Derive the closed-form expression of the impulse response $h(t)$ of the combined system consisting of systems with impulse responses $h_1(t)$ and $h_2(t)$, if the systems are connected in (i) parallel and (ii) cascade.

4.16.1 $h_1(t) = e^{-2t}u(t)$ and $h_2(t) = e^{-5t}u(t)$.
4.16.2 $h_1(t) = \delta(t) + e^{-3t}u(t)$ and $h_2(t) = \delta(t) - e^{-2t}u(t)$.
4.16.3 $h_1(t) = 2\delta(t) - e^{-4t}u(t)$ and $h_2(t) = e^{-3t}u(t)$.

Chapter 5
The Discrete Fourier Transform

In this chapter, the most often used tools for the transformation of signals from the time- to the frequency-domain and back again, the DFT and the IDFT, are presented. The frequency-domain representation of signals and systems is introduced in Sect. 5.1. In Sect. 5.2, a brief review of Fourier analysis is presented. The DFT and the IDFT are derived in Sect. 5.3. The properties of the DFT are presented in Sect. 5.4. Some applications of the DFT are presented in Sect. 5.5.

5.1 The Time-Domain and the Frequency-Domain

The independent variable, in the time-domain representation of signals and systems, is usually time. It could be anything else also, such as distance. In this domain, we analyze arbitrary signals in terms of scaled and shifted impulses. A system is characterized in terms of its impulse response (Chaps. 3 and 4). We still look for simple signals that provide more efficient signal and system analysis. This leads us to an alternate representation of signals and systems, called the frequency-domain representation. In this representation (which can be considered as the transformation of the independent variable), the variation of a signal with respect to the frequency of its constituent sinusoids is used in its characterization. At each frequency, the amplitude and phase or, equivalently, the amplitudes of the cosine and sine components of the sinusoid are used for representing a signal. Systems are characterized in terms of their responses to sinusoids. Both the time-domain and frequency-domain representations completely specify a signal or a system. In the frequency-domain, the independent variable is frequency thereby explicitly specifying the frequency components of a signal. While there are other basic signals, the sinusoid is mostly used for signal and LTI system analysis because it provides ease of signal decomposition, simpler system analysis, and more insight into the signal and system characteristics. Except for the fact that the independent variable

D. Sundararajan, *Signals and Systems*,
https://doi.org/10.1007/978-3-031-19377-4_5

is frequency, the system analysis is very similar to that used in the time-domain. That is, we decompose an input signal in terms of sinusoids, find the response of the system to each sinusoid, and, using the linearity and time-invariant properties of LTI systems, sum up all the responses to find the complete response of the system. The big advantage of the sinusoids is that the steady-state output of a stable LTI system for a sinusoidal input is of the same form. Therefore, the output of a system can be found using much simpler multiplication operation compared with the convolution operation required using the impulse signal.

A set of complex exponentials or sinusoids is used as the basis signals in the principal transforms used in signal and LTI system analysis. While sinusoidal waveforms are generated by physical devices and are easy to visualize, the complex exponential, which is a functionally equivalent mathematical representation of a sinusoid, is often used in signal and system analysis, due to its compact form and ease of manipulation. In Fourier analysis, sinusoids with constant amplitudes (or exponentials with pure imaginary exponents) are used as basis signals. Sinusoids with exponentially varying amplitudes (or exponentials with complex exponents) are used in Laplace and z-transforms. Each transform is more suitable for the analysis of certain class of signals and systems.

Fourier analysis problem is to find the coefficients $X(k)$ in the the complex exponential polynomial representation of a time-domain function $x(n)$, for example, with $N = 4$ coefficients,

$$x(n) = X(0)e^{j0\frac{2\pi}{4}n} + X(1)e^{j\frac{2\pi}{4}n} + X(2)e^{j2\frac{2\pi}{4}n} + X(3)e^{j3\frac{2\pi}{4}n}, \quad n = 0, 1, 2, 3, \tag{5.1}$$

For the given $N = 4$, $x(n)$ and the exponentials are known. The Fourier synthesis problem is to find $x(n)$, given $X(k)$ and the exponentials. In the frequency-domain representation of signals, the signals are decomposed into its constituent sinusoids or exponentials. This representation reduces the more difficult convolution operation to the much simpler multiplication, in addition to other advantages. It is similar to representing numbers by real exponentials in logarithms, which reduces the more difficult multiplication operation to the much simpler addition.

This representation is similar to decomposing a box of mixed coins of various denominations into their respective denominations. Then, finding the amount of coins reduces to counting the number of coins, multiplying by their respective values, and summing it up. This procedure is simpler than finding the amount by adding the value of the coins one by one picked up from the box. It is assumed that a coin sorting machine is available. Similarly, the availability of fast algorithms for decomposing an arbitrary waveform with adequate accuracy makes the Fourier analysis indispensable in linear signal and system analysis in all areas of science and engineering. The Laplace and the z-transforms are generalized versions of the Fourier analysis with a larger set of basis functions. In all cases, the basic decomposition, in principle, is the same as given in Eq. (5.1). Signals and their spectrums may be periodic or aperiodic and of continuous or discrete type. Still, the principle behind frequency-domain analysis remains the same. The analysis becomes simpler with sufficient paper-and-pencil and programming practice. It looks complex due to the details involved in the representation.

5.2 The Fourier Analysis

The theory of the Fourier analysis is that any periodic signal satisfying certain conditions, which are met by most signals of practical interest, can be represented uniquely as the sum of a constant value and an infinite number of harmonics. Harmonically related sinusoids, called harmonics, are a set of sinusoids consisting a fundamental harmonic with frequency f and other harmonics having frequencies those are integral multiples of f. The sum of a set of harmonically related sinusoids is not a sinusoid but is a periodic waveform with period that is the same as that of the fundamental. Given a waveform, finding the amplitude of its constituent sinusoids is called the Fourier analysis. To sum up a set of harmonically related sinusoids to synthesize an arbitrary waveform is called the Fourier synthesis. Consider the discrete periodic waveform, $x(n) = 2 + 3\sin(\frac{2\pi}{4}n) + \cos(2\frac{2\pi}{4}n)$, with period 4 samples, shown in Fig. 5.1a. The independent variable n (actually nT_s, where T_s is the sampling interval) is time, and the dependent variable is amplitude. Figure 5.1b shows the frequency-domain representation of the waveform in (a). It shows the complex amplitude, multiplied by the factor four, of its constituent complex exponentials. To find the real sinusoids, shown in Fig. 5.1c, those constitute the signal, we add up the complex exponentials.

$$
\begin{aligned}
x(n) &= \frac{1}{4}\left(8e^{j0\frac{2\pi}{4}n} - j6e^{j\frac{2\pi}{4}n} + 4e^{j2\frac{2\pi}{4}n} + j6e^{j3\frac{2\pi}{4}n}\right) \\
&= 2 + 3\sin\left(\frac{2\pi}{4}n\right) + \cos\left(2\frac{2\pi}{4}n\right)
\end{aligned}
$$

As can be seen from this example, Fourier analysis represents a signal as a linear combination of sinusoids or, equivalently, complex exponentials with pure imaginary exponents.

The Fourier reconstruction of a waveform is with respect to the least squares error criterion. That is, the mean value for power signals or the total value for the energy signals of the integral or sum of the squared magnitude of the error between the given waveform and the corresponding Fourier reconstructed waveform is guaranteed to be minimum if part of the constituent sinusoids of a waveform is used in the reconstruction and will be zero if all the constituent sinusoids are used. The reason this criterion, based on signal energy or power, is used rather than a minimum uniform deviation criterion is that (i) it is acceptable for most applications and (ii) it leads to closed-form formulas for the analytical determination of Fourier coefficients.

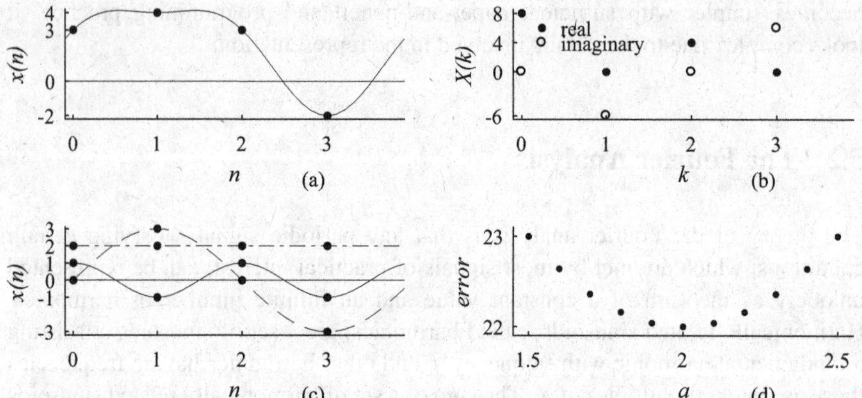

Fig. 5.1 (**a**) A periodic waveform, $x(n) = 2+3\sin(\frac{2\pi}{4}n)+\cos(2\frac{2\pi}{4}n)$, with period 4 samples and (**b**) its frequency-domain representation; (**c**) the frequency components of the waveform in (**a**); (**d**) the square error in approximating the waveform in (**a**) using only the DC component with different amplitudes

Let $x_a(n)$ be an approximation of a given waveform $x(n)$ of period N, using fewer harmonics than required. The square error between $x(n)$ and $x_a(n)$ is defined as

$$error = \sum_{n=0}^{N-1} |x(n) - x_a(n)|^2$$

For a given number of harmonics, there is no better approximation for the signal than that provided by the Fourier approximation when the least squares error criterion is applied. Assume that, we are constrained to use only the DC component to approximate the waveform in Fig. 5.1a. Let the optimal value of the DC component be a. To minimize the square error,

$$(3-a)^2 + (4-a)^2 + (3-a)^2 + (-2-a)^2$$

must be minimum. Differentiating this expression with respect to a and equating it to zero, we get

$$2(3-a)(-1) + 2(4-a)(-1) + 2(3-a)(-1) + 2(-2-a)(-1) = 0$$

Solving this equation, we get $a = 2$ as given by the Fourier analysis. The square error, for various values of a, is shown in Fig. 5.1d.

5.2.1 The Four Versions of Fourier Analysis

Fourier analysis has four versions, each of them using a set of constant amplitude sinusoids, differing in some respect, as the basis signals. Continuous periodic signals are analyzed using an infinite number of harmonically related continuous sinusoids in the FS, described in Chap. 6. Discrete aperiodic signals are analyzed using a continuum of discrete sinusoids over a finite frequency range in the DTFT, presented in Chap. 7. Continuous aperiodic signals are analyzed using a continuum of continuous sinusoids over an infinite frequency range in the FT, described in Chap. 8. The topic of the present chapter is the DFT, which analyzes the periodic extension of a finite duration discrete signal using a finite number of harmonically related discrete sinusoids. **The DFT, because of its finite and discrete nature, is the simplest of the four versions of the Fourier analysis to visualize the analysis and synthesis of waveforms. Problems in understanding the concepts in other versions of the Fourier analysis may be resolved by considering an equivalent DFT version.**

5.3 The Discrete Fourier Transform

5.3.1 The Approximation of Arbitrary Waveforms with Finite Number of Samples

We need a minimum of $2k + 1$ samples to represent a sinusoid uniquely, which completes k cycles in a period, as presented in Chap. 1. To approximate a periodic waveform in terms of DC, we need a minimum of one sample in a period. If we use the fundamental or first harmonic, which has the same period as that of the waveform to be analyzed, we need a minimum of three samples ($2k + 1 = 2(1) + 1 = 3$) in a period, since the first harmonic completes one cycle. In the frequency-domain, we need one value to represent the DC and two values (the amplitude and the phase or the amplitudes of its cosine and sine components) to specify the first harmonic. That is, three samples are required in both the time and frequency-domains. With N independent values in one domain, we can generate only N independent values in the other domain. Therefore, we need $2k + 1$ samples in both the time and frequency-domains to represent a waveform with the DC and the first k harmonically related sinusoids.

In general, an infinite number of sinusoids are required to represent an arbitrary waveform exactly. The concept of using a finite number of sinusoids is based on the fact that the waveforms encountered in practice can be approximated by a finite number of sinusoids with a finite but arbitrarily small tolerance, since, beyond some range, the spectral values become negligible. That is, all practical signals can be considered as band-limited. If the magnitude of the frequency components of a signal is identically zero outside some finite frequency range, then the signal is called band-limited. In addition, the waveforms are generally aperiodic. In order to make it finite duration, we have to truncate some part of it. Then, a periodic extension of the waveform is represented by discrete sinusoids. The truncation is acceptable because waveforms, in practice, have negligible values beyond some range. That is, all practical signals can be considered as time-limited. If the amplitude of a signal is identically zero outside some finite time interval, then the signal is called time-limited Therefore, we can represent any waveform, encountered in practice, by a finite number of samples in both the time and frequency-domains with adequate accuracy. This representation, using a finite number of samples in both the domains, is the feature of the DFT version of the Fourier analysis. That is to make the essential information, characterizing a waveform, available in any one period, in both the domains, with sufficient accuracy. The point is that, while the representation of a waveform can be made adequate, the discrete and finite nature of the DFT makes it inherently suitable for numerical analysis. And, finally, the fact that Fourier analysis plays a central part in signal and system analysis and fast algorithms are available for computing the DFT makes the DFT the heart of practical signal and system analysis. Note that, with continuous signals, both the time- and frequency-domain signals cannot be periodic. However, with uniformly sampled signals, it is possible for both the time- and frequency-domain signals to be periodic. The condition is that the ratio of the period to the sampling interval is an integer.

5.3.2 *The DFT and the IDFT*

In the DFT, a set of N samples represents a waveform in both the time and frequency-domains, whether the waveform is periodic or aperiodic and continuous or discrete. It is understood that the number of samples is adequate to represent the waveform with sufficient accuracy. The set of N samples is periodically extended and N harmonically related complex exponentials are used to represent the waveform. That is, for a real-valued signal with N samples, we are using real sinusoids with frequency indices $0, 1, 2, \ldots, \frac{N}{2}$ only. Frequency index zero represents the DC and $\frac{N}{2}$ represents a cosine waveform, assuming N is even.

The frequency components of a waveform are separated using the orthogonality property of the exponentials. For two complex exponentials $e^{j\frac{2\pi}{N}ln}$ and $e^{j\frac{2\pi}{N}kn}$ over a period of N samples, the orthogonality property is defined as

$$\sum_{n=0}^{N-1} e^{j\frac{2\pi}{N}(l-k)n} = \begin{cases} N \text{ for } l = k \\ 0 \text{ for } l \neq k \end{cases}$$

where $l, k = 0, 1, \ldots, N-1$. If $l = k$, the summation is equal to N as $e^{j\frac{2\pi}{N}(l-k)n} = e^0 = 1$. Otherwise, by using the closed-form expression for the sum of a geometric progression, we get

$$\sum_{n=0}^{N-1} e^{j\frac{2\pi}{N}(l-k)n} = \frac{1-e^{j2\pi(l-k)}}{1-e^{j\frac{2\pi(l-k)}{N}}} = 0, \text{ for } l \neq k$$

It is also known that the sine and cosine waveforms are symmetrical about the x-axis. Therefore, the sum of the equidistant samples of a sinusoid, over an integral number of periods, with a nonzero frequency index is always zero. That is, in order to find the coefficient, with a scale factor N, of a complex exponential, we multiply the samples of a signal with the corresponding samples of the complex conjugate of the complex exponential. Using each complex exponential in turn, we get the frequency coefficients of all the components of a signal as

$$X(k) = \sum_{n=0}^{N-1} x(n)W_N^{nk}, \quad k = 0, 1, \ldots, N-1 \tag{5.2}$$

where $W_N = e^{-j\frac{2\pi}{N}}$. This is the DFT equation analyzing a waveform with harmonically related discrete complex sinusoids. $X(k)$ is the coefficient, scaled by N, of the complex sinusoid $e^{j\frac{2\pi}{N}kn}$ with a specific frequency index k (frequency $\frac{2\pi}{N}k$ radians per sample). The summation of the sample values of the N complex sinusoids multiplied by their respective frequency coefficients $X(k)$ is the IDFT operation. The N-point IDFT of the frequency coefficients $X(k)$ is defined as

$$x(n) = \frac{1}{N}\sum_{k=0}^{N-1} X(k)W_N^{-nk}, \quad n = 0, 1, \ldots, N-1 \tag{5.3}$$

The sum of the sample values is divided by N in Eq. (5.3) as the coefficients $X(k)$ have been scaled by the factor N in the DFT computation.

The DFT equation can be as well written using matrices. With $N = 4$, the DFT is given by

$$\begin{bmatrix} X(0) \\ X(1) \\ X(2) \\ X(3) \end{bmatrix} = \begin{bmatrix} W_4^0 & W_4^0 & W_4^0 & W_4^0 \\ W_4^0 & W_4^1 & W_4^2 & W_4^3 \\ W_4^0 & W_4^2 & W_4^4 & W_4^6 \\ W_4^0 & W_4^3 & W_4^6 & W_4^9 \end{bmatrix} \begin{bmatrix} x(0) \\ x(1) \\ x(2) \\ x(3) \end{bmatrix}$$

The values of the transform matrix, called the twiddle factors, are equally spaced samples on the unit-circle. They are the Nth roots of unity. For example, the second row values raised to the power 4 yield 1.

$$(e^{-j\frac{2\pi}{4}n})^4 = e^{-j2\pi n} = 1, \quad n = 0, 1, 2, 3$$

The inverse and forward transform matrices are orthogonal. That is,

$$\frac{1}{4}\begin{bmatrix} 1 & 1 & 1 & 1 \\ 1 & j & -1 & -j \\ 1 & -1 & 1 & -1 \\ 1 & -j & -1 & j \end{bmatrix}\begin{bmatrix} 1 & 1 & 1 & 1 \\ 1 & -j & -1 & j \\ 1 & -1 & 1 & -1 \\ 1 & j & -1 & -j \end{bmatrix} = \begin{bmatrix} 1 & 0 & 0 & 0 \\ 0 & 1 & 0 & 0 \\ 0 & 0 & 1 & 0 \\ 0 & 0 & 0 & 1 \end{bmatrix}$$

If we use the definition of the DFT, we need N complex multiplications and $(N-1)$ complex additions for computing each of the N coefficients. The computational complexity of computing all the N coefficients is of the order of $O(N^2)$. Fast algorithms reduce this computational complexity to the order of $O(N \log_2 N)$. Because of these algorithms, the use of the DFT is more efficient in most applications compared with alternate methods.

With $N = 2$, the DFT definition is

$$\begin{bmatrix} X(0) \\ X(1) \end{bmatrix} = \begin{bmatrix} 1 & 1 \\ 1 & -1 \end{bmatrix}\begin{bmatrix} x(0) \\ x(1) \end{bmatrix}$$

The IDFT is defined as

$$\begin{bmatrix} x(0) \\ x(1) \end{bmatrix} = \frac{1}{2}\begin{bmatrix} 1 & 1 \\ 1 & -1 \end{bmatrix}\begin{bmatrix} X(0) \\ X(1) \end{bmatrix}$$

The DFT of $\{x(0) = -4, x(1) = 2\}$ is $\{X(0) = -2, X(1) = -6\}$. It can be verified that IDFT gets back the time-domain samples. With $N = 2$, the waveform is composed of 2 complex exponentials, $e^{j\frac{2\pi}{2}0n}$ and $e^{j\frac{2\pi}{2}1n}$. The two samples of these waveforms are $\{1, 1\}$ and $\{1, -1\}$. The DFT finds the coefficients of these components from $x(n)$ as -2 and -6, so that

$$\frac{-2\{1, 1\} - 6\{1, -1\}}{2} = \{-4, 2\} = x(n)$$

This summation is carried out by the IDFT. The DFT correlates the samples of the input $\{-4, 2\}$ with samples of the complex exponential $e^{j\frac{2\pi}{2}0n}$, $\{1, 1\}$, to find out $X(0) = -2$. The operation required is the sum of products of the two sequences. That is

$$X(0) = -4 \times 1 + 2 \times 1 = -2$$

Similarly,

$$X(1) = -4 \times 1 + 2 \times (-1) = -6$$

This decomposition works for any arbitrary waveform $x(n)$ of arbitrary length N due to the orthogonality of the harmonically related complex exponentials.

Let us compute the DFT of $\{x(0) = 3, x(1) = 4, x(2) = 3, x(3) = -2\}$. The DFT of this set of data is computed as

$$\begin{bmatrix} X(0) \\ X(1) \\ X(2) \\ X(3) \end{bmatrix} = \begin{bmatrix} 1 & 1 & 1 & 1 \\ 1 & -j & -1 & j \\ 1 & -1 & 1 & -1 \\ 1 & j & -1 & -j \end{bmatrix} \begin{bmatrix} 3 \\ 4 \\ 3 \\ -2 \end{bmatrix} = \begin{bmatrix} 8 \\ -j6 \\ 4 \\ j6 \end{bmatrix}$$

The DFT spectrum is $\{X(0) = 8, X(1) = -j6, X(2) = 4, X(3) = j6\}$, as shown in Fig. 5.1b. The sample values of the DC component are $\{2, 2, 2, 2\}$, as shown in Fig. 5.1c. Pointwise multiplication of these values with the first row values of the transform matrix $\{1, 1, 1, 1\}$ and summing the product yields 8. This is the scaled value of the DC component, which is 2. The same computation with the other 3 rows of the transform matrix yields 0, due to orthogonality. It is instructive to try to find the DFT coefficients of the other three frequency components.

Now, let us compute the sample values of the waveform from its DFT coefficients using the IDFT.

$$\begin{bmatrix} x(0) \\ x(1) \\ x(2) \\ x(3) \end{bmatrix} = \frac{1}{4} \begin{bmatrix} 1 & 1 & 1 & 1 \\ 1 & j & -1 & -j \\ 1 & -1 & 1 & -1 \\ 1 & -j & -1 & j \end{bmatrix} \begin{bmatrix} 8 \\ -j6 \\ 4 \\ j6 \end{bmatrix} = \begin{bmatrix} 3 \\ 4 \\ 3 \\ -2 \end{bmatrix}$$

We get back the time-domain sample values confirming that the DFT and IDFT form a transform pair. What one operation does the other undoes.

5.3.2.1 Center-Zero Format of the DFT and IDFT

The DFT coefficients $X(k)$ are periodic with period N, the number of samples of $x(n)$. Therefore, N coefficients, over any consecutive part of the spectrum, define the spectrum. However, the coefficients are usually specified in two ranges. In the standard format, the N coefficients are specified over one period starting from index $k = 0$. That is,

$$X(0), X(1), \ldots, X(N-1)$$

For example, with $N = 4$,

$$X(0), X(1), X(2), X(3)$$

In the other format, called center-zero format, the coefficient $X(0)$ is placed approximately in the middle. That is, with N even,

$$X\left(-\frac{N}{2}\right), X\left(-\frac{N}{2}+1\right), \dots, X(-1), X(0), X(1), \dots, X\left(\frac{N}{2}-2\right), X\left(\frac{N}{2}-1\right)$$

For example, with $N = 4$,

$$X(-2), X(-1), X(0), X(1)$$

One format from the other can be obtained by circularly shifting the spectrum by $N/2$ number of sample intervals. The use of the center-zero format is convenient in some derivations. Further, the display of the spectrum in this format is better for viewing. The definitions of the DFT and IDFT in this format, with N even, are

$$X(k) = \sum_{n=(-\frac{N}{2})}^{\frac{N}{2}-1} x(n)e^{-jk\frac{2\pi}{N}n}, \quad k = -\left(\frac{N}{2}\right), -\left(\frac{N}{2}-1\right), \dots, \frac{N}{2}-1 \tag{5.4}$$

$$x(n) = \frac{1}{N} \sum_{k=-(\frac{N}{2})}^{\frac{N}{2}-1} X(k)e^{jk\frac{2\pi}{N}n}, \quad n = -\left(\frac{N}{2}\right), -\left(\frac{N}{2}-1\right), \dots, \frac{N}{2}-1 \tag{5.5}$$

When the spectrum is displayed in this format, the interpretation of the spectral features is easier. The conversion of the data or spectrum from one format to another involves circular shift by half the number of samples. The spectrum

$$\{X(0) = 2, X(1) = 2 - j\sqrt{2}, X(2) = 3, X(3) = 2 + j\sqrt{2}\}$$

in center-zero format is

$$\{X(-2) = 3, X(-1) = 2 + j\sqrt{2}, X(0) = 2, X(1) = 2 - j\sqrt{2}\}$$

5.3.3 *DFT of Some Basic Signals*

While the primary purpose of the DFT is to approximate the spectra of arbitrary signals using numerical procedures, it is useful, for understanding, to derive the DFT of some simple signals analytically. The DFT of the impulse signal $x(n) = 2\delta(n)$ is simply $X(k) = 2$. As the impulse signal is nonzero only at $n = 0$, the DFT equation reduces to $x(0)$ for any value of k. A signal and its DFT form a transform pair and is denoted as $x(n) \iff X(k)$. For the specific example, the transform pair is denoted as $2\delta(n) \iff 2$. The DFT, with $N = 16$, is shown in Fig. 5.2a.

A plot of the complex coefficients $X(k)$ of the constituent complex sinusoids of a signal $x(n)$ versus k is called the complex spectrum of $x(n)$. The spectral value of two for all the frequency components implies that the impulse signal, with a value of two, is the sum of all the exponentials $\frac{2}{16}e^{j\frac{2\pi}{16}kn}$, $k = 0, 1, \ldots, 15$. In terms of real sinusoids, this impulse signal is the sum of DC component $\frac{2}{16}$, cosine waves $\frac{2}{8}\cos(\frac{2\pi}{16}kn)$, $k = 1, 2, \ldots, 7$, and $\frac{2}{16}\cos(\pi n)$.

The DFT of the DC signal $x(n) = 3$, with N samples, is $X(k) = 3N\delta(k)$. That is, $3 \Longleftrightarrow 3N\delta(k)$. As the DC signal has a constant value, its DFT evaluation essentially reduces to the summation of the sample values of the various complex exponentials. This sum is zero for all the complex exponentials with nonzero frequency index k. For $k = 0$, $X(0) = 3\sum_{n=0}^{N-1} e^{-j\frac{2\pi}{N}n0} = 3\sum_{n=0}^{N-1} 1 = 3N$. The complex exponential with $k = 0$ is the DC signal. The DFT of the DC signal $x(n) = 3$, with 16 samples, is shown in Fig. 5.2b.

The frequency range of the spectral components of a signal is called its bandwidth. The essential bandwidth of a signal is the frequency range of its spectral components containing most of its energy. The longer is the duration of a signal in the time-domain, the shorter is the essential bandwidth in its frequency-domain representation and vice versa. This is called reciprocal spreading and is well demonstrated in the case of the DC and impulse signals. The impulse signal is nonzero only at $n = 0$ in the time-domain, and its spectrum is spread with significant values throughout the whole frequency range. The reverse is the case for the DC signal.

The complex exponential signal, although of no physical significance, is the standard unit in the frequency-domain representation and analysis of signals and systems, as it is easier to manipulate and the sum of its conjugate with itself is capable of representing a physical signal. Due to the orthogonality property, the complex exponential $x(n) = e^{j\frac{2\pi}{N}np}$ with frequency index p has the transform pair $e^{j\frac{2\pi}{N}np} \Longleftrightarrow N\delta(k-p)$. The DC case presented earlier is a specific case with $p = 0$. The complex exponential signal $x(n) = e^{j\frac{2\pi}{16}n}$ with $N = 16$ and its spectrum with $X(1) = 16$ are shown in Fig. 5.2c and d, respectively.

The complex exponential signal $x(n) = e^{j(2\frac{2\pi}{16}n-\frac{\pi}{6})}$ with $N = 16$ and its spectrum are shown in Fig. 5.2e and f, respectively. This signal can be expressed as $x(n) = e^{-j\frac{\pi}{6}}e^{j2\frac{2\pi}{16}n}$. Therefore, the DFT coefficient is that of $x(n) = e^{j2\frac{2\pi}{16}n}$, which is 16 at $k = 2$, multiplied by the complex constant $e^{-j\frac{\pi}{6}} = \frac{\sqrt{3}}{2} - j\frac{1}{2}$, as shown in Fig. 5.2f.

A real sinusoid, $x(n) = \cos(\frac{2\pi}{N}np+\theta)$, is the sum of a pair of complex conjugate exponentials:

$$x(n) = \cos\left(\frac{2\pi}{N}np + \theta\right) = \frac{1}{2}(e^{j\theta}e^{j\frac{2\pi}{N}np} + e^{-j\theta}e^{-j\frac{2\pi}{N}np})$$

Using the DFT of complex exponentials, we get

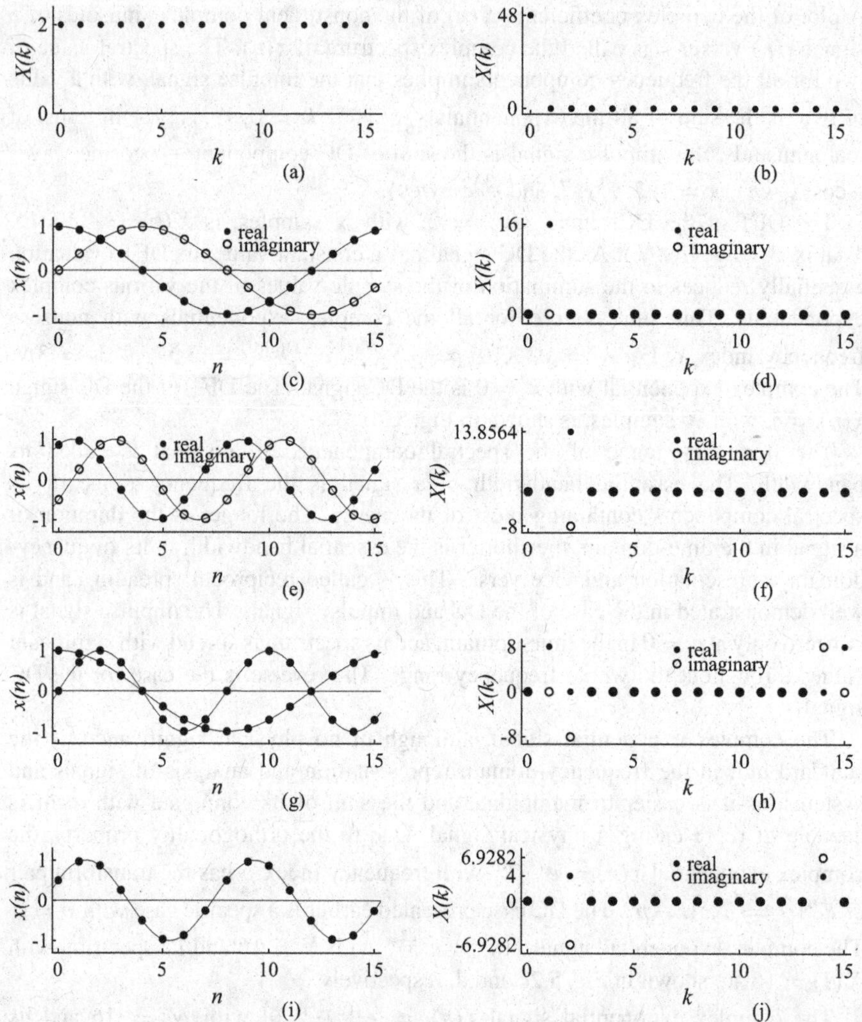

Fig. 5.2 (**a**) The spectrum of the impulse $x(n) = 2\delta(n)$ with $N = 16$; (**b**) the spectrum of the DC signal $x(n) = 3$ with $N = 16$; (**c**) the complex sinusoid $x(n) = e^{j\frac{2\pi}{16}n}$ and (**d**) its spectrum; (**e**) the complex sinusoid $x(n) = e^{j(2\frac{2\pi}{16}n-\frac{\pi}{6})}$ and (**f**) its spectrum; (**g**) the sinusoids $x(n) = \cos(\frac{2\pi}{16}n)$ and $x(n) = \sin(2\frac{2\pi}{16}n)$, and (**h**) their spectra; (**i**) The sinusoid $x(n) = \cos(2\frac{2\pi}{16}n - \frac{\pi}{3})$ and (**j**) its spectrum

$$\cos\left(\frac{2\pi}{N}np + \theta\right) \Longleftrightarrow \frac{N}{2}(e^{j\theta}\delta(k-p) + e^{-j\theta}\delta(k-(N-p)))$$

Note that, due to periodicity, $e^{-j\frac{2\pi}{N}np} = e^{j\frac{2\pi}{N}n(N-p)}$. We get the transform pairs for the cosine and sine waves, with $\theta = 0$ and $\theta = -\frac{\pi}{2}$, as

$$\cos\left(\frac{2\pi}{N}np\right) \iff \frac{N}{2}(\delta(k-p)+\delta(k-(N-p)))$$

$$\sin\left(\frac{2\pi}{N}np\right) \iff \frac{N}{2}(-j\delta(k-p)+j\delta(k-(N-p)))$$

The cosine and sine waves $x(n) = \cos(\frac{2\pi}{16}n)$ and $x(n) = \sin(2\frac{2\pi}{16}n)$ with $N = 16$ and their spectra are shown in Fig. 5.2g and h, respectively. The sinusoid $x(n) = \cos(2\frac{2\pi}{16}n - \frac{\pi}{3})$ with $N = 16$ and its spectrum are shown in Fig. 5.2i and j, respectively.

An infinite or finite sequence

$$ba^n = \{b, ba, ba^2, \ldots, \}$$

is a geometric sequence, where b and a are some fixed numbers. The first term is a constant. The rest of the terms are the product of the preceding term by the common ratio of the terms. The sum of the terms of a geometric sequence is a geometric series. For example,

$$\Sigma_N = 1 + a + a^2 + \cdots + a^{N-1}, \quad a = e^{-j\frac{2\pi}{N}}$$

is a geometric series. To find the sum in a closed form, we multiply Σ_N by a to get $a\Sigma_N$. Now,

$$\Sigma_N - a\Sigma_N = 1 - a^N \quad \text{and} \quad \Sigma_N = \frac{1-a^N}{1-a}$$

The DFT of the rectangular waveform is derived as follows.

$$x(n) = \begin{cases} 1 & \text{for } n = 0, 1, \ldots, M-1 \\ 0 & \text{for } n = M, M+1, \ldots, N-1 \end{cases}$$

$$X(k) = \sum_{n=0}^{M-1} e^{-j\frac{2\pi}{N}nk} = \frac{1-e^{-j\frac{2\pi}{N}Mk}}{1-e^{-j\frac{2\pi}{N}k}} = \frac{e^{j\frac{2\pi}{N}\frac{M}{2}k} - e^{-j\frac{2\pi}{N}\frac{M}{2}k}}{e^{j\frac{2\pi}{N}\frac{k}{2}} - e^{-j\frac{2\pi}{N}\frac{k}{2}}}$$

$$= e^{-j\frac{2\pi}{N}\frac{(M-1)}{2}k}\frac{\sin(\frac{2\pi}{N}\frac{M}{2}k)}{\sin(\frac{2\pi}{N}\frac{k}{2})} = e^{-j\frac{\pi}{N}(M-1)k}\frac{\sin(\frac{\pi}{N}Mk)}{\sin(\frac{\pi}{N}k)}$$

Verify the DFT of $\delta(n)$ and DC signals, obtained earlier, using this formula.

Example 5.1 The samples of a signal are $\{x(0) = 1, x(1) = 1, x(2) = 1, x(3) = 0\}$ and

$$\begin{bmatrix} X(0) \\ X(1) \\ X(2) \\ X(3) \end{bmatrix} = \begin{bmatrix} 1 & 1 & 1 & 1 \\ 1 & -j & -1 & j \\ 1 & -1 & 1 & -1 \\ 1 & j & -1 & -j \end{bmatrix} \begin{bmatrix} 1 \\ 1 \\ 1 \\ 0 \end{bmatrix} = \begin{bmatrix} 3 \\ -j \\ 1 \\ j \end{bmatrix}$$

$$\begin{bmatrix} x(0) \\ x(1) \\ x(2) \\ x(3) \end{bmatrix} = \frac{1}{4} \begin{bmatrix} 1 & 1 & 1 & 1 \\ 1 & j & -1 & -j \\ 1 & -1 & 1 & -1 \\ 1 & -j & -1 & j \end{bmatrix} \begin{bmatrix} 3 \\ -j \\ 1 \\ j \end{bmatrix} = \begin{bmatrix} 1 \\ 1 \\ 1 \\ 0 \end{bmatrix}$$

As the basis signals of the DFT are complex exponentials, the DFT is as well applicable to complex signals, although physical signals are real-valued. Of course, the real-valued signals can be expressed as a linear combination of complex signals.

Example 5.2 The samples of a signal are $\{x(0) = 2 + j1, x(1) = 1 - j2, x(2) = 1 + j1, x(3) = 3 + j2\}$ and

$$\begin{bmatrix} X(0) \\ X(1) \\ X(2) \\ X(3) \end{bmatrix} = \begin{bmatrix} 1 & 1 & 1 & 1 \\ 1 & -j & -1 & j \\ 1 & -1 & 1 & -1 \\ 1 & j & -1 & -j \end{bmatrix} \begin{bmatrix} 2 + j1 \\ 1 - j2 \\ 1 + j1 \\ 3 + j2 \end{bmatrix} = \begin{bmatrix} 7 + j2 \\ -3 + j2 \\ -1 + j2 \\ 5 - j2 \end{bmatrix}$$

$$\begin{bmatrix} x(0) \\ x(1) \\ x(2) \\ x(3) \end{bmatrix} = \frac{1}{4} \begin{bmatrix} 1 & 1 & 1 & 1 \\ 1 & j & -1 & -j \\ 1 & -1 & 1 & -1 \\ 1 & -j & -1 & j \end{bmatrix} \begin{bmatrix} 7 + j2 \\ -3 + j2 \\ -1 + j2 \\ 5 - j2 \end{bmatrix} = \begin{bmatrix} 2 + j1 \\ 1 - j2 \\ 1 + j1 \\ 3 + j2 \end{bmatrix}$$

5.4 Properties of the Discrete Fourier Transform

In signal and system analysis, it is often required to carry out operations such as shifting, scaling, convolution etc., in both the domains. We know the effect, in the other domain, of carrying out an operation in one domain through properties. We repeatedly use the properties in applications of the DFT and in deriving DFT algorithms. In addition, new transform pairs can be derived from available ones easily.

5.4.1 Linearity

If a sequence is a linear combination of a set of sequences, each of the same length N, then the DFT of that combined sequence is the same linear combination of the DFT of the individual sequences. That is,

$$x(n) \iff X(k), \qquad y(n) \iff Y(k), \qquad ax(n) + by(n) \iff aX(k) + bY(k),$$

where a and b are arbitrary constants.

For example,

$$\cos\left(\frac{2\pi}{4}n\right) \iff 2\left(\delta(k-1) + \delta(k-3)\right)$$

$$\sin\left(\frac{2\pi}{4}n\right) \iff 2\left(-j\delta(k-1) + j\delta(k-3)\right)$$

Then,

$$e^{j\frac{2\pi}{4}n} = \cos\left(\frac{2\pi}{4}n\right) + j\sin\left(\frac{2\pi}{4}n\right) \iff$$

$$2\left(\delta(k-1) + \delta(k-3)\right) + j2\left(-j\delta(k-1) + j\delta(k-3)\right)$$

$$= 4\delta(k-1)$$

5.4.2 Periodicity

As the complex exponential W_N^{nk} is periodic in both the variables n and k with period N ($W_N^{nk} = W_N^{n(k+N)} = W_N^{(n+N)k}$), a sequence $x(n)$ of N samples and its DFT $X(k)$ are periodic with period N. By substituting $k + aN$ for k in the DFT equation and $n + aN$ for n in the IDFT equation, we get $X(k) = X(k + aN)$ and $x(n) = x(n + aN)$, where a is any integer.

Let

$$x(n) = \{1, -3, 2, 4\} \iff X(k) = \{4, -1 + j7, 2, -1 - j7\}$$

As $-4 \bmod 4 = 0$, $x(-4) = x(0) = 1$. Similarly, $X(5) = X(1) = -1 + j7$.

5.4.3 Circular Time Reversal

Let the N-point DFT of $x(n)$ be $X(k)$. The DFT of the time reversal $x(N-n)$ of $x(n)$ is

$$\begin{bmatrix} X(0) \\ X(3) \\ X(2) \\ X(1) \end{bmatrix} = \begin{bmatrix} 1 & 1 & 1 & 1 \\ 1 & -j & -1 & j \\ 1 & -1 & 1 & -1 \\ 1 & j & -1 & -j \end{bmatrix} \begin{bmatrix} x(0) \\ x(3) \\ x(2) \\ x(1) \end{bmatrix} = X(N-k)$$

Therefore,

$$x(N-n) \Longleftrightarrow X(N-k)$$

For example,

$$x(n) = \{\check{2}, 1, 3, 4\} \Longleftrightarrow X(k) = \{\check{10}, -1 + j3, 0, -1 - j3\}$$

$$x(4-n) = \{\check{2}, 4, 3, 1\} \Longleftrightarrow X(4-k) = \{\check{10}, -1 - j3, 0, -1 + j3\}$$

5.4.4 Duality

The DFT and IDFT operations are almost similar. The differences are that the sign of the exponents in the definitions differ and there is a constant in the IDFT definition. Let the DFT of $x(n)$ be $X(k)$ with period N. Then, due to the dual nature of the definitions, we can interchange the independent variables and make the time-domain function as the DFT of the frequency-domain function with some minor changes. That is, if we compute the DFT of $X(n)$, then we get $Nx(N-k)$. That is,

$$X(n) \Longleftrightarrow Nx(N-k)$$

For example, the DFT of $\{\check{1}, 2, 3, 4\}$ is $\{\check{10}, -2 + j2, -2, -2 - j2\}$. The DFT of this is $4\{\check{1}, 4, 3, 2\}$.

5.4.5 Sum and Difference of Sequences

As the twiddle factors are all 1s with the frequency index $k = 0$, the coefficient of the DC frequency component $X(0)$ is just the sum of the time-domain samples $x(n)$. Let the transform length N be even. As the twiddle factors are all alternating 1s and

-1s with the frequency index $k = N/2$, the coefficient $X(N/2)$ is just the difference of the sum of the even and odd time-domain samples $x(n)$. These values computed this way are a check on the DFT computation.

$$X(0) = \sum_{n=0}^{N-1} x(n) \quad \text{and} \quad X\left(\frac{N}{2}\right) = \sum_{n=0,2}^{N-2} x(n) - \sum_{n=1,3}^{N-1} x(n)$$

Similarly, in the frequency-domain,

$$x(0) = \frac{1}{N}\sum_{k=0}^{N-1} X(k) \quad \text{and} \quad x\left(\frac{N}{2}\right) = \frac{1}{N}\left(\sum_{k=0,2}^{N-2} X(k) - \sum_{k=1,3}^{N-1} X(k)\right)$$

For example, let $\{\check{1}, 1, 3, 2\} \Longleftrightarrow \{\check{7}, -2 + j1, 1, -2 - j1\}$. We can verify the formulas using this transform pair.

5.4.6 Upsampling of a Sequence

Consider the sequence and its DFT

$$x(n) = \{1 \check{+} j1, 2 - j3\} \Longleftrightarrow X(k) = \{3 \check{-} j2, -1 + j4\}$$

Let us upsample $x(n)$ by a factor of 2 to get

$$x_u(n) = \{1 \check{+} j1, 0, 2 - j3, 0\} \Longleftrightarrow X_u(k) = \{3 \check{-} j2, -1 + j4, 3 - j2, -1 + j4\}$$

The spectrum is repeated.

In general, with

$$x(n) \Longleftrightarrow X(k), \; n, k = 0, 1, \ldots, N-1$$

and a positive integer upsampling factor L,

$$x_u(n) = \begin{cases} x(\frac{n}{L}) & \text{for } n = 0, L, 2L, \ldots, L(N-1) \\ 0 & \text{otherwise} \end{cases} \Longleftrightarrow X(k) = X(k \bmod N), \; k = 0, 1, \ldots, LN-1$$

The spectrum $X(k)$ is repeated L times.

The same thing happens in the upsampling of a spectrum, except for a constant factor in the amplitude of the time-domain signal. Consider the sequence and its DFT

$$x(n) = \{1 \check{+} j1, 2 - j3\} \Longleftrightarrow X(k) = \{3 \check{-} j2, -1 + j4\}$$

Let us upsample $X(k)$ by a factor of 2 to get

$$X_u(k) = \{\check{3} - j2, 0, -1 + j4, 0\} \Longleftrightarrow \{1 \check{+} j1, 2 - j3, 1 \check{+} j1, 2 - j3\}/2$$

The time-domain sequence is repeated.

5.4.7 Zero Padding the Data

Appending zeros to a sequence $x(n)$ is carried out mostly for two purposes. As the practically fast DFT algorithms are of length 2, $x(n)$ is zero padded to meet this constraint. Another purpose is to make the spectrum denser, as the frequency increment in the spectrum is inversely proportional to the length of the time-domain sequence. Sufficient number of zeros should be appended so that all the essential features, such as a peak, are adequately represented. While any number of zeros can be appended, we present the case of making the signal longer by an integer number of times, L.

Let $L = 2$ and $x(n) = \{\check{3}, 1, 2, 4\}$. $X(k) = \{\check{10}, 1 + j3, 0, 1 - j3\}$. Then

$$x_z(n) = \{\check{3}, 1, 2, 4, 0, 0, 0, 0\} \Longleftrightarrow X_z(k) = \{\check{10}, *, 1 + j3, *, 0, *, 1 - j3, *\}$$

With lengths of the sequences being 4 and 8, the DFT computes the coefficients at frequencies

$$\{0, 1, 2, 3\}/4 \quad \text{and} \quad \{0, 1, 2, 3, 4, 5, 6, 7\}/8$$

Therefore, the even-indexed DFT coefficients of $x_z(n)$ are the same as that of $X(k)$. With eight frequency components, the spectrum is denser.

In general, with $x(n) \Longleftrightarrow X(k),\ n, k = 0, 1, \ldots, N - 1,$

$$x_z(n) = \begin{cases} x(n) & \text{for } n = 0, 1, \ldots, N - 1 \\ 0 & \text{for } n = N, N + 1, \ldots, LN - 1 \end{cases} \Longleftrightarrow X_z(Lk) = X(k),\ k = 0, 1, \ldots, N - 1$$

Similarly, the zero padding of a spectrum results in a denser and scaled time-domain signal. For example, let $x(n) = \{\check{4}, 2\} \leftrightarrow \{\check{6}, 2\}$. Then,

$$x_z(n) = \{\check{4}, *, 2, *\}/2 \Longleftrightarrow \{\check{6}, 2, 0, 0\}$$

The even-indexed samples of $x_z(n)$ are the same as that of $x(n)$ divided by 2.

5.4.8 Circular Shift of a Sequence

As any periodic sequence is completely specified by its elements over a period, the shifted version of a periodic sequence can be obtained by circularly shifting its elements over a period. As the time-domain sequence $x(n)$ and its DFT $X(k)$ are considered periodic, the shift of these sequences are called as circular shift. For example, the delayed sequence $(x - 1)$ is obtained by moving the last sample of $x(n)$ to the beginning of the sequence. Similarly, the advanced sequence $(x + 2)$ is obtained by moving the first two samples of $x(n)$ to the end of the sequence. Only $(N - 1)$ unique shifts are possible for a sequence with N samples.

The distance between two samples of a sinusoid completing k cycles in its period of N samples is $\frac{2\pi}{N}k$ radians. Therefore, a shift of the sinusoid by m sample intervals to the right amounts to changing its phase by $-\frac{2\pi}{N}mk$ radians, with its amplitude unchanged. The change in the phase is $\frac{2\pi}{N}mk$ radians for a left shift. Let $x(n) \Longleftrightarrow X(k)$. Then,

$$x(n \pm m) \Longleftrightarrow e^{\pm j\frac{2\pi}{N}mk}X(k) = W_N^{\mp mk}X(k)$$

The cosine waveform $x(n) = \cos(\frac{2\pi}{16}n)$ with $N = 16$ and its DFT are shown, respectively, in Fig. 5.2g and h. By shifting $x(n)$ to the right by two sample intervals, we get $x(n) = \cos(\frac{2\pi}{16}(n - 2))$. The spectral value $X(1)$ of the delayed waveform is obtained by multiplying $X(1) = 8$ in Fig. 5.2h by $e^{-j\frac{2\pi}{16}(2)(1)} = e^{-j\frac{\pi}{4}} = \frac{1}{\sqrt{2}}(1 - j1)$. The result is $X(1) = \frac{8}{\sqrt{2}}(1 - j1)$. Similarly, $X(15) = \frac{8}{\sqrt{2}}(1 + j1)$.

5.4.9 Circular Shift of a Spectrum

The spectrum, $X(k)$, of a signal, $x(n)$, can be shifted by multiplying the signal by a complex exponential, $e^{\pm jk_0\frac{2\pi}{N}n}$, where k_0 is an integer and N is the length of $x(n)$. The new spectrum is $X(k \mp k_0)$, since a spectral component $X(k_a)e^{jk_a\frac{2\pi}{N}n}$ of the signal, multiplied by $e^{jk_0\frac{2\pi}{N}n}$, becomes $X(k_a)e^{j((k_a+k_0)\frac{2\pi}{N}n)}$ and the corresponding spectral value occurs at $k = (k_a + k_0)$, after a delay of k_0 samples. The spectrum is circularly shifted by k_0 sample intervals. For example, if $k_0 = 1$ or $k_0 = N + 1$, then the DC spectral value of the original signal appears at $k = 1$. With $k_0 = -1$ or $k_0 = N - 1$, it appears at $k = N - 1$. Let $x(n) \Longleftrightarrow X(k)$. Then,

$$e^{\mp j\frac{2\pi}{N}k_0 n}x(n) = W_N^{\pm k_0 n}x(n) \Longleftrightarrow X(k \pm k_0)$$

The complex exponential $x(n) = e^{j\frac{2\pi}{16}n}$ with $N = 16$ and its spectrum $X(1) = 16$ are shown, respectively, in Fig. 5.2c and d. By multiplying $x(n)$ with $e^{j\frac{2\pi}{16}2n}$, we get $x(n) = e^{j\frac{2\pi}{16}3n}$. Then, the spectrum becomes $X(3) = 16$.

For example,

$$x(n) = \{\check{1}, 2, 3, 4\} \Longleftrightarrow X(k) = \{\check{10}, -2 + j2, -2, -2 - j2\}$$

$$e^{j\frac{2\pi}{4}n}x(n) = \{\check{1}, j2, -3, -j4\} \Longleftrightarrow X(k) = \{-2 - j2, \check{10}, -2 + j2, -2\}$$

$$e^{j\frac{2\pi}{4}2n}x(n) = (-1)^n x(n) = \{\check{1}, -2, 3, -4\} \Longleftrightarrow X(k) = \{-2, -2 - j2, \check{10}, -2 + j2\}$$

This frequency shift by $N/2$ sample intervals is often used to find the center-zero spectrum.

5.4.10 Symmetry

Symmetry of a signal can be used to reduce its storage and computational requirements. The DFT symmetry properties for various types of signals are shown in Table 5.1. In this table, Re stands for "real part of" and Im stands for "imaginary part of." The symbol * indicates the complex conjugation operation. Note that the even-symmetry condition $x(n) = x(-n)$ is the same as $x(n) = x(N - n)$ for a periodic signal of period N.

The DFT is formulated using the complex exponentials as basis functions. Therefore, a real signal has to be reconstructed using complex conjugate pairs. Consequently, the coefficients, for each frequency component, are also conjugate symmetric. That is, the real part is even and the imaginary part is odd. Figure 5.3a shows the real signal

$$x(n) = -0.1 + \sin\left(\frac{2\pi}{16}n + \frac{\pi}{3}\right) + \cos\left(2\frac{2\pi}{16}n + \frac{\pi}{6}\right) - 0.2\sin\left(3\frac{2\pi}{16}n\right)$$

and Fig. 5.3b shows its hermitian symmetric spectrum, both with period 16. For example,

$$\sin\left(\frac{2\pi}{16}n + \frac{\pi}{3}\right) = \cos\left(\frac{2\pi}{16}n - \frac{\pi}{6}\right) \Longleftrightarrow 8\left(e^{-j\frac{\pi}{6}}\delta(k-1) + e^{j\frac{\pi}{6}}\delta(k-15)\right) = 6.9282 \mp j4$$

The nonzero spectral values are

$$\{X(0) = -1.6, X(1, 15) = 6.9282 \mp j4, X(2, 14) = 6.9282 \pm j4, X(3, 13) = \pm j1.6\}$$

Table 5.1 DFT symmetry properties

Signal $x(n)$, $n = 0, 1, \ldots, N-1$	DFT $X(k)$
Real $\mathrm{Im}(x(n)) = 0$	Hermitian $X(k) = X^*(N-k)$
Real and even $\mathrm{Im}(x(n)) = 0$ and $x(n) = x(N-n)$	Real and even $\mathrm{Im}(X(k)) = 0$ and $X(k) = X(N-k)$
Real and odd $\mathrm{Im}(x(n)) = 0$ and $x(n) = -x(N-n)$	Imaginary and odd $\mathrm{Re}(X(k)) = 0$ and $X(k) = -X(N-k)$
Real and even half-wave $\mathrm{Im}(x(n)) = 0$ and $x(n) = x(n \pm \frac{N}{2})$	Hermitian and even-indexed only $X(k) = X^*(N-k)$ and $X(2k+1) = 0$
Real and odd half-wave $\mathrm{Im}(x(n)) = 0$ and $x(n) = -x(n \pm \frac{N}{2})$	Hermitian and odd-indexed only $X(k) = X^*(N-k)$ and $X(2k) = 0$
Imaginary $\mathrm{Re}(x(n)) = 0$	Antihermitian $X(k) = -X^*(N-k)$
Imaginary and even $\mathrm{Re}(x(n)) = 0$ and $x(n) = x(N-n)$	Imaginary and even $\mathrm{Re}(X(k)) = 0$ and $X(k) = X(N-k)$
Imaginary and odd $\mathrm{Re}(x(n)) = 0$ and $x(n) = -x(N-n)$	Real and odd $\mathrm{Im}(X(k)) = 0$ and $X(k) = -X(N-k)$
Imaginary and even half-wave $\mathrm{Re}(x(n)) = 0$ and $x(n) = x(n \pm \frac{N}{2})$ Imaginary and odd half-wave $\mathrm{Re}(x(n)) = 0$ and $x(n) = -x(n \pm \frac{N}{2})$	Antihermitian and even-indexed only $X(k) = -X^*(N-k)$ and $X(2k+1) = 0$ Antihermitian and odd-indexed only $X(k) = -X^*(N-k)$ and $X(2k) = 0$
Complex and even, $x(n) = x(N-n)$	Even, $X(k) = X(N-k)$
Complex and odd, $x(n) = -x(N-n)$	Odd, $X(k) = -X(N-k)$
Complex and even half-wave $x(n) = x(n \pm \frac{N}{2})$	Even-indexed only $X(2k+1) = 0$
Complex and odd half-wave $x(n) = -x(n \pm \frac{N}{2})$	Odd-indexed only $X(2k) = 0$

A real and even-symmetric signal is composed of cosine waves only. Therefore, for each frequency component, the coefficients are real and even. Figure 5.3c shows the real and even-symmetric signal

$$x(n) = -0.2 + 0.6\cos\left(\frac{2\pi}{16}n\right) - \cos\left(2\frac{2\pi}{16}n\right) + 0.1\cos\left(8\frac{2\pi}{16}n\right)$$

and Fig. 5.3d shows its real and even-symmetric spectrum. For example,

$$-\cos\left(2\frac{2\pi}{16}n\right) \Longleftrightarrow -8\,(\delta(k-2) + \delta(k-14))$$

The nonzero spectral values are

$$\{X(0) = -3.2,\, X(1, 15) = 4.8,\, X(2, 14) = -8,\, X(8) = 1.6\}$$

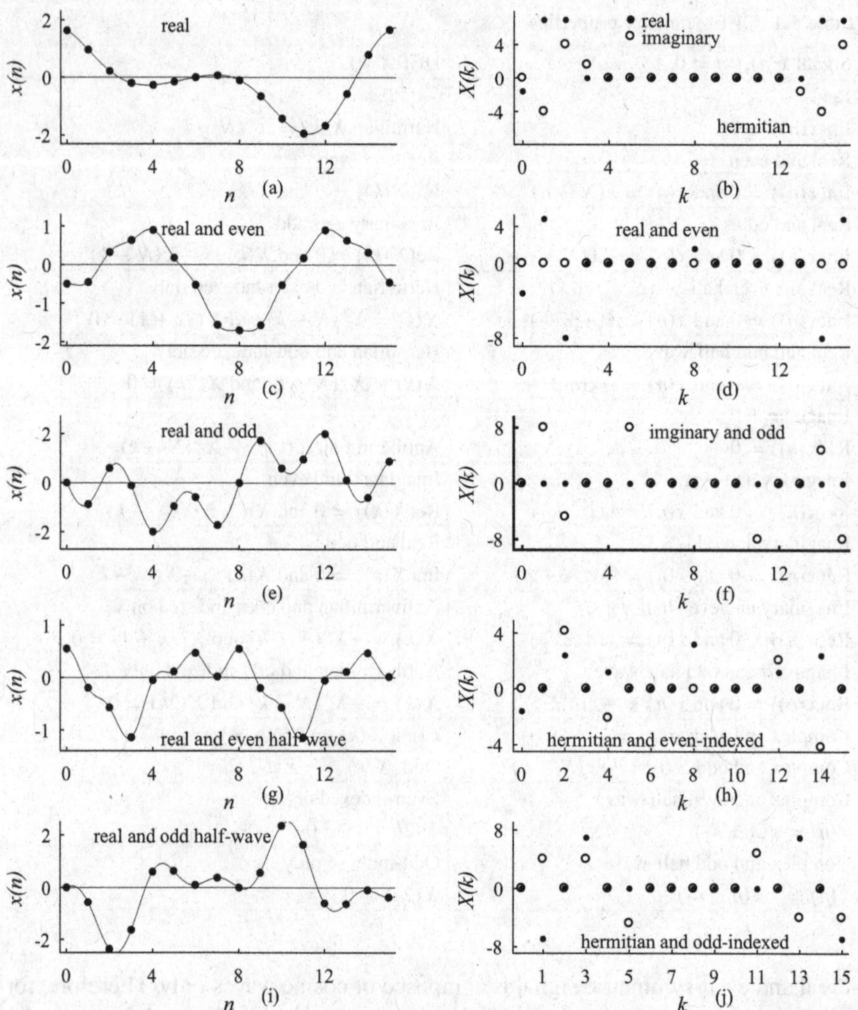

Fig. 5.3 (**a**) A real signal and (**b**) its hermitian-symmetric spectrum; (**c**) an even-symmetric real signal and (**d**) its real and even-symmetric spectrum; (**e**) an odd-symmetric real signal and (**f**) its imaginary and odd-symmetric spectrum; (**g**) a real signal with even half-wave symmetry and (**h**) its hermitian-symmetric spectrum with zero-valued odd-indexed harmonics; (**i**) a real signal with odd half-wave symmetry and (**j**) its hermitian-symmetric spectrum with zero-valued even-indexed harmonics

A real and odd-symmetric signal is composed of sine waves only. Therefore, for each frequency component, the coefficients are imaginary and odd. Figure 5.3e shows the real and odd-symmetric signal

$$x(n) = -\sin\left(\frac{2\pi}{16}n\right) + 0.6\sin\left(2\frac{2\pi}{16}n\right) - \sin\left(5\frac{2\pi}{16}n\right)$$

and Fig. 5.3f shows its imaginary and odd-symmetric spectrum. For example,

$$-\sin\left(5\frac{2\pi}{16}n\right) \Longleftrightarrow -8\left(-j\delta(k-5) + j\delta(k-11)\right)$$

The nonzero spectral values are

$$\{X(1,15) = \pm j8, X(2,14) = \mp j4.8, X(5,11) = \pm j8\}$$

A real and even half-wave symmetric signal is composed of even-indexed real frequency components only. Therefore, its spectrum is conjugate symmetric for each even-indexed frequency components and zero otherwise. Figure 5.3g shows the real and even half-wave symmetric signal

$$x(n) = -0.1 + 0.6\cos\left(2\frac{2\pi}{16}n + \frac{\pi}{3}\right) + 0.3\sin\left(4\frac{2\pi}{16}n + \frac{\pi}{6}\right) + 0.2\cos\left(8\frac{2\pi}{16}n\right)$$

and Fig. 5.3h shows its hermitian symmetric spectrum with its odd-indexed values zero. For example,

$$0.6\cos\left(2\frac{2\pi}{16}n + \frac{\pi}{3}\right) \Longleftrightarrow 8(0.6)\left(e^{j\frac{\pi}{3}}\delta(k-2) + e^{-j\frac{\pi}{3}}\delta(k-14)\right)$$

$$\{X(0) = -1.6, X(2,14) = 2.4 \pm j4.1569, X(4,12) = 1.2 \mp j2.0785, X(8) = 3.2\}$$

A real and odd half-wave symmetric signal is composed of odd-indexed real frequency components only. Therefore, its spectrum is conjugate symmetric for each odd-indexed frequency components and zero otherwise. Figure 5.3i shows the real and odd half-wave symmetric signal

$$x(n) = -\sin\left(\frac{2\pi}{16}n + \frac{\pi}{3}\right) + \cos\left(3\frac{2\pi}{16}n + \frac{\pi}{6}\right) + 0.6\sin\left(5\frac{2\pi}{16}n\right)$$

and Fig. 5.3j shows its hermitian symmetric spectrum with its even-indexed values zero. For example,

$$\cos\left(3\frac{2\pi}{16}n + \frac{\pi}{6}\right) \Longleftrightarrow 8\left(e^{j\frac{\pi}{6}}\delta(k-3) + e^{-j\frac{\pi}{6}}\delta(k-13)\right)$$

$$\{X(1,15) = -6.9282 \pm j4, X(3,13) = 6.9282 \pm j4, X(5,11) = \mp j4.8\}$$

The Basis of Fast DFT Algorithms

Any periodic function can be uniquely decomposed into its even half-wave and odd half-wave symmetric components. The even half-wave symmetric component is composed of the even-indexed frequency components, and the odd half-wave symmetric component is composed of the odd-indexed frequency components. Therefore, if an arbitrary function is decomposed into its even half-wave and odd half-wave symmetric components, then we have divided the original problem of finding the N frequency coefficients into two problems, each of them being the determination of $N/2$ frequency coefficients. This decomposition is continued until each frequency component is isolated.

5.4.11 Circular Convolution of Time-Domain Sequences

Let $x(n)$ and $h(n)$ be two periodic time-domain sequences of the same period N. Then, the circular convolution of the sequences is defined as

$$y(n) = \sum_{m=0}^{N-1} x(m)h(n-m) = \sum_{m=0}^{N-1} h(m)x(n-m),\ n = 0, 1, \ldots, N-1$$

The principal difference of this type of convolution from that of the linear convolution (Chap. 3) is that the range of the summation is restricted to a single period. Figure 5.4 shows the position of two sequences $x(n)$ and $h(n)$, of length 8, for finding their convolution output for $n = 0$ and $n = 2$ on the left and right, respectively. The process is similar to linear convolution except that the sequences are placed on a circle. Obviously, both must be of the same length. The inner sequence $x(n)$ is placed counterclockwise and is fixed. The time-reversed outer

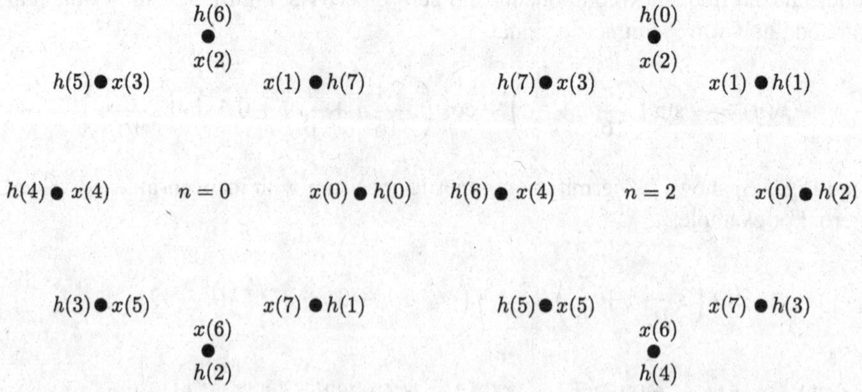

$h(6)$ ● $x(2)$ | $h(0)$ ● $x(2)$

$h(5)$ ● $x(3)$ $x(1)$ ● $h(7)$ | $h(7)$ ● $x(3)$ $x(1)$ ● $h(1)$

$h(4)$ ● $x(4)$ $n=0$ $x(0)$ ● $h(0)$ | $h(6)$ ● $x(4)$ $n=2$ $x(0)$ ● $h(2)$

$h(3)$ ● $x(5)$ $x(7)$ ● $h(1)$ | $h(5)$ ● $x(5)$ $x(7)$ ● $h(3)$

$x(6)$ ● $h(2)$ | $x(6)$ ● $h(4)$

Fig. 5.4 Circular convolution

sequence is placed clockwise. The sum of products of the corresponding elements of the sequences is the convolution output $y(0)$.

$$y(0) = x(0)h(0)+x(1)h(7)+x(2)h(6)+x(3)h(5)+x(4)h(4)+x(5)h(3)+x(6)h(2)+x(7)h(1)$$

The outer sequence is rotated counterclockwise by one sample interval and, at each point, the sum of products form the convolution output. For example,

$$y(2) = x(0)h(2)+x(1)h(1)+x(2)h(0)+x(3)h(7)+x(4)h(6)+x(5)h(5)+x(6)h(4)+x(7)h(3)$$

For N-point sequences, the output is also periodic of period N.

The convolution of $h(n)$ with a complex exponential $e^{jk_0\omega_0 n}$, $\omega_0 = \frac{2\pi}{N}$ is given as

$$\sum_{m=0}^{N-1} h(m)e^{jk_0\omega_0(n-m)} = e^{jk_0\omega_0 n}\sum_{m=0}^{N-1} h(m)e^{-jk_0\omega_0 m} = H(k_0)e^{jk_0\omega_0 n}$$

As an arbitrary $x(n)$ is reconstructed by the IDFT as

$$x(n) = \frac{1}{N}\sum_{k=0}^{N-1} X(k)W_N^{-nk},$$

the convolution of $x(n)$ and $h(n)$ is given by

$$y(n) = \frac{1}{N}\sum_{k=0}^{N-1} X(k)H(k)W_N^{-nk},$$

where $X(k)$ and $H(k)$ are, respectively, the DFT of $x(n)$ and $h(n)$. The IDFT of $X(k)H(k)$ is the circular convolution of $x(n)$ and $h(n)$.

Example 5.3 Convolve $x(n) = \{2, 1, 3, 3\}$ and $h(n) = \{1, 0, 2, 4\}$.

Solution

$$X(k) = \{9, -1 + j2, 1, -1 - j2\} \text{ and } H(k) = \{7, -1 + j4, -1, -1 - j4\}$$

$$X(k)H(k) = \{63, -7 - j6, -1, -7 + j6\}$$

The product $X(k)H(k)$ is obtained by multiplying the corresponding terms in the two sequences. The IDFT of $X(k)H(k)$ is the convolution sum, $y(n) = \{12, 19, 19, 13\}$. ■

5.4.12 *Circular Convolution of Frequency-Domain Sequences*

Let $X(k)$ and $H(k)$ be two periodic frequency-domain sequences of the same period N. Then, the circular convolution of the sequences, divided by N, is given as

$$x(n)h(n) \iff \frac{1}{N}\sum_{m=0}^{N-1} X(m)H(k-m) = \frac{1}{N}\sum_{m=0}^{N-1} H(m)X(k-m),$$

where $x(n)$ and $h(n)$ are the IDFT, respectively, of $X(k)$ and $H(k)$.

Example 5.4 Convolve $X(k) = \{9, -1 + j2, 1, -1 - j2\}$ and $H(k) = \{7, -1 + j4, -1, -1 - j4\}$.

Solution

$$x(n) = \{2, 1, 3, 3\} \text{ and } h(n) = \{1, 0, 2, 4\}$$
$$x(n)h(n) = \{2, 0, 6, 12\}$$

The product $x(n)h(n)$ is obtained by multiplying the corresponding terms in the two sequences. The DFT of $x(n)h(n)$ multiplied by four is the convolution sum of $X(k)$ and $H(k)$, $4\{20, -4 + j12, -4, -4 - j12\}$. ■

5.4.13 *Parseval's Theorem*

This theorem expresses the power of a signal in terms of its DFT spectrum. Let $x(n) \iff X(k)$ with sequence length N. The sum of the squared magnitude of the samples of a complex exponential with amplitude one, over the period N, is N. Remember that these samples occur on the unit-circle. The DFT decomposes a signal in terms of complex exponentials with coefficients $X(k)/N$. Therefore, the power of a complex exponential is $\frac{|X(k)|^2}{N^2}N = \frac{|X(k)|^2}{N}$. The power of the signal is the sum of the powers of its constituent complex exponentials and is given as

$$\sum_{n=0}^{N-1} |x(n)|^2 = \frac{1}{N}\sum_{k=0}^{N-1} |X(k)|^2$$

Example 5.5 Consider the DFT pair

$$\{2, 1, 3, 3\} \iff \{9, -1 + j2, 1, -1 - j2\}$$

The sum of the squared magnitude of the data sequence is 23 and that of the DFT coefficients divided by 4 is also 23. ■

5.5 Applications of the Discrete Fourier Transform

The DFT is extensively used to approximate the forward and inverse transforms of the other versions of the Fourier analysis as described in other chapters. In addition, important operations such as convolution, interpolation, and decimation are carried out efficiently using the DFT as presented in this section.

5.5.1 Computation of the Linear Convolution Using the DFT

Circular convolution assumes two periodic sequences of the same period and results in a periodic sequence with that period. Using the DFT, circular convolution can be efficiently carried out, as the DFT assumes a finite length sequence is periodically extended. However, the linear convolution is of prime interest in LTI system analysis. The linear convolution of two finite sequences of length N and M is a sequence of length $(N + M - 1)$.

The basis of using the DFT to evaluate the linear convolution operation, as well as approximating other versions of the Fourier analysis, is to make the period of the DFT so that all the essential information required is available in any one period with sufficient accuracy. Therefore, both the sequences to be convolved must be zero-padded to make them of length $(N + M - 1)$, at the least. This prevents the wrap-around effect of the circular convolution and makes one period output of the circular convolution the same as that of the linear convolution.

The linear convolution of $\{2, 1\}$ and $\{3, 4\}$ is $\{6, 11, 4\}$. The DFT of the sequences are, respectively, $\{3, 1\}$ and $\{7, -1\}$. The term by term product of these DFT is $\{21, -1\}$. The IDFT of this product yields the periodic convolution output $\{10, 11\}$. The last value 4 of the linear convolution is added to the first value 6 to make the first value of the circular convolution 10. The last value of the circular convolution is unaffected by aliasing in the time-domain. The DFT of the 4-point zero-padded sequences $\{2, 1, 0, 0\}$ and $\{3, 4, 0, 0\}$, respectively, are $\{3, 2-j, 1, 2+j\}$ and $\{7, 3-j4, -1, 3+j4\}$. The term by term product of these DFT is $\{21, 2-j11, -1, 2+j11\}$. The IDFT of this product yields the linear convolution output with one zero appended $\{6, 11, 4, 0\}$. We could have avoided the zero at the end by zero-padding the signals to make their length three. As fast DFT algorithms with high regularity are available only for data lengths those that are an integral power of two, the input sequences are usually zero padded to make the length of the sequences an integral power of two. Of course, this length must be greater than or equal to the sum of the lengths of the two given sequences minus one.

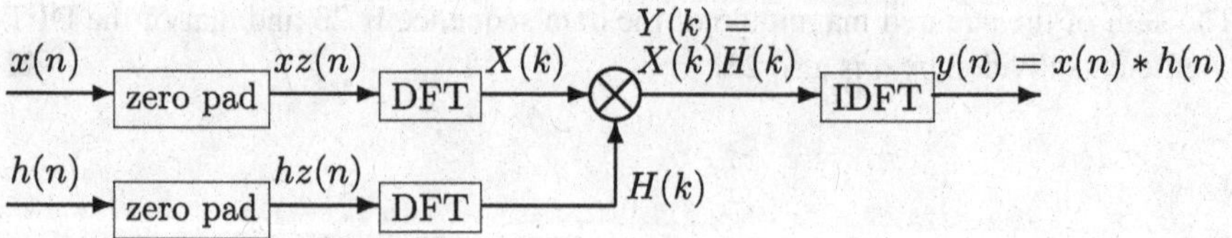

Fig. 5.5 Linear convolution using the DFT

Figure 5.5 shows the linear convolution implemented using the DFT and the IDFT. The sequences $x(n)$ and $h(n)$ are sufficiently zero-padded, as presented earlier, to form the sequences $xz(n)$ and $hz(n)$. The DFT, $X(k)$ and $H(k)$, of the zero-padded sequences are computed. The pointwise product $Y(k) = X(k)H(k)$ is the convolution output in the frequency-domain. The IDFT of $Y(k)$ is the linear convolution of the sequences $x(n)$ and $h(n)$ with some zero padding.

5.5.2 *Interpolation and Decimation*

Changing the sampling rate of a signal is required for efficient signal processing. For example, reconstructing a signal is easier with a higher sampling rate, while a lower sampling rate may be adequate for processing, requiring a shorter computation time. Changing the sampling rate of a signal by reconstructing the corresponding analog signal and resampling it at the new sampling rate introduces large errors. Therefore, sampling rate is usually changed in the discrete form itself. An analog signal sampled with an adequate sampling rate results in its proper discrete form. Sampling rate can be increased (interpolation) or decreased (decimation) to suit the processing requirements as long as the sampling theorem is not violated.

5.5.2.1 Interpolation

Increasing the sampling rate of a signal by a factor I is called interpolation. First, the signal is zero padded with $(I - 1)$ samples with value zero between successive samples. In the frequency-domain, the operation of zero-padding corresponds to duplicating the spectrum of the given waveform $(I-1)$ times, due to the periodicity of the complex exponential W_N^{nk}. This signal is passed through a lowpass filter with a cutoff frequency $\frac{\pi}{I}$ radians and a passband gain I. The resulting spectrum corresponds to that of the interpolated version of the given waveform. Note that all the frequency components of the given signal lies in the range from zero to $\frac{\pi}{I}$ radians of the duplicated spectrum. Frequency π corresponds to half the sampling frequency and the frequency with index $\frac{N}{2}$ in the DFT spectrum.

The signal, $x(n) = \cos(\frac{2\pi}{8}n - \frac{\pi}{3})$, is shown in Fig. 5.6a and its spectrum is shown in Fig. 5.6b. With the interpolation factor $I = 2$, we want twice the number

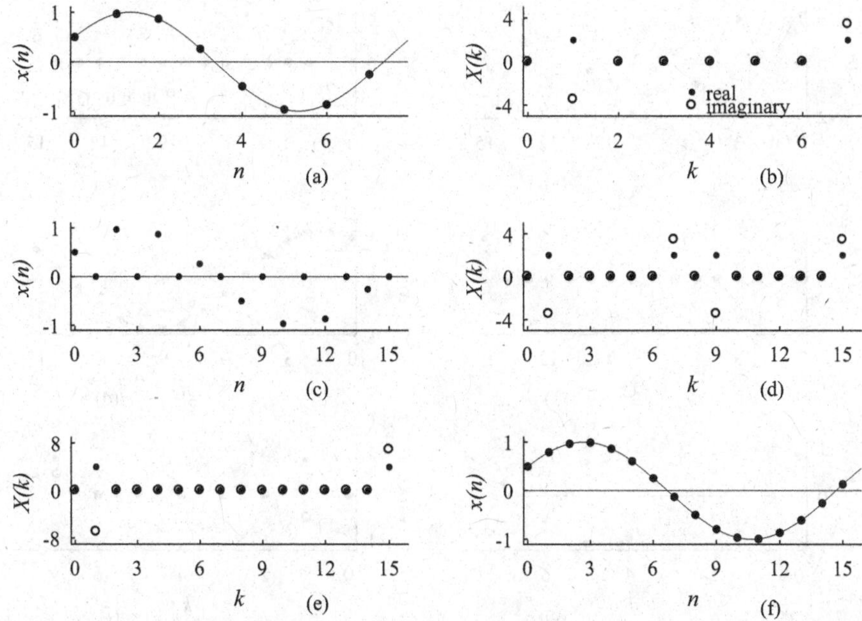

Fig. 5.6 (**a**) A real signal and (**b**) its spectrum; (**c**) the signal shown in (**a**) with zero padding in between the samples and (**d**) its spectrum, which is the same as that shown in (**b**) but repeats; (**e**) the spectrum shown in (**d**) after lowpass filtering and (**f**) the corresponding time-domain signal, which is an interpolated version of that shown in (**a**)

of samples in a cycle than that in Fig. 5.6a. This requires the insertion of one sample with zero value in between the samples, as shown in Fig. 5.6c. The DFT of the zero padded signal is shown in Fig. 5.6d. Except for the repetition, this spectrum is the same as that in Fig. 5.6b. This spectrum has two frequency components with frequency indices $k = 1$ and $k = 7$. We have to filter out the frequency component with $k = 7$. Therefore, lowpass filtering of this signal with the filter cutoff frequency $\frac{\pi}{2}$ radians and gain two yields the the spectrum shown in Fig. 5.6e and the corresponding interpolated signal, $x(n) = \cos(\frac{2\pi}{16}n - \frac{\pi}{3})$, is shown in Fig. 5.6f. The spectrum in Fig. 5.6b is the DFT of the sinusoid with eight samples in a cycle, whereas that in Fig. 5.6e is the DFT of the sinusoid with 16 samples in a cycle.

5.5.2.2 Decimation

Reducing the sampling rate of a signal by a factor D is called decimation. As we reduce the sampling rate, we have to filter the high-frequency components of the signal first, by a filter with a cutoff frequency $\frac{\pi}{D}$ and a passband gain 1, to eliminate aliasing. Then, we take every Dth sample. It is assumed that the filtered out high-

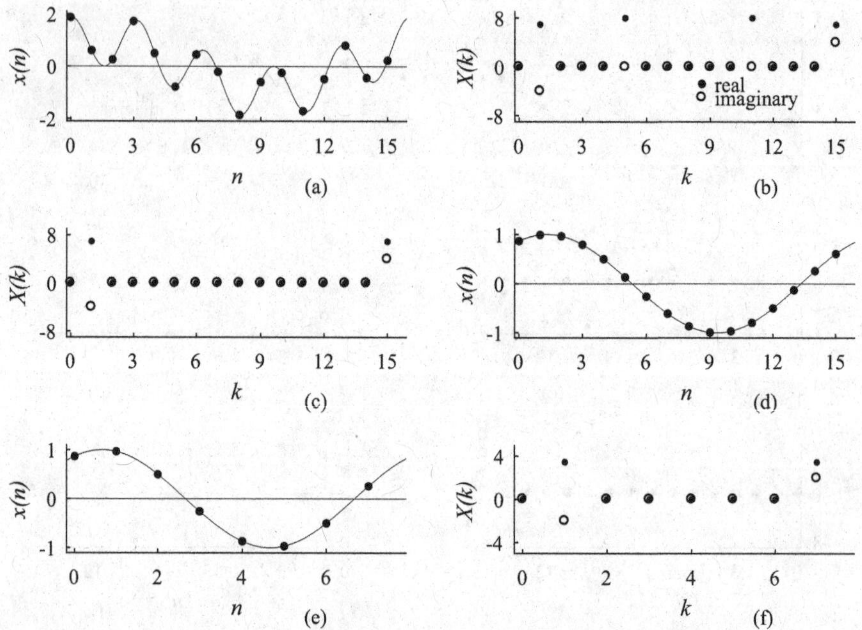

Fig. 5.7 (**a**) A real signal and (**b**) its spectrum; (**c**) the spectrum shown in (**b**) after lowpass filtering and (**d**) the corresponding time-domain signal; (**e**) the signal shown in (**d**) with decimation of alternate samples and (**f**) its spectrum, which is the same as that shown in (**c**) but compressed

frequency components are of no interest. The signal, $x(n) = \cos(\frac{2\pi}{16}n - \frac{\pi}{6}) + \cos(5\frac{2\pi}{16}n)$, is shown in Fig. 5.7a and its spectrum is shown in Fig. 5.7b. With the decimation factor $D = 2$, we want half the number of samples in a cycle than that in Fig. 5.7a. The signal is passed through a lowpass filter with cutoff frequency $\frac{\pi}{2}$ and gain of 1. The spectrum of the filter output is shown in Fig. 5.7c and the filtered signal, $x(n) = \cos(\frac{2\pi}{16}n - \frac{\pi}{6})$, is shown in Fig. 5.7d. Now, the decimated signal, $x(n) = \cos(\frac{2\pi}{8}n - \frac{\pi}{6})$, is obtained by taking every second sample. The decimated signal is shown in Fig. 5.7e, and its spectrum is shown in Fig. 5.7f.

5.5.2.3 Interpolation and Decimation

A sampling rate converter, which is a cascade of an interpolator and a decimator, can be used to convert the sampling rate by any rational factor, $\frac{I}{D}$. A single lowpass filter, with a cutoff frequency that is the smaller of $\frac{\pi}{I}$ and $\frac{\pi}{D}$, and gain of I, is adequate. The signal, $x(n) = \cos(\frac{2\pi}{4}n - \frac{\pi}{3})$, is shown in Fig. 5.8a and its spectrum is shown in Fig. 5.8b. With $I = 3$, $D = 2$, and $\frac{I}{D} = \frac{3}{2}$, we want one and a half times the number of samples in a cycle than that in Fig. 5.8a. The insertion of two samples with zero value is required, as shown in Fig. 5.8c. The spectrum of this signal, which

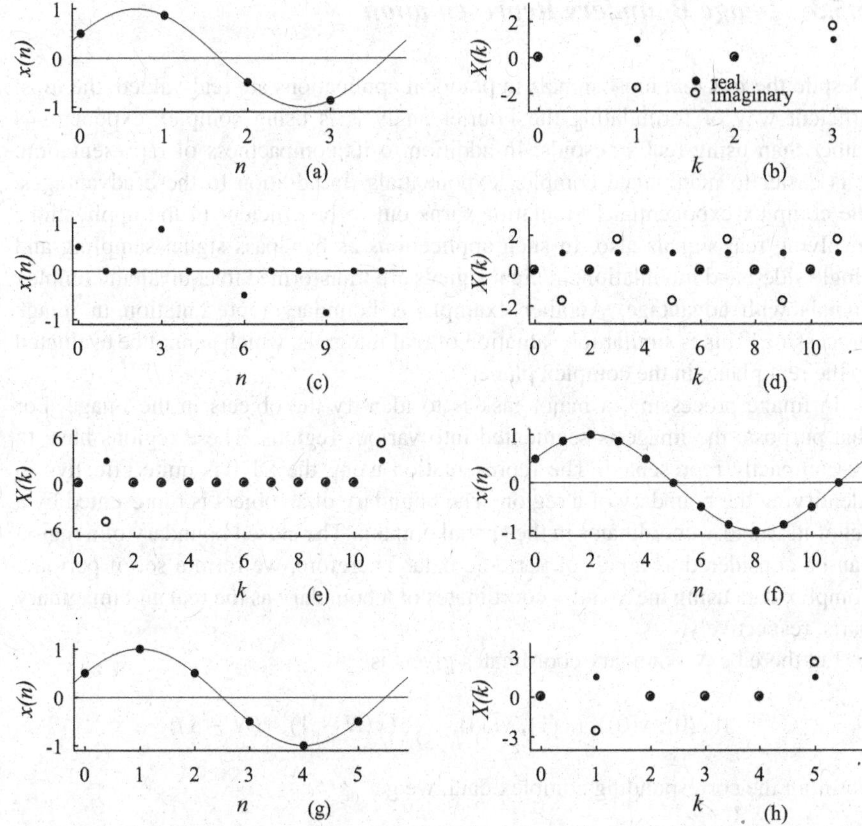

Fig. 5.8 (**a**) A real signal and (**b**) its spectrum; (**c**) the signal shown in (**a**) with zero padding in between the samples and (**d**) its spectrum, which is the same as that shown in (**b**) but repeats twice; (**e**) the spectrum shown in (**d**) after lowpass filtering and (**f**) the corresponding time-domain signal, which is an interpolated version of that shown in (**a**); (**g**) the signal shown in (**f**) with decimation of alternate samples and (**h**) its spectrum, which is the same as that shown in (**e**) but compressed

repeats twice, is shown in Fig. 5.8d. A lowpass filter, with cutoff frequency $\frac{\pi}{3}$ and gain of 3, eliminates the two high-frequency components. The resulting spectrum is shown in Fig. 5.8e, and the interpolated signal, $x(n) = \cos(\frac{2\pi}{12}n - \frac{\pi}{3})$, is shown in Fig. 5.8f. Now, by taking alternate samples, we get the decimated signal, $x(n) = \cos(\frac{2\pi}{6}n - \frac{\pi}{3})$, shown in Fig. 5.8g. Its spectrum is shown in Fig. 5.8h. Sampling rate conversion by a factor $\frac{3}{2}$ resulted in six samples in a cycle, as shown in Fig. 5.8g, compared with four samples in a cycle in Fig. 5.8a.

5.5.3 Image Boundary Representation

Despite the fact that most signals in practical applications are real-valued, the most efficient way of formulating the Fourier analysis is using complex exponentials rather than using real sinusoids. In addition to its compactness of representation, it is easier to manipulate complex exponentials. In addition to these advantages, the complex exponential formulation turns out to be efficient in the applications involving real signals also. In such applications as bandpass signal sampling and single side-band modulation, the real signals are transformed to equivalent complex signals with advantage. Another example is boundary representation in image processing. This is similar to evaluation of real integrals, which cannot be evaluated in the real plane, in the complex plane.

In image processing, a major task is to identify the objects in the image. For that purpose, the image is segmented into various regions. These regions have to be compactly represented. The representation using the DFT is quite effective in identifying the boundary of a region. The boundary of an object is represented by a set of its x and y coordinates in the spatial domain. The closed boundary of a region can be considered as a pair of periodic data. Therefore, we form a set of periodic complex data using the x and y coordinates of a boundary as the real and imaginary parts, respectively.

Let there be N boundary coordinates given as

$$\{(x(0), y(0)), (x(1), y(1)), \ldots, (x(N-1), y(N-1))\}$$

Forming the corresponding complex data, we get

$$border(n) = \{(x(0) + jy(0)), (x(1) + jy(1)), \ldots, (x(N-1) + jy(N-1))\}$$

The DFT of $border(n)$, $BORDER(k)$, represents the 2-D boundary in the frequency-domain by a 1-D data. One major advantage of this representation, as in almost all applications of Fourier analysis, is that much fewer than the required N DFT coefficients provide an adequate representation of the border, depending on the smoothness of the border.

For example, let us find the boundary representation of the 4×4 binary image $x(m,n)$.

$$x(m,n) = \begin{bmatrix} 1 & 1 & 1 & 0 \\ 1 & 0 & 1 & 0 \\ 1 & 0 & 1 & 0 \\ 1 & 1 & 1 & 0 \end{bmatrix}$$

Assume that the top left corner is the origin with coordinates $(0,0)$.

The complex data formed from the boundary coordinates of $x(m,n)$ is

$$border(n) = \{0+j0, 1+j0, 2+j0, 3+j0, 3+j1, 3+j2, 2+j2, 1+j2, 0+j2, 0+j1\}$$

The DFT of $border(n)$, $BORDER(k)$, is

$$\{15+j10, -11.8339-j8.5978, 0, 0.3031-j0.9329, 0,$$

$$-1, 0, -0.5949-j1.8310, 0, -1.8743+j1.3618\}$$

Let us use just three coefficients to reconstruct the boundary. That is, we find the IDFT of

$$\{15+j10, -11.8339-j8.5978, 0, 0, 0, 0, 0, 0, 0, -1.8743+j1.3618\}$$

to get

$$\{0.1292+j0.2764, 0.9764-j0.1708, 2.0236-j0.1708, 2.8708+j0.2764, 3.1944+j1,$$

$$2.8708+j1.7236, 2.0236+j2.1708, 0.9764+j2.1708, 0.1292+j1.7236, -0.1944+j1\}$$

By rounding these values, we get back the original coordinates of the boundary.

In reconstructing real waveforms, the DFT coefficients occur in conjugate pairs resulting in a real sinusoid for each pair of coefficients. As more and more frequency components are used, the reconstructed waveform becomes more closer to the original. In contrast, in reconstructing a closed boundary, the complex DFT coefficients of the coordinates are arbitrary. Each complex coefficient corresponds to a complex exponential in the time-domain, whose shape is a circle. The radius of the circle depends on the magnitude of the corresponding DFT coefficient. Therefore, a closed boundary is a sum of circles with different diameters traversed with different speeds. As in the case of real signals, the magnitude of the coefficients may become negligible, which depends on the smoothness of the boundary, over a considerable range of the spectrum. Therefore, a boundary can be represented with fewer coefficients with a required accuracy.

5.6 Summary

- In this chapter, the DFT, its properties, and some of its applications have been presented.
- Transform methods change a problem into another equivalent form so that it is relatively easier to interpret and solve problems.

- Frequency-domain analysis uses sinusoids or complex exponentials as basis signals to represent signals and systems, in contrast to impulse in the time-domain analysis.
- The basis functions used in Fourier analysis are constant amplitude sinusoids or exponentials with pure imaginary arguments. Fourier analysis has four versions, each version suitable for different types of signals. The sinusoidal basis functions differ, in each case, in characteristics such as discrete or continuous and finite or infinite in number.
- In all versions of Fourier analysis, the signal is represented with respect to the least squares error criterion.
- The DFT version of the Fourier analysis uses a finite number of harmonically related discrete sinusoids as basis functions. Therefore, both the input data and its spectrum are periodic and discrete. This fact makes it inherently suitable for numerical computation.
- The input to the DFT is a finite sequence of samples, and it is assumed to be periodically extended. The DFT coefficients are the coefficients of the basis complex exponentials whose superposition sum yields the periodically extended discrete signal. The IDFT carries out this sum.
- The DFT is extensively used in the approximation of the other versions of the Fourier analysis, in addition to efficient evaluation of important operations such as convolution, interpolation, and decimation.
- The periodicity property of the DFT is the key factor in deriving fast algorithms for its computation. These algorithms make the use of the DFT more efficient in most applications compared with alternate methods.
- The DFT, because of its finite and discrete nature, is the simplest of the four different versions of the Fourier analysis to visualize the analysis and synthesis of waveforms. Problems in understanding the concepts in other versions of the Fourier analysis may be resolved by considering an equivalent DFT version.

Exercises

5.1 Given the DFT spectrum $X(k)$, express the corresponding time-domain signal $x(n)$ in terms of its constituent real sinusoids.

5.1.1 $\{X(0) = 3, X(1) = \frac{1}{\sqrt{2}} - j\frac{1}{\sqrt{2}}, X(2) = -2, X(3) = \frac{1}{\sqrt{2}} + j\frac{1}{\sqrt{2}}\}$.
5.1.2 $\{X(0) = -2, X(1) = \sqrt{3} + j1, X(2) = 3, X(3) = \sqrt{3} - j1\}$.
* 5.1.3 $\{X(0) = 1, X(1) = 2 - j2\sqrt{3}, X(2) = -3, X(3) = 2 + j2\sqrt{3}\}$.
5.1.4 $\{X(0) = 3, X(1) = 4, X(2) = 1, X(3) = 4\}$.
5.1.5 $\{X(0) = -5, X(1) = j8, X(2) = 2, X(3) = -j8\}$.

5.2 Find the four samples of $x(n)$ over one period, and, then, use the DFT matrix equation to compute the spectrum $X(k)$.

5.2.1 $x(n) = 2 + 3\sin(\frac{2\pi}{4}n - \frac{\pi}{6}) - \cos(\pi n)$.
* 5.2.2 $x(n) = -1 - 2\cos(\frac{2\pi}{4}n + \frac{\pi}{3}) + 2\cos(\pi n)$.
5.2.3 $x(n) = 3 + \cos(\frac{2\pi}{4}n - \frac{\pi}{4}) - 3\cos(\pi n)$.
5.2.4 $x(n) = 1 - 2\sin(\frac{2\pi}{4}n + \frac{\pi}{4}) + 4\cos(\pi n)$.
5.2.5 $x(n) = -2 + 3\cos(\frac{2\pi}{4}n + \frac{\pi}{6}) - 2\cos(\pi n)$.

5.3 Find the IDFT of the given spectrum $X(k)$ using the IDFT matrix equation.

5.3.1 $\{X(0) = -12, X(1) = 2 - j2\sqrt{3}, X(2) = 8, X(3) = 2 + j2\sqrt{3}\}$.
5.3.2 $\{X(0) = 4, X(1) = -4\sqrt{3} + j4, X(2) = -4, X(3) = -4\sqrt{3} - j4\}$.
5.3.3 $\{X(0) = 8, X(1) = 3 - j3\sqrt{3}, X(2) = 8, X(3) = 3 + j3\sqrt{3}\}$.
* 5.3.4 $\{X(0) = -16, X(1) = -3\sqrt{2} - j3\sqrt{2}, X(2) = 8, X(3) = -3\sqrt{2} + j3\sqrt{2}\}$.
5.3.5 $\{X(0) = 12, X(1) = -2 - j2\sqrt{3}, X(2) = -12, X(3) = -2 + j2\sqrt{3}\}$.

5.4 Find the sample values of the waveform over one period first, and then use the matrix equation to find its DFT spectrum. Verify that the spectral values are the same as the corresponding coefficients of the exponentials multiplied by four.

5.4.1 $x(n) = (1 + j\sqrt{3})e^{j0\frac{2\pi}{4}n} + (2 - j2\sqrt{3})e^{j\frac{2\pi}{4}n} + (1 - j1)e^{j2\frac{2\pi}{4}n} - (1 + j1)e^{j3\frac{2\pi}{4}n}$.
5.4.2 $x(n) = (2 + j1)e^{j0\frac{2\pi}{4}n} + (3 - j2)e^{j\frac{2\pi}{4}n} + (4 - j1)e^{j2\frac{2\pi}{4}n} - (3 + j2)e^{j3\frac{2\pi}{4}n}$.
5.4.3 $x(n) = (1 - j2)e^{j0\frac{2\pi}{4}n} + (2 + j2)e^{j\frac{2\pi}{4}n} + (3 - j3)e^{j2\frac{2\pi}{4}n} + (1 - j4)e^{j3\frac{2\pi}{4}n}$.
5.4.4 $x(n) = (1 + j2)e^{j0\frac{2\pi}{4}n} + (2 + j3)e^{j\frac{2\pi}{4}n} + (4 + j4)e^{j2\frac{2\pi}{4}n} + (3 - j2)e^{j3\frac{2\pi}{4}n}$.
5.4.5 $x(n) = (2 - j2)e^{j0\frac{2\pi}{4}n} + (1 - j4)e^{j\frac{2\pi}{4}n} + (2 + j1)e^{j2\frac{2\pi}{4}n} + (1 - j2)e^{j3\frac{2\pi}{4}n}$.

5.5 Find the IDFT of the given spectrum $X(k)$ using the matrix IDFT equation.

* 5.5.1 $\{X(0) = 1 - j1, X(1) = 3 - j2, X(2) = 4 + j1, X(3) = 1 + j2\}$.
5.5.2 $\{X(0) = 3 + j3, X(1) = 1 - j1, X(2) = 2 + j3, X(3) = 1 - j4\}$.
5.5.3 $\{X(0) = 2 - j3, X(1) = 1 + j5, X(2) = 2 + j3, X(3) = 2 + j4\}$.
5.5.4 $\{X(0) = 1 - j4, X(1) = 4 + j2, X(2) = 3 + j1, X(3) = 2 + j2\}$.
5.5.5 $\{X(0) = 3 - j4, X(1) = 2 + j5, X(2) = 1 - j3, X(3) = 2 - j4\}$.

5.6 Find the DFT $X(k)$ of the given $x(n)$. Using the periodicity property of the DFT and the IDFT, find the required $x(n)$ and $X(k)$.

5.6.1 $x(n) = \{2 + j3, 1 - j2, 2 + j1, 3 + j4\}$. Find $x(13)$, $x(-22)$, $X(10)$, and $X(-28)$.
* 5.6.2 $x(n) = \{1 + j2, 2 - j3, 2 + j2, 1 - j4\}$. Find $x(-14)$, $x(43)$, $X(12)$, and $X(-7)$.

5.7 Find the DFT $X(k)$ of $x(n) = \{2 - j2, 1 + j3, 4 + j2, 1 - j2\}$. Using the time-domain shift property and $X(k)$, deduce the DFT of $x(n + 1)$, $x(n - 2)$, and $x(n + 3)$.

5.8 Find the IDFT $x(n)$ of $X(k) = \{12 + j4, 8 - j4, 4 + j4, 4 + j8\}$. Using the frequency-domain shift property, deduce the IDFT of $X(k + 1)$, $X(k - 2)$, and $X(k + 3)$.

5.9 Find the circular convolution of two frequency-domain sequences $X(k)$ and $H(k)$ using the DFT and the IDFT.

5.9.1 $X(k) = \{8 - j4, 4 + j4, 12 - j8, 8 - j12\}$ and $H(k) = \{12 - j4, 8 - j4, 4 + j8, 2 + j12\}$.
* 5.9.2 $X(k) = \{8, 4, 8, 4\}$ and $H(k) = \{12, 8, 4, 12\}$.
5.9.3 $X(k) = \{0, j4, 0, -j4\}$ and $H(k) = \{0, 4, 0, -4\}$.

5.10 Find the DFT of $x(n)$ and verify the Parseval's theorem.

5.10.1 $x(n) = \{2, 4, 3, 1\}$.
5.10.2 $x(n) = \{-2, 4, 2, 5\}$.
5.10.3 $x(n) = \{4, -1, 3, 1\}$.

5.11 Find the linear convolution of the sequences $x(n)$ and $y(n)$ using the DFT and the IDFT.

5.11.1 $\{x(0) = 2, x(1) = 4, x(2) = 3\}$ and $\{y(0) = 1, y(1) = -2\}$.
5.11.2 $\{x(0) = 2, x(1) = -4, x(2) = 3\}$ and $\{y(0) = 1, y(1) = 2\}$.
* 5.11.3 $\{x(0) = 1, x(1) = 4, x(2) = -3\}$ and $\{y(0) = -4, y(1) = 3\}$.

Chapter 6
Fourier Series

Continuous periodic signals are analyzed using an infinite set of harmonically related sinusoids and a DC component in the FS frequency-domain representation. Increasing the number of samples in a period, by decreasing the sampling interval, results in a densely sampled time-domain waveform and a broader DFT periodic spectrum. As the sampling interval tends to zero, the time-domain waveform becomes a continuous function, and the discrete spectrum becomes aperiodic. As the period of the waveform remains the same, the fundamental frequency and the harmonic spacing of the spectrum is fixed. Therefore, the discrete nature of the spectrum is unchanged. In Sect. 6.1, we derive the exponential form of the FS, starting from the defining equations of the DFT and the IDFT. Then, two equivalent trigonometric forms of the FS are deduced from the expressions of the exponential form. The properties of the FS are described in Sect. 6.2. The approximation of the FS coefficients by the DFT is presented in Sect. 6.3. Typical applications of the FS are presented in Sect. 6.4.

6.1 Fourier Series

A continuous periodic signal, $x(t)$, with period T is expressed as a sum of an infinite set of harmonically related sinusoids and a DC component in the FS. The frequency of the fundamental or first harmonic is the frequency of the waveform under analysis. That is, $\omega_0 = \frac{2\pi}{T}$. The frequency of the second harmonic is $2\omega_0$, that of the third harmonic is $3\omega_0$, and so on.

D. Sundararajan, *Signals and Systems*,
https://doi.org/10.1007/978-3-031-19377-4_6

6.1.1 FS as the Limiting Case of the DFT

While the FS can be derived using the orthogonality property of sinusoids, it is instructive to consider it as the limiting case of the DFT with the sampling interval tending to zero. Consider the continuous periodic signal $x(t)$, with period $T = 5$ s,

$$x(t) = 2 - \cos\left(\frac{2\pi}{5}t\right) + \cos\left(2\frac{2\pi}{5}t\right)$$

and its five sample values, with a sampling interval of $T_s = 1$ s, shown in Fig. 6.1a. The scaled DFT spectrum of this sample set is shown in Fig. 6.1b. The frequency increment of the spectrum is $\omega_0 = \frac{2\pi}{5}$ radians per second, and one period of the DFT spectrum corresponds to $\frac{2\pi}{5}5 = 2\pi$ radians. The samples of the same signal, with a sampling interval of $T_s = 0.125$ s, are shown in Fig. 6.1c. Reducing the sampling interval results in a densely sampled time-domain waveform. The scaled DFT spectrum of this sample set is shown in Fig. 6.1d. As the frequency increment is fixed at $\frac{2\pi}{5}$, the larger number of spectral values corresponds to a broader spectrum of width $\frac{2\pi}{5}40 = 16\pi$ radians. Eventually, as the sampling interval tends to zero, the time-domain waveform becomes continuous and the discrete spectrum becomes aperiodic.

The mathematical presentation of the foregoing argument is as follows. The IDFT of $X(k)$, $-N \leq k \leq N$ is defined as

$$x(n) = \frac{1}{2N+1} \sum_{k=-N}^{N} X(k) e^{j\frac{2\pi}{(2N+1)}nk}, \quad n = 0, \pm 1, \pm 2, \ldots, \pm N$$

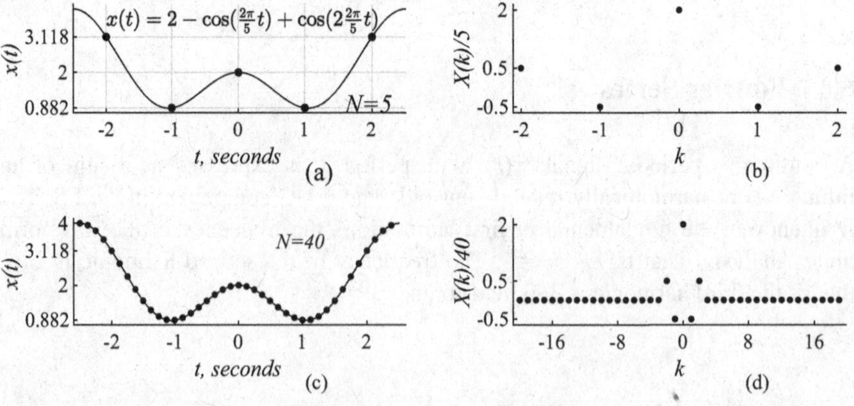

Fig. 6.1 (a) The samples, at intervals of 1 s, of a periodic continuous signal $x(t)$ with period 5 s and (b) its scaled DFT spectrum; (c) the samples of $x(t)$ at intervals of 0.125 s and (d) its scaled DFT spectrum

Substituting the DFT expression for $X(k)$, we get

$$x(n) = \frac{1}{2N+1} \sum_{k=-N}^{N} \left(\sum_{l=-N}^{N} x(l) e^{-j\frac{2\pi}{(2N+1)}lk} \right) e^{j\frac{2\pi}{(2N+1)}nk}$$

The frequency index k represents the discrete frequency $\frac{2\pi}{(2N+1)}k$. If the periodic signal, with period T, is sampled with a sampling interval of T_s seconds in order to get the samples, then the corresponding continuous frequency is given by $\frac{2\pi}{(2N+1)T_s}k = \frac{2\pi}{T}k = k\omega_0$. The time index n corresponds to nT_s seconds. The number of samples in a period is $(2N+1) = \frac{T}{T_s}$. With these substitutions, we get

$$x(nT_s) = \sum_{k=-N}^{N} \left(\frac{1}{T} \sum_{l=-N}^{N} x(lT_s) e^{-j\frac{2\pi}{T}lT_s k} T_s \right) e^{j\frac{2\pi}{T}nT_s k}$$

As T_s is reduced, the number of samples $(2N+1)$ increases, but the product $(2N+1)T_s = T$ remains constant. Hence, the fundamental frequency ω_0 also remains constant. As $T_s \rightarrow 0$, nT_s and lT_s become continuous time variables t and τ, respectively, the inner summation becomes an integral over the period T, $N \rightarrow \infty$, and differential $d\tau$ formally replaces T_s. Therefore, we get

$$x(t) = \sum_{k=-\infty}^{\infty} \left(\frac{1}{T} \int_{-\frac{T}{2}}^{\frac{T}{2}} x(\tau) e^{-j\frac{2\pi}{T}\tau k} d\tau \right) e^{j\frac{2\pi}{T}tk}$$

The exponential form of the FS for a signal $x(t)$ is

$$x(t) = \sum_{k=-\infty}^{\infty} X_{cs}(k) e^{jk\omega_0 t}, \tag{6.1}$$

where the FS coefficients $X_{cs}(k)$ are given as

$$X_{cs}(k) = \frac{1}{T} \int_{t_1}^{t_1+T} x(t) e^{-jk\omega_0 t}\, dt, \; k = 0, \pm 1, \pm 2, \ldots \tag{6.2}$$

and t_1 is arbitrary. Because of periodicity of the FS with period T, the integral from t_1 to (t_1+T) will have the same value for any value of t_1. Since sinusoids are represented in terms of exponentials, Eq. (6.1) is called the exponential form of the FS.

6.1.2 The Compact Trigonometric Form of the FS

The form of the FS, with sinusoids represented in polar form, is called the compact trigonometric form. Eq. (6.1) can be rewritten as

$$x(t) = X_{cs}(0) + \sum_{k=1}^{\infty}(X_{cs}(k)e^{jk\omega_0 t} + X_{cs}(-k)e^{-jk\omega_0 t})$$

Since $X_{cs}(k)e^{jk\omega_0 t}$ and $X_{cs}(-k)e^{-jk\omega_0 t}$ form complex conjugate pair for real signals and their sum is twice the real part of either of the terms, we get

$$x(t) = X_p(0) + \sum_{k=1}^{\infty} X_p(k)\cos(k\omega_0 t + \theta(k)), \tag{6.3}$$

where

$$X_p(0) = X_{cs}(0), \quad X_p(k) = 2|X_{cs}(k)|, \quad \theta(k) = \angle(X_{cs}(k)),\ k = 1, 2, \ldots, \infty$$

6.1.3 The Trigonometric Form of the FS

The form of the FS, with sinusoids represented in rectangular form, is called the trigonometric form. Expressing the sinusoid in Eq. (6.3) in rectangular form, we get

$$x(t) = X_c(0) + \sum_{k=1}^{\infty}(X_c(k)\cos(k\omega_0 t) + X_s(k)\sin(k\omega_0 t)), \tag{6.4}$$

where

$$X_c(0) = X_p(0) = X_{cs}(0)$$

$$X_c(k) = X_p(k)\cos(\theta(k)) = 2\,\mathrm{Re}(X_{cs}(k)),\ X_s(k) = -X_p(k)\sin(\theta(k)) = -2\,\mathrm{Im}(X_{cs}(k))$$

6.1.4 Periodicity of the FS

The FS is a periodic waveform of period that is the same as that of the fundamental, $T = \frac{2\pi}{\omega_0}$. Replacing t by $t + T$ in Eq. (6.3), we get

$$x(t) = X_p(0) + \sum_{k=1}^{\infty} X_p(k)\cos(k\omega_0(t+T) + \theta_k)$$

$$= X_p(0) + \sum_{k=1}^{\infty} X_p(k)\cos(k\omega_0 t + 2k\pi + \theta_k) = x(t)$$

If the waveform to be analyzed is defined only over the interval T, the FS represents the waveform only in that interval. On the other hand, if the waveform is periodic of period T, then the FS is valid for all t.

6.1.5 Existence of the FS

Any signal satisfying Dirichlet conditions, which are a set of sufficient conditions, can be expressed in terms of a FS. The first of these conditions is that the signal $x(t)$ is absolutely integrable over one period, that is, $\int_0^T |x(t)|dt < \infty$. From the definition of the FS, we get

$$|X_{cs}(k)| \leq \frac{1}{T}\int_{t_1}^{t_1+T} |x(t)e^{-jk\omega_0 t}|\,dt = \frac{1}{T}\int_{t_1}^{t_1+T} |x(t)||e^{-jk\omega_0 t}|\,dt$$

Since $|e^{-jk\omega_0 t}| = 1$,

$$|X_{cs}(k)| \leq \frac{1}{T}\int_{t_1}^{t_1+T} |x(t)|\,dt$$

The second condition is that the number of finite maxima and minima in one period of the signal must be finite. The third condition is that the number of finite discontinuities in one period of the signal must be finite. Most signals of practical interest satisfy these conditions.

Example 6.1 Find the three forms of the FS for the signal

$$x(t) = -1 - 2\cos\left(\frac{2\pi}{6}t - \frac{\pi}{3}\right)$$

Solution As this signal can be expressed in terms of sinusoids easily, we do not need to evaluate any integral. The fundamental frequency of the waveform is $\omega_0 = \frac{2\pi}{6}$, which is the same as that of the sinusoid. Note that the DC component is periodic with any period.

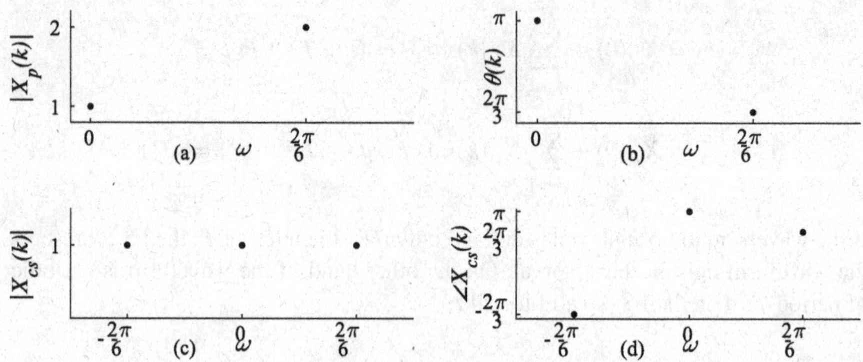

Fig. 6.2 (**a**) The FS amplitude spectrum and (**b**) the phase spectrum of the signal in compact trigonometric form; (**c**) the FS amplitude spectrum and (**d**) the phase spectrum of the signal in exponential form

Compact trigonometric form

$$x(t) = -1 - 2\cos\left(\frac{2\pi}{6}t - \frac{\pi}{3}\right) = -1 + 2\cos\left(\frac{2\pi}{6}t + \frac{2\pi}{3}\right)$$

Comparing this expression with the definition, Eq. (6.3), we get the compact trigonometric form of the FS coefficients as

$$X_p(0) = -1, \quad X_p(1) = 2, \quad \theta(1) = \frac{2\pi}{3}$$

A plot of the amplitude $X_p(k)$ of the constituent sinusoids of a signal $x(t)$ versus k or $k\omega_0$ is called the amplitude spectrum of $x(t)$. Similarly, the plot of the phase $\theta(k)$ is called the phase spectrum. The FS amplitude spectrum and the phase spectrum of the signal in compact trigonometric form are shown, respectively, in Fig. 6.2a and b.

Trigonometric form

$$x(t) = -1 - 2\cos\left(\frac{2\pi}{6}t - \frac{\pi}{3}\right) = -1 - \cos\left(\frac{2\pi}{6}t\right) - \sqrt{3}\sin\left(\frac{2\pi}{6}t\right)$$

Comparing this expression with the definition, Eq. (6.4), we get the trigonometric form of the FS coefficients as

$$X_c(0) = -1, \quad X_c(1) = -1, \quad X_s(1) = -\sqrt{3}$$

Exponential form

$$x(t) = -1 - 2\cos\left(\frac{2\pi}{6}t - \frac{\pi}{3}\right) = -1 + 2\cos\left(\frac{2\pi}{6}t + \frac{2\pi}{3}\right)$$

$$= -1 + e^{j\left(\frac{2\pi}{6}t+\frac{2\pi}{3}\right)} + e^{-j\left(\frac{2\pi}{6}t+\frac{2\pi}{3}\right)}$$

Comparing this expression with the definition, Eq. (6.1), we get the exponential form of the FS coefficients as

$$X_{cs}(0) = -1, \quad X_{cs}(1) = 1\angle\frac{2\pi}{3}, \quad X_{cs}(-1) = 1\angle-\frac{2\pi}{3}$$

The FS amplitude spectrum and the phase spectrum of the signal in exponential form are shown, respectively, in Fig. 6.2c and d. ∎

The frequencies of harmonically related continuous sinusoids must be rational numbers or rational multiples of the same transcendental or irrational number. Therefore, the ratio of frequencies of any two harmonically related sinusoids is a rational number. The fundamental frequency (of which the harmonic frequencies are integral multiples) of a combination of sinusoids is found as follows: (i) any common factors of the numerators and denominators of each of the frequencies are cancelled, and (ii) the greatest common divisor of the numerators is divided by the least common multiple of the denominators of the frequencies.

Example 6.2 Find the exponential form of the FS for the signal

$$x(t) = 2 + 4\sin\left(\frac{1}{2}t + \frac{\pi}{6}\right) + 3\cos\left(\frac{3}{5}t - \frac{\pi}{4}\right)$$

Solution The frequency of the waveforms are $\frac{1}{2}$ and $\frac{3}{5}$. There are no common factors of the numerators and denominators. The least common multiple of the denominators (2,5) is 10. The greatest common divisor of the numerators (1,3) is one. Therefore, the fundamental frequency is $\omega_0 = \frac{1}{10}$ radians per second. The fundamental period is $T = \frac{2\pi}{\omega_0} = \frac{2\pi 10}{1} = 20\pi$. The first sinusoid, the fifth harmonic shown in Fig. 6.3 (dashed line), completes five cycles, and the second sinusoid (dotted line), the sixth harmonic, completes six cycles in the period. The combined waveform (solid line) completes one cycle in the period.

$$x(t) = 2 + 2e^{j\left(\frac{1}{2}t-\frac{\pi}{3}\right)} + 2e^{-j\left(\frac{1}{2}t-\frac{\pi}{3}\right)} + \frac{3}{2}e^{j\left(\frac{3}{5}t-\frac{\pi}{4}\right)} + \frac{3}{2}e^{-j\left(\frac{3}{5}t-\frac{\pi}{4}\right)}$$

Comparing this expression with the definition, Eq. (6.1), we get the exponential form of the FS coefficients as $X_{cs}(0) = 2, \quad X_{cs}(\pm 5) = 2\angle\mp\frac{\pi}{3}, X_{cs}(\pm 6) = \frac{3}{2}\angle\mp\frac{\pi}{4}$. ∎

Example 6.3 Find the FS for a square wave defined over one period as

$$x(t) = \begin{cases} 1 & \text{for } |t| < \frac{1}{4} \\ 0 & \text{for } \frac{1}{4} < |t| < \frac{1}{2} \end{cases}$$

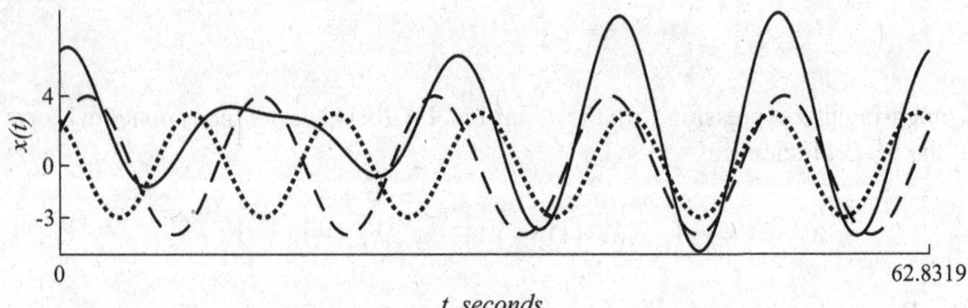

Fig. 6.3 The harmonics and the combined waveform

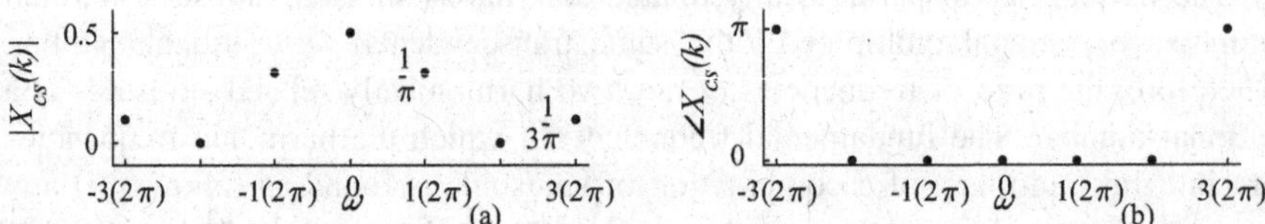

Fig. 6.4 (**a**) The FS amplitude spectrum and (**b**) the phase spectrum of the square wave in exponential form

Solution The period of the waveform is one, and the fundamental frequency is 2π. The waveform is even-symmetric and odd half-wave symmetric with a DC bias. Therefore, in addition to the DC component, the waveform is composed of odd-indexed cosine waves only.

$$X_c(0) = 2\int_0^{\frac{1}{4}} dt = \frac{1}{2}$$

$$X_c(k) = 4\int_0^{\frac{1}{4}} \cos(2\pi k\, t)dt = \begin{cases} \frac{2}{k\pi}\sin(\frac{\pi}{2}k) & \text{for } k \text{ odd} \\ 0 & \text{for } k \text{ even and } k \neq 0 \end{cases}$$

$$x(t) = \frac{1}{2} + \frac{2}{\pi}(\cos(2\pi t) - \frac{1}{3}\cos(3(2\pi t)) + \frac{1}{5}\cos(5(2\pi t)) - \cdots) \tag{6.5}$$

The FS amplitude spectrum and the phase spectrum of the signal in exponential form are shown, respectively, in Fig. 6.4a and b. ∎

6.1.6 Gibbs Phenomenon

The FS converges uniformly for waveforms with no discontinuity. At any discontinuity of a waveform, the FS converges to the average of the left- and right-hand

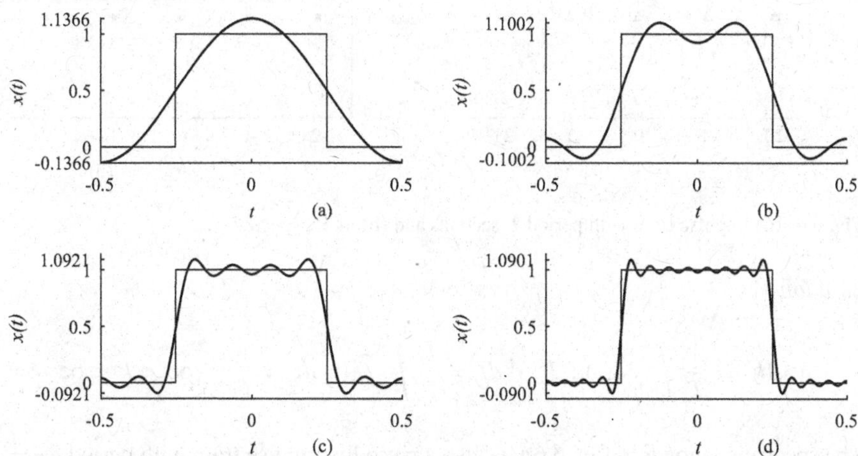

Fig. 6.5 The FS for the square wave; (**a**) using up to the first harmonic; (**b**) using up to the third harmonic; (**c**) using up to the seventh harmonic; (**d**) using up to the fifteenth harmonic

limits with overshoots and undershoots in the vicinity of the discontinuity. As the basis waveforms of the Fourier series are continuous sinusoids, they can never exactly add up to a discontinuity. This inability of the FS is referred as the Gibbs phenomenon.

Figure 6.5a, b, c, and d shows the FS for the square wave, using up to the first, third, seventh, and fifteenth harmonics, respectively. Consider the FS for the square wave using up to the first harmonic, $x(t) = \frac{1}{2} + \frac{2}{\pi}\cos(2\pi t)$. By differentiating this expression with respect to t and equating it to zero, we get $\sin(2\pi t) = 0$. The expression evaluates to zero for $t = 0$. Substituting $t = 0$ in the expression for $x(t)$, we get the value of the peak as 1.1366, as shown in Fig. 6.5a. We can find the maximum overshoots in other cases similarly.

As we use more number of harmonics, the frequency of oscillations increases, and the oscillations are confined more closer to the discontinuity. But, the largest amplitude of the oscillations settles at 1.0869 for relatively small number of harmonics. Therefore, even if we use an infinite number of harmonics to represent a waveform with discontinuity, there will be deviations of 8.69 % of the discontinuity for a moment. Of course, the area under the deviation tends to zero.

Example 6.4 Find the three forms of the FS for the periodic impulse train, shown in Fig. 6.6a, with period T seconds defined as

$$x(t) = \sum_{n=-\infty}^{\infty} \delta(t - nT)$$

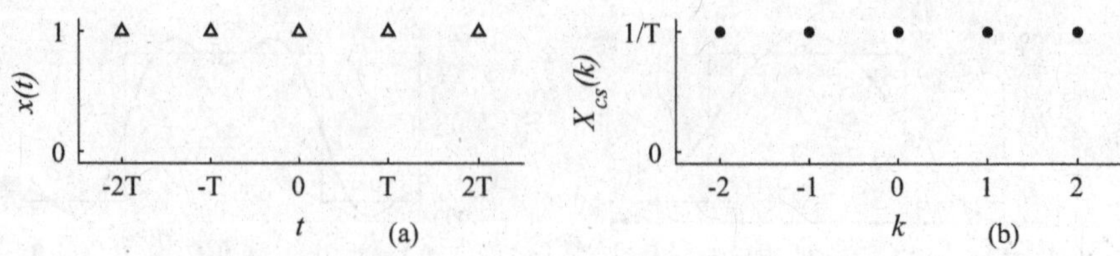

Fig. 6.6 (**a**) Impulse train with period T seconds and (**b**) its FS spectrum

Solution

$$X_{cs}(k) = \frac{1}{T}\int_{-\frac{T}{2}}^{\frac{T}{2}} \delta(t)e^{-jk\omega_0 t}\,dt = \frac{1}{T}\int_{-\frac{T}{2}}^{\frac{T}{2}} \delta(t)\,dt = \frac{1}{T}, \quad -\infty < k < \infty$$

The spectrum, shown in Fig. 6.6b, is also a periodic impulse train with period $\omega_0 = \frac{2\pi}{T}$ and constant amplitude $\frac{1}{T}$. Note that the impulses in the time-domain are of continuous type (as $x(t)$ is a function of the continuous variable t), while those of the spectrum are of discrete type (as $X_{cs}(k)$ is a function of the discrete variable k). The FS for the impulse train, in exponential form, is given by

$$x(t) = \sum_{k=-\infty}^{\infty} X_{cs}(k)e^{jk\omega_0 t} = \frac{1}{T}\sum_{k=-\infty}^{\infty} e^{jk\omega_0 t}, \; \omega_0 = \frac{2\pi}{T}$$

The FS coefficients, in compact trigonometric form, are

$$X_p(0) = X_{cs}(0) = \frac{1}{T}, \quad X_p(k) = 2|X_{cs}(k)| = \frac{2}{T}, \; \theta(k) = 0, \quad k = 1, 2, 3, \ldots$$

The FS is given by

$$x(t) = \frac{1}{T}(1 + 2(\cos(\omega_0 t) + \cos(2\omega_0 t) + \cos(3\omega_0 t) + \cdots)), \; \omega_0 = \frac{2\pi}{T}$$

As the phase $\theta(k) = 0$ is zero, the trigonometric form of the FS is the same as this form.

An alternate way of obtaining this FS is to consider the FS for a train of unit area rectangular pulses of width a and height $\frac{1}{a}$, with the width a of one pulse including the point $t = 0$. In the limiting case of $a \to 0$, the train of pulses degenerates into an impulse train, and the limiting form of its FS is the FS for the impulse train. ∎

6.2 Properties of the Fourier Series

For each operation in one domain, the properties establish the corresponding operation in the other domain, making it evident of the simpler relationship between variables for a particular operation. For example, the convolution operation in the time-domain corresponds to the much simpler multiplication operation in the frequency-domain. In addition, we can derive the FS coefficients for functions from that for related functions more easily than deriving them from the definition.

6.2.1 Linearity

The FS coefficients for a linear combination of a set of periodic signals, with the same period, is the same linear combination of their individual FS coefficients. That is,

$$x(t) \Longleftrightarrow X_{cs}(k), \quad y(t) \Longleftrightarrow Y_{cs}(k), \quad ax(t) + by(t) \Longleftrightarrow aX_{cs}(k) + bY_{cs}(k),$$

where a and b are arbitrary constants. For example, the FS coefficients for $\cos(t)$ and $\sin(t)$ are $X_{cs}(\pm 1) = \frac{1}{2}$ and $X_{cs}(\pm 1) = \mp\frac{j}{2}$, respectively. The FS coefficients for $\cos(t) + j\sin(t) = e^{jt}$ are $X_{cs}(\pm 1) = \frac{1}{2} + (j)(\mp\frac{j}{2})$. That is, the only nonzero FS coefficient is $X_{cs}(1) = 1$.

6.2.2 Symmetry

The symmetry properties simplify the evaluation of the FS coefficients. Each frequency component of a real waveform is composed of complex conjugate exponentials. Therefore, if the signal is real, then the real part of its spectrum is even, and the imaginary part is odd, called the conjugate symmetry. The FS for a real signal $x(t)$, with period T, is given by

$$X_{cs}(k) = \frac{1}{T}\int_0^T x(t)e^{-jk\frac{2\pi}{T}t}dt = \frac{1}{T}\int_0^T x(t)\left(\cos\left(k\frac{2\pi}{T}t\right) - j\sin\left(k\frac{2\pi}{T}t\right)\right)dt$$

Conjugating both sides, we get

$$X^*_{cs}(k) = \frac{1}{T}\int_0^T x(t)\left(\cos\left(k\frac{2\pi}{T}t\right) + j\sin\left(k\frac{2\pi}{T}t\right)\right)dt$$

Replacing k by $-k$, we get $X^*_{cs}(-k) = X_{cs}(k)$. For example, the FS coefficients for $4\cos(t + \frac{\pi}{3})$ are $X_{cs}(\pm 1) = 1 \pm j\sqrt{3}$.

6.2.2.1 Even Symmetry

If the signal $x(t)$ is real and even, then its product with sine basis waveforms is odd, and the $X_s(k)$ coefficients are, therefore, zero. That is, the signal is composed of cosine waveforms alone and its spectrum is real and even. As the product of the cosine basis waveforms and the signal is even, the FS defining integral can be evaluated over half the period. That is,

$$X_c(0) = \frac{2}{T}\int_{t_1}^{t_1+\frac{T}{2}} x(t)\,dt,$$

$$X_c(k) = \frac{4}{T}\int_{t_1}^{t_1+\frac{T}{2}} x(t)\cos(k\omega_0 t)\,dt, \quad k = 1, 2, \ldots, \infty$$

For example, the FS coefficients for $\cos(t)$ are $X_{cs}(\pm 1) = \frac{1}{2}$.

6.2.2.2 Odd Symmetry

If the signal $x(t)$ is real and odd, then its product with cosine basis waveforms is odd, and the $X_c(k)$ coefficients are, therefore, zero. That is, the signal is composed of sine waveforms alone and its spectrum is imaginary and odd. As the product of the sine basis waveforms and the signal is even, the FS defining integral can be evaluated over half the period. That is,

$$X_s(k) = \frac{4}{T}\int_{t_1}^{t_1+\frac{T}{2}} x(t)\sin(k\omega_0 t)\,dt, \quad k = 1, 2, \ldots, \infty$$

For example, the FS coefficients for $\sin(t)$ are $X_{cs}(\pm 1) = \mp\frac{j}{2}$.

As the FS coefficients for a real and even signal are real and even and that for a real and odd signal are imaginary and odd, it follows that the real part of the FS coefficients, $\mathrm{Re}(X_{cs}(k))$, of an arbitrary real signal $x(t)$ is the FS coefficients for its even component $x_e(t)$ and $j\,\mathrm{Im}(X_{cs}(k))$ are that for its odd component $x_o(t)$.

6.2.2.3 Half-Wave Symmetry

Even half-wave symmetry

If a periodic signal of period T satisfies the property $x(t \pm \frac{T}{2}) = x(t)$, then it is said to have even half-wave symmetry. That is, it completes two cycles of a pattern in the interval T. The FS coefficients can be expressed as

$$X_{cs}(k) = \frac{1}{T}\int_{t_1}^{t_1+\frac{T}{2}} \left(x(t) + (-1)^k x\left(t+\frac{T}{2}\right)\right) e^{-jk\omega_0 t}\, dt \tag{6.6}$$

The odd-indexed FS coefficients are zero. The even-indexed FS coefficients are given by

$$X_{cs}(k) = \frac{2}{T}\int_{t_1}^{t_1+\frac{T}{2}} x(t)e^{-jk\omega_0 t}\, dt,\ k = 0, 2, 4, \ldots$$

The full-wave rectified sine waveform, in Fig. 6.8, is an example of even half-wave symmetry.

Odd half-wave symmetry

If a periodic signal of period T satisfies the property $-x(t \pm \frac{T}{2}) = x(t)$, then it is said to have odd half-wave symmetry. That is, the values of the signal over any half period are the negatives of the values over the succeeding or preceding half period. It is obvious, from Eq. (6.6), that the even-indexed FS coefficients are zero. The odd-indexed FS coefficients are given by

$$X_{cs}(k) = \frac{2}{T}\int_{t_1}^{t_1+\frac{T}{2}} x(t)e^{-jk\omega_0 t}\, dt,\ k = 1, 3, 5, \ldots$$

The square waveform, in Fig. 6.5, is an example of odd half-wave symmetry, if the DC component is subtracted.

Any periodic signal $x(t)$, with period T, can be decomposed into its even and odd half-wave symmetric components $x_{eh}(t)$ and $x_{oh}(t)$, respectively. That is, $x(t) = x_{eh}(t) + x_{oh}(t)$, where

$$x_{eh}(t) = \frac{1}{2}\left(x(t) + x\left(t \pm \frac{T}{2}\right)\right) \quad \text{and} \quad x_{oh}(t) = \frac{1}{2}\left(x(t) - x\left(t \pm \frac{T}{2}\right)\right)$$

6.2.3 *Time Shifting*

When we shift a signal, the shape remains the same, but the signal is relocated. The shift of a typical spectral component, $X_{cs}(k_0)e^{jk_0\omega_0 t}$, by t_0 to the right results in the exponential, $X_{cs}(k_0)e^{jk_0\omega_0(t-t_0)} = e^{-jk_0\omega_0 t_0}X_{cs}(k_0)e^{jk_0\omega_0 t}$. That is, a delay of t_0 results in changing the phase of the exponential by $-k_0\omega_0 t_0$ radians without changing its amplitude.

Therefore, if the FS spectrum for $x(t)$, with the fundamental frequency $\omega_0 = \frac{2\pi}{T}$, is $X_{cs}(k)$, then

$$x(t \pm t_0) \Longleftrightarrow e^{\pm jk\omega_0 t_0} X_{cs}(k)$$

Consider the FS coefficients $X_{cs}(\pm 1) = \mp\frac{j}{2}$ for $\sin(2t)$. The FS coefficients for $\sin(2t + \frac{\pi}{2}) = \cos(2t)$, with $k = \pm 1$, $\omega_0 = 2$, and $t_0 = \frac{\pi}{4}$, are $X_{cs}(\pm 1) = e^{\pm j(1)(2)\frac{\pi}{4}}(\mp\frac{j}{2}) = \frac{1}{2}$.

6.2.4 Frequency Shifting

The spectrum, $X_{cs}(k)$, of a signal, $x(t)$, can be shifted by multiplying the signal by a complex exponential, $e^{\pm jk_0\omega_0 t}$, where k_0 is an integer and ω_0 is the fundamental frequency. The new spectrum is $X_{cs}(k \mp k_0)$, since a spectral component $X_{cs}(k_a)e^{jk_a\omega_0 t}$ of the signal, multiplied by $e^{jk_0\omega_0 t}$, becomes $X_{cs}(k_a)e^{j((k_a+k_0)\omega_0 t)}$ and the corresponding spectral value occurs at $k = (k_a + k_0)$, after a delay of k_0 samples. Therefore, we get

$$x(t)e^{\pm jk_0\omega_0 t} \Longleftrightarrow X_{cs}(k \mp k_0)$$

For example, consider the FS coefficients $X_{cs}(\pm 1) = \mp\frac{j}{2}$ for $\sin(t)$. The FS coefficients for $\cos(2t)\sin(t)$ can be computed using this property. As $\cos(2t) = \frac{1}{2}(e^{j2t} + e^{-j2t})$, the FS coefficients for the new function is the sum of the FS coefficients for $\sin(t)$ shifted to the right and left by two, in addition to the scale factor $\frac{1}{2}$. That is,

$$X_{cs}(\pm 1) = \pm\frac{j}{4} \quad \text{and} \quad X_{cs}(\pm 3) = \mp\frac{j}{4}$$

This spectrum corresponds to the time-domain function $\frac{1}{2}(\sin(3t) - \sin(t))$, which is, of course, equal to $\cos(2t)\sin(t)$.

6.2.5 Time Reversal

$$x(t) \Longleftrightarrow X_{cs}(k), \quad x(-t) \Longleftrightarrow X_{cs}(-k)$$

Time reversal of $x(t)$ results in the reversal of $X_{cs}(k)$ also. For example,

$$\sin(t) \Longleftrightarrow X_{cs}(\pm 1) = \mp\frac{j}{2}$$

and

$$\sin(-t) \Longleftrightarrow X_{cs}(\pm 1) = \pm\frac{j}{2}$$

6.2.6 Convolution in the Time-Domain

Using the FS, we get periodic or cyclic convolution as FS analyzes periodic time-domain signals. The periodic convolution is defined for two periodic signals, $x(t)$ of period T_1 and $h(t)$ of period T_2, as

$$y(t) = \int_0^T x(\tau)h(t-\tau)d\tau,$$

where T (common period of $x(t)$ and $h(t)$) is the least common multiple of T_1 and T_2. The FS coefficients for $y(t)$ are to be expressed in terms of those of $x(t)$ and $h(t)$.

The convolution of $h(t)$ of period T with a complex exponential $e^{jk_0\omega_0 t}$, $\omega_0 = \frac{2\pi}{T}$ is given as

$$\int_0^T h(\tau)e^{jk_0\omega_0(t-\tau)}d\tau = e^{jk_0\omega_0 t}\int_0^T h(\tau)e^{-jk_0\omega_0\tau}d\tau = TH_{cs}(k_0)e^{jk_0\omega_0 t}$$

As an arbitrary $x(t)$ of period T is reconstructed by the inverse FS as $x(t) = \sum_{k=-\infty}^{\infty} X_{cs}(k)e^{jk_0\omega_0 t}$, the convolution of $x(t)$ and $h(t)$ is given by $y(t) = \sum_{k=-\infty}^{\infty} TX_{cs}(k)H_{cs}(k)e^{jk\omega_0 t}$, where $X_{cs}(k)$ and $H_{cs}(k)$ are, respectively, the FS coefficients for $x(t)$ and $h(t)$. The inverse of the FS spectrum $TX_{cs}(k)H_{cs}(k)$ is the periodic convolution of $x(t)$ and $h(t)$. That is,

$$\int_0^T x(\tau)h(t-\tau)d\tau = \sum_{k=-\infty}^{\infty} TX_{cs}(k)H_{cs}(k)e^{jk\omega_0 t} \Longleftrightarrow TX_{cs}(k)H_{cs}(k)$$

Consider the convolution of $x(t) = \cos(t)$ and $h(t) = \sin(t)$ with the FS coefficients $X_{cs}(\pm 1) = \frac{1}{2}$ and $H_{cs}(\pm 1) = \mp\frac{j}{2}$, respectively. Then, with $T = 2\pi$, we get

$$TX_{cs}(\pm 1)H_{cs}(\pm 1) = 2\pi\left(\mp\frac{j}{4}\right) = \pi\left(\mp\frac{j}{2}\right)$$

These FS coefficients correspond to the time-domain function $\pi\sin(t)$. By directly evaluating the time-domain convolution, we get

$$\int_0^{2\pi} \cos(\tau)\sin(t-\tau)d\tau = \int_0^{2\pi} \cos(\tau)(\sin(t)\cos(\tau) - \cos(t)\sin(\tau))d\tau = \pi\sin(t)$$

An important application of this property is in modeling the truncation of the FS spectrum. The signal corresponding to the truncated spectrum has to be expressed in terms of the original signal $x(t)$ with fundamental frequency ω_0. The truncation operation can be considered as multiplying the spectrum of $x(t)$ with the spectrum that is one for $-N \le k \le N$ and zero otherwise. The signal corresponding to this spectrum is

$$y(t) = \frac{\sin\left(\frac{(2N+1)\omega_0 t}{2}\right)}{\sin(\frac{\omega_0 t}{2})}$$

Therefore, the signal corresponding to the truncated spectrum of $x(t)$ (using only $2N+1$ FS coefficients) is given by the convolution of $x(t)$ and $y(t)$, multiplied by the factor $\frac{1}{T}$, as

$$x_N(t) = \frac{1}{T}\int_0^T x(\tau)\frac{\sin\left(\frac{(2N+1)\omega_0(t-\tau)}{2}\right)}{\sin\left(\frac{\omega_0(t-\tau)}{2}\right)}d\tau$$

This expression is often used in explaining the Gibbs phenomenon. The alternating nature of the second function in the integrand, even in the limit as $N \to \infty$, does not change and produces deviations at any discontinuity of $x(t)$.

6.2.7 Convolution in the Frequency-Domain

Consider the FS representations of $x(t)$ and $y(t)$ with a common fundamental frequency $\omega_0 = \frac{2\pi}{T}$.

$$x(t) = \sum_{m=-\infty}^{\infty} X_{cs}(m)e^{jm\omega_0 t} \quad \text{and} \quad y(t) = \sum_{l=-\infty}^{\infty} Y_{cs}(l)e^{jl\omega_0 t}$$

The FS coefficients for $x(t)y(t)$ are to be expressed in terms of those of $x(t)$ and $y(t)$. The product of the two functions is given by

$$z(t) = x(t)y(t) = \sum_{m=-\infty}^{\infty}\sum_{l=-\infty}^{\infty} X_{cs}(m)Y_{cs}(l)e^{j(m+l)\omega_0 t}$$

Letting $m+l=k$, we get

$$z(t) = x(t)y(t) = \sum_{k=-\infty}^{\infty}\left(\sum_{m=-\infty}^{\infty} X_{cs}(m)Y_{cs}(k-m)\right)e^{jk\omega_0 t}$$

This is a FS for $z(t) = x(t)y(t)$ with coefficients $Z_{cs}(k) = \sum_{m=-\infty}^{\infty} X_{cs}(m)Y_{cs}(k-m)$. The convolution of two frequency-domain functions, with a common fundamental frequency, corresponds to the multiplication of their inverse FS in the time-domain. That is,

$$x(t)y(t) \Longleftrightarrow \frac{1}{T}\int_{0}^{T} x(t)y(t)e^{-jk\omega_0 t}\,dt = \sum_{m=-\infty}^{\infty} X_{cs}(m)Y_{cs}(k-m)$$

The convolution is aperiodic as the FS spectra are aperiodic.

Consider the convolution of the FS spectra given as $X_{cs}(\pm 1) = \frac{1}{2}$ and $Y_{cs}(\pm 2) = \mp\frac{j}{2}$, with $\omega_0 = 1$. The linear convolution of these spectra is $Z_{cs}(\pm 3) = \mp\frac{j}{4}$ and $Z_{cs}(\pm 1) = \mp\frac{j}{4}$. The corresponding time-domain function is

$$\frac{1}{2}(\sin(t) + \sin(3t)) = \cos(t)\sin(2t)$$

Note that the given FS spectra corresponds to the time-domain functions $\cos(t)$ and $\sin(2t)$.

6.2.8 Duality

The analysis equation of the FS is an integral and the synthesis equation is a summation. Therefore, there is no duality between these operations. However, as the synthesis equation of the DTFT is an integral and the analysis equation is a summation, there is duality between these two transforms. This will be presented in the next chapter.

6.2.9 Time Scaling

Scaling is the operation of replacing the independent variable t by at, where $a \neq 0$ is a constant. As we have seen in Chap. 2, the signal is compressed or expanded in the time-domain by this operation. As a consequence, the spectrum of the signal is expanded or compressed in the frequency-domain. The amplitude of the spectrum remains the same with the fundamental frequency changed to $a\omega_0$. Let the spectrum of a signal $x(t)$, with the fundamental frequency $\omega_0 = \frac{2\pi}{T}$, be $X_{cs}(k)$. Then, $x(at) \Longleftrightarrow X_{cs}(k)$ with the fundamental frequency $a\omega_0$ and $a > 0$. If $a < 0$, the spectrum, with the fundamental frequency $|a|\omega_0$, is also frequency-reversed. For example, with $a = 0.2$, the signal $\cos(t)$ becomes $\cos(0.2t)$. The spectrum remains the same, that is, $X_{cs}(\pm 1) = \frac{1}{2}$, with the fundamental frequency changed to 0.2 radians from one radian. With $a = -3$, the signal $\sin(t)$ becomes

$\sin(-3t) = -\sin(3t)$. The spectrum gets frequency-reversed, $X_{cs}(\pm 1) = \pm\frac{j}{2}$, and the fundamental frequency of the FS spectrum is changed to three radians from one radian.

6.2.10 Time-Differentiation

As the signal is decomposed in terms of exponentials of the form $e^{jk_0\omega_0 t}$, this property is essentially finding the derivative of all the constituent exponentials of a signal. The derivative of $e^{jk_0\omega_0 t}$ is $jk_0\omega_0 e^{jk_0\omega_0 t}$. Therefore, if the FS spectrum for a time-domain function $x(t)$ is $X_{cs}(k)$, then the FS spectrum for its derivative is $jk\omega_0 X_{cs}(k)$, where ω_0 is the fundamental frequency. Note that, the FS coefficient with $k = 0$ is zero, as the DC component is lost in differentiating a signal. In general,

$$\frac{d^n x(t)}{dt^n} \Longleftrightarrow (jk\omega_0)^n \, X_{cs}(k)$$

This property can be stated as the invariance of the exponentials with respect to the differentiation operation. That is, the derivative of an exponential is the same exponential multiplied by a complex scale factor. The exponentials are invariant with respect to integration and summation operations also. These properties change an integro-differential equation in the time-domain to an algebraic equation in the frequency-domain. Therefore, the analysis of linear systems is easier in the frequency-domain.

Another use of this property, in common with other properties, is to find FS spectra for signals from those of the related signals. This property can be used to find the FS for the functions represented by polynomials in terms of the FS for their derivatives. When a function is reduced to a sum of impulses, by differentiating it successively, the FS of the impulses can be found easily, and this FS is used to find the FS of $x(t)$ using the differentiation property. Consider a periodic rectangular pulse defined over one period as $x(t) = 1, \ |t| < a$ and $x(t) = 0, \ a < |t| < \frac{T}{2}$. The derivative of this signal in a period are the impulses $\delta(t + a)$ and $-\delta(t - a)$. The FS spectrum for this pair is $\frac{1}{T}(e^{jk\omega_0 a} - e^{-jk\omega_0 a}) = \frac{j2}{T}\sin(k\omega_0 a)$, where $\omega_0 = \frac{2\pi}{T}$. This spectrum is related to the spectrum of the rectangular pulse by the factor $\frac{1}{jk\omega_0}$, $k \neq 0$. Therefore, the FS spectrum for the periodic rectangular pulse is $X_{cs}(k) = 2\frac{\sin(k\omega_0 a)}{k\omega_0 T} = \frac{\sin(k\omega_0 a)}{\pi k}$. In general, use this property to obtain $X_{cs}(k)$ for $k \neq 0$ only and determine $X_{cs}(0)$ directly from the given waveform.

6.2.11 Time-Integration

For a signal $x(t) \Longleftrightarrow X_{cs}(k)$ with fundamental frequency ω_0,

$$\int_{-\infty}^{t} x(\tau)d\tau \Longleftrightarrow \frac{1}{jk\omega_0} X_{cs}(k)$$

provided the DC component of $x(t)$ is zero ($X_{cs}(0) = 0$). Consider the function $\cos(2t)$ with FS coefficients $X_{cs}(\pm 1) = \frac{1}{2}$. Then, its integral, $y(t) = \frac{\sin(2t)}{2}$, has the FS coefficients:

$$Y_{cs}(\pm 1) = \frac{1}{j(\pm 1)(2)} \frac{1}{2} = \mp \frac{j}{4}$$

6.2.11.1 Rate of Convergence of the Fourier Series

In practical problems, only the sum of a finite number of the infinite terms of the FS are used to approximate a given function $x(t)$, as no physical device can generate harmonics of infinite order. The rate of convergence of a FS indicates how rapidly the partial sums converge to $x(t)$. A smoother function has a higher rate of convergence. According to the time-integration property, each time the FS is integrated term by term, the coefficients are divided by the factor k, the index. That is, the rate of convergence of the FS is increased by the factor k, as the function becomes smoother by the integration operation. The FS for an impulse train converges slowly as all its coefficients are the same (no dependence on the index k). As the integral of a function with impulses is a function with discontinuities, the FS of such a function converges more rapidly as the magnitude of their coefficients decrease at the rate $\frac{1}{k}$, for large values of k. As the integral of a function with discontinuities results in a function with no discontinuity, the FS for such functions converges still more rapidly as the magnitude of their coefficients decrease at the rate $\frac{1}{k^2}$. The magnitude of the coefficients of a function, whose nth derivative contains impulses, decreases at the rate $\frac{1}{k^n}$.

6.2.12 *Parseval's Theorem*

As the frequency-domain representation of a signal is an equivalent representation, the power of a signal can also be expressed in terms of its spectrum. That is, the average power of a signal is the sum of the average powers of its frequency components. The average power of a complex exponential, $Ae^{j\omega_0 t}$, is

$$P = \frac{1}{T} \int_{0}^{T} |Ae^{j\omega_0 t}|^2 dt = |A|^2$$

since $|e^{j\omega_0 t}| = 1$. That is, the average power of a complex exponential is the magnitude squared of its complex amplitude (irrespective of its frequency and phase). Therefore, the total average power of a signal is

$$P = \frac{1}{T}\int_0^T |x(t)|^2 dt = \sum_{k=-\infty}^{\infty} |X_{cs}(k)|^2$$

Example 6.5 Verify Parseval's theorem for the square wave of Example 6.3. Find the sum of the powers of the DC, first harmonic, and the third harmonic components of the signal.

Solution The power using the time-domain representation is

$$P = \frac{1}{T}\int_0^T |x(t)|^2 dt = 2\int_0^{\frac{1}{4}} dt = \frac{1}{2}$$

The power using the FS is

$$P = \sum_{k=-\infty}^{\infty} |X_{cs}(k)|^2 = \left(\frac{1}{2}\right)^2 + 2\sum_{k=1,3,}^{\infty} \left(\frac{1}{k\pi}\right)^2 = \frac{1}{4} + \frac{2}{\pi^2}\frac{\pi^2}{8} = \frac{1}{2}$$

The sum of the power of the components of the signal up to the third harmonic is

$$\frac{1}{4} + \frac{2}{\pi^2} + \frac{2}{9\pi^2} = \frac{1}{4} + \frac{20}{9\pi^2} = 0.25 + 0.2252 = 0.4752$$

This example shows that the approximation of the signal by a few harmonics includes most of its power. ∎

6.3 Approximation of the Fourier Series

We approximate the integral in Eq. (6.2) by the rectangular rule of numerical integration. We take N samples of the signal

$$x(0), x\left(\frac{T}{N}\right), x\left(2\frac{T}{N}\right), \ldots, x\left((N-1)\frac{T}{N}\right),$$

by dividing the period T of the signal into N intervals. The sampling interval is $T_s = \frac{T}{N}$. Now, Eq. (6.2) is approximated as

$$X_{cs}(k) = \frac{1}{T}\sum_{n=0}^{N-1} x(nT_s)e^{-jk\omega_0 nT_s}\frac{T}{N} = \frac{1}{N}\sum_{n=0}^{N-1} x(nT_s)e^{-jk\frac{2\pi}{N}n} \qquad (6.7)$$

This is the analysis equation. The synthesis equation, Eq. (6.1), is approximated as

$$x(nT_s) = \sum_{k=0}^{N-1} X_{cs}(k)e^{j\frac{2\pi}{N}nk}, \quad n = 0, 1, \ldots, N-1.$$

Except for constant factors, the approximations of the analysis and synthesis equations are the same as the DFT and IDFT equations, respectively. Note that $X_{cs}(k)$ is aperiodic and periodicity of N selected values is assumed in the IDFT computation. For example, if we truncate the FS spectrum to $X_{cs}(-1)$, $X_{cs}(0)$, and $X_{cs}(1)$, then the periodic extension can be written, starting with $X_{cs}(0)$ and $N = 3$, as $X_{cs}(0)$, $X_{cs}(1)$, and $X_{cs}(-1)$. With $N = 4$, the periodic values are $X_{cs}(0)$, $X_{cs}(1)$, 0, and $X_{cs}(-1)$.

For N even, comparing the coefficients of the DFT with that in Eq. (6.2), we get, for real signals,

$$X_c(0) = \frac{X(0)}{N}, \quad X_c\left(\frac{N}{2}\right) = \frac{X\left(\frac{N}{2}\right)}{N}$$

$$X_c(k) = \frac{2}{N}\mathrm{Re}(X(k)), \quad X_s(k) = -\frac{2}{N}\mathrm{Im}(X(k)), \ k = 1, 2, \ldots, \frac{N}{2} - 1$$

$$X_{cs}(k) = \frac{X(k)}{N}, \quad k = 0, 1, \ldots, \frac{N}{2} - 1 \quad \text{and} \quad \mathrm{Re}\left(X_{cs}\left(\frac{N}{2}\right)\right) = \frac{X\left(\frac{N}{2}\right)}{2N}$$

For example,

$$\sin(t) \Longleftrightarrow X_{cs}(\pm 1) = \mp 0.5j$$

The four samples of $\sin(t)$ are $\{0, 1, 0, -1\}$. The DFT is $\{0, -j2, 0, j2\}$, which is $N = 4$ times of $X_{cs}(\pm 1) = \mp 0.5j$. Now, if we want to find the inverse FS of $X_{cs}(\pm 1) = \mp 0.5j$ using the IDFT, we have to multiply the FS coefficients by $N = 4$ and then take the IDFT.

6.3.1 Aliasing Effect

Let us find the FS spectrum of a sampled signal. The sampling operation can be considered as multiplying the signal ($x(t)$ with spectrum $X_{cs}(k)$) by the sampling signal ($s(t) = \sum_{n=-\infty}^{\infty} \delta(t - nT_s)$ with spectrum $\frac{1}{T_s}$). The sampled signal is $\sum_{n=-\infty}^{\infty} x(nT_s)\delta(t - nT_s)$. In the frequency-domain, sampling operation corresponds to the convolution of the spectra of the two signals. As the convolution of a signal with an impulse is just translation of the origin of the signal to the location of the impulse, we get the spectrum of the sampled signal $x(t)s(t)$ as the superposition sum of the infinite frequency-shifted spectrum of the signal, multiplied by the factor $\frac{1}{T_s}$. That is, the FS spectrum for the sampled signal is

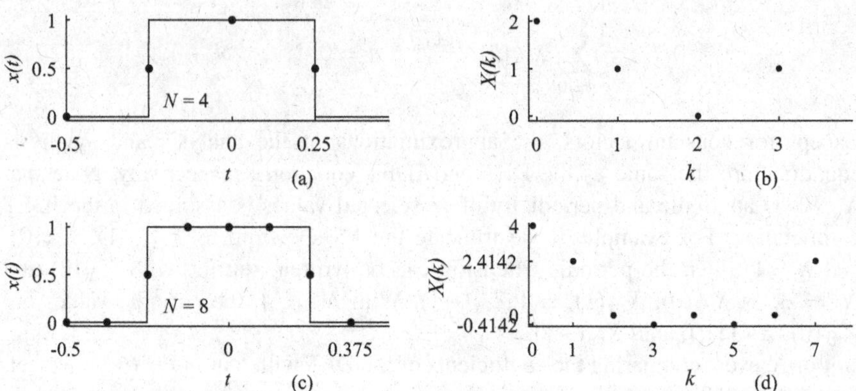

Fig. 6.7 (**a**) Four samples of the square wave and (**b**) its DFT; (**c**) eight samples of the square wave and (**d**) its DFT

$$\frac{1}{T_s}\sum_{m=-\infty}^{\infty} X_{cs}\left(k-m\frac{T}{T_s}\right),$$

where T is the period of the signal. Let us construct a sequence $x(n)$ such that its nth element has the value that is the same as the strength of the impulse $x(nT_s)\delta(t-nT_s)$ of the sampled signal. We get the DFT spectrum of $x(n)$ by multiplying the spectrum of the sampled signal by T (remember that there is no normalization factor in the definition of the DFT) and noting that $\frac{T}{T_s} = N$. Therefore, we get

$$X(k) = N\sum_{m=-\infty}^{\infty} X_{cs}(k-mN), \quad k = 0, 1, \ldots, N-1 \tag{6.8}$$

This equation shows how the DFT spectrum is corrupted due to aliasing. By sampling the signal, in order to use the DFT, to obtain a finite number of N samples in a period, we simultaneously reduce the number of distinct sinusoids and, hence, the number of distinct spectral coefficients to N. Therefore, if the signal is bandlimited, we can get the exact FS coefficients by computing the DFT of the samples of the signal. If the signal is not bandlimited or the number of samples is inadequate, we get a corrupted FS spectrum using the DFT due to the aliasing effect.

Consider the sampling of the square wave (Example 6.3) with $N = 4$ samples, shown in Fig. 6.7a. Note that, at any discontinuity, the average of the left- and right-hand limits should be taken as the sample value. The sample values, starting from $n = 0$, are $\{1, 0.5, 0, 0.5\}$. The DFT of this set of samples is $\{2, 1, 0, 1\}$, shown in Fig. 6.7b. These values can be obtained from the FS coefficients using Eq. (6.8). The DC value is $2/4 = 0.5$, which is equal to the analytical value. The coefficient of the first harmonic is $(1 + 1)/4 = 0.5$, which differs from the analytical value of 0.637. This is due to the fact that, with only four samples, all the other odd harmonics alias as the first harmonic.

$$X_c(1) = \frac{2}{\pi}\left(1 - \frac{1}{3} + \frac{1}{5} - \frac{1}{7} + \cdots\right)$$

The value of the summation can be obtained from Eq. (6.5) by substituting $t = 0$.

$$1 = \frac{1}{2} + \frac{2}{\pi}\left(1 - \frac{1}{3} + \frac{1}{5} - \frac{1}{7} + \cdots\right)$$

Therefore, we get $X_c(1) = 0.5$. As we double the number of samples, we get a better approximation of the FS coefficients by the DFT. Figure 6.7c shows the square wave with $N = 8$ samples, and its DFT is shown in Fig. 6.7d. The value of the first harmonic is $(2.4142 + 2.4142)/8 = 0.6036$, which is much closer to the actual value of 0.637. The point is that DFT should be used to approximate the FS coefficients with sufficient number of time-domain samples so that the accuracy of the approximation is adequate.

Let us compute the FS coefficients of $x(t)$ of period 4, using the DFT.

$$x(t) = 1 + 2\cos\left(\frac{2\pi}{4}t\right) + 4\sin\left(2\frac{2\pi}{4}t + \frac{\pi}{3}\right) = 1 + 2\cos\left(\frac{2\pi}{4}t\right) + 4\cos\left(2\frac{2\pi}{4}t - \frac{\pi}{6}\right)$$

$$= 1 + e^{j(\frac{2\pi}{4}t)} + e^{-j(\frac{2\pi}{4}t)} + 2e^{-j(\frac{\pi}{6})}e^{j(\frac{2\pi}{4}2t)} + 2e^{j(\frac{\pi}{6})}e^{-j(\frac{2\pi}{4}2t)}$$

The FS coefficients $X_{cs}(k)$ are

$$\{1, 1, \sqrt{3} - j1, \sqrt{3} + j1, 1\}$$

With $T_s = 1$, the four samples over the period of 4 s are

$$\{3 + 2\sqrt{3}, 1 - 2\sqrt{3}, 2\sqrt{3} - 1, 1 - 2\sqrt{3}\} = \{6.4641, -2.4641, 2.4641, -2.4641\}$$

With four samples, the DFT coefficients are

$$\{4, 4, 8\sqrt{3}, 4\}$$

Dividing $X(2)$ by 8 and the rest by 4, we get

$$\{1, 1, \sqrt{3}, 1\}$$

The 2nd harmonic is

$$4\cos\left(2\frac{2\pi}{4}t - \frac{\pi}{6}\right) = 2\sqrt{3}\cos\left(2\frac{2\pi}{4}t\right) - 2\sin\left(2\frac{2\pi}{4}t\right)$$

With $N = 4$ samples, the samples of both the $\cos(2\frac{2\pi}{4}t)$ and the DFT basis function with index 2 are $\{1, -1, 1, -1\}$, and the sum of pointwise multiplication yields 4. By multiplying with the constant $2\sqrt{3}$, we get the DFT coefficient as $8\sqrt{3}$. Dividing $8\sqrt{3}$ by $2N = 8$, we get the real part of the FS coefficient $\sqrt{3}$, as given earlier. The samples of the sine component with frequency index 2 are all zero and gets no representation.

With $T_s = 0.8$, the five samples are

$$\{6.4641, -0.0089, -1.4497, 2.3545, -2.3601\}$$

The DFT coefficients are

$$\{5, 5, 8.6603 - j5, 8.6603 + j5, 5\}$$

which are the correct FS coefficients scaled by 5.

In practice, the number of samples has to be a power of 2 to suit fast practical DFT algorithms. With $T_s = 0.5$, the eight samples are

$$\{6.4641, 4.4142, -2.4641, -2.4142, 2.4641, 1.5858, -2.4641, 0.4142\}$$

The DFT coefficients, divided by 8, are

$$\{1, 1, 1.7321 - j1, 0, 0, 0, 1.7321 + 1, 1\}$$

which are the correct FS coefficients.

The FS coefficients in the center-zero format are

$$X_{fs}(k), k = -2, -1, 0, 1, 2 = \{\sqrt{3} + j1\ , 1\ , \check{1}\ , 1, \sqrt{3} - j1\}$$

If we sum the shifted copies of the spectral values placed at a distance of four samples, we get a corrupted periodic spectrum $\{\check{1}, 1, 2\sqrt{3}, 1\}$ in the standard format.

$$\sqrt{3} + j1\ 1\ \check{1}\ 1\ \sqrt{3} - j1$$

$$\sqrt{3} + j1\ 1\ \check{1}\ 1\ \sqrt{3} - j1$$

The sum is

$$\{2\sqrt{3}, 1, \check{1}, 1\}$$

in the center-zero format. The DFT coefficients, divided by $N = 4$, are also

$$\{\check{1}, 1, 2\sqrt{3}, 1\}$$

6.4 Applications of the Fourier Series

The FS is used to analyze periodic waveforms, such as half- and full-wave rectified waveforms. The steady-state response of stable LTI systems to periodic input signals can also be found using the FS. The steady-state response is the response of a system after the transient has decayed. The transient response of a stable system always decays with time. The steady-state output of a LTI system to an input $e^{jk_0\omega_0 t}$ is the same function multiplied by the complex scale factor, $H(jk_0\omega_0)$. Therefore, the output of the system is $H(jk_0\omega_0)e^{jk_0\omega_0 t}$. The function $H(jk\omega_0)$ is the frequency response $H(j\omega)$ (Chap. 4) of the system, sampled at the discrete frequencies $\omega = k\omega_0$.

Consider the system governed by the differential equation

$$\frac{dy(t)}{dt} + y(t) = x(t)$$

The differential equation can be written, with the input $x(t) = e^{jk_0\omega_0 t}$, as

$$\frac{d(H(jk_0\omega_0)e^{jk_0\omega_0 t})}{dt} + H(jk_0\omega_0)e^{jk_0\omega_0 t} = e^{jk_0\omega_0 t}$$

Solving for $H(jk_0\omega_0)$, we get

$$H(jk_0\omega_0) = \frac{1}{1 + jk_0\omega_0}$$

For an arbitrary periodic input, as $x(t) = \sum_{k=-\infty}^{\infty} X_{cs}(k)e^{jk\omega_0 t}$, we get

$$y(t) = \sum_{k=-\infty}^{\infty} H(jk\omega_0)X_{cs}(k)e^{jk\omega_0 t} = \sum_{k=-\infty}^{\infty} \frac{X_{cs}(k\omega_0)}{1 + jk\omega_0} e^{jk\omega_0 t}$$

The more complex operation of solving a differential equation has been reduced to the evaluation of an algebraic operation.

6.4.1 Analysis of Rectified Power Supply

Most of the electronic devices require a DC supply for their operation. The input supply is usually alternating current. Therefore, two diodes or a bridge rectifier is used to convert the bipolar input voltage to DC. This way, both the positive and negative halves of the input sine wave are used. This is called a full-wave rectifier. The average value of the full-wave rectifier output is 0.637 of the peak value of the sine wave input. In addition to the required DC component, the output of the rectifier is composed of ripples. In order to design a filter circuit to reduce the ripples to negligible levels, an analysis of the output waveform of the rectifier is required. Fourier series provides such an analysis.

Let us derive the FS representation of the full-wave rectified waveform given by

$$x(t) = |\sin(2\pi t)|$$

with period 1 and the fundamental frequency $\omega_0 = 2\pi$. The waveform is shown in Fig. 6.8a. The waveform is even-symmetric, $x(-t) = x(t)$, and, hence it is composed of cosine components only. In addition, the waveform is even half-wave symmetric. That is, the first-half and second-half of the waveform are identical, $x(t + \frac{T}{2}) = x(t)$. Consequently, the components of the waveform are even-indexed cosine waves only. The DC component is the average value of the waveform, and it is

$$X_c(0) = 2\int_0^{0.5} \sin(2\pi t)\, dt = \frac{2}{\pi}$$

For even-indexed k, except $k = 0$, we get

$$\begin{aligned}
X_c(2k) &= 4\int_0^{0.5} \sin(2\pi t)\cos(2\pi(2k)t)\, dt \\
&= 2\int_0^{0.5} (\sin(2\pi(1-2k)t) + \sin(2\pi(1+2k)t))\, dt \\
&= -\frac{1}{\pi}\left(\frac{\cos(2\pi(1-2k)t)}{(1-2k)} + \frac{\cos(2\pi(1+2k)t)}{(1+2k)}\right)\Big|_0^{0.5} \\
&= \frac{2}{\pi}\left(\frac{1}{(1-2k)} + \frac{1}{(1+2k)}\right) = \frac{4}{\pi}\left(\frac{1}{1-4k^2}\right)
\end{aligned}$$

$$\begin{aligned}
x(t) &= \frac{2}{\pi} - \frac{4}{\pi}\sum_{k=1}^{\infty}\left(\frac{1}{4k^2-1}\right)\cos(2k(2\pi)t) \\
&= \frac{2}{\pi} - \frac{4}{3\pi}\cos(2(2\pi)t) - \frac{4}{15\pi}\cos(4(2\pi)t) - \frac{4}{35\pi}\cos(6(2\pi)t) - \cdots
\end{aligned}$$

Figure 6.8b shows its DC component and its reconstruction with the DC and the 2nd harmonic. Figure 6.8c shows its reconstruction with the DC and the 2nd and 4th harmonics. Figure 6.8d shows its reconstruction with the DC and the 2nd, 4th, and 6th harmonics. As the number of harmonics is increased in the synthesis process, the approximation becomes better.

There are two important aspects in the analysis of waveforms: (i) the reconstruction and (ii) the amplitude of the harmonics. Figure 6.8e and f shows, respectively,

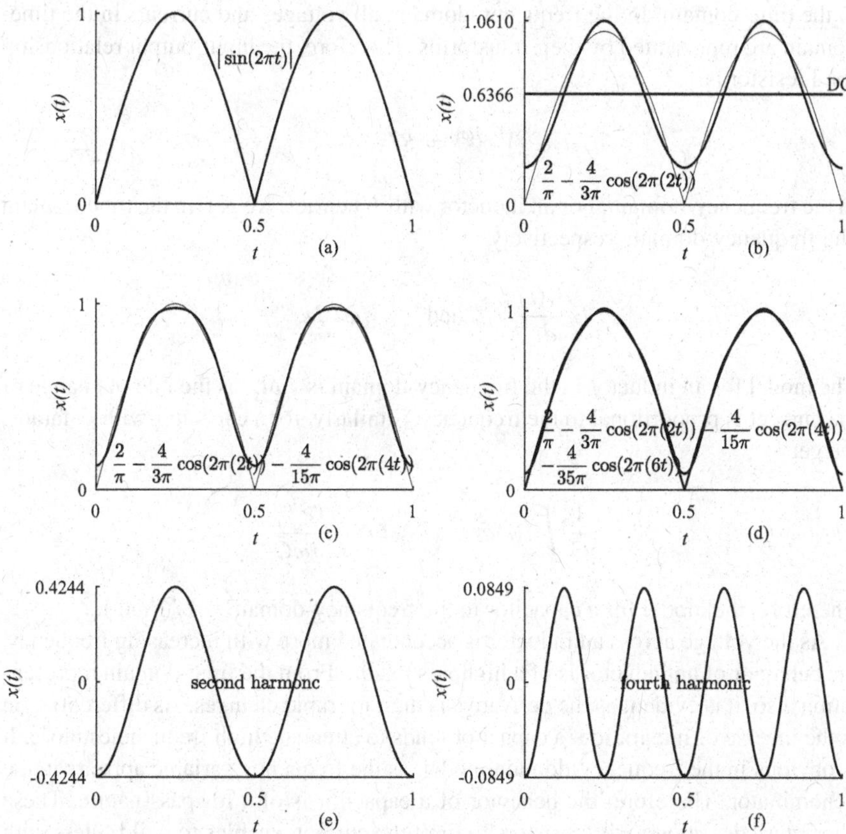

Fig. 6.8 (**a**) Full-wave rectified sine wave; (**b**) its reconstruction with DC and 2nd harmonic; (**c**) and (**d**) its reconstruction with up to 4th and 6th harmonics, respectively; (**e**) the 2nd harmonic; (**f**) the 4th harmonic

the second and fourth harmonics of the waveform in (a). The harmonics make two and four cycles in a period. In applications where the harmonics are unwanted, the amplitude of the harmonics has to be reduced below some prescribed levels. Fourier series is a tool in carrying out this task.

In the time-domain, any linear system can be modeled by integro-differential equations. It turns out that the equivalent representation in the frequency-domain provides much easier analysis. Consider the RC filter circuit shown in Fig. 3.1. This is one type of lowpass filter circuit. In general, the basic components of an electrical circuit are resistor R, inductance L, and capacitance C.

The voltage $v(t)$ across a resistor R Ohms due to current $i(t)$ flowing through it is given by

$$v(t) = Ri(t)$$

in the time-domain. In the frequency-domain, all voltages and currents in the time-domain are represented by their transforms. Therefore, the input-output relationship for a resistor is

$$Ve^{j\omega t} = RIe^{j\omega t}$$

in the frequency-domain. For an inductor with L henries, we get, in the time-domain and frequency-domain, respectively,

$$L\frac{di(t)}{dt} \quad \text{and} \quad j\omega LIe^{j\omega t}$$

The model for an inductor in the frequency-domain is $j\omega L$, as the rate of change of the current is proportional to the frequency. Similarly, for a capacitor with C farads, we get

$$\frac{1}{C}\int i(t)dt \quad \text{and} \quad \frac{Ie^{j\omega t}}{j\omega C}$$

Therefore, the model for a capacitor in the frequency-domain is $1/(j\omega C)$.

As the voltage across an inductor is accentuated more with increasing frequency, the behavior of an inductor is of a highpass nature. From the time-domain representation also, it is evident as the derivative is high for rapid changes. As differentiation is the inverse of integration, a capacitor tends to attenuate high frequencies more. It is obvious in the frequency-domain model, as the frequency variable appears in the denominator. Therefore, the behavior of a capacitor is of a lowpass nature. These characteristics, along with resistors to limit the current, enables to build filters with different frequency responses using these components. The opposition to the flow of current through a resistor is called the resistance. Similarly, the opposition to the flow of current through an inductor or capacitor is called the reactance. The impedance is the sum of the resistive and reactive part of a circuit. For example, the impedance of a series connected resistor R and inductor L is $R+j\omega L$. Similarly, the impedance of a series connected resistor R and capacitor C is $R+(1/(j\omega C))$. The frequency-domain model of a circuit makes the analysis of circuits with inductors and capacitances also similar to that of Ohm's law for DC circuit analysis.

Let the input $x(t) = |\sin(2\pi t)|$ be applied to the lowpass filter circuit, shown in Fig. 3.1. Let the values of resistor and capacitor be 10 ohms and 1/2 farads, respectively. Let us derive an expression, using the FS representation of the input waveform $x(t)$, for the output voltage across the capacitor, $y(t)$.

The FS for the full-wave rectified sine wave is

$$x(t) = \frac{2}{\pi} - \frac{4}{\pi}\sum_{k=1}^{\infty}\left(\frac{1}{4k^2-1}\right)\cos(2k(2\pi)t)$$

$$= \frac{2}{\pi} - \frac{4}{3\pi}\cos(2(2\pi)t) - \frac{4}{15\pi}\cos(4(2\pi)t) - \frac{4}{35\pi}\cos(6(2\pi)t) - \cdots$$

The voltage across the capacitor, by voltage division, is

$$y(t) = x(t)\frac{1/(j\omega C)}{R + 1/(j\omega C)} = x(t)\frac{1}{1 + j\omega RC}$$

For DC, with the fundamental frequency $\omega_0 = 2\pi$ and $k = 0$,

$$y(t) = \frac{2}{\pi}$$

since the capacitor is an open circuit to DC. For the 2nd harmonic, with the fundamental frequency $\omega_0 = 2\pi$ and $k = 2$,

$$y(t) = -\frac{4}{3\pi}\frac{1}{1 + j2(2\pi)RC}\cos((2)2\pi t)$$

With $R = 10$ and $C = 0.5$, the magnitude of the output is

$$\left|\frac{4}{3\pi}\frac{1}{1 + j20\pi}\right| = \frac{4}{3\pi}0.0159$$

The magnitude of the 2nd harmonic has been reduced by a factor of 0.0159, and the output becomes closer to the ideal output, which is DC. A large portion of the ripples drop across the series resistor. Figure 6.9a shows the rectified waveform. Figure 6.9b shows the DC component and the ripple due to 2nd harmonic. The ripple magnitude has been reduced by a factor of about 16. Figure 6.9c shows the

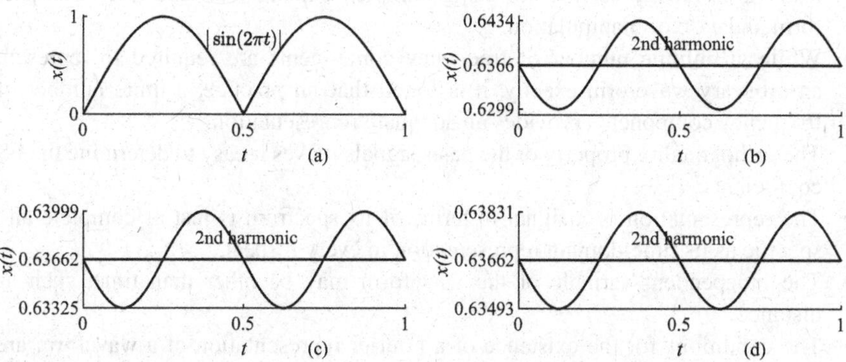

Fig. 6.9 (**a**) Full-wave rectified sine wave; (**b**) 2nd harmonic in (**a**) filtered with $R = 10$ ohms and $C = 0.5$ farads; (**c**) 2nd harmonic in (**a**) filtered with $R = 10$ ohms and $C = 1$ farads; (**d**) 2nd harmonic in (**a**) filtered with $R = 10$ ohms and $C = 2$ farads

DC component and the ripple due to 2nd harmonic with $C = 1$. The magnitude of the 2nd harmonic has been reduced by a factor of 0.0080. Figure 6.9d shows the DC component and the ripple due to 2nd harmonic with $C = 2$. The magnitude of the 2nd harmonic has been reduced by a factor of 0.0040. The sum of the responses to all the harmonic components of the input is the total response of the circuit. As it passes more readily the low-frequency components of the input compared with those of the high-frequency components, the RC circuit is a lowpass filter. As only a finite number of frequency components can be used in numerical analysis, this number has to be determined appropriately.

Similar frequency-domain analysis is applicable for other applications. For example, in mechanical engineering, friction, spring, and mass correspond to resistor, inductor, and capacitor, respectively.

6.5 Summary

- FS is one of the four versions of Fourier analysis that provides the representation of a continuous periodic time-domain waveform by a discrete aperiodic spectrum in the frequency-domain.
- FS represents a continuous periodic waveform as a linear combination of sinusoidal or, equivalently, complex exponential basis functions of harmonically related frequencies.
- Harmonics are any of the frequency components whose frequencies are integral multiples of a fundamental. The frequency of the fundamental is the same as that of the periodic waveform being analyzed.
- The FS is the limiting case of the DFT as the sampling interval of the time-domain sequence tends to zero with the period fixed.
- While physical devices generate real sinusoidal waveforms, it is found that the analysis is mostly carried out using complex exponentials due to its compact form and ease of manipulation.
- While an infinite number of frequency components are required to represent an arbitrary waveform exactly, it is found that, in practice, a finite number of frequency components provides an adequate representation.
- The orthogonality property of the basis signals makes it easy to determine the FS coefficients.
- The representation of a signal in terms of its spectrum is just as complete and specific as its time-domain representation in every respect.
- The independent variable of the waveform may be other than time, such as distance.
- The conditions for the existence of a Fourier representation of a waveform are met by signals generated by physical devices.
- The amplitude versus frequency plot of the harmonics is called the spectrum. As the spectrum is usually complex, it is represented by two plots, either the real and

imaginary parts or the magnitude and phase. While the time-domain waveform is continuous and periodic, its FS spectrum is aperiodic and discrete.

- The more smoother the waveform, the faster is the convergence of its spectrum.
- A signal can be reconstructed using its spectral components.
- The exponential and trigonometric forms of Fourier analysis are related by Euler's formula.
- The least squares error is the criterion for the representation of a signal by the Fourier analysis. With respect to this criterion, there is no better approximation than that provided by the Fourier analysis.
- The properties of Fourier analysis help to relate the effects of characteristics of signals in one domain into the other.
- LTI system analysis is simpler with the Fourier representation of signals and systems.
- The Fourier spectrum can be adequately approximated by the DFT in practical applications.
- FS is important in the analysis of signals such as acoustical, vibration, power system, communication, electrocardiogram, and frequency response.

Exercises

6.1 The FS representation of a real periodic signal $x(t)$ of period T, satisfying Dirichlet conditions, is given as

$$x(t) = X_c(0) + \sum_{k=1}^{\infty}(X_c(k)\cos(k\omega_0 t) + X_s(k)\sin(k\omega_0 t)),$$

where $\omega_0 = \frac{2\pi}{T}$ and $X_c(0)$, $X_c(k)$, and $X_s(k)$, the FS coefficients of the dc, cosine, and sine components of $x(t)$, respectively, are defined as

$$X_c(0) = \frac{1}{T}\int_{t_1}^{t_1+T} x(t)\,dt,$$

$$X_c(k) = \frac{2}{T}\int_{t_1}^{t_1+T} x(t)\cos(k\omega_0 t)\,dt, \quad k = 1, 2, \ldots, \infty$$

$$X_s(k) = \frac{2}{T}\int_{t_1}^{t_1+T} x(t)\sin(k\omega_0 t)\,dt, \quad k = 1, 2, \ldots, \infty$$

and t_1 is arbitrary. Derive the expressions for the coefficients using trigonometric identities.

6.2 Expand $x(t)$ and find the three forms of its FS coefficients without evaluating any integral. What is the fundamental frequency ω_0?

6.2.1 $x(t) = \cos(t)$.
6.2.2 $x(t) = \cos^2(t)$.
6.2.3 $x(t) = \cos^3(t)$.
* 6.2.4 $x(t) = \cos^4(t)$.
6.2.5 $x(t) = \cos^5(t)$.
6.2.6 $x(t) = \sin(t)$.
6.2.7 $x(t) = \sin^2(t)$.
6.2.8 $x(t) = \sin^3(t)$.
6.2.9 $x(t) = \sin^4(t)$.
6.2.10 $x(t) = \sin^5(t)$.

6.3 Find the three forms of the FS coefficients of $x(t)$ without evaluating any integral. What is the fundamental frequency ω_0?

6.3.1 $x(t) = 3 + \cos(2\pi t - \frac{\pi}{3}) - 2\sin(4\pi t + \frac{\pi}{6})$.
6.3.2 $x(t) = -1 - 2\sin(4\pi t - \frac{\pi}{6}) + 6\sin(8\pi t - \frac{\pi}{3})$.
6.3.3 $x(t) = 2 - 3\sqrt{2}\cos(2\pi t - \frac{\pi}{4}) + 2\sin(6\pi t + \frac{\pi}{3})$.
6.3.4 $x(t) = -3 + \sqrt{3}\cos(2\pi t) - \sin(2\pi t) + \sqrt{2}\cos(8\pi t) - \sqrt{2}\sin(8\pi t)$.
* 6.3.5 $x(t) = 1 + \frac{1}{j2}e^{j(2\pi t + \frac{\pi}{3})} + e^{j(6\pi t - \frac{\pi}{6})} - \frac{1}{j2}e^{-j(2\pi t + \frac{\pi}{3})} + e^{-j(6\pi t - \frac{\pi}{6})}$.

6.4 Find the trigonometric form of the FS coefficients. What is the fundamental frequency ω_0?

6.4.1 $x(t) = 1 + 2\sin\left(\frac{4}{7}t\right) + 3\cos\left(\frac{2}{3}t\right)$.
* 6.4.2 $x(t) = 2 - 5\cos\left(\frac{3}{7}t\right) - 2\sin\left(\frac{2}{9}t\right)$.
6.4.3 $x(t) = -3 + 2\cos\left(\frac{3}{7}t\right) - \sin\left(\frac{1}{3}t\right)$.

6.5 Find the FS of a periodic pulse train of period T, defined over one period as

$$x(t) = \begin{cases} \frac{1}{a} & \text{for } |t| < \frac{a}{2} \\ 0 & \text{for } \frac{a}{2} < |t| < \frac{T}{2} \end{cases}$$

Apply a limiting process, as $a \to 0$, to the pulse train and its FS to obtain the FS of the periodic impulse train of period T:

$$x(t) = \sum_{n=-\infty}^{\infty} \delta(t - nT)$$

6.6 Find the FS coefficients, using the time-domain convolution property, of $y(t) = x(t) * x(t)$, the convolution of $x(t)$ with itself, with $x(t)$ defined over a period as

$$x(t) = \begin{cases} A & 0 < t < \frac{T}{2} \\ 0 & \frac{T}{2} < t < T \end{cases}$$

6.7 Find the FS coefficients of $z(t) = x(t)y(t)$ with period equal to the common period of $x(t)$ and $y(t)$, where $x(t) = 2\sin(t) + 4\cos(3t)$ and $y(t) = 6\cos(2t)$, using the frequency-domain convolution property. Verify that the FS coefficients represent $z(t) = x(t)y(t)$.

6.8 Find the trigonometric FS representation of the periodic full-wave rectified sine wave defined over a period as

$$x(t) = \begin{cases} A\sin(\omega_0 t) & 0 \le t < \frac{T}{2} \\ A\sin(\omega_0 (t - \frac{T}{2})) & \frac{T}{2} \le t < T \end{cases}, \quad \omega_0 = \frac{2\pi}{T}$$

using the time-differentiation property.

6.9 Using the time-differentiation property, find the FS coefficients of the periodic signal $x(t)$ defined over a period.

6.9.1

$$x(t) = \begin{cases} -A & 0 < t < \frac{T}{2} \\ A & -\frac{T}{2} < t < 0 \end{cases}$$

6.9.2

$$x(t) = \begin{cases} A + 2 & 0 < t < \frac{T}{2} \\ -A + 2 & -\frac{T}{2} < t < 0 \end{cases}$$

6.9.3

$$x(t) = \begin{cases} 0 & 0 < t < \frac{T}{2} \\ -(t + \frac{T}{2}) & -\frac{T}{2} < t < 0 \end{cases}$$

6.9.4

$$x(t) = \frac{t}{T}, \quad 0 < t < T$$

6.9.5

$$x(t) = \begin{cases} \frac{2t}{T} & 0 \le t \le \frac{T}{2} \\ 2 - \frac{2t}{T} & \frac{T}{2} < t < T \end{cases}$$

* 6.9.6

$$x(t) = \sin(2t),\ 0 < t < \frac{\pi}{2}$$

6.10 Find the trigonometric FS representation, using Eq. (6.2), of the periodic signal defined over a period as $x(t) = \frac{3}{2}t,\ 0 < t < 2$. Using the results, find the sum of the infinite series

$$1 - \frac{1}{3} + \frac{1}{5} - \frac{1}{7} + \cdots$$

Verify Parseval's theorem.

Find the power of the frequency components of the signal up to the (i) third harmonic and (ii) fifth harmonic.

Approximate the trigonometric FS coefficients using the DFT with $N = 4$. Verify that they are the same using Eq. (6.8).

Find the location and the magnitude of a largest deviation due to Gibbs phenomenon if the signal is reconstructed using up to the third harmonic.

Deduce the trigonometric FS representation of the signals $x(t-1)$ and $2x(t)-3$.

* **6.11** Find the trigonometric FS representation, using Eq. (6.2), of the periodic signal defined over a period as

$$x(t) = \begin{cases} \frac{1}{2}t & 0 \le t < 2 \\ 2(1 - \frac{t}{4}) & 2 \le t < 4 \end{cases}$$

Using the results, find the sum of the infinite series

$$1 + \frac{1}{9} + \frac{1}{25} + \frac{1}{49} + \cdots$$

Verify Parseval's theorem. Find the power of the frequency components of the signal up to the (i) third harmonic and (ii) fifth harmonic.

Approximate the trigonometric FS coefficients using the DFT with $N = 4$. Verify that they are the same using Eq. (6.8).

Deduce the trigonometric FS representation of the signals $x(t+2)$ and $3x(t)-2$.

6.12 Find the trigonometric FS representation of the periodic half-wave rectified sine wave defined over a period as

$$x(t) = \begin{cases} A\sin(\frac{2\pi}{T}t) & 0 \le t < \frac{T}{2} \\ 0 & \frac{T}{2} \le t < T \end{cases}$$

using the frequency-domain convolution property. Verify Parseval's theorem. Find the power of the frequency components of the signal up to the third harmonic.

Approximate the trigonometric FS coefficients using the DFT with $N = 4$. Verify that they are the same using Eq. (6.8).

Deduce the FS coefficients of $x(t) + x(t - \frac{T}{2})$ and $x(t) - x(t - \frac{T}{2})$.

6.13 Find the trigonometric FS representation of the periodic half inverted cosine wave defined over a period as

$$x(t) = \begin{cases} -A\cos(\frac{2\pi}{T}t) & 0 < t < \frac{T}{2} \\ 0 & \frac{T}{2} < t < T \end{cases}$$

using the frequency-domain convolution property. Verify Parseval's theorem. Find the power of the frequency components of the signal up to the fifth harmonic. Find the FS coefficients using the DFT with $N = 4$. Verify that they are the same using Eq. (6.8). Find the location and the magnitude of a largest deviation due to Gibbs phenomenon if the signal is reconstructed using up to the third harmonic. Deduce the FS coefficients of $x(t) - x(t - \frac{T}{2})$.

6.14 Using the result of Exercise 6.13, deduce the FS representation of periodic two inverted half cosine waves defined over a period as

$$x(t) = \begin{cases} -A\cos(\frac{2\pi}{T}t) & 0 < t < \frac{T}{2} \\ A\cos(\frac{2\pi}{T}t) & \frac{T}{2} < t < T \end{cases}$$

* **6.15** Find the response of the system governed by the differential equation:

$$\frac{dy(t)}{dt} + y(t) = e^{j2t} + e^{j3t}$$

6.16 Find the response of the system governed by the differential equation:

$$\frac{dy(t)}{dt} + 2y(t) = 2 - 3\sin(t) + \cos\left(2t + \frac{\pi}{3}\right).$$

Chapter 7
The Discrete-Time Fourier Transform

A continuum of discrete sinusoids over a finite frequency range is used as the basis signals in the DTFT to analyze aperiodic discrete signals. Compared with the DFT, as the discrete aperiodic time-domain waveform contains infinite number of samples, the frequency increment of the periodic spectrum of the DFT tends to zero and the spectrum becomes continuous. The period is not affected since it is determined by the sampling interval in the time-domain. An alternate view of the DTFT is that it is the same as the FS with the roles of time- and frequency-domain functions interchanged.

In Sect. 7.1, the DTFT and its inverse and the dual relationship between the DTFT and the FS are derived. The properties of the DTFT are presented in Sect. 7.2. The approximation of the DTFT by the DFT is described in Sect. 7.3. Some typical applications of the DTFT are presented in Sect. 7.4.

7.1 The Discrete-Time Fourier Transform

7.1.1 The DTFT as the Limiting Case of the DFT

In the last chapter, we found that the FS is the limiting case of the DFT as the sampling interval in the time-domain tends to zero with the period of the waveform fixed. In this chapter, we find that the DTFT is the limiting case of the DFT as the period in the time-domain tends to infinity with the sampling interval fixed. With a predetermined sampling interval, the effective frequency range of the spectrum is fixed.

Consider the DFT magnitude spectrum $|X(k)|$ of $x(n)$ with $N = 5$ samples, shown, respectively, in Fig. 7.1b and a. The frequency increment of the spectrum is $\frac{2\pi}{5}$. Even if a signal $x(n)$ is aperiodic, in the DFT computation, periodicity is assumed. Therefore, only a set of samples of the continuous spectrum of an

D. Sundararajan, *Signals and Systems*,
https://doi.org/10.1007/978-3-031-19377-4_7

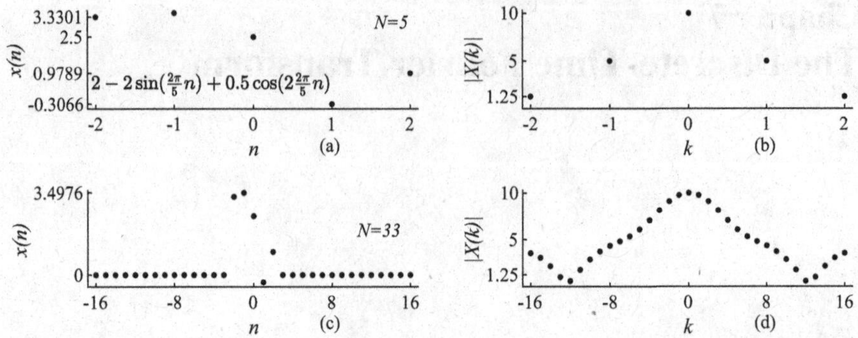

Fig. 7.1 (**a**) The time-domain function $x(n)$ with $N = 5$ and (**b**) the corresponding DFT magnitude spectrum, $|X(k)|$; (**c**) same as (**a**) with $N = 33$ and (**d**) the corresponding DFT magnitude spectrum, $|X(k)|$

aperiodic $x(n)$ is computed by the DFT. By zero-padding on either side of $x(n)$, we have made the signal longer with $N = 33$ samples, as shown in Fig. 7.1c. Its spectrum is shown in Fig. 7.1d, which is denser (frequency increment $\frac{2\pi}{33}$) compared with that in Fig. 7.1b. Eventually, as N tends to infinity, we get the aperiodic discrete signal and its periodic continuous spectrum. The spectrum is always periodic with the same period, 2π, as the sampling interval is fixed at $T_s = 1$.

The foregoing argument can be, mathematically, put as follows. The IDFT of $X(k)$, $-N \leq k \leq N$ is defined as

$$x(n) = \frac{1}{2N+1} \sum_{k=-N}^{N} X(k) e^{j\frac{2\pi}{(2N+1)}nk}, \quad n = 0, \pm 1, \pm 2, \ldots, \pm N$$

Substituting the DFT expression for $X(k)$, we get

$$x(n) = \frac{1}{2N+1} \sum_{k=-N}^{N} \left(\sum_{l=-N}^{N} x(l) e^{-j\frac{2\pi}{2N+1}lk} \right) e^{j\frac{2\pi}{2N+1}nk}$$

As N tends to ∞, due to zero-padding of $x(n)$, $\frac{2\pi}{2N+1}k$ becomes a continuous variable ω, differential $d\omega$ formally replaces $\frac{2\pi}{2N+1}$, and $2N + 1 = \frac{2\pi}{d\omega}$. The outer summation becomes an integral with limits $-\pi$ and π (actually any continuous interval of 2π). The limits of the inner summation can be written as $-\infty$ and ∞. Therefore, the DTFT $X(e^{j\omega})$ of the signal $x(n)$ is defined as

$$X(e^{j\omega}) = \sum_{n=-\infty}^{\infty} x(n) e^{-j\omega n} \tag{7.1}$$

The DTFT is commonly written as $X(e^{j\omega})$ instead of $X(j\omega)$ to emphasize the fact that it is a periodic function of ω. The inverse DTFT $x(n)$ of $X(e^{j\omega})$ is defined as

$$x(n) = \frac{1}{2\pi}\int_{-\pi}^{\pi} X(e^{j\omega})e^{j\omega n}d\omega,\ n = 0, \pm 1, \pm 2, \ldots \tag{7.2}$$

When deriving closed-form expressions for $x(n)$ or $X(e^{j\omega})$,

$$X(e^{j0}) = \sum_{n=-\infty}^{\infty} x(n),\ X(e^{j\pi}) = \sum_{n=-\infty}^{\infty} (-1)^n x(n),\ x(0) = \frac{1}{2\pi}\int_{-\pi}^{\pi} X(e^{j\omega})d\omega,$$

which can be easily evaluated, are useful to check their correctness.

The analysis equation of the DTFT is a summation, and the synthesis equation is an integral. In these equations, it is assumed that the sampling interval of the time-domain signal, T_s, is 1 s. For other values of T_s, only scaling of the frequency axis is required. However, the DTFT equations can also be expressed including T_s as

$$X(e^{j\omega T_s}) = \sum_{n=-\infty}^{\infty} x(nT_s)e^{-jn\omega T_s} \tag{7.3}$$

$$x(nT_s) = \frac{1}{\omega_s}\int_{-\frac{\omega_s}{2}}^{\frac{\omega_s}{2}} X(e^{j\omega T_s})e^{jn\omega T_s}\,d\omega,\ n = 0, \pm 1, \pm 2, \ldots, \tag{7.4}$$

where $\omega_s = \frac{2\pi}{T_s}$. The DTFT represents a discrete aperiodic signal, $x(nT_s)$, with T_s seconds between consecutive samples, as integrals of a continuum of complex sinusoids $e^{jn\omega T_s}$ (amplitude $\frac{1}{\omega_s}X(e^{j\omega T_s})d\omega$) over the finite frequency range $-\frac{\omega_s}{2}$ to $\frac{\omega_s}{2}$ (over one period of $X(e^{j\omega T_s})$). $X(e^{j\omega T_s})$ is periodic of period $\omega_s = \frac{2\pi}{T_s}$, since $e^{-jn\omega T_s} = e^{-jn(\omega+\frac{2\pi}{T_s})T_s}$. Therefore, the integration in Eq. (7.4) can be evaluated over any interval of length ω_s. As the amplitude, $\frac{1}{\omega_s}X(e^{j\omega T_s})d\omega$, of the constituent sinusoids of a signal is infinitesimal, the spectral density $X(e^{j\omega T_s})$, which is proportional to the spectral amplitude, represents the frequency content of a signal. Although the DTFT is the spectral density of a signal, it is still called the spectrum. Therefore, the DTFT spectrum is a relative amplitude spectrum.

The summation in Eq. (7.3) converges uniformly to $X(e^{j\omega T_s})$, if $x(nT_s)$ is absolutely summable, that is $\sum_{n=-\infty}^{\infty} |x(nT_s)| < \infty$. The summation converges in the least squares error sense, if $x(nT_s)$ is square summable, that is $\sum_{n=-\infty}^{\infty} |x(nT_s)|^2 < \infty$ (e.g., $x(n)$ in Example 7.2). Gibbs phenomenon is also common to all forms of Fourier analysis whenever reconstructing a continuous waveform, with one or more

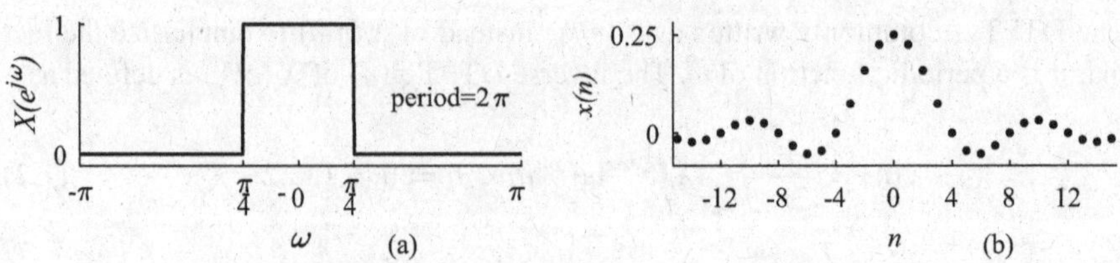

Fig. 7.2 (**a**) One period of a DTFT spectrum; (**b**) the corresponding aperiodic discrete signal

discontinuities, in either domain. In the case of the DTFT, Gibbs phenomenon occurs in the frequency-domain as the spectrum is a continuous function.

Example 7.1 Find the DTFT of the unit-impulse signal $x(n) = \delta(n)$.

$$X(e^{j\omega}) = \sum_{n=-\infty}^{\infty} \delta(n)e^{-j\omega n} = 1 \quad \text{and} \quad \delta(n) \Longleftrightarrow 1$$

That is, the unit-impulse signal is composed of complex sinusoids of all frequencies from $\omega = -\pi$ to $\omega = \pi$ in equal proportion. ∎

Example 7.2 One period of a DTFT spectrum, shown in Fig. 7.2a, is given as $X(e^{j\omega}) = u(\omega + \frac{\pi}{4}) - u(\omega - \frac{\pi}{4}),\ -\pi < \omega \leq \pi$. Find the corresponding $x(n)$.

Solution As the spectrum is even-symmetric,

$$x(n) = \frac{1}{\pi}\int_{0}^{\frac{\pi}{4}} \cos(\omega n)\, d\omega = \frac{\sin(\frac{\pi n}{4})}{n\pi}, \quad -\infty < n < \infty$$

The time-domain signal $x(n)$ is shown in Fig. 7.2b. ∎

The function of the form $x(n) = \frac{\sin(\frac{\pi n}{4})}{n\pi}$, shown in Fig. 7.2b, is called the sinc function that occurs often in signal and system analysis. It is an even function of n. At $n = 0$, the peak value is $\frac{1}{4}$, as $\lim_{\theta \to 0} \sin(\theta) = \theta$. The zeros of the sinc function occur whenever the argument of the sine function in the numerator is equal to $\pm\pi, \pm 2\pi, \pm 3\pi, \ldots$. For the specific case, the zeros occur whenever n is an integral multiple of four. As $a \to 0$, $\frac{\sin(an)}{an}$ degenerates into a DC function with amplitude one, as the zeros move to infinity. The sinc function is an energy signal, as it is square summable. However, it is not absolutely summable.

Example 7.3 Find the DTFT of the signal $x(n) = a^n u(n),\ |a| < 1$.

$$X(e^{j\omega}) = \sum_{n=-\infty}^{\infty} a^n u(n)e^{-j\omega n} = \sum_{n=0}^{\infty}(ae^{-j\omega})^n = \frac{1}{1 - ae^{-j\omega}}, \ |a| < 1$$ ∎

The DTFT of some frequently used signals, which are neither absolutely nor square summable, such as the unit-step, is obtained by applying a limiting process to appropriate signals so that they degenerate into these signals in the limit. The limit of the corresponding transform is the transform of the signal under consideration, as presented in the next example.

Example 7.4 Find the DTFT of the unit-step signal $x(n) = u(n)$.

As this signal is not absolutely or square summable, its DTFT is derived as that of the limiting form of the signal $a^n u(n)$, as $a \to 1$.

$$X(e^{j\omega}) = \lim_{a\to 1} \frac{1}{1 - ae^{-j\omega}} = \lim_{a\to 1} \left(\frac{1 - a\cos(\omega)}{1 - 2a\cos(\omega) + a^2} - j\frac{a\sin(\omega)}{1 - 2a\cos(\omega) + a^2} \right)$$

The real and imaginary parts of the DTFT spectrum of the signal $0.8^n u(n)$, shown in Fig. 7.3a, are shown, respectively, in Fig. 7.3c and e. Figure 7.3d and f shows the same for the signal $0.99^n u(n)$, shown in Fig. 7.3b. The real part of the spectrum is even, and the imaginary part is odd. The area enclosed by the real part of the spectrum is a constant (2π) independent of the value a, the base of the exponential signal. This is so because, from the inverse DTFT with $n = 0$,

$$x(0) = 1 = \frac{1}{2\pi} \int_{-\pi}^{\pi} \frac{1 - a\cos(\omega)}{1 - 2a\cos(\omega) + a^2} e^{j\omega 0} d\omega = \frac{1}{2\pi} \int_{-\pi}^{\pi} \frac{1 - a\cos(\omega)}{1 - 2a\cos(\omega) + a^2} d\omega$$

As can be seen from the figures, the real part of the spectrum becomes more peaked as $a \to 1$. Eventually, the spectrum consists of a strictly continuous component (except at $\omega = 0$) and an impulsive component. The constant area 2π is split up, as the function evaluates to 0.5 for $\omega \neq 0$ with $a \to 1$, between these components, and the spectrum becomes

$$X(e^{j\omega}) = \pi\delta(\omega) + \frac{1}{1 - e^{-j\omega}} \text{ and } u(n) \Longleftrightarrow \pi\delta(\omega) + \frac{1}{1 - e^{-j\omega}}$$ ∎

Example 7.5 Find the DTFT of the DC signal, $x(n) = 1$.

The DC signal can be written as $x(n) = u(n) + u(-n) - \delta(n)$. Due to time reversal property, if $x(n) \Longleftrightarrow X(e^{j\omega})$ then $x(-n) \Longleftrightarrow X(e^{-j\omega})$. The DTFT of $u(-n)$ is obtained from that of $u(n)$ by replacing ω by $-\omega$. Therefore, the DTFT of the DC signal is

$$\pi\delta(\omega) + \frac{1}{1 - e^{-j\omega}} + \pi\delta(-\omega) + \frac{1}{1 - e^{j\omega}} - 1 = 2\pi\delta(\omega)$$

Explicitly showing the periodicity of the DTFT spectrum, we get

$$1 \Longleftrightarrow 2\pi \sum_{k=-\infty}^{\infty} \delta(\omega + 2k\pi)$$

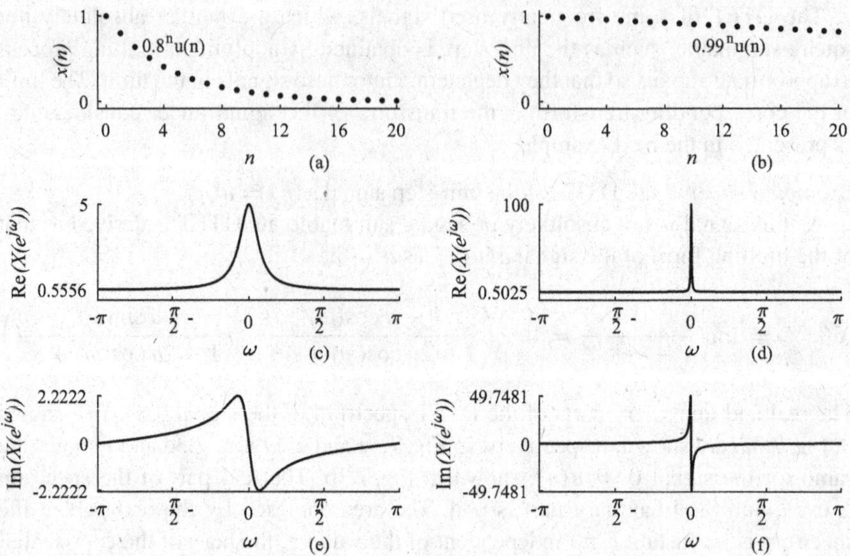

Fig. 7.3 (**a**) $x(n) = 0.8^n u(n)$; (**b**) $x(n) = 0.99^n u(n)$; (**c**) the real part of the DTFT spectrum of the signal in (**a**) and (**e**) its imaginary part; (**d**) the real part of the DTFT spectrum of the signal in (**b**) and (**f**) its imaginary part

That is, the DC signal, which is the complex exponential $x(n) = e^{j\omega n}$ with $\omega = 0$, has nonzero spectral component only at the single frequency $\omega = 0$. Note that 2π in the spectral value is a constant factor. ∎

Consider the transform pair

$$2u(n) - 1 \Longleftrightarrow 2\left(\pi\delta(\omega) + \frac{1}{1 - e^{-j\omega}}\right) - 2\pi\delta(\omega) = \frac{2}{1 - e^{-j\omega}}$$

$$2\{\ldots, 0, 0, 0, \check{1}, 1, 1, 1, \ldots\} - \{\ldots, 1, 1, 1, \check{1}, 1, 1, 1, \ldots\} = \{\ldots, -1, -1, -1, \check{1}, 1, 1, 1, \ldots\}$$

The sign or signum function $\text{sgn}(n)$ is defined as

$$\text{sgn}(n) = \begin{cases} 1 \text{ for } n \geq 0 \\ -1 \text{ for } n < 0 \end{cases}$$

There is an odd version of the sign or signum function defined as

$$sgn(n) = 2u(n) - 1 - \delta(n) \Longleftrightarrow 2\left(\pi\delta(\omega) + \frac{1}{1 - e^{-j\omega}}\right) - 2\pi\delta(\omega) - 1 = \frac{2}{1 - e^{-j\omega}} - 1$$

$$2\{\ldots, 0, 0, 0, \check{1}, 1, 1, 1, \ldots\} - \{\ldots, 1, 1, 1, \check{1}, 1, 1, 1, \ldots\} - \{\ldots, 0, 0, 0, \check{1}, 0, 0, 0, \ldots\} =$$

$$\{\ldots, -1, -1, -1, \check{0}, 1, 1, 1, \ldots\}$$

7.1.2 The Dual Relationship between the DTFT and the FS

The DTFT is the same as the FS with the roles of time- and frequency-domain functions interchanged. The analysis equation, with period of the time-domain waveform T and the fundamental frequency $\omega_0 = \frac{2\pi}{T}$, of the FS is

$$X_{cs}(k\omega_0) = \frac{1}{T}\int_{-\frac{T}{2}}^{\frac{T}{2}} x(t)e^{-jk\omega_0 t}\,dt,\ k = 0, \pm1, \pm2, \ldots$$

Replacing ω_0 by T_s, T by $\omega_s = \frac{2\pi}{T_s}$, ω by t, t by ω, and k by $-k$ in this equation, we get

$$X_{cs}(-kT_s) = \frac{1}{\omega_s}\int_{-\frac{\omega_s}{2}}^{\frac{\omega_s}{2}} x(\omega)e^{jk\omega T_s}\,d\omega,\ k = 0, \pm1, \pm2, \ldots$$

This equation is the same as the inverse DTFT with $x(kT_s) = X_{cs}(-kT_s)$ and $X(e^{j\omega T_s}) = x(\omega)$. Due to this similarity,

$$x(kT_s) \Longleftrightarrow X(e^{j\omega T_s}) \quad \text{implies} \quad X(e^{jtT_s}) = x(t) \Longleftrightarrow x(-kT_s) = X_{cs}(k\omega_0)$$

For the same periodic waveform, we get two sets of FS coefficients related by the time reversal operation because the periodic waveform occurs in the frequency-domain in the case of the DTFT and in the time-domain in the case of the FS. It is due to convention, we use complex exponential with negative exponent in the forward transform definitions of the FS and the DTFT.

Consider the signal $x(t) = \sin(3t)$ shown in Fig. 7.4a and the corresponding FS coefficients $X_{cs}(k\omega_0) = X_{cs}(\pm3) = \mp j0.5$ shown in Fig. 7.4b. From the FS synthesis equation, $-0.5je^{j3t} + 0.5je^{-j3t} = \sin(3t)$. Consider the spectrum $X(e^{j\omega T_s}) = \sin(3\omega)$ shown in Fig. 7.4c and the corresponding $x(kT_s)$, $(x(\pm3) = \pm j0.5)$, shown in Fig. 7.4d. From the DTFT analysis equation, $0.5je^{-j3\omega} - 0.5je^{j3\omega} = \sin(3\omega)$.

7.1.3 The DTFT of a Discrete Periodic Signal

A periodic signal $x(n)$ is reconstructed using its DFT coefficients $X(k)$ as

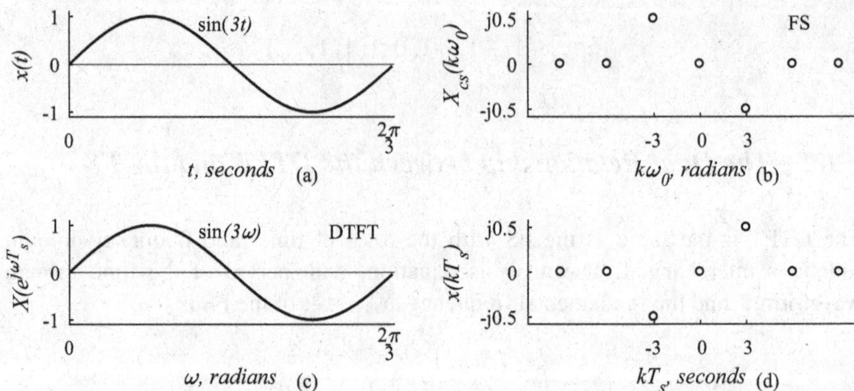

Fig. 7.4 (**a**) One period of the periodic time-domain function $x(t) = \sin(3t)$ and (**b**) the corresponding FS spectrum, $X_{cs}(k\omega_0)$; (**c**) one period of the periodic frequency-domain function $X(e^{j\omega T_s}) = \sin(3\omega)$ and (**d**) the corresponding inverse DTFT, $x(kT_s)$, which is the time reversal of $X_{cs}(k\omega_0)$ in (**b**)

$$x(n) = \frac{1}{N}\sum_{k=0}^{N-1} X(k)e^{jk\omega_0 n}, \quad \omega_0 = \frac{2\pi}{N}$$

Since the DTFT of $e^{jk\omega_0 n}$ is $2\pi\delta(\omega - k\omega_0)$, we get, from the linearity property of the DTFT, one period of the DTFT $X(e^{j\omega})$ of $x(n)$ as

$$X(e^{j\omega}) = \frac{2\pi}{N}\sum_{k=0}^{N-1} X(k)\delta(\omega - k\omega_0)$$

Therefore, the DTFT of a periodic signal is a periodic train of impulses with strength $\frac{2\pi}{N}X(k)$ at $\frac{2\pi}{N}k$ with period 2π.

For example, the DFT of $\cos(\frac{2\pi}{4}n)$ is $\{X(0) = 0, X(1) = 2, X(2) = 0, X(3) = 2\}$ with $N = 4$. One period of the DTFT $X(e^{j\omega})$ is given as $\{X(e^{j0}) = 0, X(e^{j\frac{2\pi}{4}}) = \pi\delta(\omega - \frac{2\pi}{4}), X(e^{j2\frac{2\pi}{4}}) = 0, X(e^{j3\frac{2\pi}{4}}) = \pi\delta(\omega - 3\frac{2\pi}{4})\}$.

7.1.4 Determination of the DFT from the DTFT

The DTFT of a finite sequence $x(n)$, starting from $n = n_0$, of length N is given as

$$X(e^{j\omega}) = \sum_{n=n_0}^{n_0+N-1} x(n)e^{-jn\omega}$$

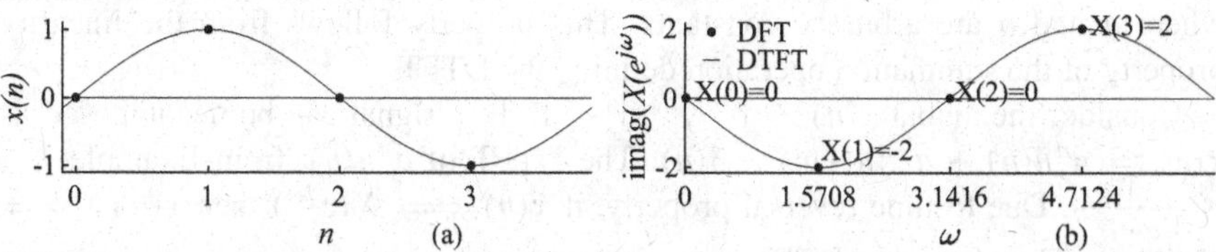

Fig. 7.5 (**a**) Sinusoid $\sin(\frac{2\pi}{4}n)$; (**b**) DFT coefficients as the samples of the DTFT

The DFT of $x(n)$ is given as

$$X(k) = \sum_{n=n_0}^{n_0+N-1} x(n)e^{-jk\omega_0 n}, \; \omega_0 = \frac{2\pi}{N}$$

Comparing the DFT and DTFT definitions of the signal, we get

$$X(k) = X(e^{j\omega})|_{\omega=k\omega_0} = X(e^{jk\omega_0})$$

The DTFT spectrum is evaluated at all frequencies along the unit-circle in the complex plane, whereas the DFT spectrum is the set of N samples of the DTFT spectrum at intervals of $\frac{2\pi}{N}$.

Let the four samples of a signal $x(n)$ be defined as $\{x(-2) = 0, x(-1) = -1, x(0) = 0, x(1) = 1\}$. The DTFT of $x(n)$ is $X(e^{j\omega}) = -e^{j\omega} + e^{-j\omega} = -j2\sin(\omega)$. The set of samples of $X(e^{j\omega})$, $\{X(0) = 0, X(1) = -j2, X(2) = 0, X(3) = j2\}$, at $\omega = 0$, $\omega = \frac{2\pi}{4}$, $\omega = 2\frac{2\pi}{4}$, and $\omega = 3\frac{2\pi}{4}$ is the DFT of $x(n) = \sin(2\pi n/4)$. Figure 7.5a and b shows, respectively, the discrete sinusoid $\sin(\frac{2\pi}{4}n)$ and its DFT coefficients as the samples of the DTFT.

7.2 Properties of the Discrete-Time Fourier Transform

Properties present the frequency-domain effect of time-domain characteristics and operations on signals and vice versa. In addition, they are used to find new transform pairs more easily.

7.2.1 Linearity

The DTFT of a linear combination of a set of signals is the same linear combination of their individual DTFT. That is,

$$x(n) \Longleftrightarrow X(e^{j\omega}), \quad y(n) \Longleftrightarrow Y(e^{j\omega}), \quad ax(n) + by(n) \Longleftrightarrow aX(e^{j\omega}) + bY(e^{j\omega}),$$

where a and b are arbitrary constants. This property follows from the linearity property of the summation operation defining the DTFT.

Consider the signal $x(n) = a^{|n|}, \ |a| < 1$. This signal can be decomposed as $x(n) = a^n u(n) + a^{-n}u(-n) - \delta(n)$. The DTFT of $a^n u(n)$, from Example 7.3, is $\frac{1}{1-ae^{-j\omega}}$. Due to time reversal property, if $x(n) \Longleftrightarrow X(e^{j\omega})$ then $x(-n) \Longleftrightarrow X(e^{-j\omega})$. Therefore, the DTFT of $a^{-n}u(-n)$ is obtained from that of $a^n u(n)$ as $\frac{1}{1-ae^{j\omega}}$. The DTFT of the signal $x(n) = a^{|n|}, \ |a| < 1$, due to linearity property, is

$$X(e^{j\omega}) = \frac{1}{1 - ae^{-j\omega}} + \frac{1}{1 - ae^{j\omega}} - 1 = \frac{1 - a^2}{1 - 2a\cos(\omega) + a^2}$$

7.2.2 *Time Shifting*

When we shift a signal, the shape remains the same, but the signal is relocated. The shift of a typical spectral component, $X(e^{j\omega_a})e^{j\omega_a n}$, by an integral number of sample intervals, n_0, to the right results in the exponential, $X(e^{j\omega_a})e^{j\omega_a(n-n_0)} = e^{-j\omega_a n_0}X(e^{j\omega_a})e^{j\omega_a n}$. That is, a delay of n_0 results in changing the phase of the exponential by $-\omega_a n_0$ radians without changing its amplitude. Therefore, if the transform of a time-domain function $x(n)$ is $X(e^{j\omega})$, then the transform of $x(n \pm n_0)$ is given by $e^{\pm j\omega n_0}X(e^{j\omega})$. That is,

$$x(n \pm n_0) \Longleftrightarrow e^{\pm j\omega n_0}X(e^{j\omega})$$

Consider the transform pair

$$(0.8)^n u(n) \Longleftrightarrow \frac{1}{1 - 0.8e^{-j\omega}}$$

Then, due to this property, we get the transform pair

$$(0.8)^{(n-2)}u(n-2) \Longleftrightarrow \frac{e^{-j2\omega}}{1 - 0.8e^{-j\omega}}$$

While this result is correct theoretically, to visualize, understand, and convince ourselves, we have to use the DFT and IDFT in this type of situations. As a checkup of this property using the IDFT, we found the 32 samples of the spectrum in the range $-\pi < \omega < \pi$ at uniform intervals and the result of the first 8 samples of the IDFT of the samples, shifted appropriately, are

$$\{1.0008, 0.8006, 0.6405, 0.5124, 0.4099, 0.3279, 0.2624, 0.2099\}$$

$$\{0.0012, 0.0010, 1.0008, 0.8006, 0.6405, 0.5124, 0.4099, 0.3279, \}$$

With more number of samples, the result will be more accurate. **For both understanding and practical use of the Fourier analysis, the DFT and IDFT are essential and should be employed to the full extent they are required.**

7.2.3 *Frequency Shifting*

The spectrum, $X(e^{j\omega})$, of a signal, $x(n)$, can be shifted by multiplying the signal by a complex exponential, $e^{\pm j\omega_0 n}$. The new spectrum is $X(e^{j(\omega \mp \omega_0)})$, since a spectral component $X(e^{j\omega_a})e^{j\omega_a n}$ of the signal multiplied by $e^{j\omega_0 n}$ becomes $X(e^{j\omega_a})e^{j(\omega_a+\omega_0)n}$ and the spectral value $X(e^{j\omega_a})$ occurs at $(\omega_a+\omega_0)$, after a delay of ω_0 radians. Therefore, we get

$$x(n)e^{\pm j\omega_0 n} \Longleftrightarrow X(e^{j(\omega \mp \omega_0)})$$

The complex exponential $e^{j\omega_0 n}$ can be considered as the product of the DC signal $x(n) = 1$ and $e^{j\omega_0 n}$. From the frequency shift property, we get the transform pair

$$1e^{j\omega_0 n} = e^{j\omega_0 n} \Longleftrightarrow 2\pi\delta(\omega - \omega_0)$$

The complex exponential is characterized by the single frequency ω_0 alone. Therefore, its spectrum is an impulse at ω_0 in the fundamental frequency range from $-\pi$ to π. As $\cos(\omega_0 n) = 0.5(e^{j\omega_0 n}+e^{-j\omega_0 n})$ and $\sin(\omega_0 n) = 0.5j(e^{-j\omega_0 n}-e^{j\omega_0 n})$,

$$\cos(\omega_0 n) \Longleftrightarrow \pi(\delta(\omega - \omega_0) + \delta(\omega + \omega_0))$$

$$\sin(\omega_0 n) \Longleftrightarrow j\pi(\delta(\omega + \omega_0) - \delta(\omega - \omega_0))$$

In Example 7.2, the frequency response of an ideal lowpass filter and its impulse response were presented. By shifting the frequency response, shown in Fig. 7.2a, by π radians, we get the frequency response of an ideal highpass filter with cutoff frequency $\pi - \frac{\pi}{4} = \frac{3\pi}{4}$, as shown in Fig. 7.6a. As the frequency response is shifted by π radians, we get the impulse response of the highpass filter by multiplying that of the lowpass filter by $e^{j\pi n} = (-1)^n$. That is, the impulse response of the highpass filter is $(-1)^n \frac{\sin(\frac{\pi n}{4})}{n\pi}$, shown in Fig. 7.6b.

7.2.4 *Convolution in the Time-Domain*

The convolution of signals $x(n)$ and $h(n)$ is defined, in Chap. 3, as

$$y(n) = \sum_{m=-\infty}^{\infty} x(m)h(n-m)$$

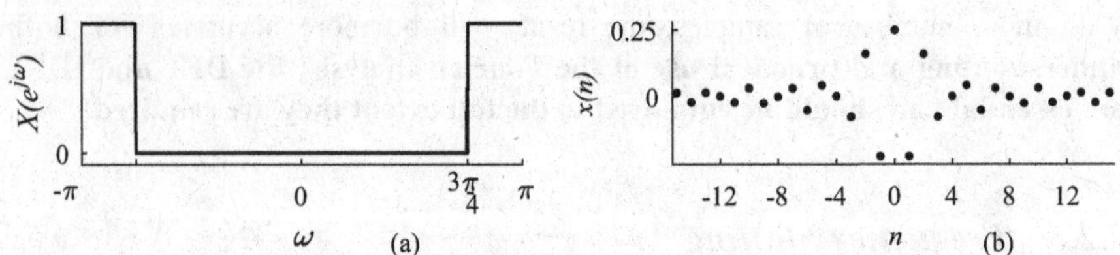

Fig. 7.6 (**a**) One period of a DTFT spectrum of a highpass filter; (**b**) the corresponding impulse response

The convolution of $h(n)$ with a complex exponential $e^{j\omega_0 n}$ is given as

$$\sum_{m=-\infty}^{\infty} h(m)e^{j\omega_0(n-m)} = e^{j\omega_0 n} \sum_{m=-\infty}^{\infty} h(m)e^{-j\omega_0 m} = H(e^{j\omega_0})e^{j\omega_0 n}$$

As an arbitrary $x(n)$ is reconstructed by the inverse DTFT as $x(n) = \frac{1}{2\pi}\int_{-\pi}^{\pi} X(e^{j\omega})e^{j\omega n}d\omega$, the convolution of $x(n)$ and $h(n)$ is given by $y(n) = \frac{1}{2\pi}\int_{-\pi}^{\pi} X(e^{j\omega})H(e^{j\omega})e^{j\omega n}d\omega$, where $X(e^{j\omega})$ and $H(e^{j\omega})$ are, respectively, the DTFT of $x(n)$ and $h(n)$. The inverse DTFT of $X(e^{j\omega})H(e^{j\omega})$ is the convolution of $x(n)$ and $h(n)$. Therefore, we get the transform pair

$$\sum_{m=-\infty}^{\infty} x(m)h(n-m) = \frac{1}{2\pi}\int_{-\pi}^{\pi} X(e^{j\omega})H(e^{j\omega})e^{j\omega n}d\omega \Longleftrightarrow X(e^{j\omega})H(e^{j\omega})$$

Consider the rectangular signal

$$x(n) = \begin{cases} 1 \text{ for } |n| \le 2 \\ 0 \text{ for } |n| > 2 \end{cases}$$

shown in Fig. 7.7a and its spectrum shown in Fig. 7.7b. The DTFT of the signal is $\frac{\sin(\frac{5\omega}{2})}{\sin(\frac{\omega}{2})}$. The DTFT of the convolution of this signal with itself is, due to the property, $\left(\frac{\sin(\frac{5\omega}{2})}{\sin(\frac{\omega}{2})}\right)^2$. As the convolution of a rectangular signal with itself is a triangular signal, this DTFT is that of a triangular signal. The triangular signal and its spectrum, which is positive for all ω, are shown, respectively, in Fig. 7.7c and d.

7.2.5 Convolution in the Frequency-Domain

The convolution of two functions in the frequency-domain corresponds to the multiplication of the inverse DTFT of the functions in the time-domain with a scale

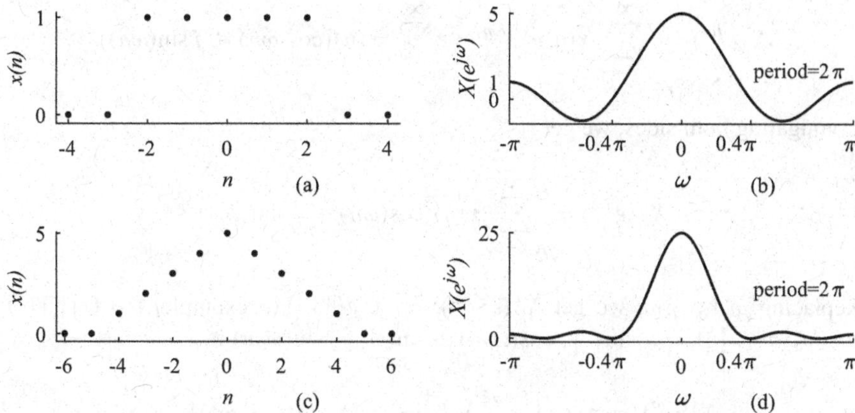

Fig. 7.7 **(a)** The rectangular signal and **(b)** its spectrum; **(c)** The triangular signal, which is the convolution of the signal in **(a)** with itself, and **(d)** its spectrum

factor. That is,

$$x(n)y(n) \Longleftrightarrow \sum_{n=-\infty}^{\infty} x(n)y(n)e^{-j\omega n} = \frac{1}{2\pi}\int_{0}^{2\pi} X(e^{jv})Y(e^{j(\omega-v)})dv$$

Note that this convolution is periodic, since the DTFT spectrum is periodic.

Consider finding the DTFT of the product of the signal $\frac{\sin(n)}{\pi n}$ with itself. One period of the DTFT of the signal is the rectangular function

$$\begin{cases} 1 \text{ for } |\omega| < 1 \\ 0 \text{ for } 1 < |\omega| < \pi \end{cases}$$

The convolution of this function with itself divided by 2π is the periodic triangular function, one period of which is defined as

$$\begin{cases} \frac{\omega+2}{2\pi} \text{ for } -2 \le \omega \le 0 \\ \frac{2-\omega}{2\pi} \text{ for } 0 < \omega \le 2 \\ 0 \quad \text{ for } -\pi \le \omega < -2 \text{ and } 2 < \omega < \pi \end{cases}$$

7.2.6 *Symmetry*

If a signal is real, then the real part of its spectrum $X(e^{j\omega})$ is even, and the imaginary part is odd, called the conjugate symmetry. The DTFT of a real signal is given by

$$X(e^{j\omega}) = \sum_{n=-\infty}^{\infty} x(n)e^{-j\omega n} = \sum_{n=-\infty}^{\infty} x(n)(\cos(\omega n) - j\sin(\omega n))$$

Conjugating both sides, we get

$$X^*(e^{j\omega}) = \sum_{n=-\infty}^{\infty} x(n)(\cos(\omega n) + j\sin(\omega n))$$

Replacing ω by $-\omega$, we get $X^*(e^{-j\omega}) = X(e^{j\omega})$. For example, the DTFT of $\cos(\omega_a(n - \frac{\pi}{4})) = \cos(\omega_a \frac{\pi}{4})\cos(\omega_a n) + \sin(\omega_a \frac{\pi}{4})\sin(\omega_a n)$ is

$$X(e^{j\omega}) = \pi\cos\left(\omega_a \frac{\pi}{4}\right)(\delta(\omega - \omega_a) + \delta(\omega + \omega_a))$$
$$+ j\pi\sin\left(\omega_a \frac{\pi}{4}\right)(\delta(\omega + \omega_a) - \delta(\omega - \omega_a))$$

If a signal is real and even, then its spectrum also is real and even. Since $x(n)\cos(\omega n)$ is even and $x(n)\sin(\omega n)$ is odd,

$$X(e^{j\omega}) = x(0) + 2\sum_{n=1}^{\infty} x(n)\cos(\omega n) \quad \text{and} \quad x(n) = \frac{1}{\pi}\int_0^{\pi} X(e^{j\omega})\cos(\omega n)d\omega$$

The DTFT of cosine function is an example of this symmetry.

If a signal is real and odd, then its spectrum is imaginary and odd. Since $x(n)\cos(\omega n)$ is odd and $x(n)\sin(\omega n)$ is even,

$$X(e^{j\omega}) = -j2\sum_{n=1}^{\infty} x(n)\sin(\omega n) \quad \text{and} \quad x(n) = \frac{j}{\pi}\int_0^{\pi} X(e^{j\omega})\sin(\omega n)d\omega$$

The DTFT of sine function is an example of this symmetry.

As the DTFT of a real and even signal is real and even and that of a real and odd is imaginary and odd, it follows that the real part of the DTFT, $\mathrm{Re}(X(e^{j\omega}))$, of an arbitrary real signal $x(n)$ is the transform of its even component $x_e(n)$ and $j\,\mathrm{Im}(X(e^{j\omega}))$ is that of its odd component $x_o(n)$.

7.2.7 *Time Reversal*

Let the spectrum of a signal $x(n)$ be $X(e^{j\omega})$. Then, $x(-n) \Longleftrightarrow X(e^{-j\omega})$. That is the time reversal of a signal results in its spectrum also reflected about the vertical axis at the origin. This result is obtained if we replace n by $-n$ and ω by $-\omega$ in the DTFT definition.

7.2.8 *Time Expansion*

As we have seen in Chap. 1, a signal is compressed or expanded by scaling operation. Consider the case of signal expansion. Let the spectrum of a signal $x(n)$ be $X(e^{j\omega})$. If we pad $x(n)$ with zeros to get $y(n)$ defined as

$$y(an) = x(n) \text{ for } -\infty < n < \infty \quad \text{and} \quad y(n) = 0 \quad \text{otherwise}$$

where $a \neq 0$ is any positive integer, then,

$$Y(e^{j\omega}) = X(e^{ja\omega})$$

The DTFT of the sequence $y(n)$ is given by

$$Y(e^{j\omega}) = \sum_{n=-\infty}^{\infty} y(n)e^{-j\omega n}$$

Since we have nonzero input values only if $n = ak,\ k = 0, \pm 1, \pm 2, \ldots$, we get

$$Y(e^{j\omega}) = \sum_{k=-\infty}^{\infty} y(ak)e^{-j\omega ak} = \sum_{k=-\infty}^{\infty} x(k)e^{-j\omega ak} = X(e^{ja\omega})$$

Therefore,

$$y(n) \Longleftrightarrow X(e^{ja\omega})$$

The spectrum is compressed. That is, the spectral value at ω in the spectrum of the signal occurs at $\frac{\omega}{a}$ in the spectrum of its expanded version. If a is negative, the spectrum is also frequency-reversed.

For example, the DTFT of the signal $x(n)$ shown in Fig. 7.7a with dots, with its only nonzero values given as $x(-1) = 1$ and $x(1) = 1$, is $X(e^{j\omega}) = e^{j\omega} + e^{-j\omega} = 2\cos(\omega)$. Using the theorem, we get the DTFT of $y(n)$ with $a = 2$, shown in Fig. 7.8a with unfilled circles, as

$$Y(e^{j\omega}) = X(e^{j2\omega}) = 2\cos(2\omega)$$

This result can be verified from the DTFT definition. The DTFT of the signal (solid line) and that of its expanded version (dashed line) are shown in Fig. 7.8b. Since the signal is expanded by a factor of two, the spectrum is compressed by a factor of two.

As an another example, consider the cosine signal $x(n) = \cos(\frac{2\pi}{8}n)$, shown in Fig. 7.8c by dots, and its DTFT

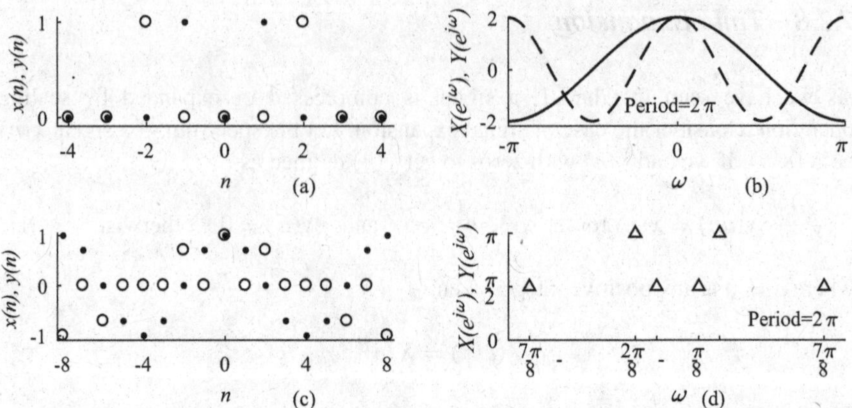

Fig. 7.8 (**a**) Signal $x(n)$ (dots) and its expanded version $y(n)$ (unfilled circles) with $a = 2$, and (**b**) the DTFT of $x(n)$ (solid line) and that of $y(n)$ (dashed line); (**c**) signal $x(n)$ (dots) and its expanded version $y(n)$ (unfilled circles) with $a = 2$, and (**d**) the DTFT of $x(n)$ (two impulses of strength π) and that of $y(n)$ (four impulses of strength $\frac{\pi}{2}$)

$$X(e^{j\omega}) = \pi \sum_{k=-\infty}^{\infty} \left(\delta\left(\omega - \frac{2\pi}{8} + 2\pi k\right) + \delta\left(\omega + \frac{2\pi}{8} + 2\pi k\right)\right),$$

shown in Fig. 7.7d with two impulses of strength π. The DTFT of $y(n)$ with $a = 2$, shown in Fig. 7.7c by unfilled circles, is

$$\begin{aligned} Y(e^{j\omega}) &= \pi \sum_{k=-\infty}^{\infty} \left(\delta\left(2\omega - \frac{2\pi}{8} + 2\pi k\right) + \delta\left(2\omega + \frac{2\pi}{8} + 2\pi k\right)\right) \\ &= \frac{\pi}{2} \sum_{k=-\infty}^{\infty} \left(\delta\left(\omega - \frac{\pi}{8} + \pi k\right) + \delta\left(\omega + \frac{\pi}{8} + \pi k\right)\right) \\ &= \frac{\pi}{2} \left(\left(\delta\left(\omega - \frac{\pi}{8}\right) + \delta\left(\omega + \frac{\pi}{8}\right)\right) + \left(\delta\left(\omega - \frac{7\pi}{8}\right) + \delta\left(\omega + \frac{7\pi}{8}\right)\right)\right), \end{aligned}$$

$-\pi < \omega \leq \pi$, shown in Fig. 7.8d with four impulses of strength $\frac{\pi}{2}$ in the fundamental frequency range from $-\pi$ to π. The expanded time-domain signal is reconstructed from its spectrum as follows:

$$y(n) = 0.5\cos\left(\frac{\pi}{8}n\right) + 0.5\cos\left(\left(\pi - \frac{\pi}{8}\right)n\right) = 0.5\cos\left(\frac{\pi}{8}n\right)(1 + (-1)^n) = \cos\left(\frac{\pi}{8}n\right)$$

for n even and $y(n)$ is zero otherwise.

7.2.9 *Frequency Differentiation*

Differentiating both sides of the DTFT defining equation, with respect to ω, we get the transform pair

$$(-jn)x(n) \Longleftrightarrow \frac{dX(e^{j\omega})}{d\omega} \quad \text{or} \quad (n)x(n) \Longleftrightarrow (j)\frac{dX(e^{j\omega})}{d\omega}$$

In general,

$$(-jn)^k x(n) \Longleftrightarrow \frac{d^k X(e^{j\omega})}{d\omega^k} \quad \text{or} \quad (n)^k x(n) \Longleftrightarrow (j)^k \frac{d^k X(e^{j\omega})}{d\omega^k}$$

This property is applicable only if the resulting signal satisfies the existence conditions of the DTFT. Consider the transform pair

$$\delta(n-2) \Longleftrightarrow e^{-j2\omega}$$

Using the property, we get the transform pair

$$n\delta(n-2) \Longleftrightarrow (j)(-j2)e^{-j2\omega} = 2e^{-j2\omega}$$

7.2.10 *Difference*

The derivative of a function is approximated by differences in the discrete case

$$y(n) = x(n) - x(n-1) \Longleftrightarrow Y(e^{j\omega}) = (1 - e^{-j\omega})X(e^{j\omega})$$

using the time shifting property.

7.2.11 *Summation*

The summation of a time-domain function, $x(n)$, can be expressed, in terms of its DTFT $X(e^{j\omega})$, as

$$y(n) = \sum_{l=-\infty}^{n} x(l) \Longleftrightarrow Y(e^{j\omega}) = \frac{X(e^{j\omega})}{(1 - e^{-j\omega})} + \pi X(e^{j0})\delta(\omega), \quad -\pi < \omega \leq \pi$$

The transform $\frac{X(e^{j\omega})}{(1-e^{-j\omega})} + \pi X(e^{j0})\delta(\omega)$ is the product of the transforms of $x(n)$ and $u(n)$ and corresponds to the convolution of $x(n)$ and $u(n)$ in the time-domain, which, of course, is equivalent to the sum of the values of $x(n)$ from $-\infty$ to n. The time-summation operation can be considered as the inverse of the time-differencing operation, if $X(e^{j0}) = 0$. This justifies the strictly continuous component of the spectrum. The impulsive component is required to take into account of the DC component of $x(n)$. This property is applicable only if the resulting signal satisfies the existence conditions of the DTFT.

Since the DTFT of unit-impulse is one and the unit-step function is a summation of the impulse, we get, using this property, the DTFT of $u(n)$, over one period, as

$$u(n) = \sum_{l=-\infty}^{n} \delta(l) \Longleftrightarrow \frac{1}{(1 - e^{-j\omega})} + \pi\delta(\omega), \quad -\pi < \omega \leq \pi$$

As an another example, consider the signal, shown in Fig. 7.9a, and the resulting signal, shown in Fig. 7.9b, obtained by summing it. The DTFT of the given signal is, from the DTFT definition, $1 + e^{-j\omega}$. Using the property, we get the DTFT of its summation as

$$\frac{1 + e^{-j\omega}}{1 - e^{-j\omega}} + 2\pi\delta(\omega), \quad -\pi < \omega \leq \pi$$

The summation of $x(n)$ is $y(n)$, shown in Fig. 7.9b along with its two components corresponding to the two terms of the transform. Note that $1 \Longleftrightarrow 2\pi\delta$.

7.2.12 Parseval's Theorem and the Energy Transfer Function

As the frequency-domain representation of a signal is an equivalent representation, the energy of a signal can also be expressed in terms of its spectrum.

$$E = \sum_{n=-\infty}^{\infty} |x(n)|^2 = \frac{1}{2\pi} \int_0^{2\pi} |X(e^{j\omega})|^2 d\omega$$

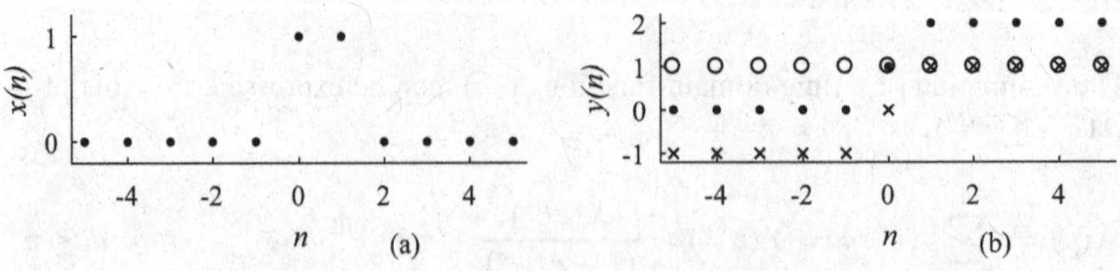

Fig. 7.9 (**a**) Signal $x(n) = u(n) - u(n-2)$; (**b**) $y(n) = \sum_{l=-\infty}^{n} x(l)$ (dotted line) and its two components

Since $x(n)$ can be considered as the FS coefficients of $X(e^{j\omega})$, this expression is the same as that corresponding to the FS with the roles of the domains interchanged. The quantity $|X(e^{j\omega})|^2$ is called the energy spectral density of the signal, since $\frac{1}{2\pi}|X(e^{j\omega})|^2 d\omega$ is the signal energy over the infinitesimal frequency band ω to $\omega + d\omega$.

Consider the signal, shown in Fig. 7.8a, and its DTFT $1 + e^{-j\omega}$. The energy of the signal, from its time-domain representation, is $1^2 + 1^2 = 2$. The energy of the signal, from its frequency-domain representation, is

$$E = \frac{1}{2\pi}\int_0^{2\pi} |1 + e^{-j\omega}|^2 d\omega = \frac{1}{2\pi}\int_0^{2\pi} (2 + 2\cos(\omega))d\omega = 2$$

The input and output of a LTI system, in the frequency-domain, is related by the transfer function $H(e^{j\omega})$ as

$$Y(e^{j\omega}) = H(e^{j\omega})X(e^{j\omega})$$

where $X(e^{j\omega})$, $Y(e^{j\omega})$, and $H(e^{j\omega})$ are the DTFT of the input, output, and impulse response of the system. The output energy spectrum is given by

$$\begin{aligned}|Y(e^{j\omega})|^2 &= Y(e^{j\omega})Y^*(e^{j\omega}) \\ &= H(e^{j\omega})X(e^{j\omega})H^*(e^{j\omega})X^*(e^{j\omega}) = |H(e^{j\omega})|^2|X(e^{j\omega})|^2\end{aligned}$$

The quantity $|H(e^{j\omega})|^2$ is called the energy transfer function, as it relates the input and output energy spectral densities of the input and output of a system.

7.3 Approximation of the Discrete-Time Fourier Transform

In the computation of the DFT, we usually use the time-domain range from $n = 0$ to $n = N - 1$. Due to periodicity of the DFT, we can always get the samples in this interval even though the data is defined in other intervals. Replacing ω by $\frac{2\pi}{N}k$ in the DTFT definition, we get

$$X(e^{j\frac{2\pi}{N}k}) = \sum_{n=0}^{N-1} x(n)e^{-j\frac{2\pi}{N}nk} = \sum_{n=0}^{N-1} x(n)W^{nk},\ k = 0, 1, \ldots, N-1$$

Let us approximate the samples of the DTFT spectrum shown in Fig. 7.2a using the DFT. The time-domain signal, shown in Fig. 7.2b, is of infinite duration, and, therefore, we have to truncate it. For example, let us take the 15 samples $x(-7), x(-6), \ldots, x(6), x(7)$. The record length of the truncated signal should be such that most of the energy of the signal is retained in the truncated

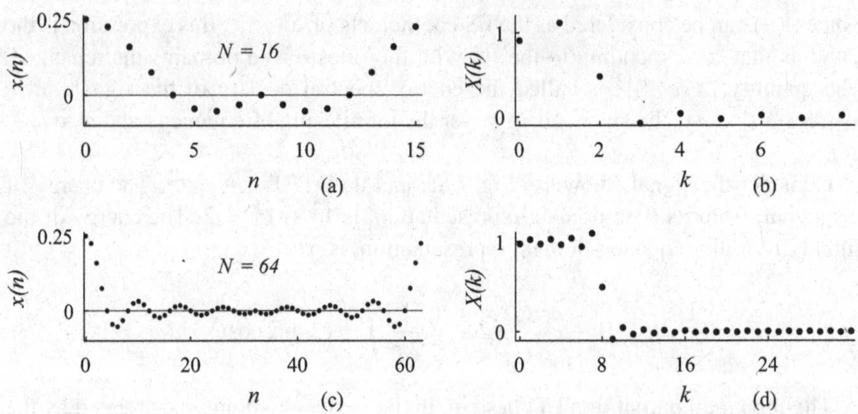

Fig. 7.10 (**a**) and (**c**) One period of the periodic extension of the truncated and zero padded aperiodic discrete signal, shown in Fig. 7.2b, with $N = 16$ and $N = 64$ samples, respectively; (**b**) and (**d**) the DFT of the signals in (**a**) and (**c**), respectively

signal. As the most efficient and regular DFT algorithms are of length that is an integral power of two, the truncated data is usually zero-padded. With one zero added and $N = 16$, the data for the DFT computation becomes $x(0), x(1), \ldots, x(7), 0, x(-7), \ldots, x(-2), x(-1)$, as shown in Fig. 7.10a. The corresponding DFT spectrum is shown in Fig. 7.10b. As the spectrum is even-symmetric, only the positive frequency half of the spectrum is shown. As the number of samples is increased, the spectral samples become more accurate, as shown in Fig. 7.10c and d with $N = 64$. Note the Gibbs phenomenon in the vicinity of the discontinuity of the spectrum.

The spectral samples obtained using the DFT are not exact because of the truncation of the input data. In effect, the actual data is multiplied by a rectangular window. Therefore, the desired spectrum is convolved with that of the rectangular window (a sinc function). This results in the distortion of the spectrum. As the level of truncation is reduced, the distortion also gets reduced. In the end, with no truncation (a rectangular window of infinite length), we get the undistorted spectrum. As an infinite data length is unacceptable for DFT computation, we start with some finite data length and keep increasing it until the difference between two successive spectra becomes negligible.

As the input signal to the DTFT is aperiodic while that of the DFT is assumed to be periodic, inevitably, truncation is required. If no truncation is necessary, we approximate the samples of the continuous DTFT spectrum by its exact samples using the DFT. We can increase the number of samples by zero padding the input data. Let

$$x(n) = \{x(-2) = 3, x(-1) = -3, x(0) = 2, x(1) = 1\}$$

and zero otherwise. The DTFT of $x(n)$ is

$$X(e^{j\omega}) = 3e^{j2\omega} - 3e^{j\omega} + 2 + e^{-j\omega}$$

The input samples $x(n)$ can be specified anywhere in the infinite range. The DFT algorithms usually assume the N-point data is in the range from 0 to $N-1$. Using the assumed periodicity of the DFT, the data can be specified in that range by periodic extension, and we get

$$xp(n) = \{x(0) = 2, x(1) = 1, x(2) = 3, x(3) = -3\}$$

The fast and practically used DFT algorithms are of length that is an integer power of 2. This requirement can be fulfilled by sufficient zero padding of the data and, at the same time, to satisfy the required number of samples of the spectrum. For example, with data length 3 and spectral samples 5, we have to zero pad the data to make the length 8. The DFT of $xp(n)$, $XP(k)$, is

$$\{XP(0) = 3, XP(1) = -1 - j4, XP(2) = 7, XP(3) = -1 + j4\},$$

The set of samples of the DTFT $X(e^{j\omega})$, at

$$\omega = 0, \omega = \frac{2\pi}{4}, \omega = \frac{2\pi}{4}, \omega = 3\frac{2\pi}{4}$$

are the same as the DFT $XP(k)$. Let us zero pad the signal $xp(n)$ to make its length 8. Then,

$$xp(n) = \{x(0) = 2, x(1) = 1, x(2) = 3, x(3) = -3, x(4) = 0, x(5) = 0, x(6) = 0, x(7) = 0\}$$

The DFT is

$$\{3, 4.8284 - j1.5858, -1 - j4, -0.8284 + j4.4142, 7, -0.8284 - j4.4142, -1 + j4, 4.8284 + j1.5858\}$$

The even-indexed samples are the same as those obtained for the $xp(n)$ with four samples.

Let us consider the effect of data truncation. The criterion for data truncation is that most of the energy of the signal is retained. Let

$$x(n) = \{x(-3) = 3, x(-2) = -3, x(-1) = 1, x(0) = 2, x(1) = 1\}$$

and zero otherwise. Then,

$$xp(n) = \{x(0) = 2, x(1) = 1, x(2) = 3, x(3) = -3, x(4) = 1\}$$

The DFT of $xp(n)$ is

$$XP(k) = \{4, 2.6180 - j3.5267, 0.3820 + j5.7063, 0.3820 - j5.7063, 2.6180 + j3.5267\}$$

The truncation operation can be considered as multiplying the signal with a window, which reduces the effective length of the data. The DFT of window

$$w(n) = \{1, 1, 1, 1, 0\}$$

is

$$W(k) = \{4, -0.3090 - j0.9511, 0.8090 - j0.5878, 0.8090 + j0.5878i - 0.3090 + j0.9511\}$$

If we multiply $xp(n)$ point-by-point by the window $w(n)$, we get the truncated signal $xp\_t(n)$. The DFT of the truncated signal $xp\_t(n)$

$$xp\_t(n) = \{2, 1, 3, -3, 0\}$$

is

$$XP\_t(k) = \{3, 2.3090 - j4.4778, 1.1910 + j5.1186, 1.1910 - j5.1186, 2.3090 + j4.4778\}$$

This is the DFT of the circular convolution of a rectangular window $W(k)$ and that of $xp(n)$, $XP(k)$, divided by 5, the convolution theorem in the frequency-domain. Let us also compute the spectrum of the truncated signal using the truncation model. The circular convolution of the two spectra can be obtained using the DFT and IDFT. The DFT of $W(k)$ is

$$\{5, 0, 5, 5, 5\}$$

The DFT of $XP(k)$ is

$$\{10, 5, -15, 15, 5\}$$

The point-by-point multiplication of the two spectra divided by 5 is

$$\{10, 0, -15, 15, 5\}$$

The IDFT of this spectrum is the spectrum of $xp\_t(n)$, $XP\_t(k)$, which is the same as that we found already. In summary, to approximate the DTFT of an arbitrary signal, start with some data length and find its DFT. Repeat this operation by doubling the data length until the difference between the energy of the signal of two consecutive iterations is negligible.

7.3.1 Approximation of the Inverse DTFT by the IDFT

Replacing ω by $\frac{2\pi}{N}k$ and $d\omega$ by $\frac{2\pi}{N}$ in the inverse DTFT definition, we get

$$x(n) = \frac{1}{N}\sum_{k=0}^{N-1} X(e^{j\frac{2\pi}{N}k})e^{j\frac{2\pi}{N}nk} = \frac{1}{N}\sum_{k=0}^{N-1} X(e^{j\frac{2\pi}{N}k})W^{-nk}, \; n = 0, 1, \ldots, N-1$$

Let us approximate the inverse DTFT of the spectrum shown in Fig. 7.2a by the IDFT. As always, at points of discontinuity, the average of the left- and right-hand limits should be taken as the sample value in Fourier analysis. The sample values of the spectrum with $N = 8$ are shown in Fig. 7.11a. The IDFT of these samples is shown in Fig. 7.11b along with the exact values. Only half of the signal is shown, as it is even-symmetric. As the number of samples is increased, as shown in Fig. 7.11c, the time-domain values become more accurate, as shown in Fig. 7.11d. As the time-domain data length is infinite, the necessary sampling interval of the spectrum is zero radians. However, as that interval is not practical with numerical analysis, we use some finite sample interval. That results in time-domain aliasing. As mentioned earlier, practical signals, with an adequate sampling interval and a sufficient record length, can be considered as both time-limited and bandlimited with a desired accuracy. This fact enables the use of the DFT and IDFT, which can be computed using fast algorithms, to approximate the other versions of Fourier analysis.

Let

$$x(n) = \{x(0) = 3, x(1) = 1, x(2) = 2, x(3) = 4\}$$

The DTFT of $x(n)$ is

$$X(e^{j\omega}) = 3 + e^{-j\omega} + 2e^{-j2\omega} + 4e^{-j3\omega}$$

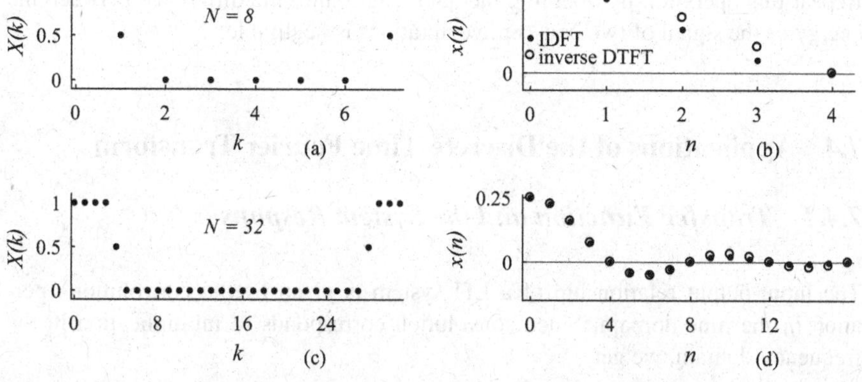

Fig. 7.11 (**a**) and (**c**) Samples of one period of the periodic DTFT spectrum, shown in Fig. 7.2a, with $N = 8$ and $N = 32$ samples, respectively; (**b**) and (**d**) The IDFT of the spectra in (**a**) and (**c**), respectively

The samples of the spectrum at $\omega = 0, \pi/2, \pi, 3\pi/2$ are

$$\{10, 1 + j3, 0, 1 - j3\}$$

The 4-point DFT of $x(n)$ is also the same. The IDFT of these samples yields $x(n)$ exactly, since the number of samples of the spectrum is sufficient to represent the data. However, if we take just two samples of the spectrum $\{10, 0\}$, and taking the IDFT, we get $\{5, 5\}$. Time-domain aliasing has occurred. That is, the time-domain samples are

$$\{3 + 2 = 5, 1 + 4 = 5\}$$

due to insufficient number of bins to hold the exact data. Aliasing is unavoidable in practice due to the finite and infinite natures, respectively, of the DFT and the DTFT in the time-domain. If we take more number of samples of the DTFT spectrum, we get the exact data with some zeros appended. With six spectral samples

$$\{10, -1.5000 - j2.5981, 5.5000 + j0.8660, 0, 5.5000 - 0.8660, -1.5000 + j2.5981\},$$

the IDFT yields

$$\{3, 1, 2, 4, 0, 0\}$$

With time-limited data, the exact time-domain samples can be obtained from adequate samples of the DTFT spectrum. Otherwise, aliasing is unavoidable. Then, it has to be ensured that time-domain aliasing is negligible by taking sufficient number of spectral samples. In summary, to approximate the inverse DTFT of an arbitrary signal, start with some number of spectral samples and find its IDFT. Repeat this operation by doubling the data length until the difference between the energy of the signal of two consecutive iterations is negligible.

7.4 Applications of the Discrete-Time Fourier Transform

7.4.1 *Transfer Function and the System Response*

The input-output relationship of a LTI system is given by the convolution operation in the time-domain. Since convolution corresponds to multiplication in the frequency-domain, we get

$$y(n) = \sum_{m=-\infty}^{\infty} x(m)h(n-m) \Longleftrightarrow Y(e^{j\omega}) = X(e^{j\omega})H(e^{j\omega}),$$

where $x(n)$, $h(n)$, and $y(n)$ are, respectively, the system input, impulse response, and output and $X(e^{j\omega})$, $H(e^{j\omega})$, and $Y(e^{j\omega})$ are their respective transforms. As input is transferred to output by multiplication with $H(e^{j\omega})$, $H(e^{j\omega})$ is called the transfer function of the system. The transfer function, which is the transform of the impulse response, characterizes a system in the frequency-domain just as the impulse response does in the time-domain.

Since the impulse function, whose DTFT is one (a uniform spectrum), is composed of complex exponentials, $e^{j\omega n}$, of all frequencies from $\omega = -\pi$ to $\omega = \pi$ with equal magnitude and zero-phase, the transform of the impulse response, the transfer function, is also called the frequency response of the system. Therefore, an exponential $Ae^{j(\omega_a n+\theta)}$ is changed to $(|H(e^{j\omega_a})|A)e^{j(\omega_a n+(\theta+\angle(H(e^{j\omega_a})))}$ at the output. A real sinusoidal input signal $A\cos(\omega_a n + \theta)$ is also changed at the output by the same amount of amplitude and phase of the complex scale factor $H(e^{j\omega_a})$. That is, $A\cos(\omega_a n + \theta)$ is changed to $(|H(e^{j\omega_a})|A)\cos(\omega_a n + (\theta + \angle(H(e^{j\omega_a})))$. The steady-state response of a stable system to the input $Ae^{j(\omega_a n+\theta)}u(n)$ is also the same.

As $H(e^{j\omega}) = \frac{Y(e^{j\omega})}{X(e^{j\omega})}$, the transfer function can also be described as the ratio of the transform $Y(e^{j\omega})$ of the response $y(n)$ to an arbitrary signal $x(n)$ to that of its transform $X(e^{j\omega})$, provided $|X(e^{j\omega})| \neq 0$ for all frequencies of interest and the system is initially relaxed.

Since the transform of a delayed signal is its transform multiplied by a factor, we can as well find the transfer function by taking the transform of the difference equation characterizing a system. Consider the difference equation of a causal LTI discrete system.

$$y(n) + a_{K-1}y(n-1) + a_{K-2}y(n-2) + \cdots + a_0 y(n-K)$$
$$= b_M x(n) + b_{M-1}x(n-1) + \cdots + b_0 x(n-M)$$

Taking the transform of both sides, we get, assuming initial conditions are all zero,

$$Y(e^{j\omega})(1 + a_{K-1}e^{-j\omega} + a_{K-2}e^{-j2\omega} + \cdots + a_0 e^{-jK\omega})$$
$$= X(e^{j\omega})(b_M + b_{M-1}e^{-j\omega} + \cdots + b_0 e^{-jM\omega})$$

The transfer function $H(e^{j\omega})$ is obtained as

$$H(e^{j\omega}) = \frac{Y(e^{j\omega})}{X(e^{j\omega})} = \frac{b_M + b_{M-1}e^{-j\omega} + \cdots + b_0 e^{-jM\omega}}{1 + a_{K-1}e^{-j\omega} + a_{K-2}e^{-j2\omega} + \cdots + a_0 e^{-jK\omega}}$$

Example 7.6 Find the response, using the DTFT, of the system governed by the difference equation

$$y(n) = x(n) + 0.6y(n-1)$$

to the input $x(n) = \cos(\frac{2\pi}{6}n + \frac{\pi}{6})$.

Solution

$$H(e^{j\omega}) = \frac{e^{j\omega}}{e^{j\omega} - 0.6}$$

Substituting $\omega = \frac{2\pi}{6}$, we get

$$H(e^{j\frac{2\pi}{6}}) = \frac{e^{j\frac{2\pi}{6}}}{e^{j\frac{2\pi}{6}} - 0.6} = 1.1471\angle(-0.6386)$$

The response of the system to the input $x(n) = \cos(\frac{2\pi}{6}n + \frac{\pi}{6})$ is $y(n) = 1.1471\cos(\frac{2\pi}{6}n + \frac{\pi}{6} - 0.6386)$. ∎

Example 7.7 Find the impulse response $h(n)$, using the DTFT, of the system governed by the difference equation

$$y(n) = x(n) - x(n-1) + 2x(n-2) + \frac{7}{12}y(n-1) - \frac{1}{12}y(n-2)$$

Solution

$$H(e^{j\omega}) = \frac{1 - e^{-j\omega} + 2e^{-j2\omega}}{(1 - \frac{7}{12}e^{-j\omega} + \frac{1}{12}e^{-j2\omega})} = \frac{1 - e^{-j\omega} + 2e^{-j2\omega}}{(1 - \frac{1}{3}e^{-j\omega})(1 - \frac{1}{4}e^{-j\omega})}$$

$$\frac{H(e^{j\omega})}{e^{j\omega}} = \frac{e^{j2\omega} - e^{j\omega} + 2}{e^{j\omega}(e^{j\omega} - \frac{1}{3})(e^{j\omega} - \frac{1}{4})}$$

The last step is required, since the degree of the numerator polynomial must be less than that of the denominator to expand into partial fractions. Expanding into partial fractions, we get

$$H(e^{j\omega}) = 24 + \frac{64e^{j\omega}}{(e^{j\omega} - \frac{1}{3})} - \frac{87e^{j\omega}}{(e^{j\omega} - \frac{1}{4})}$$

For example, putting $e^{j\omega} = 0$ on the right side of the previous expression suppressing the term $e^{j\omega}$ in the denominator, we get the first coefficient 24. Similarly, putting $e^{j\omega} = 1/3$ and $e^{j\omega} = 1/4$ in turn and suppressing the corresponding denominator term, we get the other two coefficients. Taking the inverse DTFT, we get the impulse response as

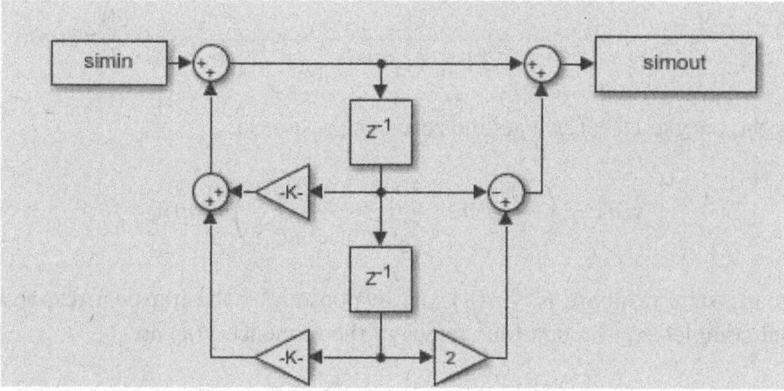

Fig. 7.12 The simulation model to find the impulse response

$$h(n) = 24\delta(n) + \left(64\left(\frac{1}{3}\right)^n - 87\left(\frac{1}{4}\right)^n\right)u(n)$$

The first four values of the impulse response $h(n)$ are

$$h(0) = 1, \quad h(1) = -0.4167, \quad h(2) = 1.6736, \quad h(3) = 1.011 \quad \blacksquare$$

The simulation model to find the impulse response is shown in Fig. 7.12. The impulse input values have to be loaded into the simin block by executing the given input program.

Example 7.8 Find the zero-state response, using the DTFT, of the system governed by the difference equation

$$y(n) = 2x(n) - x(n-1) + 3x(n-2) + \frac{9}{20}y(n-1) - \frac{1}{20}y(n-2)$$

with the input $x(n) = u(n)$, the unit-step function.

Solution

$$H(e^{j\omega}) = \frac{2 - e^{-j\omega} + 3e^{-j2\omega}}{(1 - \frac{9}{20}e^{-j\omega} + \frac{1}{20}e^{-j2\omega})} = \frac{2 - e^{-j\omega} + 3e^{-j2\omega}}{(1 - \frac{1}{5}e^{-j\omega})(1 - \frac{1}{4}e^{-j\omega})}$$

With $X(e^{j\omega}) = \frac{1}{(1-e^{-j\omega})} + \pi\delta(\omega)$,

$$Y(e^{j\omega}) = H(e^{j\omega})X(e^{j\omega}) = \frac{2 - e^{-j\omega} + 3e^{-j2\omega}}{(1 - e^{-j\omega})(1 - \frac{1}{5}e^{-j\omega})(1 - \frac{1}{4}e^{-j\omega})} + \frac{20}{3}\pi\delta(\omega)$$

Expanding into partial fractions, we get

$$Y(e^{j\omega}) = \frac{\frac{20}{3}}{(1 - e^{-j\omega})} + \frac{72}{(1 - \frac{1}{5}e^{-j\omega})} - \frac{\frac{230}{3}}{(1 - \frac{1}{4}e^{-j\omega})} + \frac{20}{3}\pi\delta(\omega)$$

Taking the inverse DTFT, we get the zero-state response.

$$y(n) = \left(\frac{20}{3} + 72\left(\frac{1}{5}\right)^n - \frac{230}{3}\left(\frac{1}{4}\right)^n\right)u(n)$$

The steady-state response is $\frac{20}{3}u(n)$, the response after the transient response has died out completely. The first four values of the sequence $y(n)$ are

$$\{y(0) = 2, \quad y(1) = 1.9, \quad y(2) = 4.755, \quad y(3) = 6.0448\}$$ ∎

The simulation model to find the step response is shown in Fig. 7.13. The step input values have to be loaded into the simin block by executing the given input program.

The transfer function concept can still be used even if the initial conditions of a system are nonzero. In that case, we have to assume that additional inputs are applied to the system at the instant the system is turned on, which will produce the same response as do the initial conditions. However, the z-transform is relatively easier for system analysis. In addition, it can handle a larger class of signals and systems than that can be analyzed by the DTFT. Wherever the DTFT is more suitable, it is better for numerical analysis as it can be approximated by the DFT using fast algorithms.

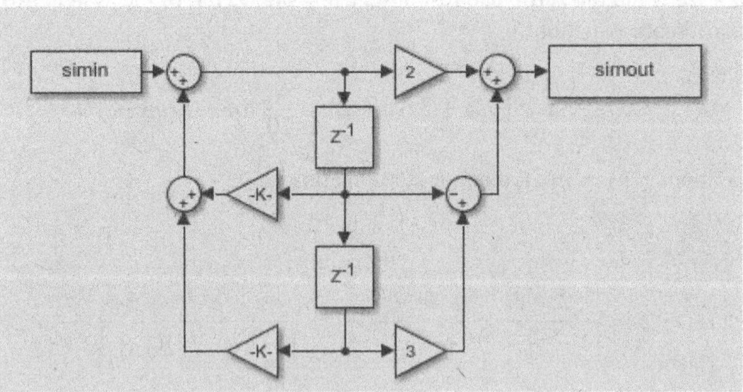

Fig. 7.13 The simulation model to find the step response

7.4.2 Digital Filter Design Using DTFT

Digital filter is widely used in signal processing applications. Usually, the specification of a filter is given in terms of its frequency response. As the filter, which is a system, is characterized by its impulse response, the design of a filter is to determine its impulse response. Therefore, one way of finding the impulse response is to find inverse DTFT of its frequency response. For example, the frequency response and the corresponding impulse response of an ideal lowpass filter are shown, respectively, in Fig. 7.2a and b, and those of an high-pass filter are shown, respectively, in Fig. 7.6a and b. A system with this type of impulse response is not practically implementable because (i) as the impulse response is not absolutely summable, it is an unstable system; and (ii) the impulse response is noncausal. The first problem is overcome by truncating part of the impulse response. The second problem is solved by shifting the impulse response to the right so that it becomes causal. With these modifications of the impulse response, of course, the filter response will not be ideal.

We prefer the response of the actual filter to uniformly converge to that of the ideal filter. But, in Fourier analysis, the convergence criteria is with respect to the square error. That is, there is a 9 % deviation of the frequency response at the band edges (discontinuities) of the filter. This problem can be reduced by using window functions to smooth the truncated impulse response. This time the price that is paid is of longer transition bands.

Figure 7.14 shows the frequency response of the ideal discrete lowpass filter. It is a periodic rectangular waveform with period 2π and even-symmetric. Therefore, the response over one-half of the period 0 to π, shown in thick lines, uniquely characterizes the filter. The expression for the frequency response is

$$H(e^{j\omega}) = \begin{cases} 1 \text{ for } 0 \leq \omega \leq \omega_c \\ 0 \text{ for } \omega_c < \omega \leq \pi \end{cases}$$

where ω_c is the cutoff frequency. The filter passes frequency components from 0 to the cutoff frequency and rejects the rest. It is evident as the magnitude of the transfer function from 0 to ω_c, called the passband of the filter, is 1 and the magnitude elsewhere, called the stopband, is zero.

The impulse response of the ideal lowpass filter, with cutoff frequency ω_c, is given by the inverse DTFT of its frequency response, and it is

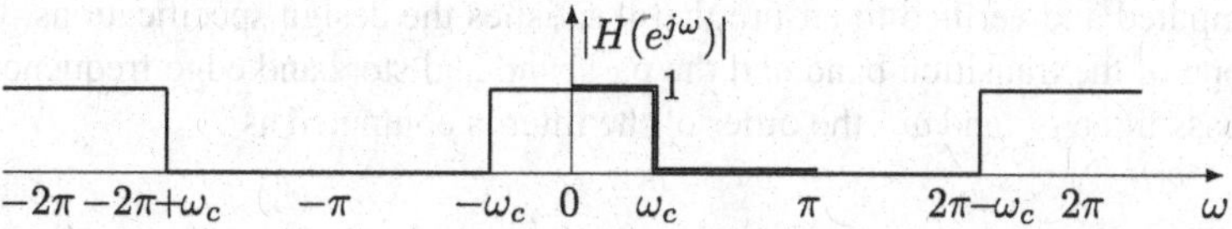

Fig. 7.14 Periodic frequency response of an ideal discrete lowpass filter

$$h(n) = \frac{\sin(\omega_c n)}{\pi n}, \quad -\infty < n < \infty$$

With $\omega_c = \frac{\pi}{4}$, it is derived in Example 7.2. The duration of the impulse response is infinity and, hence, it is not practically realizable. Further, it is unstable as the impulse response, which is a sinc function, is not absolutely summable. In order to make it realizable, we truncate the response to $N + 1$ terms and shift it to make it causal. With these modifications, the impulse response becomes

$$h(n) = \frac{\sin(\omega_c(\frac{N}{2} - n))}{\pi(\frac{N}{2} - n)}, \quad n = 0, 1, \ldots, N,$$

where N is the order of the filter.

7.4.2.1 Rectangular Window

The rectangular window, which is the truncation process itself, is given by

$$w_r(n) = \begin{cases} 1 \text{ for } n = 0, 1, \ldots, N \\ 0 \text{ otherwise} \end{cases}$$

Due to the discontinuity at the borders, the rate of convergence of the spectrum of a rectangular function is slow. Therefore, while this type of windows provides the shortest width of the transition band of the resulting filter, the attenuation possible is low due to the large side lobes of its spectrum. It is well-known that, due to Gibbs phenomenon (Chap. 6), the magnitude of the overshoot and undershoot at a discontinuity of reconstructed waveform is 0.0895. Therefore, the maximum attention possible with this type of window, irrespective of the window length, is

$$-20 \log_{10} 0.0895 = 20.96 \text{ dB}$$

In order to design a filter using window, the possible slope of the transition band of the filter is required.

The average slope of the transition band using the rectangular window is given as $0.9\frac{2\pi}{N}$ approximately. Some trial and error may be required to find the desired frequency response by changing the band edges or the filter length. Of course, after completing the filter design using any method, the frequency response must be computed and verified to ensure that it satisfies the design specifications. Using the slope of the transition band and the passband and stopband edge frequencies of a lowpass filter, ω_c and ω_r, the order of the filter is computed as

$$N \geq 0.9 \frac{2\pi}{\omega_r - \omega_c}$$

Then, the impulse response is specified as

$$h(n) = \frac{\sin(\omega_c(\frac{N}{2} - n))}{\pi(\frac{N}{2} - n)}, \quad n = 0, 1, \ldots, N, \quad \omega_c = \frac{\omega_r + \omega_c}{2}$$

Example 7.9 Given that the passband and stopband edge frequencies of a lowpass filter are 0.25π radians and 0.35π radians, respectively. The minimum attenuation in the stopband is to be 20 dB. Assuming that the sampling frequency is $f_s = 512$ Hz, design the lowpass filter using the rectangular window.

Solution The passband edge frequency 0.25π corresponds to $\frac{0.25\pi}{2\pi}512 = 64$ Hz. The stopband edge frequency 0.35π corresponds to $\frac{0.35\pi}{2\pi}512 = 89.6$ Hz. As the maximum attenuation possible using the rectangular window is about 21 dB, the specification of 20 dB is attainable. From the given specifications, the order of the filter is computed as

$$N \geq 0.9\frac{2\pi}{0.35\pi - 0.25\pi} = 18$$

The cutoff frequency of the corresponding ideal filter is the average of the band edges, and it is

$$\omega_c = \frac{\omega_r + \omega_c}{2} = \frac{0.35\pi + 0.25\pi}{2} = 0.3\pi$$

The impulse response of the filter is given by

$$h(n) = \frac{\sin(\omega_c(\frac{N}{2} - n))}{\pi(\frac{N}{2} - n)} = \frac{\sin(0.3\pi(9 - n))}{\pi(9 - n)}, \quad n = 0, 1, \ldots, 18$$

The impulse response values, with a precision of four digits after the decimal point, are

$$\begin{aligned}
\{h(n), n &= 0, 1, \ldots, 18\} \\
&= \{0.0286, 0.0378, 0.0141, -0.0312, -0.0637, -0.0468, \\
&\quad 0.0328, 0.1514, 0.2575, 0.3, 0.2575, 0.1514, 0.0328, \\
&\quad -0.0468, -0.0637, -0.0312, 0.0141, 0.0378, 0.0286\}
\end{aligned}$$

The impulse response and the magnitude and phase of the frequency response of the filter are shown in Fig. 7.15a, b, and d, respectively. The passband in (b) is shown in expanded linear scale in Fig. 7.15c. The passband and stopband ripples are equal in the filters designed using the window method. ∎

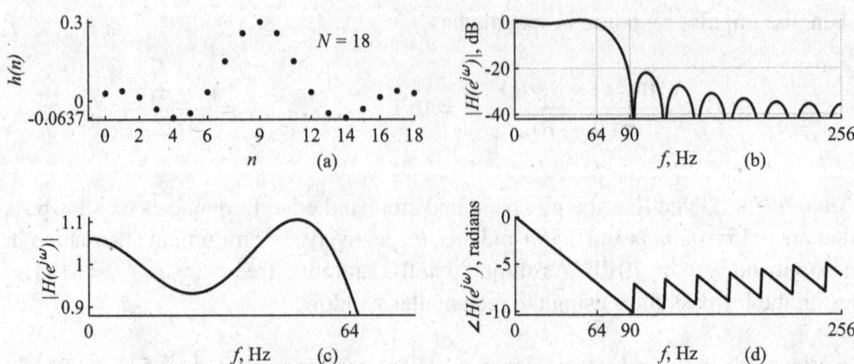

Fig. 7.15 (**a**) Filter impulse response; (**b**) the magnitude of the frequency response; (**c**) the passband in (**b**) shown in expanded linear scale; (**d**) the phase response

Filters can be designed with arbitrary frequency specifications using other methods. However, there are certain basic types of filters, which are often used in practice. These filters are lowpass, highpass, bandpass, and bandstop. While each type can be designed independently, usually, the other types of filters are expressed as a linear combination of lowpass filters of various parameters, or they are designed using some transformation. Let us design two lowpass filters with one having a cutoff frequency π radians and the other having a cutoff frequency $\omega_c < \pi$. Then, by subtracting the impulse response of the second filter from that of the first, we get a highpass filter with cutoff frequency ω_c. That is, the impulse response of a highpass filter is given by

$$h(n) = \frac{\sin(\pi(\frac{N}{2} - n))}{\pi(\frac{N}{2} - n)} - \frac{\sin(\omega_c(\frac{N}{2} - n))}{\pi(\frac{N}{2} - n)}, \qquad n = 0, 1, \ldots, N$$

The order of the filter N must be even for highpass and bandstop filters.

Now, the steps involved in designing FIR filters using the window method are listed.

1. The ideal frequency response is specified.
2. Compute the corresponding impulse response.
3. An appropriate window is selected.
4. By multiplying the impulse response with the window function, the filter coefficients are obtained.
5. A trial-and-error procedure is followed, typically by varying the window length, so that the filter satisfies the given specifications.

7.4.2.2 Hamming Window

Several windows are available with different characteristics for filter design. The Hamming window is defined as

$$w_{ham}(n) = \begin{cases} 0.54 - 0.46\cos(\frac{2\pi}{N}n) & \text{for } n = 0, 1, \ldots, N \\ 0 & \text{otherwise} \end{cases}$$

This window is a linear combination of three time-shifted rectangular windows, resulting in the reduction of the amplitude of the side lobes at the cost of increased width of the transition band. The window provides attenuation up to about 53 dB. The average slope of the transition band obtained using this window is $3.3\frac{2\pi}{N}$. In common with filter design using windows, some trial and error is required to get the required filter. Given the passband and stopband edge frequencies of a lowpass filter, ω_c and ω_r, we find the order of the filter as

$$N \geq 3.3\frac{2\pi}{\omega_r - \omega_c}$$

Then, the impulse response is specified as

$$h(n) = w_{ham}(n)\frac{\sin(\omega_c(\frac{N}{2} - n))}{\pi(\frac{N}{2} - n)}, \quad n = 0, 1, \ldots, N, \quad \omega_c = \frac{\omega_r + \omega_c}{2}$$

Example 7.10 The passband and stopband edge frequencies of a highpass filter are 0.55π and 0.25π, respectively. The minimum attenuation required in the stopband is 49 dB. Design the highpass filter using the Hamming window. The sampling frequency is $f_s = 512$ Hz.

Solution The maximum attenuation provided by filters using the Hamming window is about 53 dB. Therefore, the specification of 49 dB is realizable. Now, we find the order of the filter as

$$N \geq 3.3\frac{2\pi}{0.55\pi - 0.25\pi} = 22$$

The cutoff frequency of the corresponding ideal filter is computed as

$$\omega_c = \frac{\omega_r + \omega_c}{2} = \frac{0.25\pi + 0.55\pi}{2} = 0.4\pi$$

The impulse response of the highpass filter is given by

$$h(n) = w_{ham}(n)\left(\frac{\sin(\pi(11 - n))}{\pi(11 - n)} - \frac{\sin(0.4\pi(11 - n))}{\pi(11 - n)}\right), \quad n = 0, 1, \ldots, 22$$

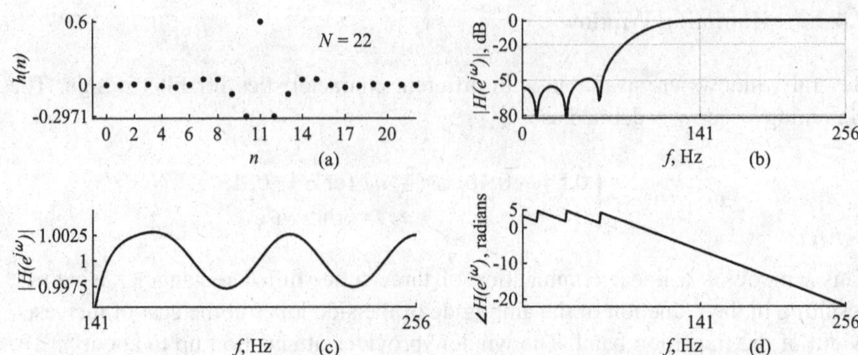

Fig. 7.16 (**a**) Filter impulse response; (**b**) the magnitude of the frequency response; (**c**) the passband in (**b**) shown in expanded linear scale; (**d**) the phase response

The impulse response values, with a precision of four digits after the decimal point, are

$$\{h(n), n = 0, 1, \ldots, 22\}$$
$$= -0.0022, 0, 0.0051, 0.0056, -0.0093, -0.0239, 0, 0.0553,$$
$$0.0525, -0.0867, -0.2971, 0.6, -0.2971, -0.0867, 0.0525,$$
$$0.0553, 0, -0.0239, -0.0093, 0.0056, 0.0051, 0, -0.0022\}$$

The impulse response and the magnitude and phase of the frequency response of the filter are shown in Fig. 7.16a, b, and d, respectively. The passband in (b) is shown in expanded linear scale in Fig. 7.16c. ∎

7.4.3 Digital Differentiator

In this subsection, we derive the impulse response of the digital differentiator from its frequency response. This differentiator takes the samples of a continuous signal $x(t)$ and produces the samples of its derivative. The periodic frequency response, shown in Fig. 7.17a over one period, of the ideal digital differentiator is defined as

$$H(e^{j\omega}) = j\omega, \ -\pi < \omega < \pi$$

For example, the input and the output of the differentiator are

$$\sin(\omega_0 n) \Longleftrightarrow j\pi(\delta(\omega + \omega_0) - \delta(\omega - \omega_0))$$
$$j\pi(j)(-\omega_0\delta(\omega + \omega_0) - \omega_0\delta(\omega - \omega_0)) \Longleftrightarrow \omega_0 \cos(\omega_0 n)$$

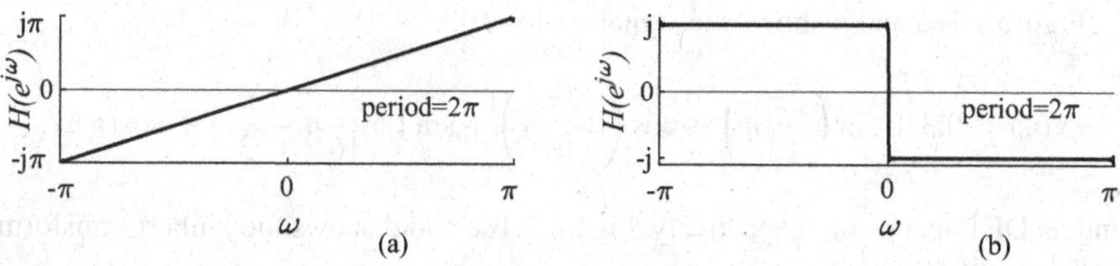

Fig. 7.17 (**a**) The frequency response of the ideal digital differentiator; (**b**) the frequency response of the ideal Hilbert transformer

The impulse response of the ideal differentiator is obtained by finding the inverse DTFT of its frequency response.

$$h(n) = \frac{1}{2\pi}\int_{-\pi}^{\pi} j\omega e^{j\omega n}d\omega = \frac{\cos(\pi n)}{n} = \begin{cases} \frac{(-1)^n}{n} & \text{for } n \neq 0 \\ 0 & \text{for } n = 0 \end{cases}, -\infty < n < \infty$$

As the frequency response of the differentiator is imaginary and odd-symmetric, the impulse response is real and odd-symmetric.

7.4.4 *Hilbert Transform*

Although most practical signals are real-valued, we need, in applications such as the sampling of bandpass signals and single-sideband amplitude modulation, a complex signal whose real part is the given real signal $x(n)$ and the imaginary part is the Hilbert transform of $x(n)$. In the Hilbert transform, every real frequency component of a real signal $x(n)$ is shifted to the right by $-\frac{\pi}{2}$ radians. That is, a phase of $-\frac{\pi}{2}$ radians is added. For example, the Hilbert transform of $\sin(\omega n)$ is $\sin(\omega n - \frac{\pi}{2}) = -\cos(\omega n)$. Most of the transforms have two domains, whereas there is only one domain in the Hilbert transform. Consider the complex signal formed with the real part being a real signal and the imaginary part being its Hilbert transform. The spectral values of this complex signal are zero for negative frequencies (a one-sided spectrum). The complex signal formed by the sine signal and its Hilbert transform is

$$\sin(\omega n) - j\cos(\omega n) = -je^{j\omega n}$$

The DFT of $\sin(\omega n)$, with N samples in a cycle, is $-j\frac{N}{2}$ at ω and $j\frac{N}{2}$ at $-\omega$, whereas that of $-je^{j\omega n}$ is $-jN$ at ω only. Similarly, a transform with its imaginary part being the Hilbert transform of its real part, for example, the transfer function of a causal system, corresponds to a one-sided time-domain signal. In this subsection, the impulse response of the Hilbert transformer is derived from its frequency response.

Figure 7.18a and b shows the signal

$$x(n) = 0.3 + \sin\left(\frac{2\pi}{16}n\right) + \cos\left(3\frac{2\pi}{16}n\right) + \sin\left(5\frac{2\pi}{16}n - \frac{\pi}{3}\right) + \cos(\pi n)$$

and its DFT spectrum, respectively. Figure 7.18c and d shows the Hilbert transform of the signal in (a)

$$x_H(n) = \sin\left(\frac{2\pi}{16}n - \frac{\pi}{2}\right) + \cos\left(3\frac{2\pi}{16}n - \frac{\pi}{2}\right) + \sin\left(5\frac{2\pi}{16}n - \frac{\pi}{3} - \frac{\pi}{2}\right)$$

and its spectrum, respectively. The DC component 0.3 and the component with frequency π, $\cos(\pi n)$, become sine terms with frequencies 0 and π radians. At these frequencies, the samples of the sine wave are all zero. The differences between the spectra in (b) and (d) are that the values at index $k = 0$ and at $k = \frac{N}{2} = 8$ are zero in (d), the values of the other positive frequency components in (b) are multiplied by $-j$, and those of the negative frequency components in (b) are multiplied by j. Therefore, the spectrum of a real signal modified in this way is the DFT of its Hilbert transform, and its IDFT gives the Hilbert transform of the signal.

The signal $jx_H(n)$ and its spectrum are shown in Fig. 7.18e and f, respectively. Compared with the spectrum in Fig. 7.18b, the coefficients at index $k = 0$ and at $k = \frac{N}{2} = 8$ are zero, the coefficients of the positive frequency components are modified by $j(-j) = 1$, and those of the negative frequency components are modified by $j(j) = -1$. Therefore, the spectrum is the same as in (b) with the values of the negative frequency components negated and the values with indices 0 and 8 zero. The complex signal $x(n) + jx_H(n)$ and its spectrum are shown in Fig. 7.18g and h, respectively. The spectral values in (h) with indices from 1 to 7 are twice of those in the first half of (b). Values with indices 0 and 8 are the same and the rest of the values are zero.

The periodic frequency response, shown in Fig. 7.17b over one period, of the ideal Hilbert transformer is defined as

$$H(e^{j\omega}) = \begin{cases} -j \text{ for } 0 < \omega < \pi \\ j \text{ for } -\pi < \omega < 0 \end{cases}$$

The impulse response of the ideal Hilbert transformer is obtained by finding the inverse DTFT of its frequency response.

$$\begin{aligned} h(n) &= \frac{1}{2\pi}\int_0^{\pi} -je^{j\omega n}d\omega + \frac{1}{2\pi}\int_{-\pi}^{0} je^{j\omega n}d\omega \\ &= \begin{cases} \frac{2\sin^2(\frac{\pi n}{2})}{\pi n} & \text{for } n \neq 0 \\ 0 & \text{for } n = 0 \end{cases}, -\infty < n < \infty \end{aligned}$$

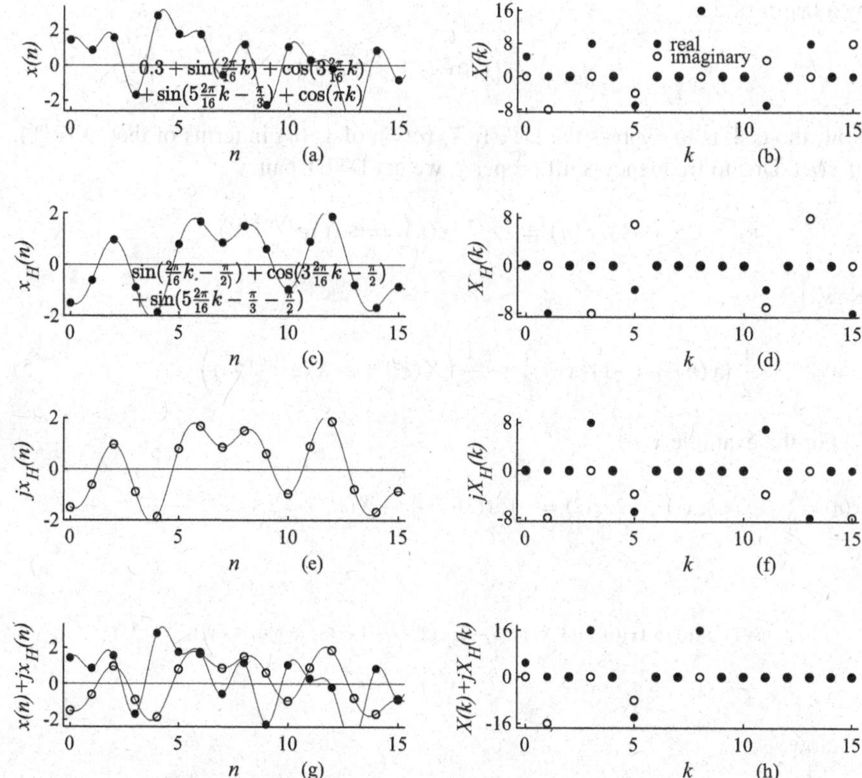

Fig. 7.18 (**a**) An arbitrary signal and (**b**) its DFT; (**c**) the Hilbert transform of signal in (**a**) and (**d**) its DFT; (**e**) the signal in (**c**) multiplied by j and (**f**) its DFT; (**g**) the sum of signals in (**a**) and (**e**), and (**h**) its one-sided DFT spectrum

7.4.5 Downsampling

In multirate digital signal processing, two more basic operations, upsampling and downsampling, are required in addition to addition, multiplication, and delaying. The upsampling operation has been presented in the properties section. Downsampling a signal is reducing its sampling rate by discarding the samples whose indices are not integer multiples of M, where M is the downsampling factor. The nth sample of the downsampled version, $x_d(n)$, of a signal, $x(n)$, is defined as

$$x_d(n) = x(Dn)$$

For example, the downsampled version of the signal

$$x(n) = \{x(0) = 3, x(1) = 2, x(2) = 1, x(3) = 4\}$$

by a factor of 2 is

$$x_d(n) = \{x(0) = 3, x(1) = 1\}$$

Now, the task is to express the DTFT, $X_d(e^{j\omega})$, of $x_d(n)$ in terms of that, $X(e^{j\omega})$, of $x(n)$. Due to frequency-shift property, we get DTFT pair

$$(-1)^n x(n) = e^{-j\pi n} x(n) \Longleftrightarrow X(e^{j(\omega+\pi)})$$

Now,

$$\frac{1}{2}\left(x(n) + (-1)^n x(n)\right) \leftrightarrow \frac{1}{2}\left(X(e^{j\omega}) + X(e^{j(\omega+\pi)})\right) \tag{7.5}$$

For the example $x(n)$,

$$x(n) = \{x(0) = 3, x(1) = 2, x(2) = -1, x(3) = 4\} \leftrightarrow X(e^{j\omega}) = 3 + 2e^{-j\omega} - e^{-j2\omega} + 4e^{-j3\omega}$$

$$\begin{aligned}(-1)^n x(n) = \{x(0) = 3, x(1) = -2, x(2) = -1, x(3) = -4\} &\leftrightarrow X(e^{j(\omega+\pi)})\\ &= 3 - 2e^{-j\omega} - e^{-j2\omega} - 4e^{-j3\omega}\end{aligned}$$

Now,

$$\begin{aligned}\frac{1}{2}\left(x(n) + (-1)^n x(n)\right) &= \{x(0) = 3, x(1) = 0, x(2) = -1, x(3) = 0\}\\ &\leftrightarrow \frac{1}{2}\left(X(e^{j\omega}) + X(e^{j(\omega+\pi)})\right) = 3 - e^{-j2\omega}\end{aligned}$$

The DTFT spectrum is periodic with period π since

$$3 - e^{-j2(\omega+\pi)} = 3 - e^{-j2\omega}$$

The spectral samples at

$$\omega = 0, \frac{\pi}{2}, \pi, \frac{3\pi}{2}$$

are, respectively,

$$\{2, 4, 2, 4\}$$

The IDFT of one period of the spectral samples $\{2, 4\}$ is $\{3, -1\}$, which is the downsampled version of $x(n)$. Therefore, we compress the spectrum by replacing

ω by $\omega/2$ in Equation (7.5), which is the spectrum $X_d(e^{j\omega})$ of the downsampled signal

$$x_d(n) = x(2n) \leftrightarrow X_d(e^{j\omega}) = \frac{1}{2}\left(X(e^{j\frac{\omega}{2}}) + X(e^{j(\frac{\omega}{2}+\pi)})\right), \quad 0 < \omega < 2\pi \tag{7.6}$$

7.5 Summary

- In this chapter, the DTFT, its properties, its applications, and its approximation by the DFT have been presented.
- The DTFT analyzes aperiodic discrete signals in terms of a continuum of discrete sinusoids over a finite frequency range. Due to the discrete nature of the signal with an infinite range, the DTFT spectrum is periodic and continuous.
- The DTFT is the limiting case of the DFT as the period of the time-domain sequence tends to infinity with the sampling interval fixed.
- The DTFT spectrum is continuous and the magnitude of the infinite frequency components is infinitesimal. The spectral density $X(e^{j\omega})$ represents $x(n)$ as a relative amplitude spectrum.
- The DFT spectrum is the samples of the DTFT spectrum at equal intervals of $2\pi/N$.
- There is a dual relationship between the FS and the DTFT.
- The spectral analysis of discrete signals, design of filters, and LTI discrete system analysis are typical applications of the DTFT.
- As is the case with the other versions of the Fourier analysis, the DTFT is also approximated by the DFT.

Exercises

7.1 Find the DTFT of

$$x(n) = \begin{cases} 1 \text{ for } 0 \le n < N \\ 0 \text{ otherwise} \end{cases}$$

With $N = 5$, compute the values of $X(e^{j\omega})$ of $x(n)$ at $\omega = 0, \pi$.

* **7.2** Find the DTFT of

$$x(n) = \begin{cases} 1 \text{ for } -N \le n \le N \\ 0 \text{ otherwise} \end{cases}$$

With $N = 5$, compute the values of $X(e^{j\omega})$ of $x(n)$ at $\omega = 0, \pi$.

7.3 Find the DTFT of $x(n) = (a)^n \cos(\omega_0 n)u(n)$, $a < 1$. With $\omega_0 = \frac{\pi}{2}$ and $a = 0.9$, compute the values of $X(e^{j\omega})$ of $x(n)$ at $\omega = 0, \pi$.

* **7.4** Find the DTFT of $x(n) = (a)^n \sin(\omega_0 n)u(n)$, $a < 1$. With $\omega_0 = \frac{\pi}{2}$ and $a = 0.7$, compute the values of $X(e^{j\omega})$ of $x(n)$ at $\omega = 0, \pi$.

7.5 Apply a limiting process, as $N \to \infty$, so that

$$x(n) = \begin{cases} \cos(\omega_0 n) & \text{for } |n| \le N \\ 0 & \text{for } |n| > N \end{cases}$$

degenerates into the cosine function and, hence, derive the DTFT of the signal $\cos(\omega_0 n)$, $-\infty < n < \infty$.

7.6 Apply a limiting process, as $a \to 1$, so that $a^{|n|}\cos(\omega_0 n)$, $a < 1$ degenerates into $\cos(\omega_0 n)$ and, hence, derive the DTFT of the signal $\cos(\omega_0 n)$, $-\infty < n < \infty$.

7.7 Apply a limiting process so that $x(n)$ degenerates into the dc function and, hence, derive the DTFT of the dc function, $x(n) = 1$.

7.7.1 $x(n) = \begin{cases} 1 & \text{for } |n| \le M \\ 0 & \text{for } |n| > M \end{cases}$ as $M \to \infty$.
7.7.2 $x(n) = a^{|n|}$, $0 < a < 1$ as $a \to 1$.
7.7.3 $x(n) = \frac{\sin(an)}{an}$ as $a \to 0$.

7.8 Given the description of the periodic signal $x(t)$ over one period, find its FS. Then, using the duality property, find the corresponding DTFT pair. Verify the DTFT pair using the inverse DTFT equation.

7.8.1

$$x(t) = \begin{cases} 2 & \text{for } 0 < t < 2 \\ -2 & \text{for} 2 < t < 4 \end{cases}$$

7.8.2 $x(t) = 1.5t$, $0 \le t < 2$.
7.8.3

$$x(t) = \begin{cases} \frac{4}{3}t & \text{for } 0 \le t < 1.5 \\ \frac{4}{3}(3 - t) & \text{for } 1.5 \le t < 3 \end{cases}$$

7.9 Find the DTFT of $x(n)$.

7.9.1 $x(n) = 2\cos(\frac{2\pi}{8}n + \frac{\pi}{3})$.
7.9.2 $x(n) = j4\sin(\frac{2\pi}{6}n - \frac{\pi}{6})$.
7.9.3 $x(n) = 2e^{j(\frac{2\pi}{9}n+\frac{\pi}{4})}$.
7.9.4 $x(n) = u(n-2)$.
* 7.9.5 $x(n) = (0.6)^n u(n-2)$.

7.10 Given the sample values over a period of a periodic sequence, find its DTFT using the DFT.

7.10.1 $\{x(0) = 2, x(1) = 3, x(2) = 1, x(3) = 4\}$.
* 7.10.2 $\{x(0) = 4, x(1) = 1, x(2) = 2, x(3) = 3\}$.
7.10.3 $\{x(0) = 3, x(1) = 4, x(2) = -2, x(3) = 1\}$.

7.11 Find the DTFT, $X(e^{j\omega})$, of $x(n)$. Find also the DFT, $X(k)$, of $x(n)$ with $N = 4$. Verify that the DFT values correspond to the samples of $X(e^{j\omega})$ at $\omega = 0, \frac{\pi}{2}, \pi, \frac{3\pi}{2}$.

7.11.1 $\{x(n), n = 0, 1, 2, 3\} = \{2, 3, -1, 4\}$ and $x(n) = 0$ otherwise.
7.11.2 $\{x(n), n = 0, 1, 2, 3\} = \{4, 0, 0, 0\}$ and $x(n) = 0$ otherwise.
7.11.3 $\{x(n), n = 0, 1, 2, 3\} = \{0, -2, 0, 0\}$ and $x(n) = 0$ otherwise.
7.11.4 $\{x(n), n = 0, 1, 2, 3\} = \{3, 3, 3, 3\}$ and $x(n) = 0$ otherwise.
7.11.5 $\{x(n), n = 0, 1, 2, 3\} = \{2, -2, 2, -2\}$ and $x(n) = 0$ otherwise.

7.12 Find the DTFT of the signal

$$x(n) = \begin{cases} 1 \text{ for } n \geq 0 \\ -1 \text{ for } n < 0 \end{cases}$$

using the linearity property.

7.13 Find the DTFT of the signal $x(n) = (n+1)(0.7)^n u(n)$ using the linearity property. Find the spectral values at $\omega = 0, \pi$.

7.14 Find the DTFT of the signal $x(n) = 0,\ n < 0, x(0) = 2, x(1) = 2$, and $x(n) = 5,\ n > 1$ using the transform of $u(n)$, and linearity and time shifting properties.

7.15 Find the DTFT of the signal with its nonzero values defined as $x(n) = (0.6)^n,\ 0 \leq n \leq 7$ using the transform of $(0.6)^n u(n)$, and the linearity and time shifting properties.

7.16 Find the inverse DTFT of $X(e^{j\omega})$ using the linearity property.

7.16.1 $X(e^{j\omega}) = \frac{1}{(1-0.5e^{-j\omega})(1-0.4e^{-j\omega})}$.
* 7.16.2 $X(e^{j\omega}) = \frac{1}{(1-0.5e^{-j\omega})(1-0.25e^{-j\omega})}$.

7.17 Find the impulse response $h_l(n)$ of an ideal lowpass filter with cutoff frequency $\frac{\pi}{3}$ radians. Using the frequency shifting property and the $h_l(n)$ obtained, find the impulse response $h_h(n)$ of an ideal highpass filter with cutoff frequency $\frac{2\pi}{3}$ radians.

7.18 Find the DTFT of the signal

$$x(n) = \begin{cases} 1 \text{ for } -N \leq n \leq N \\ 0 \text{ otherwise} \end{cases}$$

using the DTFT of shifted unit-step signals.

7.19 Using the frequency shifting property, find the inverse DTFT of $X(e^{j\omega}) = \frac{1}{(1-0.6e^{-j(\omega-\frac{\pi}{3})})}$.

7.20 Find the DTFT of the signal $x(n) = e^{j\omega_0 n}u(n)$ using the frequency shifting property.

7.21 Find the convolution of the finite sequences $x(n)$ and $h(n)$ using the DTFT.

* 7.21.1 $\{x(n),\ n = 0, 1, 2, 3\} = \{1, 0, 2, 3\}$ and $\{h(n),\ n = 1, 2, 3\} = \{-2, 1, -4\}$.

7.21.2 $\{x(n),\ n = -4, -3, -2, -1\} = \{3, 1, 0, -4\}$ and $\{h(n),\ n = -4, -3, -2, -1\} = \{1, 0, -1, 3\}$.

7.21.3 $\{x(n),\ n = -1, 0, 1\} = \{2, 0, 3\}$ and $\{h(n),\ n = -1, 0, 1\} = \{-3, 2, 2\}$.

7.22 Using the time-domain convolution property, find the DTFT of the convolution of $x(n)$ and $h(n)$.

7.22.1 $x(n) = (0.5)^n u(n)$ and $h(n) = x(n)$.

7.22.2 $x(n) = (0.6)^n u(n)$ and $h(n) = u(n)$.

7.22.3 $x(n) = (0.7)^n u(n)$ and $h(n) = (0.3)^n u(n)$.

7.23 Using the frequency-domain convolution property, find the DTFT of the product of $x(n)$ and $h(n)$.

7.23.1 $x(n) = 2\sin(n)$ and $h(n) = \cos(n)$.

7.23.2 $x(n) = e^{(j\omega_0 n)}$ and $h(n) = u(n)$.

7.24 Using the time expansion property, find the DTFT of the signal $y(n)$ defined as

$$y(an) = x(n) \text{ for } -\infty < n < \infty \quad \text{and} \quad y(n) = 0 \quad \text{otherwise}$$

7.24.1 $x(n) = 3, \ |n| \leq 2$ and $x(n) = 0$ otherwise, and $a = 2$.
* 7.24.2 $x(n) = (0.6)^n u(n)$ and $a = -4$.
7.24.3 $x(n) = \frac{\sin(\frac{\pi n}{3})}{\pi n}$ and $a = 2$.
7.24.4 $x(n) = u(n)$ and $a = 3$.
7.24.5 $x(n) = u(n - 2)$ and $a = 2$.
7.24.6 $x(n) = \cos(\frac{2\pi}{8}(n - 1))$ and $a = 2$.

7.25 Find the DTFT of the signal $x(n) = n(0.8)^n u(n)$ using the multiplication by n property.

7.26 Using the time-summation property, find the DTFT of the summation

$$y(n) = \sum_{l=-\infty}^{n} x(l)$$

7.26.1 $x(-1) = 2, x(1) = -2$ and $x(n) = 0$ otherwise.
7.26.2 $x(n) = \delta(n + 2)$.
7.26.3 $x(n) = u(n + 2)$.
* 7.26.4 $x(n) = (0.6)^n u(n)$.

7.27 Verify Parseval's theorem.

7.27.1 $x(-1) = 1, x(1) = -1$, and $x(n) = 0$ otherwise.
7.27.2 $x(n) = \frac{\sin(n)}{\pi n}$.

7.28 Find the DTFT of $x(n) = (0.4)^n u(n)$. Compute the samples of $X(e^{j\omega})$ of $x(n)$ using the DFT with $N = 4$. Compare the DFT values with the exact sample values of $X(e^{j\omega})$.

* **7.29** Find the DTFT of $x(n) = (0.3)^n u(n)$. Approximate the values of $x(n)$, using the IDFT with $N = 4$, from the samples of the DTFT of $x(n)$. Compare the IDFT values with the exact values of $x(n)$.

7.30 Using the DTFT, find the impulse response $h(n)$ of the system governed by the difference equation

$$y(n) = 2x(n) - 3x(n - 1) + 2x(n - 2) + \frac{5}{6}y(n - 1) - \frac{1}{6}y(n - 2)$$

with input $x(n)$ and output $y(n)$. List the first four values of $h(n)$.

* **7.31** Using the DTFT, find the impulse response $h(n)$ of the system governed by the difference equation

$$y(n) = x(n) - 4x(n-1) - \frac{11}{12}y(n-1) - \frac{1}{6}y(n-2)$$

with input $x(n)$ and output $y(n)$. List the first four values of $h(n)$.

7.32 Using the DTFT, find the impulse response $h(n)$ of the system governed by the difference equation

$$y(n) = x(n) + \frac{11}{15}y(n-1) - \frac{2}{15}y(n-2)$$

with input $x(n)$ and output $y(n)$. List the first four values of $h(n)$.

7.33 Using the DTFT, find the frequency response of the system governed by the difference equation

$$y(n) = x(n) + 0.8y(n-1)$$

Deduce the steady-state response of the system to the input $x(n) = \cos(\frac{2\pi}{8}n - \frac{\pi}{6})u(n)$.

7.34 Using the DTFT, find the zero-state response of the system governed by the difference equation

$$y(n) = x(n) - 2x(n-1) + 3x(n-2) + \frac{7}{12}y(n-1) - \frac{1}{12}y(n-2)$$

with the input $x(n) = u(n)$, the unit-step function.

* **7.35** Using the DTFT, find the zero-state response of the system governed by the difference equation

$$y(n) = 3x(n) + 2x(n-1) + x(n-2) + \frac{8}{15}y(n-1) - \frac{1}{15}y(n-2)$$

with the input $x(n) = (\frac{1}{2})^n u(n)$.

7.36 Find the Hilbert transform $x_H(n)$ of the signal.

8.36.1 $x(n) = 2 - \cos^2(2\pi n/8)$.
* 8.36.2 $x(n) = (-1)^n + \sin^2(2\pi n/12)$.

7.37 Find the Hilbert transform $x_H(n)$ of the signal $\{x(n),\ n = 0, 1, 2, 3\} = \{4, 5, 4, 3\}$ using the DFT and the IDFT.

Chapter 8
The Fourier Transform

The FT is the frequency-domain representation of continuous aperiodic signals in terms of a continuum of sinusoids over an infinite frequency range. Compared with the FS, as the period of the periodic waveform tends to infinity, the waveform becomes aperiodic, and the interval between the spectral points tends to zero resulting in a continuous aperiodic spectrum. Compared with the DTFT, as the sampling interval of the time-domain waveform tends to zero, the waveform becomes continuous, and the period of the spectrum tends to infinity resulting in the continuous periodic spectrum of the DTFT becoming a continuous aperiodic spectrum.

In Sect. 8.1, we derive the FT starting from the definition of the DTFT. The properties of the FT are presented in Sect. 8.2. The FT of mixed class of signals is derived in Sect. 8.3. In Sect. 8.4, the approximation of the samples of the FT by those of the DFT is described. Some typical applications of the FT are presented in Sect. 8.5.

8.1 The Fourier Transform

8.1.1 The FT as a Limiting Case of the DTFT

The FT is the same as the DTFT with the sampling interval of the time-domain waveform tending to zero. Consider the samples of the continuous sinc function, $\frac{\sin(\frac{\pi}{3}t)}{\pi t}$, with sampling interval $T_s = 1$ s and its DTFT spectrum, multiplied by T_s, shown, respectively, in Fig. 8.1a and b. The DTFT spectrum is periodic with period $\frac{2\pi}{T_s} = \frac{2\pi}{1}$ radians. Reducing the sampling interval by a factor of 2 results in the doubling of the period of the spectrum, as shown in Fig. 8.1c and d. As the number of samples is increased, the amplitude of the spectrum will also increase. But the product of the amplitude and the sampling interval approaches a finite

D. Sundararajan, *Signals and Systems*,
https://doi.org/10.1007/978-3-031-19377-4_8

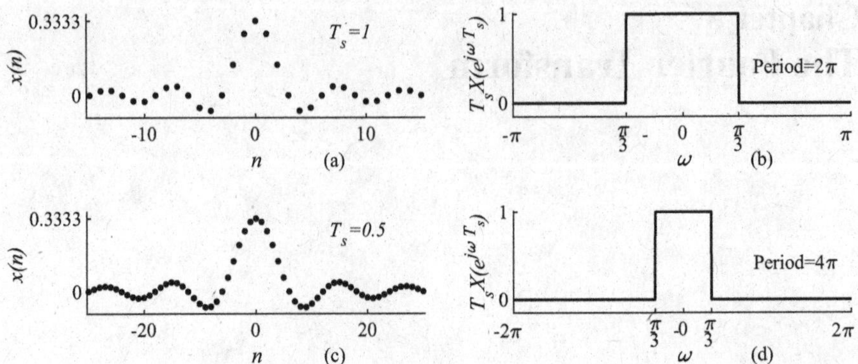

Fig. 8.1 (**a**) Samples of the sinc function $\frac{\sin(\frac{\pi}{3}t)}{\pi t}$, with $T_s = 1$ s and (**b**) its DTFT spectrum, multiplied by T_s, with period 2π radians; (**c**) sinc function with $T_s = 0.5$ s and (**d**) its DTFT spectrum, multiplied by T_s, with period 4π radians

limiting function. As the sampling interval tends to zero, the time-domain waveform becomes continuous with a corresponding aperiodic spectrum.

The foregoing argument can be, mathematically, put as follows. Substituting for $X(e^{j\omega T_s})$ and $\frac{1}{\omega_s}$ replaced by $\frac{T_s}{2\pi}$ in Eq. (7.4), we get

$$x(nT_s) = \frac{T_s}{2\pi} \int_{-\frac{\omega_s}{2}}^{\frac{\omega_s}{2}} e^{j\omega nT_s} \left(\sum_{l=-\infty}^{\infty} x(lT_s)e^{-j\omega lT_s} \right) d\omega$$

As T_s tends to 0, ω_s tends to ∞; nT_s and lT_s become, respectively, continuous time variables t and τ; differential $d\tau$ formally replaces T_s; and the summation becomes an integral. Therefore, we get

$$x(t) = \frac{1}{2\pi} \int_{-\infty}^{\infty} \left(\int_{-\infty}^{\infty} x(\tau)e^{-j\omega\tau} d\tau \right) e^{j\omega t} d\omega = \frac{1}{2\pi} \int_{-\infty}^{\infty} X(j\omega)e^{j\omega t} d\omega$$

The FT $X(j\omega)$ of $x(t)$ is defined as

$$X(j\omega) = \int_{-\infty}^{\infty} x(t)e^{-j\omega t} dt \tag{8.1}$$

The inverse FT $x(t)$ of $X(j\omega)$ is defined as

$$x(t) = \frac{1}{2\pi} \int_{-\infty}^{\infty} X(j\omega)e^{j\omega t} d\omega \tag{8.2}$$

The FT represents a continuous aperiodic signal $x(t)$ as integrals of a continuum of complex sinusoids (amplitude $\frac{1}{2\pi}X(j\omega)d\omega$) over an infinite frequency range.

Although the amplitudes are infinitesimal, the spectrum $X(j\omega)$ (actually the spectral density) gives the relative variations of the amplitudes of the constituent complex sinusoids of a signal. When deriving closed-form expressions for $X(j\omega)$ or $x(t)$,

$$X(j0) = \int_{-\infty}^{\infty} x(t)dt \quad \text{and} \quad x(0) = \frac{1}{2\pi}\int_{-\infty}^{\infty} X(j\omega)d\omega,$$

which can be easily evaluated, are useful to check their correctness. By replacing ω by $2\pi f$ and since $d\omega = 2\pi df$, Eqs. (8.1) and (8.2) can be expressed in terms of the cyclic frequency f as

$$X(j2\pi f) = \int_{-\infty}^{\infty} x(t)e^{-j2\pi ft}dt \quad \text{and} \quad x(t) = \int_{-\infty}^{\infty} X(j2\pi f)e^{j2\pi ft}df$$

Gibbs phenomenon is common to all forms of Fourier analysis, whenever a continuous function, with one or more discontinuities, is reconstructed in either domain.

8.1.2 *Existence of the FT*

Any signal satisfying Dirichlet conditions, which are a set of sufficient conditions, can be expressed in terms of a FT. The first of these conditions is that the signal $x(t)$ is absolutely integrable, that is, $\int_{-\infty}^{\infty} |x(t)|dt < \infty$. From the definition of the FT, we get

$$|X(j\omega)| \le \int_{-\infty}^{\infty} |x(t)e^{-j\omega t}|\,dt = \int_{-\infty}^{\infty} |x(t)||e^{-j\omega t}|\,dt$$

Since $|e^{-j\omega t}| = 1$,

$$|X(j\omega)| \le \int_{-\infty}^{\infty} |x(t)|\,dt$$

Hence, the condition $\int_{-\infty}^{\infty} |x(t)|dt < \infty$ implies that $X(j\omega)$ will exist. The second condition is that the number of finite maxima and minima of $x(t)$ in any finite interval must be finite. The third condition is that the number of finite discontinuities of $x(t)$ in any finite interval must be finite. Most signals of practical interest satisfy these conditions.

As Fourier analysis approximates a signal in the least squares error sense,

$$\int_{-\infty}^{\infty} |x(t)|^2dt - \frac{1}{2\pi}\int_{-\infty}^{\infty} |X(j\omega)|^2d\omega = 0,$$

the FT $X(j\omega)$ of a square integrable signal, $\int_{-\infty}^{\infty} |x(t)|^2dt < \infty$, also exists.

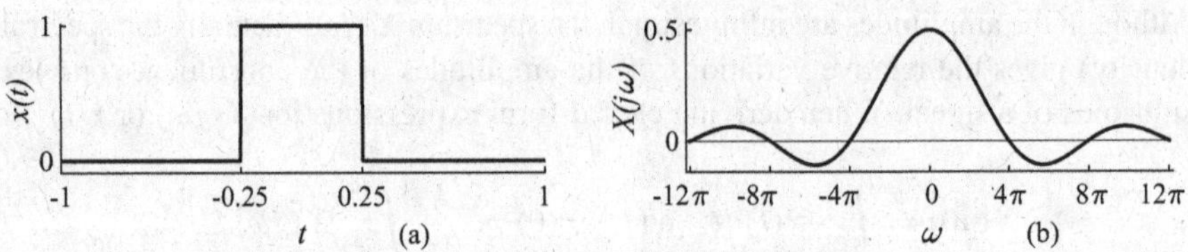

Fig. 8.2 (**a**) The pulse $x(t) = u(t + 0.25) - u(t - 0.25)$ and (**b**) its FT spectrum

Example 8.1 Find the FT of the rectangular pulse $x(t) = u(t + a) - u(t - a)$.

Solution

$$X(j\omega) = \int_{-a}^{a} e^{-j\omega t} dt = 2\int_{0}^{a} \cos(\omega t) dt = \frac{2\sin(\omega a)}{\omega}$$

$$u(t + a) - u(t - a) \iff \frac{2\sin(\omega a)}{\omega}$$

The pulse and its FT are shown, respectively, in Fig. 8.2a and b with $a = 0.25$. ∎

The function of the form $\frac{\sin(\omega a)}{\omega}$, a specific case shown in Fig. 8.2b, is called the sinc function that occurs often in signal and system analysis. It is an even function of ω. At $\omega = 0$, the peak value is a, as $\lim_{\theta \to 0} \sin(\theta) = \theta$. The zeros of the sinc function occur whenever the numerator argument (ωa) of the sine function is equal to $\pm\pi, \pm 2\pi, \ldots$. That is, at $\omega = \pm\frac{\pi}{a}, \pm\frac{2\pi}{a}, \ldots$. For the specific case, the zeros occur whenever ω equals a multiple of 4π. The area enclosed by the sinc function is π irrespective of the value of a, as, by finding the inverse FT of $X(j\omega)$ in Example 8.1 with $t = 0$,

$$x(0) = \frac{1}{2\pi}\int_{-\infty}^{\infty} \frac{2\sin(\omega a)}{\omega} d\omega = 1$$

It is also known that the area enclosed by the function is equal to the area of the triangle inscribed within its main hump. The sinc function is not absolutely integrable. But, it is square integrable and, hence, is an energy signal.

As $a \to 0$, the function $\frac{\sin(\omega a)}{a\omega}$ is expanded and, eventually, degenerates into a DC function. The first pair of zeros at $\omega = \pm\frac{\pi}{a}$ move to infinity, and the function becomes a horizontal line with amplitude 1. As a becomes larger, the numerator sine function $\sin(\omega a)$ of $\frac{\sin(\omega a)}{\pi\omega}$ alone is compressed (frequency of oscillations is increased). As a consequence, the amplitudes of all the ripples along with that of the main hump increase with fixed ratios to one another. While the ripples and the main hump become taller and narrower, the area enclosed by each and the total area enclosed by the function remain fixed. In the limit, as $a \to \infty$, the main hump and all the ripples of significant amplitude are concentrated at $\omega = 0$, and $\frac{\sin(\omega a)}{\pi\omega}$ degenerates into a unit impulse.

Example 8.2 Find the FT $X(j\omega)$ of the real, causal, and decaying exponential signal $x(t) = e^{-at}u(t),\ a > 0$. Find the value of $x(0)$ from $X(j\omega)$.

Solution

$$X(j\omega) = \int_0^{\infty} e^{-at}e^{-j\omega t}dt = \int_0^{\infty} e^{-(a+j\omega)t}dt = -\frac{e^{-(a+j\omega)t}}{a+j\omega}\bigg|_0^{\infty} = \frac{1}{a+j\omega}$$

$$e^{-at}u(t),\ a > 0 \Longleftrightarrow \frac{1}{a+j\omega}$$

$$x(0) = \frac{1}{2\pi}\int_{-\infty}^{\infty}\frac{1}{a+j\omega}d\omega = \frac{1}{2\pi}\int_{-\infty}^{\infty}\frac{a}{\omega^2+a^2}d\omega - \frac{j}{2\pi}\int_{-\infty}^{\infty}\frac{\omega}{\omega^2+a^2}d\omega$$

As the imaginary part of $X(j\omega)$ is odd, its integral evaluates to zero. Therefore,

$$x(0) = \frac{1}{2\pi}\int_{-\infty}^{\infty}\frac{a}{\omega^2+a^2}d\omega = \frac{1}{2\pi}\int_{-\infty}^{\infty}\frac{d\left(\frac{\omega}{a}\right)}{\left(\frac{\omega}{a}\right)^2+1} = \frac{1}{2\pi}\tan^{-1}\left(\frac{\omega}{a}\right)\bigg|_{-\infty}^{\infty} = \frac{1}{2}$$

The value of $x(t)$ at $t = 0$ is always $\frac{1}{2}$ for any value of a. Note that the Fourier reconstructed waveform converges to the average of the right- and left-hand limits at any discontinuity. ∎

For some signals, such as a step signal or a sinusoid, which are neither absolutely nor square integrable, the FT is obtained by applying a limiting process to appropriate signals so that they degenerate into these signals in the limit. The limit of the corresponding transform is the transform of the signal under consideration, as presented in the next example.

Example 8.3 Find the FT of $x(t) = u(t)$, the unit step function.

Solution As $u(t)$ is not absolutely or square integrable, we consider it as the limiting form of the decaying exponential, $e^{-at}u(t),\ a > 0$, as $a \to 0$. Therefore, as the FT of the exponential is $\frac{1}{j\omega+a}$, the FT of $u(t)$ is given by

$$X(j\omega) = \lim_{a\to 0}\frac{1}{j\omega+a} = \lim_{a\to 0}\frac{a}{\omega^2+a^2} - \lim_{a\to 0}\frac{j\omega}{\omega^2+a^2} = \lim_{a\to 0}\frac{a}{\omega^2+a^2} + \frac{1}{j\omega}$$

The area under the real part of $X(j\omega)$ is π regardless of the value of a, as found in Example 8.2. As $a \to 0$, the value of this function tends to zero at all values of ω except when $\omega = 0$, where its area is π. Therefore, $\lim_{a\to 0}\frac{a}{\omega^2+a^2} = \pi\delta(\omega)$, and

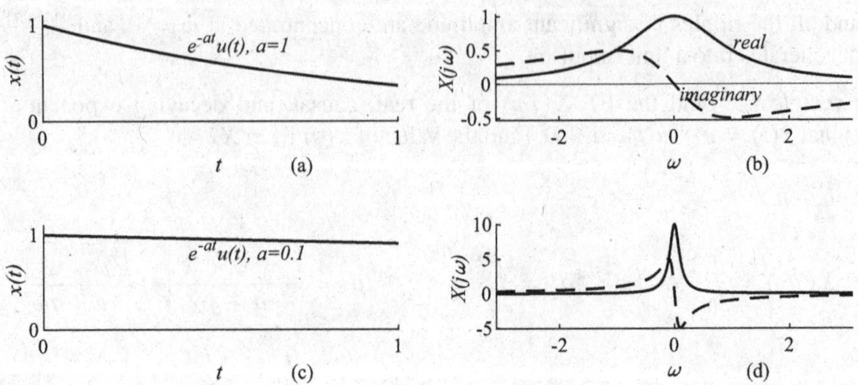

Fig. 8.3 (**a**) $x(t) = e^{-at}u(t) and\ a = 1$ and (**b**) its FT spectrum. The real part of the FT is shown by the continuous line, and the imaginary part is shown by the dashed line; (**c**) $x(t) = e^{-at}u(t) and\ a = 0.1$ and (**d**) its FT spectrum

$$u(t) \Longleftrightarrow \pi\delta(\omega) + \frac{1}{j\omega}$$

That is, the spectrum of the unit step function is composed of an impulsive component $\pi\delta(\omega)$ (an impulse of strength π at $\omega = 0$) and a strictly continuous (except at $\omega = 0$) component $\frac{1}{j\omega}$. The real part of the transform $\pi\delta(\omega)$ corresponds to the even component $u_e(t) = 0.5$ of $u(t)$, and the imaginary part $\frac{1}{j\omega}$ corresponds to the odd component $u_o(t) = -0.5,\ t < 0$ and $u_o(t) = 0.5,\ t > 0$.

Figure 8.3 depicts the limiting process by which a real exponential function degenerates into a unit step function. Figure 8.3a and c shows, respectively, the signal $e^{-at}u(t)$ with $a = 1$ and $a = 0.1$. Figure 8.3b and d shows, respectively, their corresponding spectra. The real part of the spectrum (continuous line) is an even function with a peak value of $\frac{1}{a}$ at $\omega = 0$, and the imaginary part (dashed line) is an odd function with peaks of value $\pm\frac{1}{2a}$ at $\omega = \mp a$. As $a \to 0$, the real part becomes more peaked and, eventually, degenerates into an impulse of strength π, that is, $\pi\delta(\omega)$. The imaginary part becomes a rectangular hyperbola in the limit. ∎

Example 8.4 Find the FT of the unit impulse signal $x(t) = \delta(t)$.

Solution Using the sampling property of the impulse, we get

$$X(j\omega) = \int_{-\infty}^{\infty} \delta(t)e^{-j\omega t}dt = e^{-j\omega 0}\int_{-\infty}^{\infty} \delta(t)dt = 1 \quad \text{and} \quad \delta(t) \Longleftrightarrow 1$$

The unit impulse signal is composed of complex sinusoids, with zero phase shift, of all frequencies from $\omega = -\infty$ to $\omega = \infty$ in equal proportion. That is,

$$\delta(t) = \frac{1}{2\pi}\int_{-\infty}^{\infty} e^{j\omega t}d\omega = \frac{1}{2\pi}\int_{-\infty}^{\infty} \cos(\omega t)d\omega = \frac{1}{\pi}\int_{0}^{\infty} \cos(\omega t)d\omega$$

∎

Example 8.5 Find the inverse FT of $X(j\omega) = \delta(\omega)$.

Solution

$$x(t) = \frac{1}{2\pi}\int_{-\infty}^{\infty} \delta(\omega)e^{j\omega t}d\omega = \frac{1}{2\pi} \quad \text{and} \quad 1 \Longleftrightarrow 2\pi\delta(\omega)$$

An impulse at $\omega = 0$ properly represents the DC signal, since it is characterized by the single frequency $\omega = 0$ alone. That is, $x(t) = e^{j\omega_0 t}$ with $\omega_0 = 0$. Similar to the DFT, the scale factor $\frac{1}{2\pi}$ is included in the inverse transform. Therefore, the spectrum of DC is an impulse at $\omega = 0$ with strength 2π rather than one. The placement of the constant in the forward or inverse definition of a transform is a matter of convention. ∎

Example 8.6 Find the inverse FT of $X(j\omega) = \delta(\omega - \omega_0)$.

Solution

$$x(t) = \frac{1}{2\pi}\int_{-\infty}^{\infty} \delta(\omega-\omega_0)e^{j\omega t}d\omega = \frac{1}{2\pi}e^{j\omega_0 t} \quad \text{and} \quad e^{j\omega_0 t} \Longleftrightarrow 2\pi\delta(\omega-\omega_0)$$

That is, the spectrum of the complex sinusoid $e^{j\omega_0 t}$ is an impulse at $\omega = \omega_0$ with strength 2π. ∎

8.2 Properties of the Fourier Transform

Properties present the frequency-domain effect of time-domain characteristics and operations on signals and vice versa. In addition, they are used to find new transform pairs more easily.

8.2.1 Linearity

The FT of a linear combination of a set of signals is the same linear combination of their individual FT. That is,

$$x(t) \Longleftrightarrow X(j\omega), \quad y(t) \Longleftrightarrow Y(j\omega), \quad ax(t) + by(t) \Longleftrightarrow aX(j\omega) + bY(j\omega),$$

where a and b are arbitrary constants. This property follows from the linearity property of the integral defining the FT. Consider the sign, $sgn(t)$ (pronounced as $signum(t)$) signal, defined as

$$x(t) = \begin{cases} 1 & \text{for } t > 0 \\ -1 & \text{for } t < 0 \end{cases}$$

This signal can be expressed as $(2u(t) - 1)$. Substituting the respective FT, we get the FT of $x(t)$ as $2(\pi\delta(\omega) + \frac{1}{j\omega}) - 2\pi\delta(\omega) = \frac{2}{j\omega}$.

8.2.2 *Duality*

The forward and inverse FT definitions differ only by the reversed algebraic sign in the exponent of the complex exponential, the interchange of the variables t and ω, and the constant $\frac{1}{2\pi}$ in the inverse FT. Due to this similarity, there exists a dual relationship between time- and frequency-domain functions. Consider the inverse FT defined as

$$x(t) = \frac{1}{2\pi}\int_{-\infty}^{\infty} X(j\omega)e^{j\omega t}d\omega$$

By replacing t by $-t$, we get

$$x(-t) = \frac{1}{2\pi}\int_{-\infty}^{\infty} X(j\omega)e^{-j\omega t}d\omega \quad \text{and} \quad 2\pi x(-t) = \int_{-\infty}^{\infty} X(j\omega)e^{-j\omega t}d\omega$$

This is a forward transform with $2\pi x(-t)$ being the FT of $X(j\omega)$. To put it another way, we get $2\pi x(-t)$ by taking the FT of $x(t)$ twice in succession, $2\pi x(-t) = \text{FT}(\text{FT}(x(t)))$. Let $x(t) \Longleftrightarrow X(j\omega)$. If we replace the variable ω in the frequency-domain function by $\pm t$, then the corresponding frequency-domain function is obtained by replacing the variable t by $\mp\omega$ in the original time-domain function multiplied by 2π. For an even $x(t)$, as $X(j\omega)$ is also even, the sign change of either t or ω is not required. For example, consider the FT pairs

$$\cos(2t) \Longleftrightarrow \pi(\delta(\omega+2)+\delta(\omega-2)) \quad \text{and} \quad \sin(3t) \Longleftrightarrow j\pi(\delta(\omega+3)-\delta(\omega-3))$$

Using the property, we get the transform pairs

$$2\cos(2(-\omega)) = 2\cos(2\omega) \Longleftrightarrow \delta(t+2) + \delta(t-2)$$

$$2\sin(3\omega) \Longleftrightarrow j(\delta(-t+3) - \delta(-t-3)) = j(\delta(t-3) - \delta(t+3))$$

8.2.3 Symmetry

If a signal $x(t)$ is real, then the real part of its spectrum $X(j\omega)$ is even, and the imaginary part is odd, called the conjugate symmetry. The FT of $x(t)$ is given by

$$X(j\omega) = \int_{-\infty}^{\infty} x(t)e^{-j\omega t}dt = \int_{-\infty}^{\infty} x(t)(\cos(\omega t) - j\sin(\omega t))dt$$

Conjugating both sides, we get

$$X^*(j\omega) = \int_{-\infty}^{\infty} x(t)(\cos(\omega t) + j\sin(\omega t))dt$$

Replacing ω by $-\omega$, we get $X^*(-j\omega) = X(j\omega)$. This is expected since a real sinusoid is composed of a pair of complex exponentials. An example is

$$x(t) = e^{-t}u(t) \Longleftrightarrow X(j\omega) = \frac{1}{j\omega + 1} = \frac{1}{\omega^2 + 1} - \frac{j\omega}{\omega^2 + 1}$$

If a signal $x(t)$ is real and even, then its spectrum also is real and even. Since $x(t)\cos(\omega t)$ is even and $x(t)\sin(\omega t)$ is odd,

$$X(j\omega) = 2\int_{0}^{\infty} x(t)\cos(\omega t)dt \quad \text{and} \quad x(t) = \frac{1}{\pi}\int_{0}^{\infty} X(j\omega)\cos(\omega t)d\omega$$

The FT $\pi(\delta(\omega+1)+\delta(\omega-1))$ of $\cos(t)$ is an example of the FT of an even function.

Similarly, if a signal $x(t)$ is real and odd, then its spectrum is imaginary and odd.

$$X(j\omega) = -j2\int_{0}^{\infty} x(t)\sin(\omega t)dt \quad \text{and} \quad x(t) = \frac{j}{\pi}\int_{0}^{\infty} X(j\omega)\sin(\omega t)d\omega$$

The FT $j\pi(\delta(\omega+1)-\delta(\omega-1))$ of $\sin(t)$ is an example of the FT of an odd function.

As the FT of a real and even signal is real and even and that of a real and odd is imaginary and odd, it follows that the real part of the FT, $\mathrm{Re}(X(j\omega))$, of an arbitrary real signal $x(t)$ is the transform of its even component $x_e(t)$ and $j\,\mathrm{Im}(X(j\omega))$ is that of its odd component $x_o(t)$. For example,

$$\begin{aligned} x(t) &= e^{-t}u(t) = x_e(t) + x_o(t) = 0.5e^{-|t|} + (0.5e^{-t}u(t) - 0.5e^{t}u(-t)) \\ &\Longleftrightarrow X(j\omega) = \frac{1}{j\omega + 1} = \frac{1}{\omega^2 + 1} - \frac{j\omega}{\omega^2 + 1} \end{aligned}$$

8.2.4 Time Shifting

When we shift a signal, the shape remains the same, but the signal is relocated. The shift of a typical spectral component, $X(j\omega_a)e^{j\omega_a t}$, by t_0 to the right results in the exponential, $X(j\omega_a)e^{j\omega_a(t-t_0)} = e^{-j\omega_a t_0}X(j\omega_a)e^{j\omega_a t}$. That is, a delay of t_0 results in changing the phase of the exponential by $-\omega_a t_0$ radians without changing its amplitude. Therefore, if the FT of $x(t)$ is $X(j\omega)$, then

$$x(t \pm t_0) \Longleftrightarrow e^{\pm j\omega t_0}X(j\omega)$$

Consider the FT of $\cos(2t)$, $\pi(\delta(\omega+2)+\delta(\omega-2))$. Now, the FT of $\cos(2(t-\frac{\pi}{4})) = \cos(2t-\frac{\pi}{2}) = \sin(2t)$ is

$$\pi(e^{-j(-2)\frac{\pi}{4}}\delta(\omega+2) + e^{-j2\frac{\pi}{4}}\delta(\omega-2)) = j\pi(\delta(\omega+2) - \delta(\omega-2))$$

8.2.5 Frequency Shifting

The spectrum, $X(j\omega)$, of a signal, $x(t)$, can be shifted by multiplying the signal by a complex exponential, $e^{\pm j\omega_0 t}$. The new spectrum is $X(j(\omega \mp \omega_0))$, since a spectral component $X(j\omega_a)e^{j\omega_a t}$ of the signal multiplied by $e^{j\omega_0 t}$ becomes $X(j\omega_a)e^{j(\omega_a+\omega_0)t}$ and the spectral value $X(j\omega_a)$ occurs at $(\omega_a + \omega_0)$, after a delay of ω_0 radians. That is,

$$x(t)e^{\pm j\omega_0 t} \Longleftrightarrow X(j(\omega \mp \omega_0))$$

Duality applies for both transform pairs and properties. This property is the dual of the time-shifting property.

Consider the FT pair $e^{-2t}u(t) \Longleftrightarrow \frac{1}{2+j\omega}$. The FT of $e^{-2t}\cos(3t)u(t) = e^{-2t}\frac{(e^{j3t}+e^{-j3t})}{2}u(t)$ is

$$X(j\omega) = \frac{1}{2}\left(\frac{1}{2+j(\omega-3)} + \frac{1}{2+j(\omega+3)}\right) = \frac{2+j\omega}{(2+j\omega)^2+9}$$

For example,

$$\frac{1}{2+j\omega}|_{\omega=0} = \frac{1}{2}$$

and

$$\frac{1}{2+j(\omega-3)}|_{\omega=3} = \frac{1}{2} \quad \text{and} \quad \frac{1}{2+j(\omega+3)}|_{\omega=-3} = \frac{1}{2}$$

8.2.6 Convolution in the Time Domain

The convolution $x(t) * h(t)$ of signals $x(t)$ and $h(t)$ is defined, in Chap. 4, as

$$y(t) = x(t) * h(t) = \int_{-\infty}^{\infty} x(\tau)h(t-\tau)d\tau$$

The convolution of $h(t)$ with a complex exponential $e^{j\omega_0 t}$ is given as

$$\int_{-\infty}^{\infty} h(\tau)e^{j\omega_0(t-\tau)}d\tau = e^{j\omega_0 t}\int_{-\infty}^{\infty} h(\tau)e^{-j\omega_0\tau}d\tau = H(j\omega_0)e^{j\omega_0 t}$$

As an arbitrary signal $x(t)$ is reconstructed by the inverse FT as $x(t) = \frac{1}{2\pi}\int_{-\infty}^{\infty} X(j\omega)e^{j\omega t}d\omega$, the convolution of $x(t)$ and $h(t)$ is given by $y(t) = \frac{1}{2\pi}\int_{-\infty}^{\infty} X(j\omega)H(j\omega)e^{j\omega t}d\omega$. The inverse FT of $X(j\omega)H(j\omega)$ is the convolution of $x(t)$ and $h(t)$. That is,

$$\int_{-\infty}^{\infty} x(\tau)h(t-\tau)d\tau = \frac{1}{2\pi}\int_{-\infty}^{\infty} X(j\omega)H(j\omega)e^{j\omega t}d\omega \Longleftrightarrow X(j\omega)H(j\omega)$$

Therefore, convolution in the time domain corresponds to multiplication in the frequency domain. This property is one of the major reasons for the dominant role of the frequency-domain analysis in the study of signals and systems.

The convolution of a rectangular pulse, centered at the origin, of width a and height $\frac{1}{a}$ with itself yields a triangular waveform, centered at the origin, with width $2a$ and height $\frac{1}{a}$. Figure 8.4a and b shows, respectively, these waveforms with $a = 2$. Since convolution in the time domain corresponds to multiplication in the

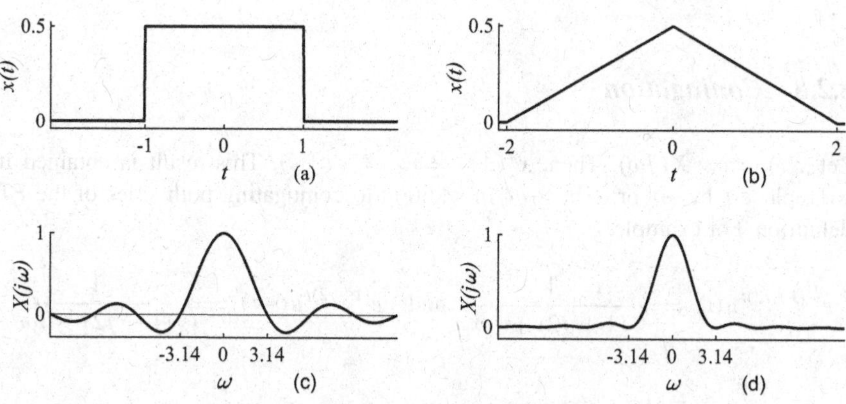

Fig. 8.4 (**a**) The rectangular pulse with width 2 and height 0.5; (**b**) the triangular waveform with width 4 and height 0.5, which is the convolution of the pulse in (**a**) with itself; (**c**) and (**d**) their corresponding FT spectra

frequency domain and the FT of the rectangular pulse is $\frac{2\sin(\frac{\omega a}{2})}{a\omega}$, we get the FT of the triangular waveform as

$$X(j\omega) = \frac{2\sin(\frac{\omega a}{2})}{a\omega}\frac{2\sin(\frac{\omega a}{2})}{a\omega} = \left(\frac{2\sin(\frac{\omega a}{2})}{a\omega}\right)^2$$

Figure 8.4c and d shows, respectively, their FT spectra.

8.2.7 Convolution in the Frequency Domain

Similar to the time-domain convolution, we find that the convolution of two frequency-domain functions corresponds to the multiplication of the inverse FT of the functions in the time domain with a scale factor. That is,

$$x(t)y(t) \Longleftrightarrow \int_{-\infty}^{\infty} x(t)y(t)e^{-j\omega t}dt = \frac{1}{2\pi}\int_{-\infty}^{\infty} X(jv)Y(j(\omega - v))dv$$

The FT of $\sin(t)\cos(t)$ is the convolution of the FT of $\sin(t)$ and $\cos(t)$ divided by 2π. That is,

$$\begin{aligned}
&\frac{1}{2\pi}(j\pi(\delta(\omega+1) - \delta(\omega-1)) * \pi(\delta(\omega+1) + \delta(\omega-1))) \\
&= \frac{j\pi^2}{2\pi}((\delta(\omega+1) * \delta(\omega+1)) - (\delta(\omega-1) * \delta(\omega-1))) \\
&= \frac{j\pi}{2}(\delta(\omega+2) - \delta(\omega-2)) \Longleftrightarrow \frac{1}{2}\sin(2t) = \sin(t)\cos(t)
\end{aligned}$$

8.2.8 Conjugation

Let $x(t) \Longleftrightarrow X(j\omega)$. Then, $x^*(\pm t) \Longleftrightarrow X^*(\mp j\omega)$. This result is obtained if we replace t by $-t$ or ω by $-\omega$, in addition to conjugating both sides of the FT definition. For example,

$$e^{-(1+j2)t}u(t) \Longleftrightarrow \frac{1}{(1+j2)+j\omega} \quad \text{and} \quad e^{(1-j2)t}u(-t) \Longleftrightarrow \frac{1}{(1-j2)-j\omega}$$

8.2.9 *Time Reversal*

Let $x(t) \Longleftrightarrow X(j\omega)$. Then, $x(-t) \Longleftrightarrow X(-j\omega)$. That is, the time reversal of a signal results in its spectrum also reflected about the vertical axis at the origin. This result is obtained if we replace t by $-t$ and ω by $-\omega$ in the FT definition. For example,

$$e^{-3t}u(t) \Longleftrightarrow \frac{1}{3+j\omega} \quad \text{and} \quad e^{3t}u(-t) \Longleftrightarrow \frac{1}{3-j\omega}$$

8.2.10 *Time Scaling*

Scaling is the operation of replacing the independent variable t by at, where $a \neq 0$ is a real constant. As we have seen in Chap. 2, the signal is compressed ($|a| > 1$) or expanded ($|a| < 1$) in the time domain by this operation. As a consequence, the spectrum of the signal is expanded or compressed in the frequency domain. With a negative, the signal is also time-reversed.

Let the spectrum of a signal $x(t)$ be $X(j\omega)$. By replacing at by τ, t by $\frac{\tau}{a}$, and dt by $\frac{d\tau}{a}$, with $a > 0$, in the FT definition of $x(at)$, we get

$$\int_{-\infty}^{\infty} x(at)e^{-j\omega t}dt = \frac{1}{a}\int_{-\infty}^{\infty} x(\tau)e^{-j\omega\frac{\tau}{a}}d\tau = \frac{1}{a}X\left(j\left(\frac{\omega}{a}\right)\right)$$

The FT of $x(-at)$, due to the time-reversal property, becomes

$$\frac{1}{a}X\left(j\left(\frac{-\omega}{a}\right)\right) = \frac{1}{a}X\left(j\left(\frac{\omega}{-a}\right)\right)$$

By combining both the results, we get

$$x(at) \Longleftrightarrow \frac{1}{|a|}X\left(j\left(\frac{\omega}{a}\right)\right), \qquad a \neq 0$$

The factor $\frac{1}{|a|}$ ensures that the scaled waveforms in both the domains have the same energy or power. A compressed signal varies more rapidly and, hence, requires higher-frequency components to synthesize. Therefore, the spectrum is expanded. The reverse is the case for signal expansion.

Consider the transform pair $\sin(2t) \Longleftrightarrow (j\pi)(\delta(\omega+2) - \delta(\omega-2))$. $\sin(6t)$ is a time-compressed version of $\sin(2t)$ with $a = 3$. Using the property, the transform of $\sin(6t)$ is obtained from that of $\sin(2t)$ as follows.

$$\frac{1}{3}(j\pi)\left(\delta\left(\frac{\omega}{3}+2\right)-\delta\left(\frac{\omega}{3}-2\right)\right)=\frac{1}{3}(j\pi)\left(\delta\left(\frac{\omega+6}{3}\right)-\delta\left(\frac{\omega-6}{3}\right)\right)$$
$$=(j\pi)(\delta(\omega+6)-\delta(\omega-6))$$

Note that $\delta(a\omega)=\frac{1}{|a|}\delta(\omega)$.

Consider the transform pair

$$\cos(t)\Longleftrightarrow\pi(\delta(\omega+1)+\delta(\omega-1))$$

$\cos(-2t)$ is a time-compressed version of $\cos(t)$ with $a=-2$. Using the property, the transform of $\cos(-2t)=\cos(2t)$ is obtained from that of $\cos(t)$ as follows.

$$\frac{1}{2}(\pi)\left(\delta\left(\frac{\omega}{-2}+1\right)+\delta\left(\frac{\omega}{-2}-1\right)\right)=\frac{1}{2}(\pi)\left(\delta\left(\frac{\omega-2}{-2}\right)+\delta\left(\frac{\omega+2}{-2}\right)\right)$$
$$=\pi(\delta(\omega+2)+\delta(\omega-2))$$

8.2.11 Time Differentiation

The derivative of a typical spectral component $X(j\omega_a)e^{j\omega_a t}$ is $j\omega_a X(j\omega_a)e^{j\omega_a t}$. Therefore, if the transform of a time-domain function $x(t)$ is $X(j\omega)$, then the transform of its derivative is given by $j\omega X(j\omega)$. That is,

$$\frac{dx(t)}{dt}\Longleftrightarrow j\omega X(j\omega)$$

Note that the spectral value with $\omega=0$ is zero, as the DC component is lost in differentiating a signal. The factor ω implies that the magnitude of the high-frequency components is enhanced more and, hence, rapid time variations of the signal are accentuated. The property is valid only if the derivative function is Fourier transformable. For example,

$$e^{-t}u(t)\Longleftrightarrow\frac{1}{j\omega+1}\quad\text{and}\quad\frac{d(e^{-t}u(t))}{dt}=(\delta(t)-e^{-t}u(t))\Longleftrightarrow\frac{j\omega}{j\omega+1}$$

In general,

$$\frac{d^n x(t)}{dt^n}\Longleftrightarrow(j\omega)^n X(j\omega)$$

Consider finding the FT, shown in Fig. 8.5b, of the triangular waveform $x(t)=0.5(t+2)u(t+2)-tu(t)+0.5(t-2)u(t-2)$, shown in Fig. 8.5a. This problem was solved using the convolution property. Now, we use the differentiation property. The

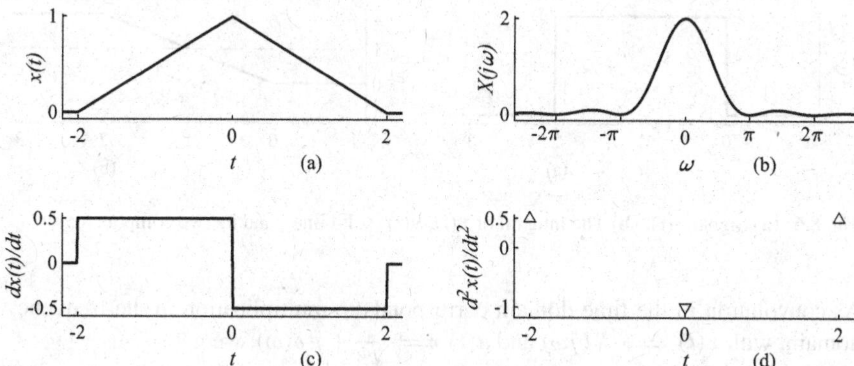

Fig. 8.5 (**a**) The triangular waveform and (**b**) its spectrum; (**c**) the first derivative of the triangular waveform; (**d**) the second derivative of the triangular waveform

FT of scaled and shifted impulse function can be found easily. Therefore, the idea is to reduce the given function to a set of impulses by differentiating it successively. (This method is applicable to signals that are characterized or approximated by any piecewise polynomial function with finite energy.) Then, the FT of the impulses can be related to the FT of the given function by the differentiation property. The first and second derivatives of the triangular waveform, $\frac{dx(t)}{dt} = 0.5u(t+2) - u(t) + 0.5u(t-2)$ and $\frac{d^2x(t)}{dt^2} = 0.5\delta(t+2) - \delta(t) + 0.5\delta(t-2)$, are shown, respectively, in Fig. 8.5c and d. Let the FT of the triangular waveform be $X(j\omega)$. Then, the FT of the impulses of $\frac{d^2x(t)}{dt^2}$ shown in Fig. 8.5d, $(0.5e^{j2\omega} - 1 + 0.5e^{-j2\omega})$, must be equal to $(j\omega)^2X(j\omega)$. That is,

$$(0.5e^{j2\omega} - 1 + 0.5e^{-j2\omega}) = \cos(2\omega) - 1 = -2\sin^2(\omega) = -\omega^2X(j\omega)$$

Solving for $X(j\omega)$, we get the FT of the triangular waveform as

$$X(j\omega) = 2\left(\frac{\sin(\omega)}{\omega}\right)^2$$

8.2.12 *Time Integration*

The definite integral, $y(t)$, of a time-domain signal, $x(t)$, can be expressed as the convolution of $x(t)$ and the unit-step signal, $u(t)$, as

$$y(t) = \int_{-\infty}^{t} x(\tau)d\tau = \int_{-\infty}^{\infty} x(\tau)u(t-\tau)d\tau = x(t) * u(t)$$

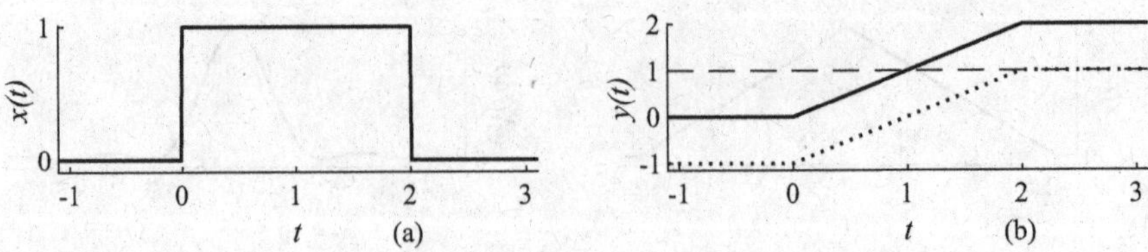

Fig. 8.6 (**a**) Signal $x(t)$. (**b**) The integral of $x(t)$, $y(t)$ (solid line), and its two components

As convolution in the time domain corresponds to multiplication in the frequency domain, with $x(t) \Longleftrightarrow X(j\omega)$ and $u(t) \Longleftrightarrow \frac{1}{j\omega} + \pi\delta(\omega)$, we get

$$\int_{-\infty}^{t} x(\tau)d\tau \Longleftrightarrow X(j\omega)\left(\frac{1}{j\omega} + \pi\delta(\omega)\right) = \frac{X(j\omega)}{j\omega} + \pi X(j0)\delta(\omega)$$

Note that, if $X(j0) = 0$, the integration operation can be considered as the inverse of the differentiation operation. The property is valid only if $y(t)$ is Fourier transformable. The factor ω in the denominator implies that the magnitude of the high-frequency components is reduced more and, hence, rapid time variations of the signal are reduced, resulting in a smoother signal.

Consider the signal $x(t) = u(t) - u(t-2)$, shown in Fig. 8.6a, with the FT $X(j\omega) = \frac{1}{j\omega}(1 - e^{-j2\omega})$ and $X(j0) = 2$. Now, using the property,

$$y(t) = \int_{-\infty}^{t} x(\tau)d\tau \Longleftrightarrow Y(j\omega) = \frac{X(j\omega)}{j\omega} + 2\pi\delta(\omega) = 2\pi\delta(\omega) + \frac{(e^{-j2\omega} - 1)}{\omega^2}$$

The integral of $x(t)$ is $y(t) = tu(t) - (t-2)u(t-2)$, shown in Fig. 8.6b along with its two components corresponding to the two terms of its transform.

8.2.13 *Frequency Differentiation*

Differentiating both sides of the FT definition with respect to ω yields

$$(-jt)x(t) \Longleftrightarrow \frac{dX(j\omega)}{d\omega} \quad \text{or} \quad tx(t) \Longleftrightarrow j\frac{dX(j\omega)}{d\omega}$$

The property is valid only if the resulting function is Fourier transformable. In general,

$$(-jt)^n x(t) \Longleftrightarrow \frac{d^n X(j\omega)}{d\omega^n} \quad \text{or} \quad (t)^n x(t) \Longleftrightarrow (j)^n \frac{d^n X(j\omega)}{d\omega^n}$$

For example,

$$e^{-2t}u(t) \Longleftrightarrow \frac{1}{j\omega+2} \quad \text{and} \quad te^{-2t}u(t) \Longleftrightarrow \frac{1}{(j\omega+2)^2}$$

8.2.14 *Parseval's Theorem and the Energy Transfer Function*

As the frequency-domain representation of a signal is an equivalent representation, energy E of a signal can also be expressed in terms of its spectrum. Note that this theorem is only applicable to FT of energy signals. From the frequency-domain convolution property, we get

$$\int_{-\infty}^{\infty} x(t)y(t)e^{-j\omega t}dt = \frac{1}{2\pi}\int_{-\infty}^{\infty} X(jv)Y(j(\omega - v))dv$$

Letting $\omega = 0$ and then replacing v by ω, we get

$$\int_{-\infty}^{\infty} x(t)y(t)dt = \frac{1}{2\pi}\int_{-\infty}^{\infty} X(j\omega)Y(-j\omega)d\omega$$

Assuming $x^*(t) = y(t)$, $X^*(-j\omega) = Y(j\omega)$ and $X^*(j\omega) = Y(-j\omega)$. Therefore, we get

$$\int_{-\infty}^{\infty} x(t)x^*(t)dt = \frac{1}{2\pi}\int_{-\infty}^{\infty} X(j\omega)X^*(j\omega)d\omega$$

$$E = \int_{-\infty}^{\infty} |x(t)|^2 dt = \frac{1}{2\pi}\int_{-\infty}^{\infty} |X(j\omega)|^2 d\omega$$

This relationship is called Parseval's theorem. This expression is the limiting form of the corresponding expression for DTFT as the sampling interval of the time-domain signal tends to zero. Alternately, this expression can also be considered as the limiting form of the corresponding expression for FS as the period of the signal tends to infinity. For real signals, as $|X(j\omega)|$ is even, we get

$$E = \int_{-\infty}^{\infty} |x(t)|^2 dt = \frac{1}{\pi}\int_{0}^{\infty} |X(j\omega)|^2 d\omega$$

The quantity $|X(j\omega)|^2$ is called the energy spectral density of the signal, since $\frac{1}{2\pi}|X(j\omega)|^2 d\omega$ is the signal energy over the infinitesimal frequency band ω to $\omega + d\omega$.

Example 8.7 Find the energy of the signal $x(t) = e^{-t}u(t)$. Find the value of T such that 99% of the signal energy lies in the range $0 \le t \le T$. What is the corresponding signal bandwidth B, where B is such that 99% of the spectral energy lies in the range $0 \le \omega \le B$?

Solution From the transform pair of Example 8.2, we get

$$e^{-t}u(t) \Longleftrightarrow \frac{1}{1 + j\omega}$$

The energy E of the signal is

$$E = \int_{-\infty}^{\infty} |x(t)|^2 dt = \int_{0}^{\infty} e^{-2t} dt = \frac{1}{2}$$

By changing the upper limit to T, we get

$$\int_{0}^{T} e^{-2t} dt = -\frac{1}{2}(e^{-2T} - 1) = \frac{0.99}{2} = 0.495$$

Solving for T, we get $T = 2.3026$ s. This value is required in order to truncate the signal for numerical analysis.

Using the spectrum,

$$\frac{1}{\pi}\int_{0}^{B} \frac{d\omega}{1 + \omega^2} = \frac{1}{\pi}\tan^{-1}(B) = 0.495 \quad \text{or} \quad B = \tan(0.495\pi) = 63.6567$$

Using this value, we can determine the sampling interval required to sample this signal. As the sampling frequency must be greater than twice of that of the highest-frequency component, the sampling frequency must be greater than (2)(63.6567) radians/second. Therefore, the sampling interval must be smaller than $\frac{2\pi}{(2)(63.6567)} = 0.0494$ s. ∎

Since $|X(j\omega)|^2 = X(j\omega)X^*(j\omega) = X(j\omega)X(-j\omega)$ for real signals, $x(t)*x(-t) \Longleftrightarrow |X(j\omega)|^2$. The convolution $x(t)*x(-t)$, called the autocorrelation of $x(t)$, is defined as

$$x(t) * x(-t) = \int_{-\infty}^{\infty} x(\tau)x(\tau - t)d\tau \Longleftrightarrow |X(j\omega)|^2$$

The input and output of a LTI system, in the frequency domain, are related by the transfer function $H(j\omega)$ as $Y(j\omega) = H(j\omega)X(j\omega)$, where $X(j\omega)$, $Y(j\omega)$, and $H(j\omega)$ are the FT of the input, output, and impulse response of the system. The output energy spectrum is given by

$$\begin{aligned} |Y(j\omega)|^2 &= Y(j\omega)Y^*(j\omega) \\ &= H(j\omega)X(j\omega)H^*(j\omega)X^*(j\omega) = |H(j\omega)|^2|X(j\omega)|^2 \end{aligned}$$

The quantity $|H(j\omega)|^2$ is called the energy transfer function, as it relates the input and output energy spectral densities of a system.

8.3 Fourier Transform of Mixed Class of Signals

As the most general version of the Fourier analysis, the FT is capable of representing all types of signals. Therefore, the relation between the FT and other versions of the Fourier analysis is important in dealing with mixed class of signals. The signal $x(t)$ and its FT $X(j\omega)$ are, in general, continuous and aperiodic. The inverse FT of a sampled spectrum $X_s(j\omega)$ yields a periodic signal, which is the sum of a periodic repetition of $x(t)$. This version corresponds to the FS. On the other hand, the FT of a sampled signal $x_s(t)$ yields a periodic spectrum, which is the sum of a periodic repetition of $X(j\omega)$. This version corresponds to the DTFT. Sampling in both the domains corresponds to the DFT with both the signal and its spectrum sampled and periodic.

8.3.1 The FT of a Continuous Periodic Signal

A periodic signal $x(t)$ is reconstructed using its FS coefficients $X_{cs}(k)$ as

$$x(t) = \sum_{k=-\infty}^{\infty} X_{cs}(k)e^{jk\omega_0 t},$$

where ω_0 is the fundamental frequency. Since the FT of $e^{jk\omega_0 t}$ is $2\pi\delta(\omega - k\omega_0)$, we get, from the linearity property of the FT,

$$x(t) = \sum_{k=-\infty}^{\infty} X_{cs}(k)e^{jk\omega_0 t} \Longleftrightarrow X(j\omega) = 2\pi \sum_{k=-\infty}^{\infty} X_{cs}(k)\delta(\omega - k\omega_0)$$

Therefore, the FT of a periodic signal is a sum of impulses with strength $2\pi X_{cs}(k)$ occurring at intervals of ω_0.

Example 8.8 Find the FT of the signal $x(t) = \cos(\omega_0 t)$.

Solution The FS spectrum for $\cos(\omega_0 t)$ is $\frac{1}{2}(\delta(k-1) + \delta(k+1))$. Multiplying this result by 2π and with the discrete impulse $\delta(k-1)$ corresponding to the continuous impulse $\delta(\omega - \omega_0)$, we get the FT as $\pi(\delta(\omega - \omega_0) + \delta(\omega + \omega_0))$. Hence,

$$\cos(\omega_0 t) \Longleftrightarrow \pi(\delta(\omega - \omega_0) + \delta(\omega + \omega_0))$$

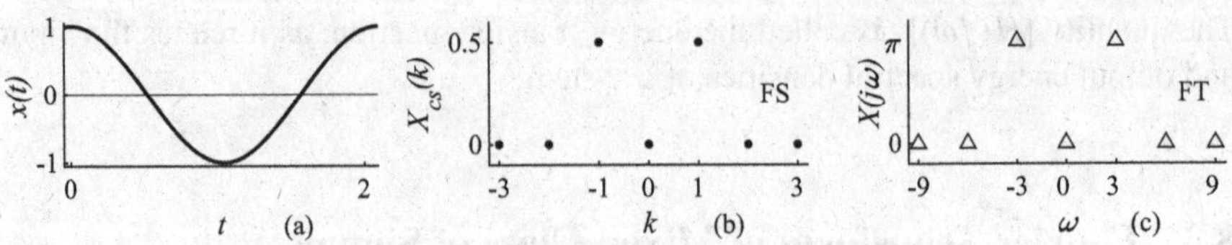

Fig. 8.7 (**a**) The sinusoid $\cos(3t)$; (**b**) its FS spectrum, $X_{cs}(k)$; (**c**) its FT, $X(j\omega)$

Similarly,

$$\sin(\omega_0 t) \Longleftrightarrow (j\pi)(\delta(\omega+\omega_0)-\delta(\omega-\omega_0))$$

In general,

$$\cos(\omega_0 t+\phi) \Longleftrightarrow \pi(e^{j\phi}\delta(\omega-\omega_0)+e^{-j\phi}\delta(\omega+\omega_0))$$

For example, the FS and FT spectra of $\cos(3t)$, shown in Fig. 8.7a, are shown in Fig. 8.7b and c, respectively. ∎

The spectra in Fig. 8.7b and c are the equivalent representations of a single sinusoid by the FS and the FT. In Fig. 8.7b, the discrete spectrum $X_{cs}(k)$ consists of two nonzero discrete impulses of value 0.5. In Fig. 8.7c, the continuous spectrum $X(j\omega)$ consists of two continuous impulses with the value of their integrals being π, which, after dividing by the scale factor 2π, becomes 0.5. The amplitude of a constituent complex exponential of a signal $x(t)$ is $X_{cs}(k)$ in the case of the FS and $\frac{1}{2\pi}X(j\omega)d\omega$ in the case of the FT. Note that $(\delta(\omega-\omega_0)d\omega)|_{\omega=\omega_0}=1$. Remember that both the spectra in Fig. 8.7b and c represent the same waveform and, from either spectra, we get $0.5(e^{j3t}+e^{-j3t})=\cos(3t)$.

$$\cos(3t) \Longleftrightarrow \pi(\delta(\omega-3)+\delta(\omega+3))$$

The conclusion is that, in this case, the FT includes impulses on the imaginary axis, which are the equivalent of coefficients in a FS.

$$\cos(3t) \Longleftrightarrow 0.5(\delta(k-1)+\delta(k+1))$$

8.3.2 *Determination of the FS from the FT*

Let $x(t)$ be a periodic signal of period T. Let us define an aperiodic signal $x_p(t)$ that is identical with $x(t)$ over its one period from t_1 to t_1+T and is zero otherwise, where t_1 is arbitrary. The FT of this signal is

$$X_p(j\omega) = \int_{-\infty}^{\infty} x_p(t)e^{-j\omega t}dt = \int_{t_1}^{t_1+T} x(t)e^{-j\omega t}dt$$

The FS spectrum for $x(t)$ is

$$X_{cs}(k) = \frac{1}{T}\int_{t_1}^{t_1+T} x(t)e^{-jk\omega_0 t}dt, \quad \omega_0 = \frac{2\pi}{T}$$

Comparing the FS and FT definitions of the signals, we get

$$X_{cs}(k) = \frac{1}{T}X_p(j\omega)|_{\omega=k\omega_0} = \frac{1}{T}X_p(jk\omega_0)$$

The discrete samples of $\frac{1}{T}X_p(j\omega)$, at intervals of ω_0, constitute the FS spectrum for the periodic signal $x(t)$. While the spectral values at discrete frequencies are adequate to reconstruct one period of the periodic waveform using the inverse FS, spectral values at continuum of frequencies are required to reconstruct one period of the periodic waveform and the infinite extent zero values of the aperiodic waveform using the inverse FT. Note the similarity of this relationship to that between the DTFT and the DFT.

Example 8.9 Find the FS spectrum for the periodic signal $x(t)$, one period of which is defined as

$$x(t) = \begin{cases} 1 & \text{for } |t| < 1 \\ 0 & \text{for } 1 < |t| < 2 \end{cases}$$

Solution Using the derivative method, the FT of $x_p(t)$ is obtained as follows:

$$j\omega X_p(j\omega) = e^{j\omega} - e^{-j\omega} \quad \text{and} \quad X_p(j\omega) = 2\frac{\sin(\omega)}{\omega}$$

Since $X_{cs}(k) = \frac{1}{T}X_p(jk\omega_0)$, with $T = 4$ and $\omega = k\omega_0 = k\frac{2\pi}{4} = \frac{\pi}{2}k$, we get

$$X_{cs}(k) = \frac{2}{4}\frac{\sin(\frac{\pi}{2}k)}{\frac{\pi}{2}k} = \frac{\sin(\frac{\pi}{2}k)}{\pi k}$$ ∎

Figure 8.8a and b shows, respectively, one period of a periodic square wave and its FS spectrum, $X_{cs}(k)$, as the samples of its FT, $X(j\omega)$, divided by its period T.

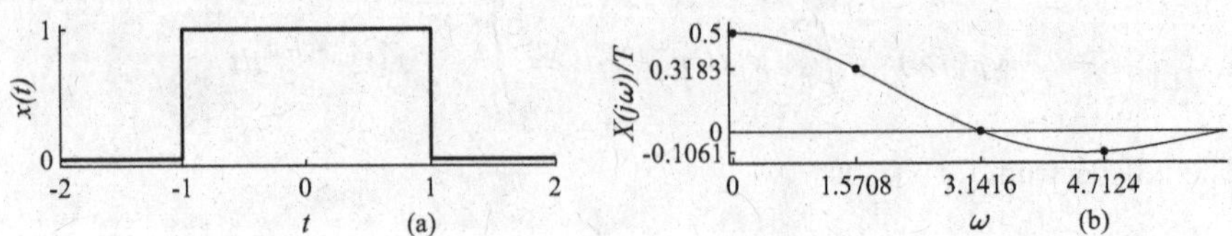

Fig. 8.8 (**a**) One period of a periodic square wave; (**b**) its FS spectrum, $X_{cs}(k)$, as the samples of its FT, $X(j\omega)$, divided by its period T

8.3.3 The FT of a Sampled Signal and the Aliasing Effect

Two equivalent expressions for the FT of a sampled signal are derived. Once we sample a signal, its spectrum becomes periodic due to the reduction in the range of the frequencies from infinity to some finite value. The periodic spectrum has a FS representation. Another point of view is that the sampled signal is considered as obtained by multiplying a continuous signal by an impulse train. Then, the FT of the sampled signal is the convolution of the FT of the continuous signal and the FT of the impulse train.

Let the FT of a signal $x(t)$ be $X(j\omega)$. The sampled version of this signal, $x_s(t)$, is obtained by multiplying it with an impulse train, $s(t) = \sum_{n=-\infty}^{\infty} \delta(t - nT_s)$, where T_s is the period and n is an integer. That is,

$$x_s(t) = x(t)s(t) = x(t) \sum_{n=-\infty}^{\infty} \delta(t - nT_s) = \sum_{n=-\infty}^{\infty} x(nT_s)\delta(t - nT_s)$$

The FS representation of the impulse train, from Chap. 6, is given as

$$s(t) = \frac{1}{T_s} \sum_{k=-\infty}^{\infty} e^{jk\omega_s t},$$

where $\omega_s = \frac{2\pi}{T_s}$. Therefore, the sampled signal $x_s(t)$ is also given by

$$x_s(t) = \frac{1}{T_s} \sum_{k=-\infty}^{\infty} x(t)e^{jk\omega_s t}$$

$$= \frac{1}{T_s}(\cdots + x(t)e^{-j\omega_s t} + x(t) + x(t)e^{j\omega_s t} + \cdots)$$

Let the FT of $x_s(t)$ be $X_s(j\omega)$. Then, using the linearity and frequency-shifting properties of the Fourier transform, we get

$$X_s(j\omega) = \frac{1}{T_s}(\cdots + X(j(\omega+\omega_s)) + X(j\omega) + X(j(\omega-\omega_s)) + \cdots)$$
$$= \frac{1}{T_s}\sum_{k=-\infty}^{\infty} X(j(\omega - k\omega_s))$$

This expression represents the convolution of the spectra of $x(t)$ and $s(t)$ (since it is the FT of their product), and we could as well have obtained the result through the frequency-domain convolution property, as we shall see later. As the FT of the sampled signal is expressed as a sum of the shifted versions of that of the corresponding continuous signal, it is easy to visualize the form of $X_s(j\omega)$ if we know $X(j\omega)$. The sampling of a signal has made the resulting spectrum periodic with period ω_s, the sampling frequency, in addition to scaling the amplitude by the factor $\frac{1}{T_s}$, where T_s is the sampling interval. The periodicity is the result of the reduction of the range of frequencies, due to sampling, over which sinusoids can be distinguished. The factor $\frac{1}{T_s}$ arises from the fact that

$$x(t) = \int_{-\infty}^{\infty} x(\tau)\delta(t-\tau)d\tau = \lim_{T_s\to 0}\sum_{n=-\infty}^{\infty} x(nT_s)T_s\delta(t-nT_s) = \lim_{T_s\to 0} T_s x_s(t)$$

Figure 8.9a and b shows, respectively, the continuous sinc function and its aperiodic FT spectrum.

$$x(t) = \frac{\sin\left(\frac{2\pi}{3}t\right)}{\pi t} \Longleftrightarrow X(j\omega) = \left(u\left(\omega + \frac{2\pi}{3}\right) - u\left(\omega - \frac{2\pi}{3}\right)\right)$$

With $T_s = 0.1$, period $\frac{2\pi}{0.1} = 20\pi$ radians and amplitude $\frac{1}{0.1} = 10$. Then,

$$x_s(t) = \sum_{n=-\infty}^{\infty} \frac{\sin\left(\frac{2\pi}{3}(0.1n)\right)}{\pi(0.1n)}\delta(t-0.1n) \Longleftrightarrow$$

$$X_s(j\omega) = \sum_{k=-\infty}^{\infty} 10\left(u\left(\omega + \frac{2\pi}{3} - 20k\pi\right) - u\left(\omega - \frac{2\pi}{3} - 20k\pi\right)\right)$$

At any discontinuity of the time-domain function, the strength of the sample should be equal to the average value of the right- and left-hand limits.

While the FT $X(j\omega)$ of $x(t)$ uniquely determines the FT $X_s(j\omega)$ of $x_s(t)$, the converse is not necessarily true. By sampling the signal, we simultaneously reduce the effective frequency range of the sinusoids available to represent the signal, and, hence, the FT of the sampled signal becomes periodic due to aliasing effect. Therefore, if the signal is bandlimited and the sampling frequency is greater than twice the highest-frequency component of the signal, we can recover its exact FT

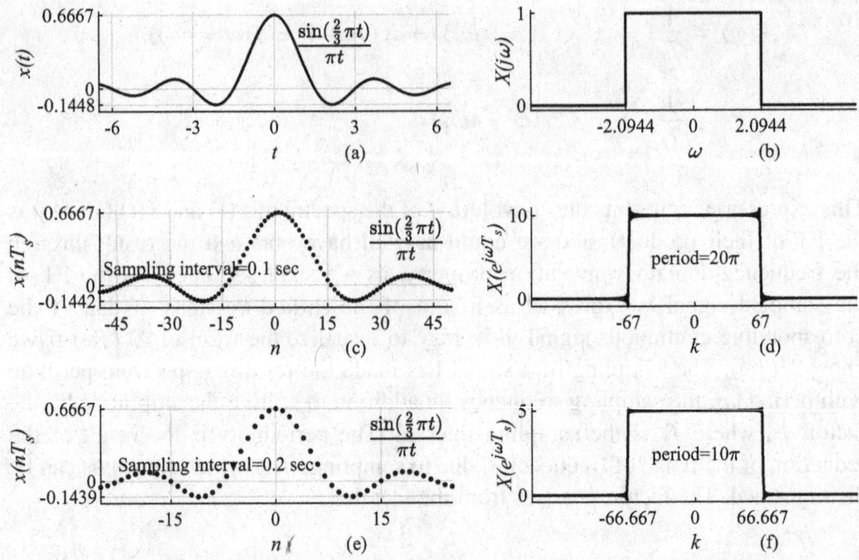

Fig. 8.9 (**a**) The sinc function $\frac{\sin\left(\frac{2\pi}{3}t\right)}{\pi t}$ and (**b**) its FT spectrum; (**c**) discrete samples of (**a**) with $T_s = 0.1$ s and (**d**) its periodic DTFT spectrum with period 20π radians; (**e**) discrete samples of (**a**) with $T_s = 0.2$ s and (**f**) its DTFT spectrum with period 10π radians

from that of its sampled version by lowpass filtering (since the periodic repetition of $X(j\omega)$, yielding $X_s(j\omega)$, does not result in the overlapping of its nonzero portions). If the sampling frequency is not sufficiently high, we can only recover a corrupted version of its FT spectrum, since the periodic repetition of $X(j\omega)$ results in the overlapping of its nonzero portions. For example, with $T_s = 2$,

$$x_s(t) = \sum_{n=-\infty}^{\infty} \frac{\sin\left(\frac{2\pi}{3}(2n)\right)}{\pi(2n)}\delta(t-2n) \Longleftrightarrow$$

$$X_s(j\omega) = \sum_{k=-\infty}^{\infty} 0.5(u(\omega + \frac{2\pi}{3} - k\pi) - u(\omega - \frac{2\pi}{3} - k\pi))$$

As the FT of $\delta(t - nT_s)$ is $e^{-jn\omega T_s}$ and due to the linearity property of the FT, the FT of the sampled signal $x_s(t) = \sum_{n=-\infty}^{\infty} x(nT_s)\delta(t - nT_s)$ is also given by

$$X_s(j\omega) = \sum_{n=-\infty}^{\infty} x(nT_s)e^{-jn\omega T_s}$$

This expression, which, of course, is completely equivalent to that derived earlier for $X_s(j\omega)$, reminds us that the relation is a FS with the roles of the domains interchanged and corresponds to the DTFT. The time-domain samples $x(nT_s)$ are the FS coefficients of the corresponding continuous periodic spectrum $X_s(j\omega)$ of period $2\pi/T_s$.

8.3.4 The FT and the DTFT of Sampled Aperiodic Signals

Let us construct a sequence with the discrete sample values, at intervals of T_s, of the signal $x(t)$. These sample values are the same as the strengths (areas) of the corresponding impulses $x(nT_s)\delta(t - nT_s)$ of the sampled signal. The DTFT of $x(nT_s)$ is defined as

$$X(e^{j\omega T_s}) = \sum_{n=-\infty}^{\infty} x(nT_s)e^{-jn\omega T_s}$$

That is, the DTFT of a sequence $x(nT_s)$ and the FT of the corresponding sampled signal,

$$\sum_{n=-\infty}^{\infty} x(nT_s)\delta(t - nT_s),$$

are the same when the DTFT version includes the sampling interval, T_s. Figure 8.9c and d shows, respectively, the discrete samples of the sinc function $x(0.1n) = \frac{\sin\left(\frac{2\pi}{3}(0.1n)\right)}{\pi(0.1n)}$ with $T_s = 0.1$ s and 2010 samples. The DTFT spectrum with period 20π radians is the same as the corresponding $X_s(j\omega)$. The DTFT spectrum is approximated by the DFT with 2010 samples. The edge frequency is

$$\frac{67 \times 2\pi}{2010 \times 0.1} = 2.0944 = \frac{2\pi}{3} \text{ radians}$$

Note the overshoots at the edges of the spectrum due to Gibbs phenomenon. Figure 8.9e and f shows, respectively, the discrete samples of the sinc function $x(0.2n) = \frac{\sin\left(\frac{2\pi}{3}(0.2n)\right)}{\pi(0.2n)}$ with $T_s = 0.2$ s and its DTFT spectrum with period 10π radians, which is the same as the corresponding $X_s(j\omega)$.

Usually, the DTFT spectrum is computed with the assumption of $T_s = 1$ s. The FT of the corresponding sampled continuous signal $x_s(t)$ is obtained by scaling the frequency axis of this DTFT spectrum so that the period of the spectrum becomes $\frac{2\pi}{T_s}$.

8.3.5 The FT and the DFT of Sampled Periodic Signals

The FT of a bandlimited periodic signal $x(t)$, from earlier results, is

$$x(t) = \sum_{k=-N}^{N} X_{cs}(k)e^{jk\omega_0 t} \Longleftrightarrow X(j\omega) = 2\pi \sum_{k=-N}^{N} X_{cs}(k)\delta(\omega - k\omega_0),$$

where $\omega_0 = \frac{2\pi}{T}$, the fundamental frequency of $x(t)$. Let us sample the periodic signal by multiplying it with an impulse train

$$s(t) = \sum_{n=-\infty}^{\infty} \delta(t - nT_s) \Longleftrightarrow S(j\omega) = \frac{2\pi}{T_s} \sum_{m=-\infty}^{\infty} \delta(\omega - m\omega_s)$$

with the interval between impulses being $T_s = \frac{2\pi}{\omega_s}$. Then, as multiplication in the time domain corresponds to convolution in the frequency domain, the FT $X_s(j\omega)$ of the sampled signal $x_s(t) = x(t)s(t)$ is $\frac{1}{2\pi}X(j\omega) * S(j\omega)$. The FT of the sampled signal, as convolution of a signal with an impulse is the relocation of the origin of the signal at the location of the impulse, is

$$X_s(j\omega) = \frac{2\pi}{T_s} \sum_{m=-\infty}^{\infty} \sum_{k=-N}^{N} X_{cs}(k)\delta(\omega - k\omega_0 - m\omega_s)$$

where $\omega_s = \frac{2\pi}{T_s}$. As $X(k) = (2N+1)X_{cs}(k)$, where $X(k)$ is the DFT of the $2N+1$ discrete samples of $x(t)$ over one period, we get

$$X_s(j\omega) = \frac{2\pi}{(2N+1)T_s} \sum_{m=-\infty}^{\infty} \sum_{k=-N}^{N} X(k)\delta(\omega - k\omega_0 - m\omega_s)$$

This FT corresponds to the sampled periodic time-domain signal

$$x_s(t) = \sum_{n=-\infty}^{\infty} x(nT_s)\delta(t - nT_s)$$

The period of the time-domain signal $x(n)$ of the DFT is $2N+1$ samples, and that of corresponding sampled continuous signal $x_s(t)$ is $(2N+1)T_s = T$ s. The period of the FT spectrum is $\omega_s = \frac{2\pi}{T_s}$ radians, and the spectral samples are placed at intervals of $\omega_0 = \frac{2\pi}{(2N+1)T_s} = \frac{2\pi}{T}$ radians.

Consider the discrete samples, over two periods, of the continuous cosine wave $\cos(2\frac{2\pi}{48}t)$ with sampling interval $T_s = 3$ s and its DFT spectrum shown, respectively, in Fig. 8.10a and b. Both the waveform and its spectrum are periodic

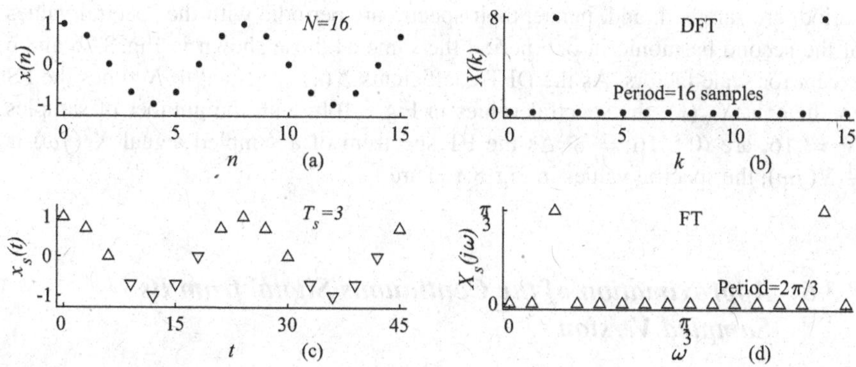

Fig. 8.10 (**a**) The discrete samples, over two periods, of the continuous cosine wave $\cos(2\frac{2\pi}{48}t)$ with sampling interval $T_s = 3$ s and (**b**) its DFT spectrum; (**c**) the sampled version of the cosine wave $\cos(2\frac{2\pi}{48}t)$ and (**d**) its periodic FT spectrum

with period $N = 16$ samples. The sampled version of the cosine wave is shown in Fig. 8.10c. The waveform is periodic with period $NT_s = T = 48$ s. The FT spectrum of the waveform in Fig. 8.10c is shown in Fig. 8.10d. The spectrum is periodic with period $\frac{2\pi}{T_s} = \frac{2\pi}{3}$ radians. The spectral samples are placed at intervals of $\omega_s = \frac{2\pi}{NT_s} = \frac{2\pi}{48} = 0.1309$ radians.

$$x_s(t) = \sum_{n=-\infty}^{\infty} \cos(2\frac{2\pi}{48}n(3))\delta(t - 3n) \Longleftrightarrow$$

$$X_s(j\omega) = \frac{\pi}{3} \sum_{m=-\infty}^{\infty} \left(\delta(\omega - 2\frac{2\pi}{48} - \frac{2m\pi}{3}) + \delta(\omega + 2\frac{2\pi}{48} - \frac{2m\pi}{3}) \right)$$

The point is that we should mean the same waveform by looking at DFT and FT spectra. The term $\frac{2m\pi}{3}$ indicates that the spectrum is periodic with period $\frac{2\pi}{3}$ radians and, hence, the time-domain waveform is sampled with a sampling interval of 3 s. The two impulse terms, with strength $\frac{\pi}{3}$, indicate a cosine waveform with frequency $2\frac{2\pi}{48}$ radians and amplitude 1. The DFT spectrum indicates a cosine waveform $\cos(2\frac{2\pi}{16}n)$. With a sampling interval of 3 s, this waveform corresponds to $\cos(2\frac{2\pi}{48}t)$. $x_s(t)$ repeats with a period equal to the total sample time NT_s, and $X_s(j\omega)$ repeats with a period equal to the total bandwidth $N\omega_s$. With $T_s = 1$ sec, the usual way the DFT is defined, the total bandwidth is $N\omega_s = 2\pi$ radians. Then, we rescale the frequency axis, with $T_s = 3$. That is, 2π becomes $2\pi/3$ radians.

Consider the differences between the cosine waveforms with amplitude 1 and their spectra in Figs. 8.7 and 8.10. The waveform is continuous in Fig. 8.7a and makes one cycle in the fundamental period. The FS and FT spectra in Fig. 8.7b and c are aperiodic. The waveforms in Fig. 8.10a and c, with two cycles in the fundamental period, are sampled, and, hence, their spectra are periodic with the spectral values of the second harmonic in one period the same as those shown in Fig. 8.7b and c except for scale factors. As the DFT coefficients $X(k)$ are equal to N times the FS coefficients $X_{cs}(k)$, the spectral values in Fig. 8.10b, with the number of samples $N = 16$, are $(0.5)16 = 8$. As the FT spectrum of a sampled signal $X_s(j\omega)$ is $\frac{1}{T_s}X(j\omega)$, the spectral values in Fig. 8.10d are $\frac{\pi}{3}$.

8.3.6 Approximation of the Continuous Signal from Its Sampled Version

The zero-order hold filter is commonly used to approximate a continuous signal $x(t)$ from its sampled version $x_s(t)$. The impulse response of this filter is a rectangular pulse of unit height and width T_s, $h(t) = u(t) - u(t - T_s)$, where T_s is the sampling interval of $x_s(t)$. By passing $x_s(t)$ through this filter, we get an output signal, which is the convolution of $x_s(t)$ and $h(t)$, that is a staircase approximation of $x(t)$. The convolution of $x_s(t)$, which is a sum of impulses, with $h(t)$ results in replacing each impulse of $x_s(t)$ by a pulse of width T_s and height equal to its strength (holding the current sample value until the next sample arrives).

8.4 Approximation of the Fourier Transform

In approximating the FS by the DFT, we determine the appropriate sampling interval and take samples over one period. In approximating the FT by the DFT, we have to fix the record length as well. These two parameters have to be fixed so that most of the energy of the signal is included in the selected record length and the continuous spectrum of the FT is represented by a sufficiently accurate and dense set of spectral samples.

The integral in Eq. (8.1) is approximated by the rectangular rule of numerical integration. The summation interval can start from zero, since the truncated signal, of length T, is assumed periodic by the DFT, although the input signal can be nonzero in any interval. We divide the period T into N intervals of width $T_s = \frac{T}{N}$ and represent the signal at N points as $x(0), x(\frac{T}{N}), x(2\frac{T}{N}), \ldots, x((N-1)\frac{T}{N})$. The sampling interval in the time domain is T_s seconds, and that in the frequency domain is $\frac{2\pi}{NT_s} = \frac{2\pi}{T}$ radians per second. Now, Eq. (8.1) is approximated as

$$X\left(j\frac{2\pi k}{NT_s}\right) = T_s \sum_{n=0}^{N-1} x(nT_s)e^{-j\frac{2\pi}{N}nk}, \quad k = 0, 1, \ldots, N-1 \tag{8.3}$$

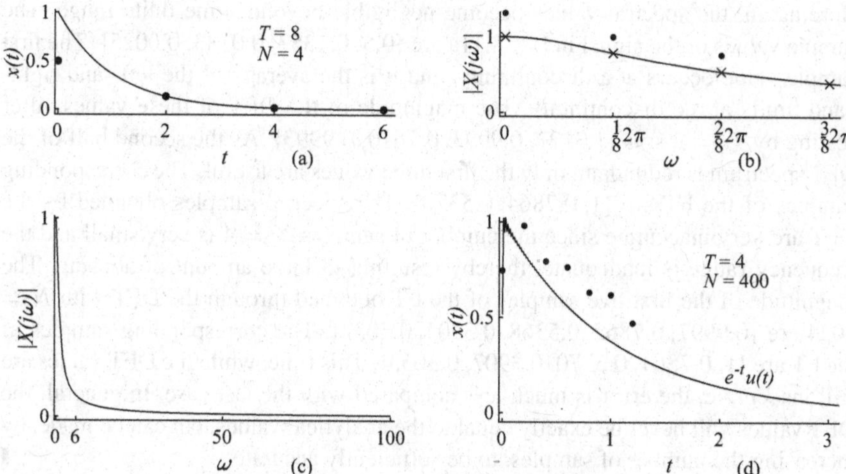

Fig. 8.11 (**a**) The exponential waveform $x(t) = e^{-t}u(t)$, with four samples over the range $0 \leq t < 8$; (**b**) the magnitude of the FT (solid line) and the samples of the FT obtained through the DFT with $N = 4$ (dots) and $N = 1024$ (crosses) samples; (**c**) the magnitude of the FT spectrum; (**d**) the partially (dotted) and fully reconstructed waveforms

Equation (8.2) is approximated as

$$x(nT_s) = \frac{1}{NT_s}\sum_{k=0}^{N-1} X\left(j\frac{2\pi k}{NT_s}\right)e^{j\frac{2\pi}{N}nk}, \quad n = 0, 1, \ldots, N-1 \tag{8.4}$$

Except for the scale factors, Eqs. (8.3) and (8.4) are, respectively, the DFT and the IDFT of N samples. By multiplying the DFT coefficients by the sampling interval T_s, we get the approximate samples of the FT. By multiplying the IDFT values by $\frac{1}{T_s}$, we get the approximate samples of the time-domain signal.

Example 8.10 Approximate the magnitude of the FT of the signal $x(t) = e^{-t}u(t)$ using the DFT.

Solution From the transform pair of Example 8.7, we get

$$X(j\omega) = \frac{1}{1 + j\omega} \quad \text{and} \quad |X(j\omega)| = \frac{1}{\sqrt{1 + \omega^2}}$$

Figure 8.11a shows the exponential signal $e^{-t}u(t)$ with four samples over a period of $T = 8$ s. Figure 8.11b shows the magnitude of the FT and the samples of the FT obtained through the DFT with $N = 4$ and $N = 1024$ samples. While the signal is of infinite duration, we have truncated it to 8 s duration. The truncated signal has most of the energy of the untruncated signal. This signal has also an infinite bandwidth.

Here again, the spectral values become negligible beyond some finite range. The sample values of the signal in Fig. 8.11a are $\{0.5, 0.1353, 0.0183, 0.0025\}$. The first sample value occurs at a discontinuity, and it is the average of the left- and right-hand limits at the discontinuity. The magnitude of the DFT of these values, after scaling by $T_s = 2$ s, is $\{1.3123, 0.9993, 0.7610, 0.9993\}$. As the second half of the DFT spectrum is redundant, only the first three values are useful. The corresponding samples of the FT are $\{1, 0.7864, 0.5370\}$. The spectral samples obtained by the DFT are very inaccurate since the number of samples $N = 4$ is very small and the frequency range is inadequate, thereby resulting in large amount of aliasing. The magnitude of the first five samples of the FT obtained through the DFT with $N = 1024$ are $\{0.9997, 0.7862, 0.5368, 0.3905, 0.3032\}$. The corresponding samples of the FT are $\{1, 0.7864, 0.5370, 0.3907, 0.3033\}$. This time, while the DFT values are still inaccurate, the error is much less compared with the last case. In general, the DFT values will never be exactly equal to the analytical values, but can be made, by increasing the number of samples, to be sufficiently accurate. ∎

In order to approximate the FT of an arbitrary signal by the DFT, a trial-and-error procedure is used. A set of samples over a reasonable record length of the signal with an initial sampling interval is taken, and the DFT is computed. Then, keeping the record length the same, we double the number of samples. That is, we reduce the sampling interval by one-half, and the DFT is computed. This process is repeated until the spectral values near the middle of the spectrum for real signals (at the end of the spectrum for complex signals) become negligibly small, which ensures very little aliasing. Now, the sampling interval is fixed. Truncation of a signal is multiplying it with a rectangular window. As the window becomes longer, the truncation becomes less. In the frequency domain, the spectrum of the window becomes more closer to an impulse from that of a sinc function. The convolution of the spectra of the untruncated signal and the window distorts the spectrum of the signal to a lesser extent. Therefore, keeping the sampling interval the same, we keep doubling the record length and use the DFT to compute the spectral samples. When truncation becomes negligible, the spectral values with two successive lengths will be almost the same. Now, the record length is fixed.

A similar procedure for the approximation of the IFT is required. Now, we have to fix the record length of the spectrum. The spectrum, as shown in Fig. 8.11c for positive values of ω (the spectrum is conjugate symmetric), is slowly decaying and is of infinite extent. The cutoff frequency has to be selected to suit the accuracy requirements. The magnitude of the spectrum is approximately $1/\omega$ for large values of ω. The peak value of the spectrum is 1. If we want to discard the values of the spectrum that are less than two-hundredth of the peak, then $\omega = 200$ radians is the cutoff frequency. Of course, the cutoff frequency can also be fixed based on signal energy.

In contrast to most signals in the time domain, the spectra of signals are almost always two-sided. For real signals, the spectrum is conjugate symmetric, and, usually, the positive frequency side is shown, as can be seen in Fig. 8.11b and c.

For the example waveform, the record length T in the time domain is 4 s, and ω_s, the sampling frequency of the spectrum, is $\pi/2$ radians.

Let the sampling interval of the reconstructed signal be T_s seconds. Then, the record length of the reconstructed signal is NT_s seconds, where N is the number of samples. Let the number of samples N be 7 and the sampling interval T_s be 0.2 s. Then, the record length T in the time domain is 1.4 s. Let the frequency increment be 2.5π. Then, the four samples of the FT spectrum at

$$\omega = \{0, 7.8540, 15.7080, 23.5619\}$$

are

$$\{1, 0.0160 - j0.1253, 0.0040 - j0.0634, 0.0018 - j0.042\}$$

Conjugating the last three spectral samples and concatenating in the reverse order, we get

$$\{1, 0.0160 - j0.1253, 0.0040 - j0.0634, 0.0018 - j0.042, 0.0018 + j0.042, 0.0040 + j0.0634, 0.0160 + j0.1253\}$$

The IDFT of these values divided by $T_s = 0.2$ are

$$\{0.7454, 0.9794, 0.7935, 0.7626, 0.6309, 0.6177, 0.4704\}$$

shown in Fig. 9.10d. The actual samples of e^{-t} at

$$t = \{0, 0.2, 0.4, 0.6, 0.8, 1.0, 1.2\}$$

are

$$\{1, 0.8187, 0.6703, 0.5488, 0.4493, 0.3679, 0.3012\}$$

The fully reconstructed waveform, with $N = 400$ and $T = 4$, is also shown in Fig. 8.11d.

8.5 Applications of the Fourier Transform

8.5.1 *Transfer Function and the System Response*

The input-output relationship of a LTI system is given by the convolution operation in the time domain. Since convolution corresponds to multiplication in the frequency domain, we get

$$y(t) = \int_{-\infty}^{\infty} x(\tau)h(t-\tau)d\tau \Longleftrightarrow Y(j\omega) = X(j\omega)H(j\omega),$$

where $x(t)$, $h(t)$, and $y(t)$ are, respectively, the system input, impulse response, and output and $X(j\omega)$, $H(j\omega)$, and $Y(j\omega)$ are their respective transforms. As input is transferred to output by multiplication with $H(j\omega)$, $H(j\omega)$ is called the transfer function of the system. The transfer function, which is the transform of the impulse response, characterizes a system in the frequency domain just as the impulse response does in the time domain.

Since the impulse function, whose FT is one (a uniform spectrum), is composed of complex exponentials, $e^{j\omega t}$, of all frequencies with equal magnitude and zero phase, the transform of the impulse response, the transfer function, is also called the frequency response of the system. Therefore, an exponential $Ae^{j(\omega_a t+\theta)}$ is changed to $(|H(j\omega_a)|A)e^{j(\omega_a t+(\theta+\angle(H(j\omega_a)))}$ at the output. A real sinusoidal input signal $A\cos(\omega_a t+\theta)$ is also changed at the output by the same amount of amplitude and phase of the complex scale factor $H(j\omega_a)$. That is, $A\cos(\omega_a t+\theta)$ is changed to $(|H(j\omega_a)|A)\cos(\omega_a t+(\theta+\angle(H(j\omega_a)))$. The steady-state response of a stable system to the input $Ae^{j(\omega_a t+\theta)}u(t)$ is also the same.

As $H(j\omega) = \frac{Y(j\omega)}{X(j\omega)}$, the transfer function can also be described as the ratio of the transform $Y(j\omega)$ of the response $y(t)$ to an arbitrary signal $x(t)$ to that of its transform $X(j\omega)$, provided $|X(j\omega)| \neq 0$ for all frequencies and the system is initially relaxed.

As the transform of the derivative of a signal is its transform multiplied by a factor, we can readily find the transfer function from the differential equation. Consider the second-order differential equation of a stable and initially relaxed LTI continuous system.

$$\frac{d^2y(t)}{dt^2} + a_1\frac{dy(t)}{dt} + a_0y(t) = b_2\frac{d^2x(t)}{dt^2} + b_1\frac{dx(t)}{dt} + b_0x(t)$$

Taking the FT of both sides, we get

$$(j\omega)^2Y(j\omega) + a_1(j\omega)Y(j\omega) + a_0Y(j\omega) =$$
$$(j\omega)^2b_2X(j\omega) + b_1(j\omega)X(j\omega) + b_0X(j\omega)$$

The transfer function $H(j\omega)$ is obtained as

$$H(j\omega) = \frac{Y(j\omega)}{X(j\omega)} = \frac{(j\omega)^2b_2 + (j\omega)b_1 + b_0}{(j\omega)^2 + a_1(j\omega) + a_0}$$

Example 8.11 Find the response, using the FT, of the system governed by the differential equation

$$\frac{dy(t)}{dt} + y(t) = x(t)$$

to the input $x(t) = 2\cos(t + \frac{\pi}{4})$.

Solution

$$H(j\omega) = \frac{1}{1 + j\omega}$$

Substituting $\omega = 1$, we get

$$H(j1) = \frac{1}{1 + j1} = \frac{1}{\sqrt{2}} \angle\left(-\frac{\pi}{4}\right)$$

The response of the system to the input $x(t) = 2\cos(t + \frac{\pi}{4})$ is $y(t) = \frac{2}{\sqrt{2}}\cos(t + \frac{\pi}{4} - \frac{\pi}{4}) = \sqrt{2}\cos(t)$. ∎

Example 8.12 Find the impulse response, using the FT, of the system governed by the differential equation

$$\frac{dy(t)}{dt} + 3y(t) = x(t)$$

Solution

$$H(j\omega) = \frac{1}{3 + j\omega}$$

The impulse response of the system, which is the inverse FT of $H(j\omega)$, is $h(t) = e^{-3t}u(t)$. ∎

Example 8.13 Find the zero-state response of the system governed by the differential equation

$$\frac{d^2y(t)}{dt^2} + 4\frac{dy(t)}{dt} + 4y(t) = \frac{d^2x(t)}{dt^2} + \frac{dx(t)}{dt} + 2x(t)$$

with the input $x(t) = u(t)$, the unit-step function.

Solution

$$H(j\omega) = \frac{(j\omega)^2 + (j\omega) + 2}{(j\omega)^2 + 4(j\omega) + 4}$$

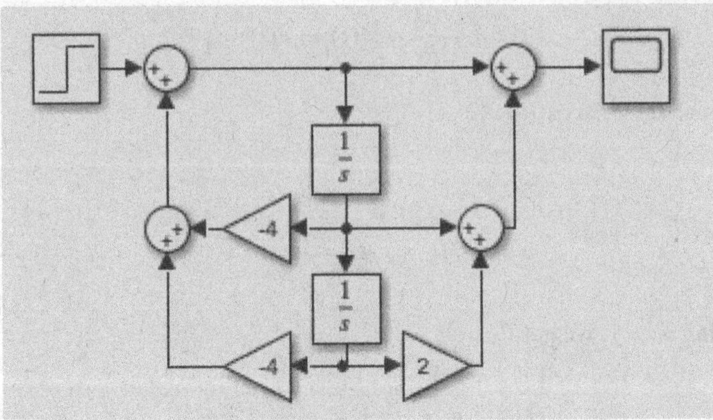

Fig. 8.12 The simulation model for step input

With $X(j\omega) = \pi\delta(\omega) + \frac{1}{j\omega}$,

$$Y(j\omega) = H(j\omega)X(j\omega) = \frac{(j\omega)^2 + (j\omega) + 2}{(j\omega)((j\omega)^2 + 4(j\omega) + 4)} + \frac{\pi\delta(\omega)((j\omega)^2 + (j\omega) + 2)}{(j\omega)^2 + 4(j\omega) + 4}$$

Expanding into partial fractions, we get

$$Y(j\omega) = \frac{0.5}{j\omega} + \frac{0.5}{j\omega + 2} - \frac{2}{(j\omega + 2)^2} + 0.5\pi\delta(\omega)$$

Taking the inverse FT, we get the zero-state response

$$y(t) = (0.5 + 0.5e^{-2t} - 2te^{-2t})u(t)$$

The steady-state response is $0.5u(t)$, and the transient response due to the input is $(0.5e^{-2t} - 2te^{-2t})u(t)$. The simulation model for step input is shown in Fig. 8.12. Running this model will yield the same response as that obtained analytically. ∎

Systems with nonzero initial conditions cannot be directly analyzed with FT. Further, handling of the frequency variable $j\omega$ is relatively more difficult. For these reasons, the Laplace transform is preferable for system analysis. However, the FT, wherever it is more suitable, is efficient, as it can be approximated by the DFT using fast algorithms.

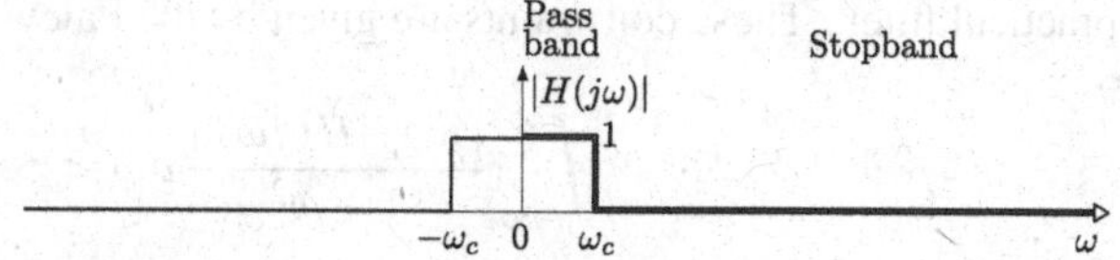

Fig. 8.13 Frequency response of an ideal lowpass filter

8.5.2 *Ideal Filters and Their Unrealizability*

Filters are prominent examples of LTI systems for signal analysis, manipulation, and processing. Common applications of filters include removing noise from signals and selection of individual channels in radio or television receivers. We present, in this subsection, the constraints involved in the realization of practical filters.

The frequency response of an ideal lowpass filter is shown in Fig. 8.13. As it is even-symmetric, the specification of the response over the interval from $\omega = 0$ to $\omega = \infty$, shown in thick lines, characterizes a filter.

$$H(j\omega) = \begin{cases} 1 \text{ for } 0 \leq \omega < \omega_c \\ 0 \text{ for } \omega > \omega_c \end{cases}$$

From $\omega = 0$ to $\omega = \omega_c$, the filter passes frequency components of a signal with a gain of 1 and rejects the other frequency components, since the output of the filter, in the frequency domain, is given by $Y(j\omega) = H(j\omega)X(j\omega)$. The magnitudes of the frequency components of the signal, $X(j\omega)$, with frequencies up to ω_c are multiplied by 1 and the rest by 0. The range of frequencies from 0 to ω_c is called the passband, and the range from ω_c to ∞ is called the stopband. This ideal filter model is practically unrealizable since its impulse response (inverse of $H(j\omega)$) extends from $t = -\infty$ to $t = \infty$, which requires a noncausal system. Practical filters approximate this model.

The impulse response of practical systems must be causal. The even and odd components, for $t > 0$, of a causal time function $x(t)$ are given as

$$x_e(t) = \frac{x(t) + x(-t)}{2} = \frac{x(t)}{2} \quad \text{and} \quad x_o(t) = \frac{x(t) - x(-t)}{2} = \frac{x(t)}{2}$$

That is, $x(t) = 2x_e(t) = 2x_0(t)$, $t > 0$ and $x_e(t) = -x_0(t)$, $t < 0$. As the FT of an even signal is real and that of an odd signal is imaginary, $x(t)$ can be obtained by finding the inverse FT of either the real part or the imaginary part of its spectrum $X(j\omega)$. That is,

$$x(t) = \frac{2}{\pi}\int_0^{\infty} \text{Re}(X(j\omega))\cos(\omega t)d\omega = -\frac{2}{\pi}\int_0^{\infty} \text{Im}(X(j\omega))\sin(\omega t)d\omega, \; t > 0$$

The point is that the real and imaginary parts or, equivalently, the magnitude and the phase of the FT of a causal signal are related. This implies that there are constraints, for the realizability, on the magnitude of the frequency response, $H(j\omega)$, of a practical filter. These constraints are given by the Paley-Wiener criterion as

$$\int_{-\infty}^{\infty} \frac{|\log_e |H(j\omega)||}{1+\omega^2} d\omega < \infty$$

To satisfy this criterion, the magnitude of the frequency response $|H(j\omega)|$ can be zero at discrete points but not over any continuous band of frequencies. If $H(j\omega)$ is zero over a band of frequencies, $|\log_e |H(j\omega)|| = \infty$, and the condition is violated. On the other hand, if $H(j\omega)$ is zero at a finite set of discrete frequencies, the value of the integral may still be finite, although the integrand is infinite at these frequencies. In addition, any transition of this function cannot vary more rapidly than by exponential order. The $H(j\omega)$ of the ideal filter shown in Fig. 8.11 does not meet the Paley-Wiener criterion. Further, the order of the filter must be infinite to have a constant gain all over the passband. Therefore, neither the flatness of the bands nor the sharpness of the transition between the bands of ideal filters is realizable by practical filters.

8.5.3 Modulation and Demodulation

Modulation and demodulation operations are fundamental to communication applications. These operations are required in signal communication because of different frequency ranges required for the signals to be communicated and for efficient transmission of signals. As the antenna size is inversely proportional to the frequency of the signal, the lower the frequency of the signal, the larger is the required antenna size. For example, an antenna of size about 30 km is required to transmit the audio signal efficiently. Therefore, it is a necessity to embed the audio signal, called the message signal, in a much higher-frequency signal, called the carrier signal, which can be transmitted more efficiently. The operation of embedding the message signal in a carrier signal is called modulation. The embedding involves the variation of some property of the carrier signal in accordance with the message signal. At the receiving end, the message signal has to be extracted from the modulated carrier signal. This operation is called demodulation. There are different types of these operations with distinct characteristics. We understand these operations using the property of the FT that the multiplication of two signals in the time domain corresponds to convolution in the frequency domain.

8.5.3.1 Double Sideband, Suppressed Carrier (DSB-SC), Amplitude Modulation

In this type of modulation, the amplitude A of the carrier signal, $A\cos(\omega_c(t)+\theta_c)$, is varied in some manner with the message signal, $m(t)$, where ω_c and θ_c are constants. Let the FT of $m(t)$ be $M(j\omega)$. Then, the FT of the product of the message and carrier

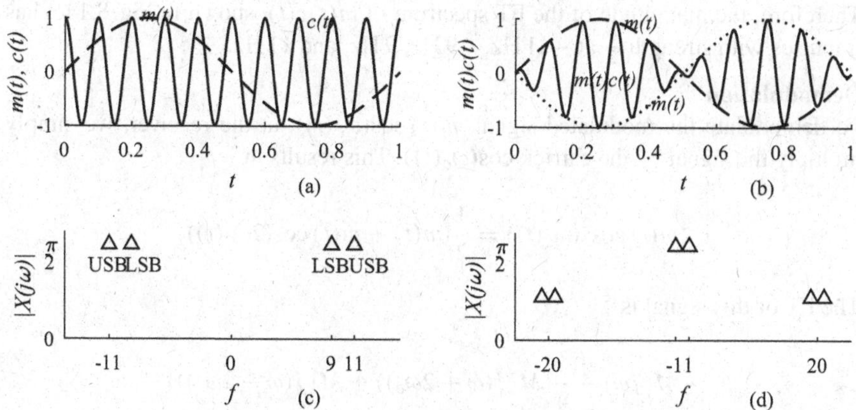

Fig. 8.14 **(a)** $m(t) = \sin(2\pi t)$ and $c(t) = \cos(20\pi t)$; **(b)** $m(t)c(t)$; **(c)** the magnitude of the FT spectrum of $m(t)c(t)$; **(d)** the magnitude of the FT spectrum of $(m(t)c(t))c(t)$

signals, with $A = 1$ and $\theta_c = 0$, is given as

$$m(t)\cos(\omega_c(t)) \iff \frac{1}{2}(M(j(\omega + \omega_c)) + M(j(\omega - \omega_c)))$$

After modulation, a copy of the spectrum of the message signal is placed at ω_c, and another copy is placed at $-\omega_c$. Each copy of the spectrum of the message signal has the upper sideband (USB) portion (the right half of the spectrum centered at ω_c and the left half of the spectrum centered at $-\omega_c$) and the lower sideband (LSB) portion (the left half of the spectrum centered at ω_c and the right half of the spectrum centered at $-\omega_c$). As there are two sidebands and no carrier in the spectrum, this form of modulation is called double sideband, suppressed carrier, amplitude modulation. Note that the message signal can be recovered from either sideband.

Let the message signal be $m(t) = \sin(2\pi t)$ and the carrier signal be $c(t) = \cos(20\pi t)$, as shown in Fig. 8.14a. For illustration, we are using a sine wave of 1 Hz as the message signal and cosine wave of 10 Hz as the carrier signal. However, it should be noted that, in practice, the message signal will have a finite bandwidth and the carrier frequency will be much higher. For example, the bandwidth of a message signal could be 3 kHz with a carrier frequency 3000 kHz. The product $m(t)c(t)$ is shown in Fig. 8.14b. The envelopes of $m(t)c(t)$ are $m(t)$ and $-m(t)$, since $m(t)\cos(20\pi t) = m(t)$ when $\cos(20\pi t) = 1$ and $m(t)\cos(20\pi t) = -m(t)$ when $\cos(20\pi t) = -1$. For this specific example, the FT $X(j\omega)$ of $m(t)c(t)$ is

$$\frac{j\pi}{2}(\underbrace{-\delta(f + (10 + 1)) + \delta(f + (-10 - 1))}_{\text{USB}} + \underbrace{\delta(f + (10 - 1)) - \delta(f + (-10 + 1))}_{\text{LSB}})$$

Therefore, the magnitude of the FT spectrum of $m(t)c(t)$, shown in Fig. 8.14c, has impulses with strength $\frac{\pi}{2}$ at -11 Hz, -9 Hz, 9 Hz, and 11 Hz.

Demodulation

To demodulate the modulated signal, $m(t)\cos(\omega_c(t))$, at the receiver, we simply multiply the signal by the carrier, $\cos(\omega_c(t))$. This results in

$$m(t)\cos^2(\omega_c(t)) = \frac{1}{2}(m(t) + m(t)\cos(2\omega_c(t))$$

The FT of this signal is

$$\frac{1}{2}M(j\omega) + \frac{1}{4}(M(j(\omega + 2\omega_c)) + M(j(\omega - 2\omega_c)))$$

The spectrum of the message signal is centered at $\omega = 0$ and can be recovered by lowpass filtering. The other two spectra are the transform of $m(t)$ modulated by a carrier with frequency $2\omega_c$. For the specific example, the magnitude of the spectrum of $(m(t)c(t))c(t)$ is shown in Fig. 8.14d.

To use this type of demodulation, we have to generate the carrier signal with the same frequency and phase. This requires a complex receiver. While this form is used in certain applications, for commercial radio broadcasting, another type of modulation and demodulation, described next, is most commonly used.

8.5.3.2 Double Sideband, with Carrier (DSB-WC), Amplitude Modulation

In this type of modulation, the amplitude of the carrier signal, $\cos(\omega_c(t))$, is varied in some manner with the modulating signal, $(1+km(t))$, where ω_c and k are constants. Let the FT of $m(t)$ be $M(j\omega)$. Then, the FT of the product of the message and carrier signal is given as

$$(1 + km(t))\cos(\omega_c t) \Longleftrightarrow$$
$$\frac{k}{2}(M(j(\omega + \omega_c)) + M(j(\omega - \omega_c))) + \pi(\delta(\omega + \omega_c) + \delta(\omega - \omega_c))$$

After modulation, a copy of the spectrum of the message signal is placed at ω_c, and another copy is placed at $-\omega_c$. As there are two sidebands and the carrier in the spectrum of the transmitted signal, this form of modulation is called double sideband, with carrier, amplitude modulation. This form of modulation is intended for simple receivers without the need for generating the carrier signal. For example, let the signals $m(t)$ and $c(t)$ be the same as shown in Fig. 8.14a. Figure 8.15a and b shows, respectively, $(1 + 0.8m(t))c(t)$ and the magnitude of the FT spectrum of $(1 + 0.8m(t))c(t)$. The signal can be demodulated by a simple envelope detector circuit or a rectifier followed by a lowpass filter, if the message signal rides on the carrier signal. That is, $(1 + km(t)) \geq 0$ for all values of t.

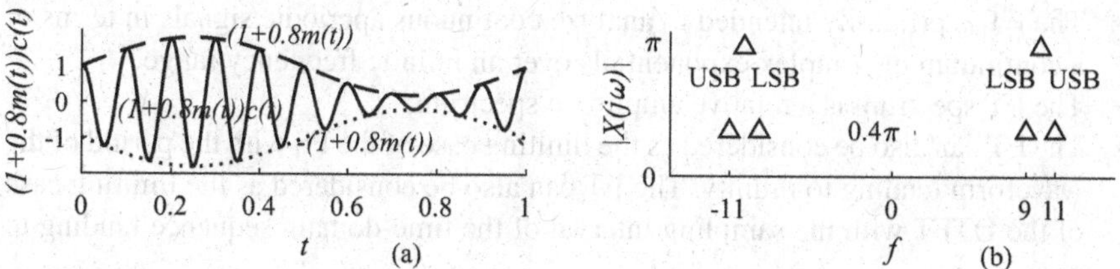

Fig. 8.15 **(a)** $(1 + 0.8m(t))c(t)$; **(b)** the magnitude of the FT spectrum of $(1 + 0.8m(t))c(t)$

The basis of modulation in the two cases studied is frequency shifting. One advantage of this type of modulation is the transmission of several signals over the same channel using the frequency-division multiplexing method. The signals share portions of the bandwidth of the channel with adequate separation between them.

8.5.3.3 Pulse Amplitude Modulation (PAM)

In the modulation types so far presented, the carrier is a sinusoid. The use of a pulse train as the carrier and modulating its amplitude in accordance with the message signal is called the pulse amplitude modulation (PAM). The pulse train consists of constant width and amplitude pulses with uniform spacing between them. The message signal modulates the amplitude of the pulses. This is essentially the same as that of sampling of continuous signals using an impulse train, presented earlier. The difference is that the sampling pulse, unlike the impulse, has a finite width. The FS spectrum of this signal is $\frac{\sin(a\,k\omega_s)}{k\pi}$, where a is half the width of the pulse, $\omega_s = \frac{2\pi}{T_s}$, and T_s is the sampling interval. Proceeding as in the case of the impulse sampling, we get the FT of the modulated signal as

$$X_s(j\omega) = \left(\cdots + \frac{\sin(a\omega_s)}{\pi} X(j(\omega + \omega_s)) + \frac{2a}{T_s} X(j\omega) + \cdots\right)$$

The spectrum, centered at $\omega = 0$, is unaltered except for a scale factor compared with that of the signal. Therefore, we can recover the original spectrum using a lowpass filter. Using this type of modulation, several message signals can be transmitted over the same channel by the method called time-division multiplexing. The time between two pulses of a modulated signal can be used by other modulated signals.

8.6 Summary

- In this chapter, we studied the FT, its properties, its approximation by the DFT, and some of its applications.

- The FT is primarily intended to analyze continuous aperiodic signals in terms of a continuum of complex exponentials over an infinite frequency range.
- The FT spectrum is a relative amplitude spectrum.
- The FT can also be considered as the limiting case of the FS with the period of the waveform tending to infinity. The FT can also be considered as the limiting case of the DTFT with the sampling interval of the time-domain sequence tending to zero.
- The FT is the most general type of Fourier analysis, and, hence, it can be used to analyze a mixed class of signals.
- The FT can be approximated by the DFT to a desired accuracy with proper choice of the record length and the number of samples.
- The FT has wide applications in signal and system analysis.

Exercises

8.1 Starting from the defining equations of the exponential form of the FS and the inverse FS, derive the defining equations of the FT and the inverse FT as the period of the time-domain waveform tends to infinity.

8.2 Derive an expression, using the defining integral, for the FT of the signal $x(t) = e^{-at}\cos(\omega_0 t)u(t),\ a > 0$. With $a = 0.4$ and $\omega_0 = 3$, compute $X(j0)$.

8.3 Derive an expression, using the defining integral, for the FT of the signal $x(t) = e^{-at}\sin(\omega_0 t)u(t),\ a > 0$. With $a = 0.1$ and $\omega_0 = 2$, compute $X(j0)$.

8.4 Derive an expression, using the defining integral, for the FT of the signal

$$x(t) = \begin{cases} 1 - |t| & \text{for } |t| < 1 \\ 0 & \text{elsewhere} \end{cases}$$

Compute $X(j0)$ and $X(j(2\pi))$.

8.5 Derive an expression, using the defining integral, for the FT of the signal $x(t) = e^{-3|t|}$. Compute $X(j0)$.

8.6 Derive an expression, using the defining integral, for the FT of the signal

$$x(t) = \begin{cases} \cos(10t) & \text{for } |t| < 1 \\ 0 & \text{for } |t| > 1 \end{cases}$$

Compute $X(j0)$.

* **8.7** Derive an expression, using the defining integral, for the FT of the signal $x(t) = te^{-2t}u(t)$. Compute $X(j0)$.

8.8 Apply a limiting process so that $x(t)$ degenerates into the impulse function, and, hence, derive the FT of the impulse function $\delta(t)$.

8.8.1 $x(t) = ae^{-at}u(t), \ a > 0, \text{ as } a \to \infty.$

8.8.2 $x(t) = \begin{cases} \frac{1}{2a} & \text{for } |t| < a \\ 0 & \text{for } |t| > a \end{cases}, \ a > 0, \text{ as } a \to 0.$

8.8.3 $x(t) = \begin{cases} \frac{1}{a} & \text{for } 0 < t < a \\ 0 & \text{elsewhere} \end{cases}, \ a > 0, \text{ as } a \to 0.$

8.8.4 $x(t) = \begin{cases} \frac{1}{a} & \text{for } -a < t < 0 \\ 0 & \text{elsewhere} \end{cases}, \ a > 0, \text{ as } a \to 0.$

8.8.5 $x(t) = \begin{cases} \frac{1}{a^2}(a - |t|) & \text{for } |t| < a \\ 0 & \text{for } |t| > a \end{cases}, \ a > 0, \text{ as } a \to 0.$

8.8.6 $x(t) = \begin{cases} \frac{1}{2a^2}(2a + t) & \text{for } -2a < t < 0 \\ 0 & \text{elsewhere} \end{cases}, \ a > 0, \text{ as } a \to 0.$

8.9 Apply a limiting process so that $x(t)$ degenerates into the DC function, and, hence, derive the FT of the DC function, $x(t) = 1$.

$$x(t) = \begin{cases} 1 & \text{for } |t| < a \\ 0 & \text{for } |t| > a \end{cases}, \ a > 0, \text{ as } a \to \infty.$$

8.10 Derive the FT of the function $y(t) = \begin{cases} -1 & \text{for } t < 0 \\ 1 & \text{for } t > 0 \end{cases}$ by applying a limiting process to the signal $x(t) = e^{-at}u(t) - e^{at}u(-t)$, as $a \to 0$.

8.11 Apply a limiting process so that $x(t)$ degenerates into the cosine function, and, hence, derive the FT of the cosine function, $\cos(t)$.

$$x(t) = \begin{cases} \cos(t) & \text{for } |t| < a \\ 0 & \text{for } |t| > a \end{cases}, \ a > 0, \text{ as } a \to \infty$$

8.12 Apply a limiting process so that $e^{-a|t|}\sin(t), \ a > 0$ degenerates into $\sin(t)$, as $a \to 0$, and, hence, derive the FT of $\sin(t)$.

8.13 Derive the FT of the unit-step function $u(t)$ using the FT of the functions

$$x(t) = \begin{cases} 1 & \text{for } t > 0 \\ -1 & \text{for } t < 0 \end{cases} \quad \text{and} \quad y(t) = 1$$

8.14 Using the duality property, find the FT of the signal $x(t)$.

8.14.1 $x(t) = \frac{1}{2+jt}$.

* 8.14.2 $x(t) = 2\frac{\sin(3t)}{t}$.

8.14.3 $x(t) = \pi\delta(t) + \frac{1}{jt}$.

8.15 Using the linearity and frequency-shifting properties, find the FT of $x(t)$.

8.15.1 $x(t) = \cos(\omega_0 t)u(t)$.

8.15.2 $x(t) = \sin(\omega_0 t)u(t)$.

* 8.15.3 $x(t) = \begin{cases} \cos(\omega_0 t) & \text{for } |t| < a \\ 0 & \text{for } |t| > a \end{cases}, \quad a > 0.$

8.15.4 $x(t) = \begin{cases} \sin(\omega_0 t) & \text{for } |t| < a \\ 0 & \text{for } |t| > a \end{cases}, \quad a > 0.$

* **8.16** Derive the inverse FT of the function

$$X(j\omega) = \frac{1}{\omega^2}\left(e^{-j4\omega} - 1\right)$$

using the time-domain convolution property.

8.17 Using the time-domain convolution property, find the FT of the convolution of $x(t)$ and $h(t)$.

8.17.1 $x(t) = \begin{cases} -2 & \text{for } 0 < t < 4 \\ 0 & \text{for } t < 0 \text{ and } t > 4 \end{cases}$ and $h(t) = \begin{cases} 3 & \text{for } 0 < t < 5 \\ 0 & \text{for } t < 0 \text{ and } t > 5 \end{cases}$

8.17.2 $x(t) = e^{-2t}u(t)$ and $h(t) = e^{-3t}u(t)$.

8.17.3 $x(t) = e^{-t}u(t)$ and $h(t) = \begin{cases} 1 & \text{for } 0 < t < 1 \\ 0 & \text{for } t < 0 \text{ and } t > 1 \end{cases}$

8.17.4 $x(t) = \begin{cases} (1 - |t|) & \text{for } |t| < 1 \\ 0 & \text{otherwise} \end{cases}$

and $h(t) = \begin{cases} 1 & \text{for } 0 < t < 1 \\ 0 & \text{for } t < 0 \text{ and } t > 1 \end{cases}$

8.17.5 $x(t) = e^{-at}u(t)$, $a > 0$ and $h(t) = x(t)$.

8.18 Using the frequency-domain convolution property, find the FT of the product of $x(t)$ and $h(t)$.

8.18.1 $x(t) = \cos(\omega_0 t)$ and $h(t) = u(t)$.

* 8.18.2 $x(t) = \sin(\omega_0 t)$ and $h(t) = u(t)$.

8.18.3 $x(t) = \cos(\omega_0 t)$ and $h(t) = \begin{cases} 1 & \text{for } |t| < a \\ 0 & \text{for } |t| > a \end{cases}, \quad a > 0.$

8.18.4 $x(t) = \sin(\omega_0 t)$ and $h(t) = \begin{cases} 1 & \text{for } |t| < a \\ 0 & \text{for } |t| > a \end{cases}, \quad a > 0.$

8.19 Derive the FT of the function $x(t) = e^{-a|t|}$, $a > 0$ using the linearity and time-reversal properties.

8.20 Using the time-scaling property, find the FT of the signal $x(at)$.

8.20.1 $x(t) = \cos(t)$ and $a = -2$.
8.20.2 $x(t) = e^{-2t}u(t)$ and $a = 2$.
8.20.3 $x(t) = e^{-2t}u(t)$ and $a = \frac{1}{2}$.
8.20.4 $x(t) = e^{-2t}u(t)$ and $a = -\frac{1}{2}$.
8.20.5 $x(t) = \begin{cases} 1 & \text{for } |t| < 2 \\ 0 & \text{for } |t| > 2 \end{cases}$ and $a = 2$.
8.20.6 $x(t) = \begin{cases} 1 & \text{for } |t| < 2 \\ 0 & \text{for } |t| > 2 \end{cases}$ and $a = -2$.
8.20.7 $x(t) = \begin{cases} 1 & \text{for } |t| < 2 \\ 0 & \text{for } |t| > 2 \end{cases}$ and $a = \frac{1}{2}$.
8.20.8 $x(t) = u(t)$ and $a = 3$.
* 8.20.9 $x(t) = u(t)$ and $a = -2$.
8.20.10 $x(t) = u(t - 4)$ and $a = 2$.

8.21 Using the time-differentiation property, find the FT of the derivative of the signal $x(t) = \sin(4t)$.

8.22 Using the time-differentiation property, find the FT of the signal $x(t)$.

8.22.1 $x(t) = \begin{cases} (1 - t) & \text{for } 0 < t < 1 \\ 0 & \text{for } t < 0 \text{ and } t \geq 1 \end{cases}$
8.22.2 $x(t) = \begin{cases} t & \text{for } 0 < t < 1 \\ 0 & \text{for } t \leq 0 \text{ and } t > 1 \end{cases}$
8.22.3 $x(t) = \begin{cases} 1 & \text{for } -1 < t < 0 \\ -1 & \text{for } 0 < t < 1 \\ 0 & \text{for } t < -1 \text{ and } t > 1 \end{cases}$
* 8.22.4 $x(t) = \begin{cases} t & \text{for } 0 \leq t < 1 \\ 1 & \text{for } 1 \leq t < 2 \\ (3 - t) & \text{for } 2 \leq t < 3 \\ 0 & \text{for } t < 0 \text{ and } t > 3 \end{cases}$
8.22.5 $x(t) = e^{-2|t|}$.

8.23 Using the time-integration property, find the FT of $y(t)$, where

$$y(t) = \int_{-\infty}^{t} x(\tau)d\tau$$

8.23.1 $x(t) = \delta(t-3)$.

8.23.2 $x(t) = \begin{cases} 2 \text{ for } -1 < t < 0 \\ -2 \text{ for } 0 < t < 1 \\ 0 \text{ for } t < -1 \text{ and } t > 1 \end{cases}$

8.23.3 $x(t) = \cos(3t)$.

8.23.4 $x(t) = \begin{cases} \sin(t) \text{ for } 0 \le t < \frac{\pi}{2} \\ 0 \text{ for } t < 0 \text{ and } t > \frac{\pi}{2} \end{cases}$

8.23.5 $x(t) = \begin{cases} \cos(t) \text{ for } 0 < t < \pi \\ 0 \text{ for } t < 0 \text{ and } t > \pi \end{cases}$

* 8.23.6 $x(t) = e^{-t}u(t)$.

8.23.7 $x(t) = u(t)$.

8.24 Using the frequency-differentiation property, find the FT of the signal $x(t)$.

8.24.1 $x(t) = t^2 e^{-t} u(t)$.

8.24.2 $x(t) = tu(t)$.

8.24.3 $x(t) = te^{-2|t|}$.

* 8.24.4 $x(t) = \begin{cases} t \text{ for } 0 < t < 1 \\ 0 \text{ for } t < 0 \text{ and } t > 1 \end{cases}$

8.24.5 $x(t) = \begin{cases} t \text{ for } -1 < t < 1 \\ 0 \text{ for } t < -1 \text{ and } t > 1 \end{cases}$

8.24.6 $x(t) = \begin{cases} t\sin(t) \text{ for } 0 < t < \pi \\ 0 \text{ for } t < 0 \text{ and } t > \pi \end{cases}$

8.25 Using the linearity, time-shifting, frequency-differentiation properties and the FT of $u(t)$, find the FT of the signal

$$x(t) = \begin{cases} 0 \text{ for } t < 0 \\ t \text{ for } 0 \le t \le 3 \\ 3 \text{ for } t > 3 \end{cases}$$

8.26 Find the energy of the signal $x(t) = e^{-2t}u(t)$. Find the value of T such that 90% of the signal energy lies in the range $0 \le t \le T$. What is the corresponding signal bandwidth?

8.27 Derive Parseval's theorem for aperiodic signals from that for the Fourier series of periodic signals, as the period tends to infinity.

8.28 Using the complex FS coefficients of the periodic signal $x(t)$, find its FT.

8.28.1 $x(t) = \sum_{n=-\infty}^{\infty} \delta(t - nT)$.

8.28.2 $x(t) = 2 + 3\cos(2t) + 4\sin(4t) - 5e^{-j6t} + 6e^{j10t}$.

* 8.28.3 $x(t) = -1 - 3\sin(3t) + 2\cos(5t) + 6e^{-j7t}$.

8.28.4 $x(t) = 3 - 2\cos(10t) + 3\sin(15t) - e^{j25t}$.

8.29 Using the FT, find the complex FS coefficients of the periodic signal $x(t)$.

8.29.1 $x(t) = 5e^{-j(t+\frac{\pi}{3})}$.
8.29.2 $x(t) = 2\cos(2t - \frac{\pi}{4})$.
8.29.3 $x(t) = 3\sin(3t - \frac{\pi}{6})$.
* 8.29.4 $x(t) = \sum_{n=-\infty}^{\infty}(t)(u(t) - u(t-2))$.

8.30 Find the inverse FT, $x(t)$, of $X(j\omega)$. Find the sampled signal $x_s(t)$ and its transform $X_s(j\omega)$ for the sampling interval $T_s = 0.25, 0.5, 1, 2$, and 3 s.

8.30.1

$$X(j\omega) = \begin{cases} \cos(\omega) & \text{for } |\omega| < \pi \\ 0 & \text{elsewhere} \end{cases}$$

8.30.2

$$X(j\omega) = \begin{cases} \sin(2\omega) & \text{for } |\omega| < \pi \\ 0 & \text{elsewhere} \end{cases}$$

8.31 Find the FT of $x(t)$ and its sampled versions with the sampling interval $T_s = 0.01, 0.1, 1$, and 10 s. What are the spectral values of $x(t)$ and its sampled versions at $\omega = 0$?

8.31.1 $x(t) = e^{-t}u(t)$.
* 8.31.2 $x(t) = e^{-|t|}$.

8.32 Find the FT of $x(t)$ and its sampled versions with the sampling interval $T_s = 0.1, 0.5, 1$, and 2 s.

8.32.1 $x(t) = 2\cos(\frac{2\pi}{32}t) + \sin(3\frac{2\pi}{32}t)$.
8.32.2 $x(t) = 4\sin(\frac{2\pi}{24}t) + \cos(5\frac{2\pi}{24}t)$.

* **8.33** Approximate the samples of the FT of the signal

$$x(t) = \begin{cases} 1 - |t| & \text{for } |t| < 1 \\ 0 & \text{elsewhere} \end{cases}$$

using the DFT with $N = 4$ samples. Assume that the signal is periodically extended with period $T = 2$ s. Compare the first two samples of the FT obtained using the DFT with that of the exact values.

8.34 Approximate the samples of the FT of the signal

$$x(t) = \begin{cases} 1 & \text{for } |t| < 2 \\ 0 & \text{for } |t| > 2 \end{cases}$$

using the DFT with $N = 4$ samples. Assume that the signal is periodically extended with period $T = 8$ s. Compare the first two samples of the FT obtained using the DFT with that of the exact values.

8.35 Find the response $y(t)$, using the FT, of the system governed by the differential equation

$$\frac{dy(t)}{dt} + y(t) = e^{jt}$$

Verify your solution by substituting it into the differential equation.

8.36 Using the FT, find the zero-state response $y(t)$ of the system governed by the differential equation

$$2\frac{dy(t)}{dt} + 3y(t) = \delta(t)$$

Verify your solution by substituting it into the differential equation.

8.37 Using the FT, find the zero-state response $y(t)$ of the system governed by the differential equation

$$\frac{dy(t)}{dt} + 2y(t) = u(t)$$

Verify your solution by substituting it into the differential equation.

8.38 Using the FT, find the zero-state response $y(t)$ of the system governed by the differential equation

$$3\frac{dy(t)}{dt} + 2y(t) = 4e^{-2t}u(t)$$

Verify your solution by substituting it into the differential equation.

Chapter 9
The z-Transform

In the Fourier analysis, we decompose a signal in terms of its constituent constant amplitude sinusoids. Systems are modeled in terms of their responses to sinusoids. This representation provides an insight into the signal and system characteristics and makes the evaluation of important operations, such as convolution, easier. The general constraint on the signal to be analyzed is that it is absolutely or square integrable/summable. Even with this constraint, the use of Fourier analysis is extensive in signal and system analysis. However, we still need the generalization of the Fourier analysis so that a larger class of signals and systems could be analyzed in the frequency domain, retaining all the advantages of the frequency-domain methods. The generalization of the Fourier analysis for discrete signals, called the z-transform, is described in this chapter.

The differences between the z-transform and the Fourier analysis are presented in Sect. 9.1. In Sect. 9.2, the z-transform is derived starting from the DTFT definition. In Sect. 9.3, the properties of the z-transform are described. In Sect. 9.4, the inverse z-transform is derived, and two frequently used methods to find the inverse z-transform are presented. Typical applications of the z-transform are described in Sect. 9.5.

9.1 Fourier Analysis and the z-Transform

In Fourier analysis, we analyze a waveform in terms of constant amplitude sinusoids $A\cos(\omega n + \theta)$ (shown in Fig. 1.4). The Fourier analysis is generalized by making the basis signals a larger set of sinusoids, by including sinusoids with exponentially varying amplitudes $Ar^n \cos(\omega n + \theta)$ (shown in Fig. 1.7). This extension enables us to analyze a larger set of signals and systems than that is possible with the Fourier analysis. The sinusoids, whether they have constant amplitude or varying amplitude, have the key advantages to be the basis signals in terms of ease of signal

D. Sundararajan, *Signals and Systems*,
https://doi.org/10.1007/978-3-031-19377-4_9

decomposition and efficient signal and system analysis. In Fourier analysis, we use fast algorithms to obtain the frequency-domain representation of signals. In the case of the transforms that use sinusoids with exponentially varying amplitudes, it is found that a short table of transform pairs is adequate for most practical purposes.

In the Fourier representation, the spectrum of a one-dimensional signal is also one-dimensional, the spectral coordinates being the frequency ω and the complex amplitude of the complex sinusoids. In the case of the generalized transforms, the rate of change of the amplitude of the exponentially varying amplitude sinusoids is also a parameter. This makes the spectrum of a one-dimensional signal two-dimensional, a surface. The spectrum provides infinite spectral representations of the signal, that is, the spectral values along any appropriate closed contour of the two-dimensional spectrum could be used to reconstruct the signal. Therefore, a signal may be reconstructed using constant amplitude sinusoids or exponentially decaying sinusoids or exponentially growing sinusoids or an infinite combination of these types of sinusoids.

The advantages of the z-transform include the pictorial description of the behavior of the system obtained by the use of the complex frequency; the ability to analyze unstable systems or systems with exponentially growing inputs; automatic inclusion of the initial conditions of the system in finding the output; and easier manipulation of the expressions involving the variable z than those with $e^{j\omega}$.

9.2 The z-Transform

We assume, in this chapter, that all signals are causal, that is, $x(n) = 0, n < 0$, unless otherwise specified. This leads to the one-sided or unilateral version of the z-transform, which is mostly used for practical system analysis. If a signal $x(n)u(n)$ is not Fourier transformable, then its exponentially weighted version, $(x(n)r^{-n})$, may be Fourier transformable for the positive real quantity $r > 1$. If $x(n)u(n)$ is Fourier transformable, $(x(n)r^{-n})$ may still be transformable for some values of $r < 1$. The DTFT of this signal is

$$\sum_{n=0}^{\infty}(x(n)r^{-n})e^{-j\omega n}$$

By combining the exponential factors, we get

$$X(re^{j\omega}) = \sum_{n=0}^{\infty} x(n)(re^{j\omega})^{-n}$$

This equation can be interpreted as the generalized Fourier analysis of the signal $x(n)$ using exponentials with complex exponents or sinusoids with varying amplitudes as the basis signals. By substituting $z = re^{j\omega}$, we get the defining equation of the one-sided or unilateral z-transform of $x(n)$ as

$$X(z) = \sum_{n=0}^{\infty} x(n)z^{-n} \tag{9.1}$$

Expanding the summation, we get

$$X(z) = x(0) + x(1)z^{-1} + x(2)z^{-2} + x(3)z^{-3} + \cdots$$

where z is a complex variable. Therefore, the basis functions used in the z-transform are of the form $z^n = e^{(\sigma+j\omega)n} = r^n e^{j\omega n} = r^n(\cos(\omega n) + j\sin(\omega n)) = (a + jb)^n$. While $X(e^{j\omega})$ is the DTFT of $x(n)$, $X(z) = X(re^{j\omega})$ is the DTFT of $x(n)r^{-n}$ for all values of r for which $\sum_{n=0}^{\infty} |x(n)r^{-n}| < \infty$. If the value one is included in these values of r, then $X(e^{j\omega})$ can be obtained from $X(z)$ by the substitution $z = e^{j\omega}$. The z-transform of a signal $x(n)$, $X(z)$, exists for $|z| > r_0$ if $|x(n)| \leq r_0^n$ for some constant r_0. For example, $x(n) = a^{n^2}$ does not have a z-transform. In essence, the z-transform of a signal, whether it is converging or not, is the DTFT of all its versions, obtained by multiplying it with a real exponential of the form r^{-n}, so that the modified signal is guaranteed to converge.

The z-transform, $X(z)$, represents a sequence only for the set of values of z for which it converges, that is, the magnitude of $X(z)$ is finite. The region that comprises this set of values in the z-plane (a complex plane used for displaying the z-transform) is called the region of convergence (ROC). For a given positive number c, the equation $|z| = |a + jb| = c$ or $a^2 + b^2 = c^2$ describes a circle in the z-plane with center at the origin and radius c. Consequently, the condition $|z| > c$ for ROC specifies the region outside this circle. If the ROC of the z-transform of a sequence includes the unit circle, then its DTFT can be obtained from $X(z)$ by replacing z with $e^{j\omega}$.

Example 9.1 Find the z-transform of the unit-impulse signal, $\delta(n)$.

Solution Using the definition, we get

$$X(z) = 1, \text{ for all } z \quad \text{and} \quad \delta(n) \Longleftrightarrow 1, \text{ for all } z$$

The transform pair for a delayed impulse $\delta(n - m)$ is

$$\delta(n - m) \Longleftrightarrow z^{-m}, \ |z| > 0,$$

where m is positive. ∎

Example 9.2 Find the z-transform of the finite sequence with its only nonzero samples specified as $\{x(0) = 5, x(2) = 4, x(5) = -2\}$.

Solution Using the definition, we get

$$X(z) = 5 + 4z^{-2} - 2z^{-5} = \frac{5z^5 + 4z^3 - 2}{z^5}, \quad |z| > 0$$ ∎

The geometric sequence, $a^n u(n)$, is fundamental to the study of linear discrete systems, as the natural response of systems is in that form.

Example 9.3 Find the z-transform of the geometric sequence, $a^n u(n)$.

Solution Substituting $x(n) = a^n$ in the defining equation of the z-transform, we get

$$X(z) = \sum_{n=0}^{\infty} (a^{-1}z)^{-n} = 1 + (a^{-1}z)^{-1} + (a^{-1}z)^{-2} + (a^{-1}z)^{-3} + \cdots$$

$$= \frac{1}{1 - (a^{-1}z)^{-1}} = \frac{z}{z - a}, \quad |z| > |a|$$

It is known that the geometric series $1 + r + r^2 + \cdots$ converges to $\frac{1}{1-r}$, if $|r| < 1$. If $|z| > |a|$, the common ratio of the series $r = \frac{a}{z}$ has magnitude that is less than one. Therefore, the ROC of the z-transform is given as $|z| > |a|$, and we get the transform pair

$$a^n u(n) \Longleftrightarrow \frac{z}{z - a}, \quad |z| > |a|$$ ∎

Note that the DTFT of $a^n u(n)$ does not exist for $a > 1$, whereas the z-transform exists for all values of a as long as $|z| > |a|$. The z-transform spectrum of a sequence is usually displayed by the locations of zeros and poles of the z-transform and its magnitude. The pole-zero plot and the magnitude of the z-transform $\frac{z}{z-0.8}$ of the signal $a^n u(n)$ with $a = 0.8$ are shown, respectively, in Fig. 9.1a and b. When $z = 0.8$, $|X(z)| = \infty$. This point marked by the symbol $\times$ in Fig. 9.1a is called a pole of $X(z)$ (the peak in Fig. 9.1b). When $z = 0$, $X(z) = 0$. This point marked by the symbol o in Fig. 9.1a is called a zero of $X(z)$ (the valley in Fig. 9.1b). The pole-zero plot specifies a transform $X(z)$, except for a constant factor. In the region outside the circle with radius 0.8, $X(z)$ exists and is a valid frequency-domain representation of the signal. In general, the ROC of a z-transform is the region in the z-plane that is exterior to the smallest circle, centered at the origin, enclosing all its poles.

Example 9.4 Find the z-transform of the signal $e^{j\omega n}u(n)$. Deduce the z-transform of $\sin(\omega n)u(n)$.

Solution Using the transform of $a^n u(n)$ with $a = e^{j\omega}$, we get

$$e^{j\omega n}u(n) \Longleftrightarrow \frac{z}{z - e^{j\omega}}, \quad |z| > 1$$

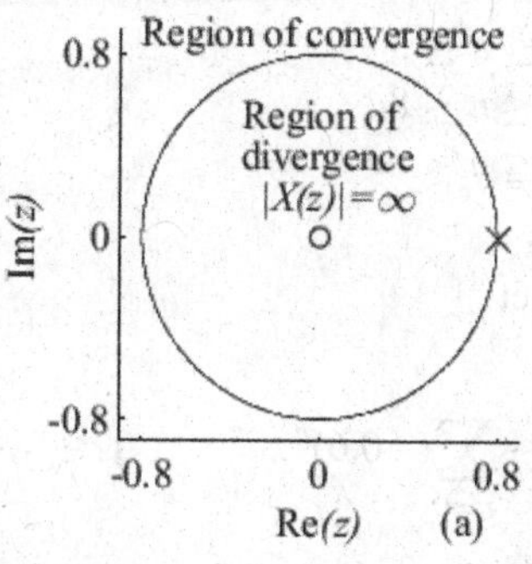

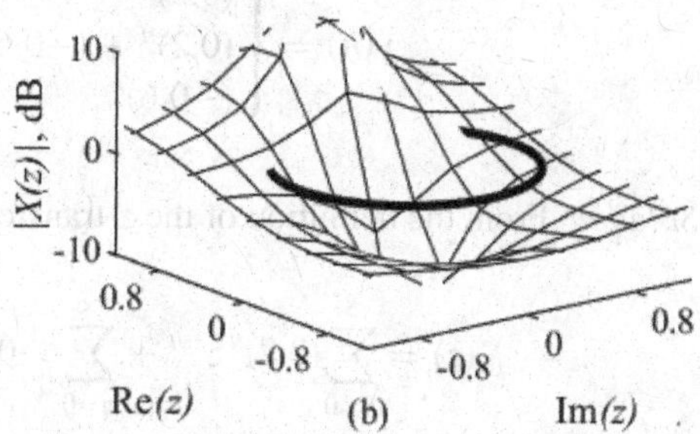

Fig. 9.1 (**a**) The pole-zero plot of the z-transform $\frac{z}{z-0.8}$ of $(0.8)^n u(n)$; (**b**) the magnitude of the z-transform

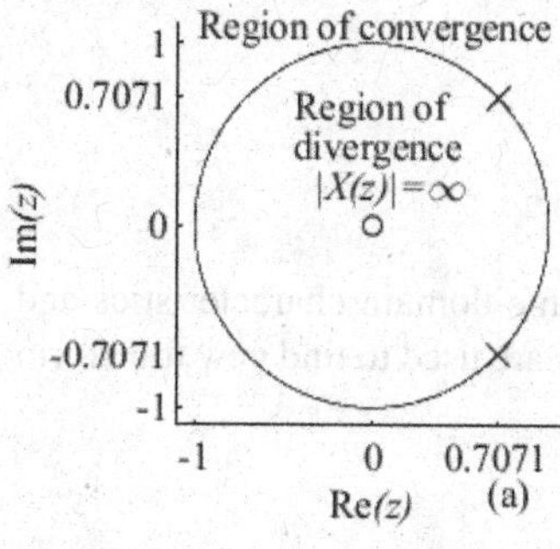

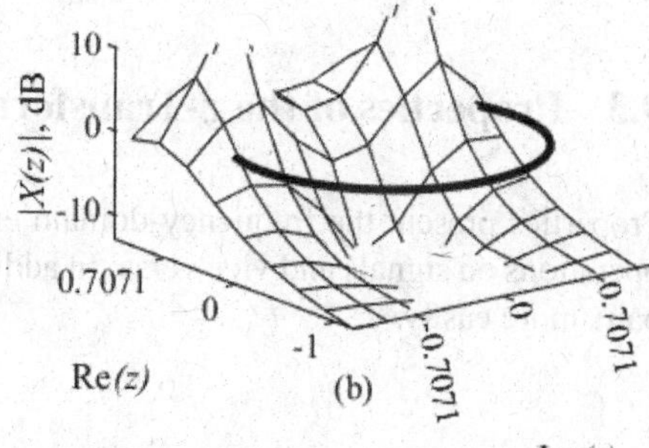

Fig. 9.2 (**a**) The pole-zero plot of the z-transform $\frac{z\sin(\frac{\pi}{4})}{(z-e^{j\frac{\pi}{4}})(z-e^{-j\frac{\pi}{4}})}$ of $\sin(\frac{\pi}{4}n)u(n)$; (**b**) the magnitude of the z-transform

Since the magnitude of $a = e^{j\omega}$ is 1, the convergence condition is $|z| > 1$. Using the fact that $j2\sin(\omega n) = (e^{j\omega n} - e^{-j\omega n})$, we get

$$j2X(z) = \frac{z}{z - e^{j\omega}} - \frac{z}{z - e^{-j\omega}}, \quad |z| > 1$$

$$\sin(\omega n)u(n) \Longleftrightarrow \frac{z\sin(\omega)}{(z - e^{j\omega})(z - e^{-j\omega})} = \frac{z\sin(\omega)}{z^2 - 2z\cos(\omega) + 1}, \quad |z| > 1$$ ∎

Figure 9.2a shows the pole-zero plot, and Fig. 9.2b shows the magnitude of the z-transform $\frac{z\sin(\frac{\pi}{4})}{z^2-2z\cos(\frac{\pi}{4})+1}$ of the signal $\sin(\frac{\pi}{4}n)u(n)$. There are a zero at $z = 0$ and poles at $z = e^{j\frac{\pi}{4}}$ and $z = e^{-j\frac{\pi}{4}}$, a pair of complex conjugate poles.

Example 9.5 Find the z-transform of the signal defined as

$$x(n) = \begin{cases} (0.2)^n & \text{for } 0 \le n \le 5 \\ (0.2)^n + (-0.6)^n & \text{for } 6 \le n \le 8 \\ (-0.6)^n & \text{for } 9 \le n < \infty \end{cases}$$

Solution From the definition of the z-transform, we get

$$X(z) = \sum_{n=0}^{8} (0.2)^n z^{-n} + \sum_{n=0}^{\infty} (-0.6)^n z^{-n} - \sum_{n=0}^{5} (-0.6)^n z^{-n}$$

$$X(z) = \frac{z^9 - (0.2)^9}{z^8(z - 0.2)} + \frac{z}{z + 0.6} - \frac{z^6 - (-0.6)^6}{z^5(z + 0.6)}, \ |z| > 0.6$$ ∎

9.3 Properties of the z-Transform

Properties present the frequency-domain effect of time-domain characteristics and operations on signals and vice versa. In addition, they are used to find new transform pairs more easily.

9.3.1 *Linearity*

It is often advantageous to decompose a complex sequence into a linear combination of simpler sequences (as in Example 9.4) in the manipulation of sequences and their transforms. If $x(n) \Longleftrightarrow X(z)$ and $y(n) \Longleftrightarrow Y(z)$, then

$$ax(n) + by(n) \Longleftrightarrow aX(z) + bY(z),$$

where a and b are arbitrary constants. The z-transform of a linear combination of sequences is the same linear combination of the z-transforms of the individual sequences. This property is due to the linearity of the defining summation operation of the transform.

9.3.2 *Left Shift of a Sequence*

The shift property is used to express the transform of the shifted version, $x(n + m)u(n)$, of a sequence $x(n)$ in terms of its transform $X(z)$. If $x(n)u(n) \Longleftrightarrow X(z)$ and m is a positive integer, then

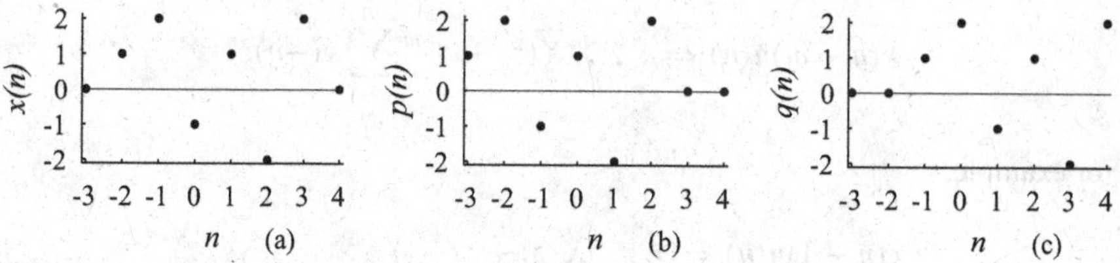

Fig. 9.3 **(a)** $x(n)$; **(b)** $p(n) = x(n+1)$; **(c)** $q(n) = x(n-1)$

$$x(n+m)u(n) \Longleftrightarrow z^m X(z) - z^m \sum_{n=0}^{m-1} x(n)z^{-n}$$

Let the z-transform of the sequence $x(n+m)u(n)$ be $Y(z)$. Then,

$$z^{-m}Y(z) = x(m)z^{-m} + x(m+1)z^{-m-1} + \cdots$$

By adding m terms, $\sum_{n=0}^{m-1} x(n)z^{-n}$, to both sides of the equation, we get

$$z^{-m}Y(z) + x(m-1)z^{-m+1} + x(m-2)z^{-m+2} + \cdots + x(0) = X(z)$$

$$Y(z) = z^m X(z) - z^m \sum_{n=0}^{m-1} x(n)z^{-n}$$

For example,

$$x(n+1)u(n) \Longleftrightarrow zX(z) - zx(0) \text{ and } x(n+2)u(n) \Longleftrightarrow z^2X(z) - z^2x(0) - zx(1)$$

Consider the sequence $x(n)$ with $x(-2) = 1, x(-1) = 2, x(0) = -1, x(1) = 1, x(2) = -2, x(3) = 2$, and $x(n) = 0$ otherwise, shown in Fig. 9.3a. The transform of $x(n)$ is $X(z) = -1 + z^{-1} - 2z^{-2} + 2z^{-3}$. The sequence $p(n)$, shown in Fig. 9.3b, is the left-shifted sequence $x(n+1)$. The transform of $p(n)u(n) = x(n+1)u(n)$ is

$$P(z) = 1 - 2z^{-1} + 2z^{-2} = zX(z) - zx(0) = z(-1 + z^{-1} - 2z^{-2} + 2z^{-3}) + z$$

9.3.3 Right Shift of a Sequence

If $x(n)u(n) \Longleftrightarrow X(z)$ and m is a positive integer, then

$$x(n-m)u(n) \Longleftrightarrow z^{-m}X(z) + z^{-m}\sum_{n=1}^{m} x(-n)z^{n}$$

For example,

$$x(n-1)u(n) \Longleftrightarrow z^{-1}X(z) + x(-1)$$
$$x(n-2)u(n) \Longleftrightarrow z^{-2}X(z) + z^{-1}x(-1) + x(-2)$$

The sequence $q(n)$, shown in Fig. 9.3c, is the right-shifted sequence $x(n-1)$. The transform of $q(n)u(n) = x(n-1)u(n)$ is

$$Q(z) = 2 - z^{-1} + z^{-2} - 2z^{-3} + 2z^{-4}$$
$$= 2 + z^{-1}(-1 + z^{-1} - 2z^{-2} + 2z^{-3}) = x(-1) + z^{-1}X(z)$$

In finding the response $y(n)$ of a system for $n \geq 0$, the initial conditions, such as $y(-1)$ and $y(-2)$, must be taken into account. The shift properties provide the way for the automatic inclusion of the initial conditions. The left shift property is more convenient for solving difference equations in advance operator form. Consider solving the difference equation $y(n) = x(n) + \frac{1}{2}y(n-1)$ with the initial condition $y(-1) = 3$ and $x(n) = 0$. The solution is $y(n) = \frac{3}{2}\left(\frac{1}{2}\right)^{n} u(n)$, using time-domain method. Taking the z-transform of the difference equation, we get $Y(z) = X(z) + \frac{1}{2}(z^{-1}Y(z) + 3)$. Solving for $Y(z)$, $Y(z) = \frac{\frac{3}{2}z}{z-\frac{1}{2}}$. The inverse transform of $Y(z)$ is $y(n) = \frac{3}{2}\left(\frac{1}{2}\right)^{n} u(n)$, which is the same as that obtained earlier.

9.3.4 Convolution

If $x(n)u(n) \Longleftrightarrow X(z)$ and $h(n)u(n) \Longleftrightarrow H(z)$, then

$$y(n) = \sum_{m=0}^{\infty} h(m)x(n-m) \Longleftrightarrow Y(z) = H(z)X(z)$$

The DTFT of $x(n)r^{-n}$ is the z-transform $X(z)$ of $x(n)$. The convolution of $x(n)r^{-n}$ and $h(n)r^{-n}$ corresponds to $X(z)H(z)$ in the frequency domain. The inverse DTFT of $X(z)H(z)$, therefore, is the convolution of $x(n)r^{-n}$ and $h(n)r^{-n}$ given by

$$\sum_{m=0}^{\infty} x(m)r^{-m}h(n-m)r^{(-n+m)} = r^{-n}\sum_{m=0}^{\infty} x(m)h(n-m) = r^{-n}(x(n) * h(n))$$

As finding the inverse z-transform is the same as finding the inverse DTFT in addition to multiplying the signal by r^n, as will be seen later, we get the convolution of $x(n)$ and $h(n)$ by finding the inverse z-transform of $X(z)H(z)$.

Consider the two sequences and their transforms $x(n) = (\frac{1}{2})^n u(n) \Longleftrightarrow X(z) = \frac{z}{z-\frac{1}{2}}$ and $h(n) = (\frac{1}{3})^n u(n) \Longleftrightarrow H(z) = \frac{z}{z-\frac{1}{3}}$. The convolution of the sequences, in the transform domain, is given by the product of their transforms,

$$X(z)H(z) = \frac{z}{z-\frac{1}{2}}\frac{z}{z-\frac{1}{3}} = \frac{3z}{z-\frac{1}{2}} - \frac{2z}{z-\frac{1}{3}}$$

The inverse transform of $X(z)H(z)$ is the convolution of the sequences in the time domain, and it is $\left(3(\frac{1}{2})^n - 2(\frac{1}{3})^n\right) u(n)$.

9.3.5 Multiplication by *n*

If $x(n)u(n) \Longleftrightarrow X(z)$, then

$$nx(n)u(n) \Longleftrightarrow -z\frac{d}{dz}X(z)$$

Differentiating the defining expression for $X(z)$ with respect to z and multiplying it by $-z$, we get

$$-z\frac{d}{dz}X(z) = -z\frac{d}{dz}\sum_{n=0}^{\infty} x(n)z^{-n} = \sum_{n=0}^{\infty} nx(n)z^{-n} = \sum_{n=0}^{\infty}(nx(n))z^{-n}$$

For example,

$$\delta(n) \Longleftrightarrow 1 \quad \text{and} \quad n\delta(n) = 0 \Longleftrightarrow 0$$

$$u(n) \Longleftrightarrow \frac{z}{z-1} \quad \text{and} \quad nu(n) \Longleftrightarrow \frac{z}{(z-1)^2}$$

9.3.6 Multiplication by a^n

If $x(n)u(n) \Longleftrightarrow X(z)$, then

$$a^n x(n)u(n) \Longleftrightarrow X\left(\frac{z}{a}\right)$$

From the z-transform definition, we get

$$X(z) = \sum_{n=0}^{\infty} a^n x(n) z^{-n} = \sum_{n=0}^{\infty} x(n) \left(\frac{z}{a}\right)^{-n} = X\left(\frac{z}{a}\right)$$

Multiplication of $x(n)$ by a^n corresponds to scaling the frequency variable z. For example,

$$u(n) \Longleftrightarrow \frac{z}{z-1} \quad \text{and} \quad (2)^n u(n) \Longleftrightarrow \frac{\frac{z}{2}}{(\frac{z}{2}-1)} = \frac{z}{z-2}$$

The pole at $z = 1$ in the transform of $u(n)$ is shifted to the point $z = 2$ in the transform of $(2)^n u(n)$.

With $a = -1$ and $x(n)u(n) \Longleftrightarrow X(z)$, $(-1)^n x(n)u(n) \Longleftrightarrow X(-z)$. For example, $u(n) \Longleftrightarrow \frac{z}{z-1}$ and $(-1)^n u(n) \Longleftrightarrow \frac{-z}{-z-1} = \frac{z}{z+1}$.

9.3.7 *Summation*

If $x(n)u(n) \Longleftrightarrow X(z)$, then $y(n) = \sum_{m=0}^{n} x(m) \Longleftrightarrow Y(z) = \frac{z}{z-1} X(z)$. The product $\frac{z}{z-1} X(z)$ corresponds to the convolution of $x(n)$ and $u(n)$ in the time domain, which, of course, is equivalent to the sum of the first $n+1$ values of $x(n)$.

For example, $x(n) = (-1)^n u(n) \Longleftrightarrow \frac{z}{z+1}$. Then, $Y(z) = \frac{z}{z-1}\frac{z}{z+1} = \frac{1}{2}\left(\frac{z}{z-1} + \frac{z}{z+1}\right)$. Taking the inverse z-transform, we get $y(n) = \frac{1}{2}(1 + (-1)^n)$.

9.3.8 *Initial Value*

Using this property, the initial value of $x(n)$, $x(0)$, can be determined directly from $X(z)$. If $x(n)u(n) \Longleftrightarrow X(z)$, then

$$x(0) = \lim_{z\to\infty} X(z) \quad \text{and} \quad x(1) = \lim_{z\to\infty} (z(X(z) - x(0)))$$

From the definition of the transform, we get

$$\lim_{z\to\infty} X(z) = \lim_{z\to\infty} (x(0) + x(1)z^{-1} + x(2)z^{-2} + x(3)z^{-3} + \cdots) = x(0)$$

As $z \to \infty$, each term, except $x(0)$, tends to zero. Let $X(z) = \frac{(z^2-2z+5)}{(z^2+3z-2)}$. Then,

$$x(0) = \lim_{z\to\infty} \frac{(z^2 - 2z + 5)}{(z^2 + 3z - 2)} = 1$$

Note that, when $z \to \infty$, only the terms of the highest power are significant.

9.3.9 Final Value

Using this property, the final value of $x(n)$, $x(\infty)$, can be determined directly from $X(z)$. If $x(n)u(n) \Longleftrightarrow X(z)$, then

$$\lim_{n\to\infty} x(n) = \lim_{z\to 1} ((z-1)X(z))$$

provided the ROC of $(z-1)X(z)$ includes the unit circle (otherwise, $x(n)$ has no limit as $n \to \infty$). Let $X(z) = \frac{(z^2-2z+5)}{(z^2+3z-2)}$. The property does not apply since the ROC of $(z-1)X(z)$ does not include the unit circle. Let $X(z) = \frac{(z^2-2z+5)}{(z^2-1.5z+0.5)}$. Then,

$$\lim_{n\to\infty} x(n) = \lim_{z\to 1} (z-1)\frac{(z^2 - 2z + 5)}{(z^2 - 1.5z + 0.5)} = \lim_{z\to 1} \frac{(z^2 - 2z + 5)}{(z - 0.5)} = 8$$

The value $\lim_{n\to\infty} x(n)$, if it is nonzero, is solely due to the scaled unit-step component of $x(n)$. Multiplying $X(z)$ by $(z-1)$ and setting $z = 1$ is just finding the partial fraction coefficient of the unit-step component of $x(n)$.

9.3.10 Transform of Semiperiodic Functions

Consider the function $x(n)u(n)$ that is periodic of period N for $n \geq 0$, that is, $x(n+N) = x(n)$, $n \geq 0$. Let $x_1(n) = x(n)u(n) - x(n-N)u(n-N) \Longleftrightarrow X_1(z)$. $x_1(n)$ is equal to $x(n)u(n)$ over its first period and is zero elsewhere. Then,

$$x(n)u(n) = x_1(n) + x_1(n-N) + x_1(n-2N) + \cdots$$

Using the right shift property, the transform of $x(n)u(n)$ is

$$X(z) = X_1(z)(1 + z^{-N} + z^{-2N} + \cdots) = \frac{X_1(z)}{1 - z^{-N}} = X_1(z)\left(\frac{z^N}{z^N - 1}\right)$$

Let us find the transform of $x(n) = (-1)^n u(n)$ with period $N = 2$. $X_1(z) = 1 - z^{-1} = \frac{z-1}{z}$. From the property,

$$X(z) = \frac{z^2}{(z^2-1)}\frac{(z-1)}{z} = \frac{z}{(z+1)}$$

9.4 The Inverse z-Transform

Consider the transform pair $x(n)u(n) \Longleftrightarrow \frac{z}{z-2},\ |z| > 2$. Multiplying the signal by $(\frac{1}{4})^n u(n)$ gives $x(n)(\frac{1}{4})^n u(n) \Longleftrightarrow \frac{z}{z-0.5},\ |z| > 0.5$, due to the multiplication by a^n property. Now, the ROC includes the unit circle in the z-plane. Let us substitute $z = e^{j\omega}$ in $\frac{z}{z-0.5}$ to get $\frac{1}{1-0.5e^{-j\omega}}$. The inverse DTFT of this transform is the signal $0.5^n u(n) = x(n)(\frac{1}{4})^n u(n)$. Now, multiplying both sides by $4^n u(n)$ gives the original time-domain signal $x(n)u(n) = 2^n u(n)$. This way of finding the inverse z-transform gives us a clear understanding of how the z-transform is the generalized version of the DTFT.

The inverse z-transform relation enables us to find the corresponding sequence from its z-transform. The DTFT of $x(n)r^{-n}$ can be written as

$$X(re^{j\omega}) = \sum_{n=0}^{\infty} x(n)(re^{j\omega})^{-n}$$

The inverse DTFT of $X(re^{j\omega})$ is

$$x(n)r^{-n} = \frac{1}{2\pi}\int_{-\pi}^{\pi} X(re^{j\omega})e^{j\omega n}d\omega$$

Multiplying both sides by r^n, we get

$$x(n) = \frac{1}{2\pi}\int_{-\pi}^{\pi} X(re^{j\omega})(re^{j\omega})^n d\omega$$

Let $z = re^{j\omega}$. Then, $dz = jre^{j\omega}d\omega = jzd\omega$. Now, the inverse z-transform of $X(z)$, in terms of the variable z, is defined as

$$x(n) = \frac{1}{2\pi j}\oint_C X(z)z^{n-1}dz \tag{9.2}$$

with the integral evaluated, in the counterclockwise direction, along any simply connected closed contour C, encircling the origin, that lies in the ROC of $X(z)$. As ω varies from $-\pi$ to π, the variable z traverses the circle of radius r in the counterclockwise direction once. We can use any appropriate contour of integration in evaluating the inverse z-transform because the transform values corresponding to the contour are taken in the inverse process. As can be seen from Figs. 9.1b and

9.2b, the z-transform values vary with each of the infinite choices for the contour of integration.

9.4.1 *Finding the Inverse z-Transform*

While the most general way of finding the inverse z-transform is to evaluate the contour integral Eq. (9.2), for most practical purposes, two other simpler methods are commonly used.

9.4.1.1 The Partial Fraction Method

In LTI system analysis, we are mostly encountered with the problem of inverting a z-transform that is a rational function (a ratio of two polynomials in z). In the partial fraction method, the rational function of the z-transform is decomposed into a linear combination of transforms such as those of $\delta(n)$, $a^n u(n)$, and $na^n u(n)$. Then, it is easy to find the inverse transform from a short table of transform pairs.

Consider finding the partial fraction expansion of $X(z) = \frac{z}{(z-\frac{1}{5})(z-\frac{1}{4})}$. As the partial fraction of the form $\frac{kz}{(z-p)}$ is more convenient, we first expand $\frac{X(z)}{z}$ and then multiply both sides by z.

$$\frac{X(z)}{z} = \frac{1}{(z-\frac{1}{5})(z-\frac{1}{4})} = \frac{A}{(z-\frac{1}{5})} + \frac{B}{(z-\frac{1}{4})}$$

Multiplying all the expressions by $(z-\frac{1}{5})$, we get

$$\left(z-\frac{1}{5}\right)\frac{X(z)}{z} = \frac{1}{(z-\frac{1}{4})} = A + \frac{B(z-\frac{1}{5})}{(z-\frac{1}{4})}$$

Letting $z = \frac{1}{5}$, we get $A = (z-\frac{1}{5})\frac{X(z)}{z}|_{z=\frac{1}{5}} = -20$. Similarly, $B = (z-\frac{1}{4})\frac{X(z)}{z}|_{z=\frac{1}{4}} = 20$. Therefore,

$$X(z) = \frac{-20z}{(z-\frac{1}{5})} + \frac{20z}{(z-\frac{1}{4})}$$

The time-domain sequence $x(n)$ corresponding to $X(z)$ is given by

$$x(n) = \left(-20\left(\frac{1}{5}\right)^n + 20\left(\frac{1}{4}\right)^n\right)u(n)$$

The first four values of the sequence $x(n)$ are

$$x(0) = 0, \quad x(1) = 1, \quad x(2) = 0.45, \quad x(3) = 0.1525$$

As the sum of the terms of a partial fraction will always produce a numerator polynomial whose order is less than that of the denominator, the order of the numerator polynomial of the rational function must be less than that of the denominator. This condition is satisfied by $\frac{X(z)}{z}$, as the degree of the numerator polynomial, for z-transforms of practical interest, is at the most equal to that of the denominator.

Example 9.6 Find the inverse z-transform of

$$X(z) = \frac{z^2}{(z - \frac{1}{2})(z + \frac{1}{3})}$$

Solution

$$\frac{X(z)}{z} = \left(\frac{z}{(z - \frac{1}{2})(z + \frac{1}{3})}\right) \quad \text{and} \quad X(z) = \frac{\frac{3}{5}z}{(z - \frac{1}{2})} + \frac{\frac{2}{5}z}{(z + \frac{1}{3})}$$

$$x(n) = \left(\frac{3}{5}\left(\frac{1}{2}\right)^n + \frac{2}{5}\left(-\frac{1}{3}\right)^n\right) u(n)$$

The first four values of the sequence $x(n)$ are

$$x(0) = 1, \quad x(1) = 0.1667, \quad x(2) = 0.1944, \quad x(3) = 0.0602$$ ∎

The partial fraction method applies for complex poles also. Of course, the complex poles and their coefficients will always appear in conjugate pairs for $X(z)$ with real coefficients. Therefore, finding one of the coefficients of each pair of poles is sufficient.

Example 9.7 Find the inverse z-transform of

$$X(z) = \frac{z}{(z^2 - 2z + 4)}$$

Solution Factorizing the denominator of $X(z)$ and finding the partial fraction, we get

$$\frac{X(z)}{z} = \left(\frac{1}{(z - (1 + j\sqrt{3}))(z - (1 - j\sqrt{3}))}\right)$$

$$X(z) = \left(\frac{\frac{z}{j2\sqrt{3}}}{z-(1+j\sqrt{3})} + \frac{\frac{z}{-j2\sqrt{3}}}{z-(1-j\sqrt{3})}\right)$$

$$x(n) = \left(\frac{1}{j2\sqrt{3}}\right)(1+j\sqrt{3})^n + \left(\frac{1}{-j2\sqrt{3}}\right)(1-j\sqrt{3})^n,\ n=0,1,\ldots$$

The two terms of $x(n)$ form a complex conjugate pair. The conjugate of a complex number $z = x + jy$, denoted by z^*, is defined as $z^* = x - jy$, that is, the imaginary part is negated. Now, $z + z^* = 2x$ (twice the real part of z or z^*). This result is very useful in simplifying expressions involving complex conjugate poles. Let $(a + jb)$ and $(a - jb)$ be a pair of complex conjugate poles and $(c + jd)$ and $(c - jd)$ be their respective partial fraction coefficients. Then, the poles combine to produce the time-domain response $2A(r)^n \cos(\omega n + \theta)$, where $r = \sqrt{a^2+b^2}$ and $\omega = \tan^{-1}(\frac{b}{a})$, and $A = \sqrt{c^2+d^2}$ and $\theta = \tan^{-1}(\frac{d}{c})$. For the specific example, twice the real part of $\left(\frac{1}{j2\sqrt{3}}\right)(1+j\sqrt{3})^n$ or $\left(\frac{1}{-j2\sqrt{3}}\right)(1-j\sqrt{3})^n$ is

$$x(n) = \frac{1}{\sqrt{3}}(2)^n \cos\left(\frac{\pi}{3}n - \frac{\pi}{2}\right)u(n)$$

The first four values of the sequence $x(n)$ are

$$x(0) = 0,\quad x(1) = 1,\quad x(2) = 2,\quad x(3) = 0$$ ∎

Example 9.8 Find the inverse z-transform of

$$X(z) = \frac{(z^2 - 2z + 2)}{(z^2 - \frac{7}{12}z + \frac{1}{12})}$$

Solution

$$\frac{X(z)}{z} = \left(\frac{(z^2-2z+2)}{z(z-\frac{1}{3})(z-\frac{1}{4})}\right) \quad \text{and} \quad X(z) = \left(24 + \frac{52z}{(z-\frac{1}{3})} - \frac{75z}{(z-\frac{1}{4})}\right)$$

$$x(n) = 24\delta(n) + \left(52\left(\frac{1}{3}\right)^n - 75\left(\frac{1}{4}\right)^n\right)u(n)$$

The first four values of the sequence $x(n)$ are

$$x(0) = 1,\quad x(1) = -1.4167,\quad x(2) = 1.0903,\quad x(3) = 0.7541$$ ∎

For a pole of order m, there must be m partial fraction terms corresponding to poles of order $m, m-1, \ldots, 1$.

Example 9.9 Find the inverse z-transform of

$$X(z) = \frac{z^2}{(z-\frac{1}{3})^2(z-\frac{1}{2})}$$

Solution

$$\frac{X(z)}{z} = \left(\frac{z}{(z-\frac{1}{3})^2(z-\frac{1}{2})}\right) = \left(\frac{A}{(z-\frac{1}{3})^2} + \frac{B}{(z-\frac{1}{3})} + \frac{C}{(z-\frac{1}{2})}\right)$$

Now, A can be found to be -2 by substituting $z = \frac{1}{3}$ in the expression $\frac{z}{(z-\frac{1}{2})}$. C can be found to be 18 by substituting $z = \frac{1}{2}$ in the expression $\frac{z}{(z-\frac{1}{3})^2}$. One method to determine the value of B is to substitute a value for z, which is not equal to any of the poles. For example, by substituting $z = 0$ in the expression, the only unknown B is evaluated to be -18. Another method is to subtract the term $\frac{-2}{(z-\frac{1}{3})^2}$ from the expression $\frac{z}{(z-\frac{1}{3})^2(z-\frac{1}{2})}$ to get $\frac{3}{(z-\frac{1}{2})(z-\frac{1}{3})}$. Substituting $z = \frac{1}{3}$ in the expression $\frac{3}{(z-\frac{1}{2})}$, we get $B = -18$. Therefore,

$$X(z) = \left(-\frac{2z}{(z-\frac{1}{3})^2} - \frac{18z}{(z-\frac{1}{3})} + \frac{18z}{(z-\frac{1}{2})}\right)$$

$$x(n) = \left(-2n\left(\frac{1}{3}\right)^{n-1} - 18\left(\frac{1}{3}\right)^n + 18\left(\frac{1}{2}\right)^n\right)u(n)$$

The first four values of the sequence $x(n)$ are

$$x(0) = 0, \quad x(1) = 1, \quad x(2) = 1.1667, \quad x(3) = 0.9167$$ ∎

The next example is similar to Example 9.9 with the difference that a second-order pole occurs at $z = 0$.

Example 9.10 Find the inverse z-transform of

$$X(z) = \frac{z^2+1}{z^2(z-\frac{1}{3})}$$

Solution

$$\frac{X(z)}{z} = \left(\frac{z^2+1}{z^3(z-\frac{1}{3})}\right) = \left(\frac{A}{z^3} + \frac{B}{z^2} + \frac{C}{z} + \frac{D}{(z-\frac{1}{3})}\right)$$

$$X(z) = \left(\frac{-3}{z^2} + \frac{-9}{z} - 30 + \frac{30z}{(z - \frac{1}{3})} \right)$$

$$x(n) = -30\delta(n) - 9\delta(n-1) - 3\delta(n-2) + 30\left(\frac{1}{3}\right)^n u(n)$$

The first four values of the sequence $x(n)$ are

$$x(0) = 0, \quad x(1) = 1, \quad x(2) = 0.3333, \quad x(3) = 1.1111$$

■

9.4.1.2 The Long Division Method

By dividing the numerator polynomial by the denominator polynomial, we can express a z-transform in a form that is similar to that of the defining series. Then, from inspection, the sequence values can be found. For example, the inverse z-transform of $X(z) = \frac{z}{z-0.8}$ is obtained dividing z by $z - 0.8$. The quotient is

$$X(z) = 1 + 0.8z^{-1} + 0.64z^{-2} + 0.512z^{-3} + \cdots$$

Comparing with the definition of the z-transform, the time-domain values are $x(0) = 1$, $x(1) = 0.8$, $x(2) = 0.64$, $x(3) = 0.512$, and so on. These values can be verified from $x(n) = (0.8)^n u(n)$, which is the closed-form solution of the inverse z-transform (Example 9.3). This method is particularly useful when only the first few values of the time-domain sequence are required.

9.5 Applications of the z-Transform

9.5.1 *Transfer Function*

The input-output relationship of a LTI system is given by the convolution operation in the time domain. Since convolution corresponds to multiplication in the frequency domain, we get

$$y(n) = \sum_{m=0}^{\infty} x(m)h(n-m) \Longleftrightarrow Y(z) = X(z)H(z),$$

where $x(n)$, $h(n)$, and $y(n)$ are, respectively, the system input, impulse response, and output and $X(z)$, $H(z)$, and $Y(z)$ are their respective transforms. As input is transferred to output by multiplication with $H(z)$, $H(z)$ is called the transfer function of the system. The transfer function, which is the transform of the impulse response, characterizes a system in the frequency domain just as the impulse response does in the time domain. For stable systems, the frequency response $H(e^{j\omega})$ is obtained from $H(z)$ by replacing z by $e^{j\omega}$.

We can as well apply any input, with nonzero spectral amplitude for all values of z in the ROC, to the system and find the response, and the ratio of the z-transforms $Y(z)$ of the output and $X(z)$ of the input is $H(z) = \frac{Y(z)}{X(z)}$. Consider the system governed by the difference equation

$$y(n) = 2x(n) - 3y(n-1)$$

The impulse response of the system is $h(n) = 2(-3)^n$. The transform of $h(n)$ is $H(z) = \frac{2z}{z+3}$. The output of this system, with initial condition zero, to the input $x(n) = 3u(n)$ is $y(n) = \frac{3}{2}(1 + 3(-3)^n)u(n)$, using time-domain method. The transform of $y(n)$ is

$$Y(z) = \frac{3}{2}\left(\frac{z}{z-1} + \frac{3z}{z+3}\right)$$

The transform of the input $x(n) = 3u(n)$ is $X(z) = \frac{3z}{z-1}$. Now,

$$H(z) = \frac{Y(z)}{X(z)} = \frac{\frac{3}{2}\left(\frac{z}{z-1} + \frac{3z}{z+3}\right)}{\frac{3z}{z-1}} = \frac{2z}{z+3},$$

which is the same as the transform of the impulse response.

Since the transform of a delayed signal is its transform multiplied by a factor, we can as well find the transfer function by taking the transform of the difference equation characterizing a system. Consider the difference equation of a causal LTI discrete system.

$$\begin{aligned}y(n) + a_{K-1}y(n-1) + a_{K-2}y(n-2) + \cdots + a_0 y(n-K)\\ = b_M x(n) + b_{M-1}x(n-1) + \cdots + b_0 x(n-M)\end{aligned}$$

Taking the z-transform of both sides, we get, assuming initial conditions are all zero,

$$\begin{aligned}Y(z)(1 + a_{K-1}z^{-1} + a_{K-2}z^{-2} + \cdots + a_0 z^{-K})\\ = X(z)(b_M + b_{M-1}z^{-1} + \cdots + b_0 z^{-M})\end{aligned}$$

The transfer function $H(z)$ is obtained as

$$H(z) = \frac{Y(z)}{X(z)} = \frac{b_M + b_{M-1}z^{-1} + \cdots + b_0 z^{-M}}{1 + (a_{K-1}z^{-1} + a_{K-2}z^{-2} + \cdots + a_0 z^{-K})}$$

$$= \frac{\sum_{l=0}^{M} b_{M-l} z^{-l}}{1 + \sum_{l=1}^{K} a_{K-l} z^{-l}}$$

The transfer function written in positive powers of z,

$$H(z) = \frac{z^{K-M}(b_M z^M + b_{M-1} z^{M-1} + \cdots + b_0)}{z^K + (a_{K-1} z^{K-1} + a_{K-2} z^{K-2} + \cdots + a_0)},$$

is more convenient for manipulation.

9.5.2 Characterization of a System by Its Poles and Zeros

By using the pole-zero representation of the z-transform, the transfer function can be written as

$$H(z) = B \frac{z^{K-M}(z - z_1)(z - z_2) \cdots (z - z_M)}{(z - p_1)(z - p_2) \cdots (z - p_K)} = B z^{K-M} \frac{\prod_{i=1}^{M} (z - z_i)}{\prod_{i=1}^{K} (z - p_i)}$$

where B is a constant. As the coefficients of the polynomials of $H(z)$ are real for practical systems, the zeros and poles are real-valued, or they always occur as complex conjugate pairs.

The numerator and denominator polynomials of $H(z)$, when equated to zero, have M and K roots, respectively. The roots

$$\{z_1, z_2, \ldots, z_M\} \quad \text{and} \quad \{p_1, p_2, \ldots, p_K\}$$

are complex frequencies. As $H(z)$ evaluates to zero at $\{z_1, z_2, \ldots, z_M\}$, these frequencies are called the *zeros* of $H(z)$. As $H(z)$ evaluates to infinity at $\{p_1, p_2, \ldots, p_M\}$, these frequencies are called the *poles* of $H(z)$. Any linear system is completely specified by its poles and zeros and a constant factor. The poles determine the time variation of the response of the system, while the zeros determine the magnitude. When a pole or zero repeats m times, it is said to be of multiplicity m. They are also referred to as second-order or third-order poles or zeros. A nonrepetitive pole or zero is called a simple pole or zero.

The pole-zero plot of the transfer function $H(z)$ of a system is a pictorial description of its characteristics, such as speed of response, frequency selectivity, and stability. Poles with magnitudes much smaller than one result in a fast-responding system with its transient response decaying rapidly. On the other hand, poles with magnitudes closer to one result in a sluggish system. Complex conjugate poles located inside the unit circle result in an oscillatory transient response that decays with time. The frequency of oscillation is higher for poles located in the second and third quadrants of the unit circle. Complex conjugate poles located on the unit circle result in a steady oscillatory transient response. Poles located on the positive real axis inside the unit circle result in exponentially decaying transient response. An alternating positive and negative sample is the transient response due to poles located on the negative real axis. The frequency components of an input signal with frequencies close to a zero will be suppressed, while those close to a pole will be readily transmitted. Poles located symmetrically about the positive real axis inside the unit circle and close to the unit circle in the passband result in a lowpass system that more readily transmits low-frequency signals than high-frequency signals. Zeros located symmetrically about the negative real axis in the stopband further enhance the lowpass character of the frequency response. On the other hand, poles located symmetrically about the negative real axis inside the unit circle and close to the unit circle in the passband result in a highpass system that more readily transmits high-frequency signals than low-frequency signals. For example, a system with its pole-zero plot such as that shown in Fig. 9.1 is a lowpass filter. The stability of a system can also be determined from its pole-zero plot, as presented later.

9.5.3 Frequency Response and the Locations of the Poles and Zeros

The frequency response of a system characterizes its performance. For example, a high-fidelity amplifier should have distortion-free response up to 20 kHz. Even for system design, the required frequency response is usually specified. Mathematically, it is obtained from $H(z)$ by replacing z by $e^{j\omega}$. Note that $z = re^{j\omega}$ and on the unit circle $r = 1$. Frequency response is the response of the system for an everlasting complex exponential with unit magnitude and zero phase. As that type of signal starts at time $-\infty$, the response of the system observed at any finite time is that of steady state. In practice, the frequency response is approximated by applying a sinusoidal signal to the system, starting at a finite time, and measuring the response after the transient response decays to a negligible level.

Replacing z by $e^{j\omega}$ in $H(z)$, we get

$$H(z)|_{z=e^{j\omega}} = a(0)\frac{(e^{j(K-M)\omega})(e^{j\omega} - z_1)(e^{j\omega} - z_2)\cdots(e^{j\omega} - z_M)}{(e^{j\omega} - p_1)(e^{j\omega} - p_2)\cdots(e^{j\omega} - p_K)}$$

Let the magnitude and phase of the terms in the numerator be

$$(1\angle\theta_{KM}), (r_1\angle\theta_1), (r_2\angle\theta_2), \ldots, (r_M\angle\theta_M)$$

Let the magnitude and phase of the terms in the denominator be

$$(d_1\angle\phi_1), (d_2\angle\phi_2), \ldots, (d_K\angle\phi_K)$$

Then, the frequency response $H(e^{j\omega})$ is given by

$$H(e^{j\omega}) = |a(0)|\angle(a(0))\frac{(1\angle\theta_{KM})(r_1\angle\theta_1)(r_2\angle\theta_2)\cdots(r_M\angle\theta_M)}{(d_1\angle\phi_1)(d_2\angle\phi_2)\cdots(d_K\angle\phi_K)}$$

Therefore, the magnitude of the frequency response is given as

$$|H(e^{j\omega})| = |a(0)|\frac{(r_1)(r_2)\cdots(r_M)}{(d_1)(d_2)\cdots(d_K)}$$

The phase of the frequency response is given as

$$\angle H(e^{j\omega}) = (\angle(a(0)) + \theta_{KM} + \theta_1 + \theta_2 + \cdots + \theta_M) - (\phi_1 + \phi_2 + \cdots + \phi_K)$$

These expressions are useful for understanding and can be used to find the response using graphical methods. However, in practice, DFT is commonly used to approximate the frequency response.

Let us find the frequency response of the transfer function

$$H(z) = \frac{z+1}{z-0.8}$$

Replacing z by $e^{j\omega}$, we get

$$H(e^{j\omega}) = \frac{e^{j\omega}+1}{e^{j\omega}-0.8}$$

Let us use the DFT to evaluate the frequency response. DFT is polynomial evaluation at roots of unity. The DFTs of the numerator and denominator polynomials of $H(z)$ are separately computed, and term-by-term division yields the frequency response. The frequency response is continuous. But, in practice, it is computed at finite number of points. Let use compute at 512 points. The 512-point DFTs of the numerator and denominator polynomials are separately computed, and, then, term-by-term division is carried out. To make the length 512, both the numeration and denominator coefficients are sufficiently zero-padded. Figure 9.4a and b shows, respectively, the magnitude and phase of the frequency response. The magnitude plot is even symmetric about the midpoint, while the phase plot is odd symmetric. Therefore, only positive half of the spectra are shown. Let us check the plot at $\pi/2$. Then,

$$H(e^{j\frac{2\pi 128}{512}}) = \frac{j+1}{j-0.8} = 1.1043\angle - 83.6598^\circ$$

At $\omega = 0$, the magnitude is 10 and the phase is 0.

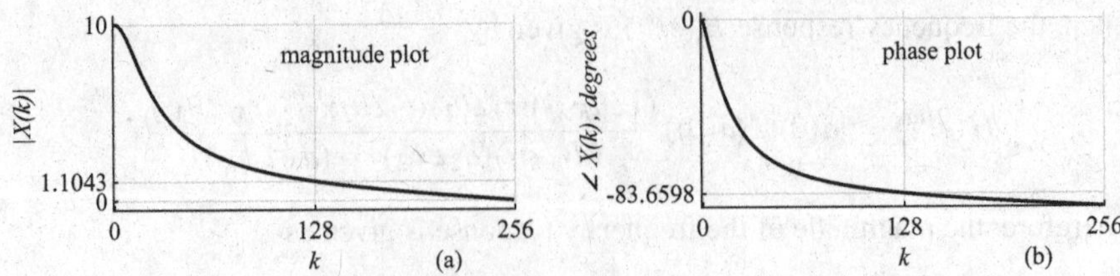

Fig. 9.4 (**a**) Magnitude of the frequency response; (**b**) phase response

9.5.4 Design of Digital Filters

Normalized lowpass analog filter design procedures are well established. Further, transformation methods, which operate on the independent variable, are available to design other types of filters. However, digital filters are preferred in practice. Now, the task reduces to finding a suitable transformation to obtain the transfer function of the corresponding digital filter from that of the analog filter. The criterion in the transformation is to preserve the desired characteristics of the analog filter as much as possible. We obtained the corresponding difference equation from that of the differential equation of a RC lowpass filter in Chap. 3.

9.5.4.1 The Bilinear Transformation

One of the transformations often used in the transformation of an analog transfer function to the corresponding digital transfer function is the bilinear transformation. Basically, it is an approximation of a differential equation into a difference equation using the trapezoidal algorithm of numerical integration. The resulting digital filter preserves the steady-state response of the analog filter.

Let the transfer function of the analog filter be

$$H(s) = \frac{2}{s+2}$$

The corresponding differential equation is

$$\frac{dy(t)}{dt} + 2y(t) = 2x(t)$$

Now, the derivative can be replaced by a finite difference, as shown in Chap. 3. Instead, the trapezoidal algorithm approximates the derivative by several trapezoids of sufficiently small width. The difference equation characterizing this algorithm is

$$y(n) = y(n-1) + \frac{T_s}{2}(x(n) + x(n-1)),$$

where T_s is the sampling interval. The z-transform of this equation is

$$Y(z) = z^{-1}Y(z) + \frac{T_s}{2}(X(z) + z^{-1}X(z))$$

Multiplying both sides by z, we get

$$zY(z) = Y(z) + \frac{T_s}{2}(zX(z) + X(z))$$

Finding the transfer function, we get

$$H(z) = \frac{Y(z)}{X(z)} = \frac{T_s}{2}\frac{(z+1)}{(z-1)}$$

As the transfer function of the integrator in the s-plane is $1/s$, we get the transformation by the substitution

$$s = \frac{2}{T_s}\frac{(z-1)}{(z+1)}$$

For the $H(s)$ given earlier, the corresponding $H(z)$ is

$$H(z) = \frac{2}{s+2}\Big|_{s=\frac{2}{T_s}\frac{(z-1)}{(z+1)}} = \frac{T_s(z+1)}{(T_s+1)z + (T_s-1)}$$

Frequency Warping

The frequency response in the s-plane is aperiodic and extends from $-\infty < \omega_a < \infty$. On the other hand, the frequency response in the z-plane is periodic with a finite range, $-\pi < \omega_d T_s \le \pi$. Substituting $s = j\omega_a$ and $z = e^{j\omega_d T_s}$ in

$$s = \frac{2}{T_s}\frac{(z-1)}{(z+1)}$$

we get

$$\begin{aligned} j\omega_a &= \frac{2}{T_s}\frac{(e^{j\omega_d T_s}-1)}{(e^{j\omega_d T_s}+1)} \\ &= \frac{2}{T_s}\frac{(e^{0.5j\omega_d T_s} - e^{-0.5j\omega_d T_s})}{(e^{0.5j\omega_d T_s} + e^{-0.5j\omega_d T_s})} = \frac{j2}{T_s}\frac{-0.5j(e^{0.5j\omega_d T_s} - e^{-0.5j\omega_d T_s})}{0.5(e^{0.5j\omega_d T_s} + e^{-0.5j\omega_d T_s})} \\ &= \frac{2}{T_s} j \tan\left(\frac{T_s}{2}\omega_d\right) \end{aligned}$$

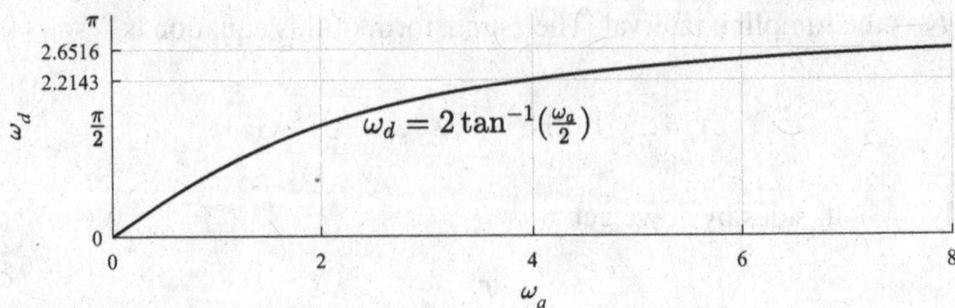

Fig. 9.5 The nonlinear relationship between analog, ω_a, and discrete, ω_d, frequencies in the bilinear transformation

Therefore, the relationship between the frequency variables ω_a and ω_d is given by

$$\omega_d = \frac{2}{T_s}\tan^{-1}\left(\frac{T_s}{2}\omega_a\right) \quad \text{and} \quad \omega_a = \frac{2}{T_s}\tan\left(\frac{T_s}{2}\omega_d\right)$$

The relationship is obviously nonlinear. The bilinear transformation is a conformal mapping, which preserves the angles between oriented curves with respect to both magnitude and direction, that maps the $j\omega$ axis in the Laplace plane onto the unit circle in the z-plane only once. The region to the left-half Laplace plane is mapped inside the unit circle, and the right half is mapped to the outside, between ω_a and ω_d. Figure 9.5 shows the nonlinear relationship between analog, ω_a, and discrete, ω_d, frequencies in the bilinear transformation, with $T_s = 1$ s. The entire frequency range from 0 to ∞ in the s-plane is mapped into the range 0 to π/T_s in the z-plane only once. Therefore, there is no aliasing problem in using this transformation. However, as an infinite range of frequencies is compressed to a finite range, the frequency scale is compressed. For example, $\omega_a = 2$ radians is mapped to

$$\omega_d = 2\tan^{-1}\left(\frac{\omega_a}{2}\right) = 2\tan^{-1}\left(\frac{2}{2}\right) = 2\tan^{-1}(1) = 2\frac{\pi}{4} = \frac{\pi}{2} = 1.5708 \text{ radians},$$

as shown in the figure. Initially, the relationship is almost linear for a short range. That is, for $\omega_d < 0.3/T_s$, $\omega_d \approx \omega_a$. With increasing frequency, the corresponding range in the z-plane becomes progressively shorter compared with that in the s-plane.

The transfer functions of the analog and the corresponding digital filter, obtained using the bilinear transformation with $T_s = 0.2$ s, are

$$H(s) = \frac{2}{s+2} \quad \text{and} \quad H(z) = \frac{0.1667(z+1)}{z - 0.6667}$$

The details of obtaining the digital filter coefficients are given later. Figure 9.6a shows the magnitude of the frequency response in dB of the analog and digital (dotted line) filters. The responses are obtained by replacing $s = j\omega$ in $H(s)$ and $z = e^{j\omega}$ in $H(z)$. That is,

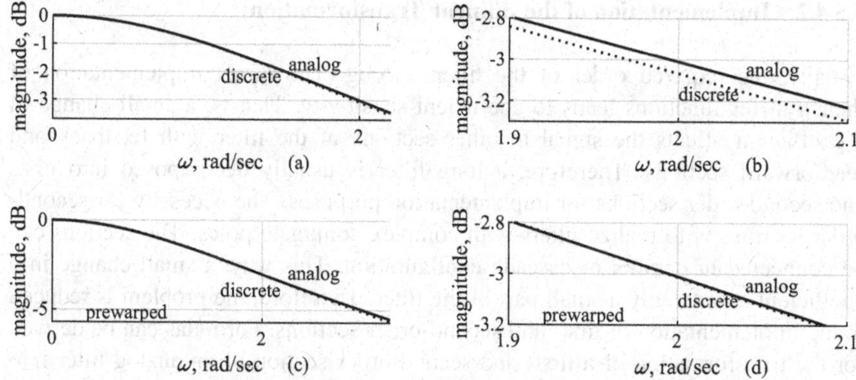

Fig. 9.6 **(a)** The frequency response; **(b)** response in expanded scale around the cutoff frequency; **(c)** the frequency response with prewarping; **(b)** response in expanded scale around the cutoff frequency

$$H(j\omega) = \frac{2}{j\omega + 2} \quad \text{and} \quad H(e^{j\omega}) = \frac{0.1667(e^{j\omega} + 1)}{e^{j\omega} - 0.6667}$$

For example, with $\omega = 0$, $e^{j\omega} = 1$, and the magnitude of both the transfer functions is 1 or $20 \log 10(1) = 0$ dB. With $\omega = 2$, the magnitude of the analog filter is

$$|H(j2)| = \left|\frac{2}{j2 + 2}\right| = \frac{1}{\sqrt{2}} \text{ or } 20 \log 10 \left(\frac{1}{\sqrt{2}}\right) = -3.0103 \text{ dB}$$

With $\omega = 2$, $20 \log 10(|H(e^{j2})|) = -3.0671$ dB, which is not equal to that of the analog filter due to warping. The divergence of the two responses begins before $\omega = 2$, and the expanded responses around the cutoff frequency are shown Fig. 9.6b.

Let us prewarp the cutoff frequency. That is, we design an analog filter with a higher cutoff frequency. With $T_s = 0.2$,

$$\omega_a = (2/T_s) \tan((0.5)(2)T_s) = 10 \tan(0.2) = 2.0271$$

The transfer functions of the analog and the corresponding digital filter, obtained using the bilinear transformation with $T_s = 0.2$ s and prewarping, are

$$H(s) = \frac{2.0271}{s + 2.0271} \quad \text{and} \quad H(z) = \frac{0.1685(z + 1)}{z - 0.0.6629}$$

With $\omega = 2$, $20 \log 10(|H(e^{j2})|) = -3.0127$ dB, which is much closer to that of the analog filter due to prewarping. Figure 9.6c shows the magnitude of the frequency response in dB of the analog and digital (dotted line) filters. The divergence of the two responses begins after $\omega = 2$, and the expanded responses around the cutoff frequency are shown Fig. 9.6d.

9.5.4.2 Implementation of the Bilinear Transformation

Usually, the required order of the filter is long. The direct implementation of their transfer functions leads to coefficient sensitivity. That is, a small change in a coefficient affects the signal to other sections of the filter with feedback and feedforward sections. Therefore, a long filter is usually decomposed into first- and second-order sections for implementation purposes. The necessity for second-order sections is to realize filters with complex conjugate poles. The sections can be connected in parallel or cascade configuration. This way, a small change in a coefficient affects only a small part of the filter. Therefore, the problem is reduced to the implementation of first- and second-order sections. Formulas can be derived for the transformation of a first- and second-order section of an analog filter into the corresponding digital filter. Let us find the formulas for the transformation of a first-order filter and use them to find the digital filter coefficients of the example filter. The general form of analog and discrete first-order transfer function are

$$\frac{c_1 s + c_0}{d_1 s + d_0} \quad \text{and} \quad \frac{a_1 z + a_0}{z + b_0}$$

$$H(s) = \frac{2}{s+2} \quad \text{and} \quad H(z) = 0.1667\frac{z+1}{z-0.6667}$$

The coefficients of the digital filter, designed using the bilinear transformation with $T_s = 0.2$, have to be obtained. Using the transformation

$$s = \frac{2}{T_s}\frac{(z-1)}{(z+1)} = k\frac{(z-1)}{(z+1)},$$

we get, with $k = 2/T_s = 2/0.2 = 10$,

$$a_1 = \frac{(c_0 + c_1 k)}{D}, \; a_0 = \frac{(c_0 - c_1 k)}{D}, \; b_0 = \frac{(d_0 - d_1 k)}{D}, \; \text{where } D = (d_0 + d_1 k)$$

Further simplifications may be possible where some of the values are 0 or 1. For the filter

$$H(s) = \frac{2}{s+2},$$

with $T_s = 0.2$, we get

$$a_1 = \frac{2}{D}, \; a_0 = \frac{2}{D}, \; b_0 = \frac{(2-k)}{D}, \; \text{where } D = (2+k)$$

With $D = 2 + 10 = 12$,

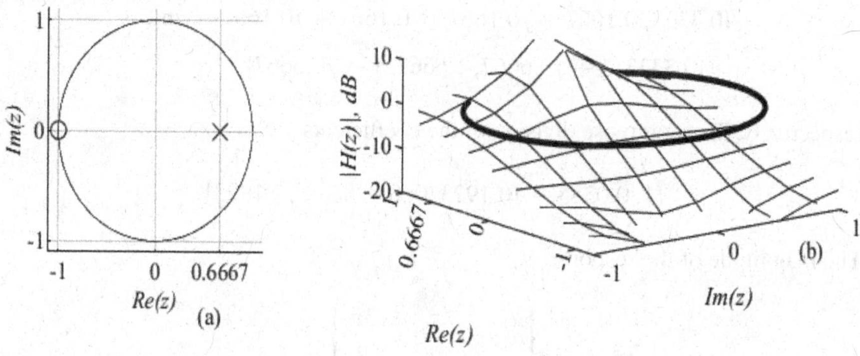

Fig. 9.7 (**a**) Pole-zero plot of a lowpass filter; (**b**) the magnitude of its frequency response

$$a(0) = a(1) = \frac{2}{12} = \frac{1}{6} = 0.1667, \quad b(0) = \frac{(2-10)}{12} = -\frac{8}{12} = -\frac{2}{3} = -0.6667$$

Therefore, the transfer function of the digital filter is

$$H(z) = \frac{0.1667(e^{j\omega} + 1)}{e^{j\omega} - 0.6667}$$

The pole-zero plot of the filter is shown in Fig. 9.7a. The pole is located at $\omega = 0.6667$, and the zero is located at $\omega = -1$. Therefore, the magnitude of the frequency response is high for low frequencies, which are near (1,0) point on the unit circle. The zero suppresses the response for high frequencies, which are near (−1,0) point on the unit circle. The magnitude of the frequency response is shown in Fig. 9.7b, which clearly depicts the lowpass nature of the response on the unit circle (shown in a thick line). The z-plane is like a rubber sheet. Zeros pin the sheet down at zero level, and poles push it up to infinite level. The net response due to all the zeros and poles is the frequency response.

Let us approximate the frequency response using the DFT. The numerator coefficients of the filter are $\{0.1667, 0.1667\}$. The denominator coefficients are $\{1, -0.6667\}$. The 2-point DFT of the coefficients are $\{0.3333, 0\}$ and $\{0.3333, 1.6667\}$, respectively. The pointwise division of the coefficients yields

$$(0.3333, 0)/(0.3333, 1.6667) = (1, 0)$$

The magnitude of the response is 1 at $\omega = 0$ radian and 0 at $\omega = \pi$.

Typically, the frequency response is computed at 256 or 512 points in the frequency range 0 to π radians by computing the DFT of zero-padded coefficients. For illustration, let us use points from 0 to 2π radians. The zero-padded numerator coefficients of the filter are $\{0.1667, 0.1667, 0, 0\}$. The zero-padded denominator coefficients are $\{1, -0.6667, 0, 0\}$. The 4-point DFT of the coefficients are

$$\{0.3333, 0.1667 - j0.1667, 0, 0.1667 + j0.1667\} \quad \text{and}$$

$$\{0.3333, 1 + j0.6667, 1.6667, 1 - j0.6667\},$$

respectively. The pointwise division of the coefficients yields

$$\{1, 0.0385 - j0.1923, 0, 0.0385 + j0.1923\}$$

The magnitude of the response at

$$\left\{\omega = 0, \frac{\pi}{2}, \pi, \frac{3\pi}{2}\right\}$$

radians is

$$\{1, 0.1961, 0, 0.1961\}$$

The corresponding angles are, in degrees,

$$\{0, -78.6901, 0, 78.6901\}$$

At $\omega = \pi$, the phase angle jumps from −90 to 90, and the average is 0.

The transfer functions of the analog highpass filter and the corresponding digital filter, obtained using the bilinear transformation with $T_s = 0.1$ s, are

$$H(s) = \frac{s}{s+5} \quad \text{and} \quad H(z) = \frac{0.8(z-1)}{z-0.6}$$

The pole-zero plot of the filter is shown in Fig. 9.8a. The pole is located at $\omega = 0.6$, and the zero is located at $\omega = 1$. Therefore, the magnitude of the frequency response is small for low frequencies, which are near (1,0) point on the unit circle. The zero suppresses the response for low frequencies. The magnitude of the frequency response is shown in Fig. 9.8b, which clearly depicts the highpass nature of the response on the unit circle (shown in a thick line).

9.5.5 System Response

Example 9.11 Find the zero-input, zero-state, transient, steady-state, and complete responses of the system governed by the difference equation

$$y(n) = 2x(n) - x(n-1) + 3x(n-2) + \frac{9}{20}y(n-1) - \frac{1}{20}y(n-2)$$

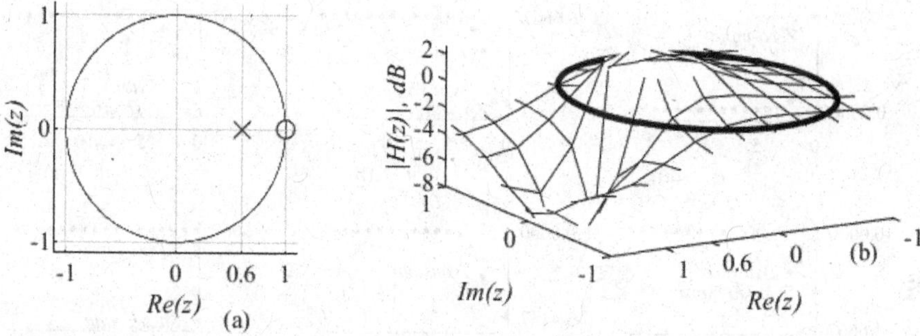

Fig. 9.8 (**a**) Pole-zero plot of a highpass filter; (**b**) the magnitude of its frequency response

with the initial conditions $y(-1) = 3$ and $y(-2) = 2$ and the input $x(n) = u(n)$, the unit-step function.

Solution The z-transforms of the terms of the difference equation are

$$x(n) \Longleftrightarrow \frac{z}{z-1}, \qquad x(n-1) \Longleftrightarrow \frac{1}{z-1}, \qquad x(n-2) \Longleftrightarrow \frac{1}{z(z-1)}$$

$$y(n) \Longleftrightarrow Y(z), \qquad y(n-1) \Longleftrightarrow y(-1) + z^{-1}Y(z) = z^{-1}Y(z) + 3$$

$$y(n-2) \Longleftrightarrow y(-2) + z^{-1}y(-1) + z^{-2}Y(z) = z^{-2}Y(z) + 3z^{-1} + 2$$

Substituting the corresponding transform for each term in the difference equation and factoring, we get

$$\frac{Y(z)}{z} = \frac{2z^2 - z + 3}{(z-1)(z-\frac{1}{5})(z-\frac{1}{4})} + \frac{\left(\frac{5}{4}z - \frac{3}{20}\right)}{(z-\frac{1}{5})(z-\frac{1}{4})}$$

The first term on the right-hand side is $H(z)X(z)/z$ and corresponds to the zero-state response. The second term is due to the initial conditions and corresponds to the zero-input response.

Expanding into partial fractions, we get

$$\frac{Y(z)}{z} = \frac{\frac{20}{3}}{(z-1)} + \frac{72}{(z-\frac{1}{5})} - \frac{\frac{230}{3}}{(z-\frac{1}{4})} - \frac{2}{(z-\frac{1}{5})} + \frac{\frac{13}{4}}{(z-\frac{1}{4})}$$

Taking the inverse z-transform, we get the complete response.

$$y(n) = \overbrace{\frac{20}{3} + 72\left(\frac{1}{5}\right)^n - \frac{230}{3}\left(\frac{1}{4}\right)^n}^{\text{zero-state}} \overbrace{-2\left(\frac{1}{5}\right)^n + \frac{13}{4}\left(\frac{1}{4}\right)^n}^{\text{zero-input}}, \quad n = 0, 1, \ldots$$

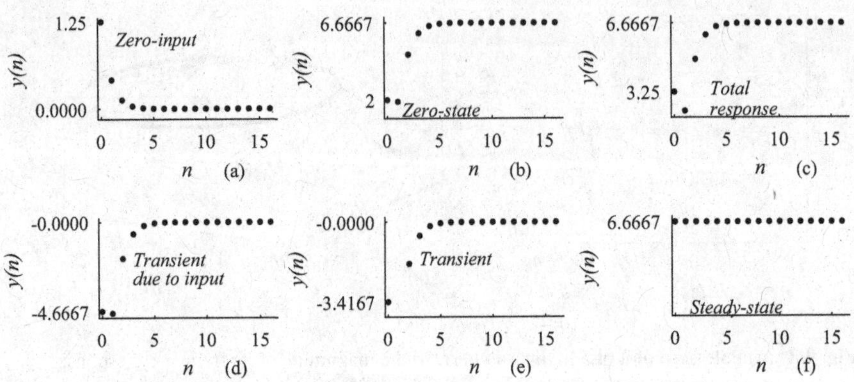

Fig. 9.9 Various components of the response of the system in Example 9.11

$$y(n) = \frac{20}{3} + 70\left(\frac{1}{5}\right)^n - \frac{881}{12}\left(\frac{1}{4}\right)^n, \quad n = 0, 1, \ldots$$

The first four values of $y(n)$ are

$$y(0) = 3.2500, \quad y(1) = 2.3125, \quad y(2) = 4.8781, \quad y(3) = 6.0795$$

The responses are shown in Fig. 9.9. The zero-input response (a) is $-2(\frac{1}{5})^n + \frac{13}{4}(\frac{1}{4})^n$, the response due to initial conditions alone. The zero-state response (b) is $\frac{20}{3} + 72\left(\frac{1}{5}\right)^n - \frac{230}{3}\left(\frac{1}{4}\right)^n$, the response due to input alone. The transient response (e) is $72\left(\frac{1}{5}\right)^n - \frac{230}{3}\left(\frac{1}{4}\right)^n - 2\left(\frac{1}{5}\right)^n + \frac{13}{4}\left(\frac{1}{4}\right)^n = 70\left(\frac{1}{5}\right)^n - \frac{881}{12}\left(\frac{1}{4}\right)^n$, the response that decays with time. The steady-state response (f) is $\frac{20}{3}u(n)$, the response after the transient response has died out completely. Either the sum of the zero-input and zero-state components (a) and (b) or the sum of the transient and steady-state components (e) and (f) of the response is the complete response (c) of the system. Either the difference of the transient and zero-input components (e) and (a) or the difference of the zero-state and steady-state components (b) and (f) of the response is the transient response (d) of the system due to input alone. The initial and final values of $y(n)$ are 3.25 and $\frac{20}{3}$, respectively. These values can be verified by applying the initial and final value properties to $Y(z)$. We can also verify that the initial conditions at $n = -1$ and at $n = -2$ are satisfied by the zero-input component of the response. Figure 9.10 shows the simulation diagram of the system with initial conditions producing the total response. The initial conditions are set in the delay units on the output side. The unit-step input values have to be loaded into the simin block by executing the given input program. ∎

The same set of coefficients is involved in both the difference equation and transfer function models of a system. Therefore, either of the models can be used to

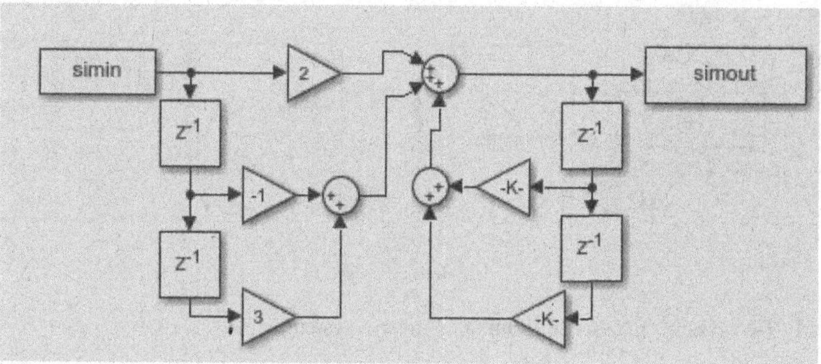

Fig. 9.10 The simulation diagram of the system

determine the complete response of a system. In formulating the transfer function model, we have assumed that the initial conditions are zero. However, it should be noted that, with appropriately chosen input that yields the same output as the initial conditions, we can use the transfer function concept even for problems with nonzero initial conditions. Consider the transform of the output obtained in the example in presenting the right-shift property, $Y(z) = \frac{\frac{3}{2}z}{z-\frac{1}{2}}$. This equation can be considered as $Y(z) = H(z)X'(z)$ with $H(z) = \frac{z}{z-\frac{1}{2}}$ and $X'(z) = \frac{3}{2}$. $X'(z)$ corresponds to the time-domain input $\frac{3}{2}\delta(n)$, which produces the same response that results from the initial condition.

9.5.5.1 Inverse Systems

Two systems with impulse responses h_1 and h_2 form an inverse system, when connected in cascade, if $h_1 * h_2 = \delta(n)$. That is, input remains the same at the output. This implies that, in the frequency domain, $H_1(z)H_2(z) = 1$.

The difference equation characterizing the trapezoidal numerical algorithm, used in the bilinear transform, is

$$y(n) = y(n-1) + \frac{T_s}{2}(x(n) + x(n-1)),$$

where T_s is the sampling interval. The z-transform of this equation is

$$Y(z) = z^{-1}Y(z) + \frac{T_s}{2}(X(z) + z^{-1}X(z))$$

Multiplying both sides by z, we get

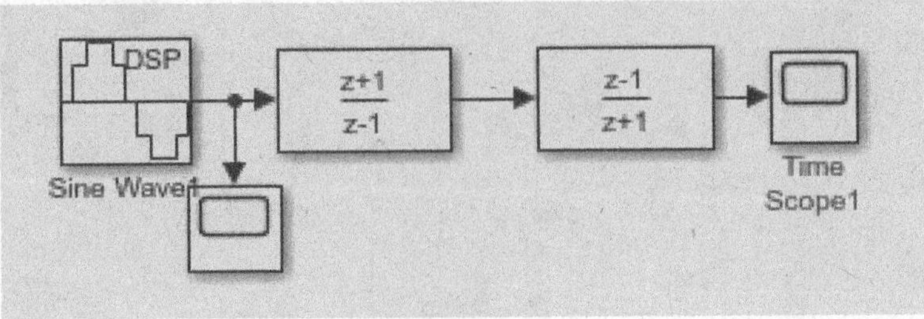

Fig. 9.11 Two systems in cascade forming an inverse system

$$zY(z) = Y(z) + \frac{T_s}{2}(zX(z) + X(z))$$

Finding the transfer function, we get

$$H_1(z) = \frac{Y(z)}{X(z)} = \frac{T_s}{2}\frac{(z+1)}{(z-1)}$$

The transfer function of the inverse of this system is

$$H_2(z) = \frac{2}{T_s}\frac{(z-1)}{(z+1)}$$

The difference equation characterizing the inverse system is

$$y(n) = \frac{2}{T_s}(x(n) - x(n-1)) - y(n-1)),$$

which is a numerical differentiator. Figure 9.11 shows the simulation of the two systems in cascade forming an inverse system

9.5.6 System Stability

The zero-input response of a system depends solely on the locations of its poles. A system is considered stable if its zero-input response, due to finite initial conditions, converges, marginally stable if its zero-input response tends to a constant value or oscillates with a constant amplitude, and unstable if its zero-input response diverges. Commonly used marginally stable systems are oscillators, which produce a bounded zero-input response. The response corresponding to each pole p of a system is of the form $r^n e^{jn\theta}$, where the magnitude and phase of the pole are r and θ, respectively. If $r < 1$, then r^n tends to zero as n tends to ∞. If $r > 1$, then r^n tends to ∞ as n tends to ∞. If $r = 1$, then $r^n = 1$ for all n. However, the response tends to infinity, for poles of order more than 1 lying on the unit circle, as the expression for the response includes a factor that is a function of n. Poles of any order lying inside the unit circle do not cause instability. Therefore, we conclude that, from the locations of the poles of a system,

- All the poles, of any order, of a stable system must lie inside the unit circle. That is, the ROC of $H(z)$ must include the unit circle.
- Any pole lying outside the unit circle or any pole of order more than 1 lying on the unit circle makes a system unstable.
- A system is marginally stable if it has no poles outside the unit circle and has poles of order 1 on the unit circle.

Figure 9.12 shows pole locations of some transfer functions and the corresponding impulse responses. If all the poles of a system lie inside the unit circle, the bounded-input bounded-output stability condition (Chap. 3) is satisfied. However, the converse is not necessarily true, since the impulse response is an external description of a system and may not include all its poles. The bounded-input bounded-output stability condition is not satisfied by a marginally stable system.

9.5.7 Realization of Systems

To implement a system, a realization diagram has to be derived. Several realizations of a system are possible, each realization differing in such characteristics as the amount of arithmetic required, sensitivity to coefficient quantization, etc. The z-transform of the output of a Nth-order system is given as

$$Y(z) = X(z)H(z) = X(z)\frac{N(z)}{D(z)} = \frac{X(z)(b_N + b_{N-1}z^{-1} + \cdots + b_0 z^{-N})}{1 + a_{N-1}z^{-1} + \cdots + a_0 z^{-N}}$$

Let $R(z) = \frac{X(z)}{D(z)}$. Then, $Y(z) = R(z)N(z)$. Now, the system structure can be realized as a cascade of two systems. The first system, $R(z) = \frac{X(z)}{D(z)}$, has only poles with input $x(n)$ and output $r(n)$. The second system, $Y(z) = R(z)N(z)$, has only zeros with input $r(n)$ and output $y(n)$, where

$$r(n) = x(n) - \sum_{k=1}^{N} a_{N-k} r(n-k) \quad \text{and} \quad y(n) = \sum_{k=0}^{N} b_{N-k} r(n-k)$$

Both the systems can share a set of delay units as the term $r(n-k)$ is common. The realization of a second-order system is shown in Fig. 9.13. This realization is known as the canonical form I realization, implying the use of the minimum number of delay elements. A transposed form of a system structure is obtained by (i) reversing the directions of all the signal flow paths, (ii) replacing the junction

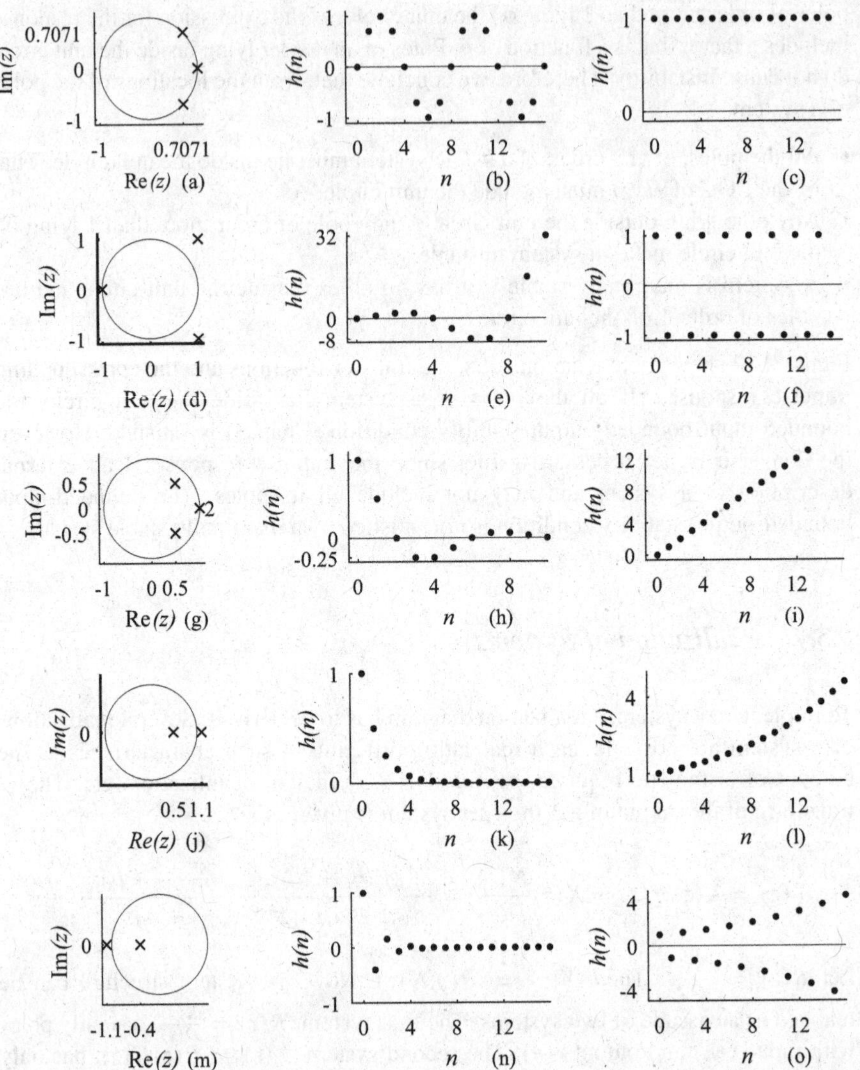

Fig. 9.12 The poles of some transfer functions $H(z)$ and the corresponding impulse responses $h(n)$: **(a)** $H(z) = \frac{(z/\sqrt{2})}{z^2-\sqrt{2}z+1} = \frac{0.5jz}{z-((1/\sqrt{2})-(j/\sqrt{2}))} - \frac{0.5jz}{z-((1/\sqrt{2})+(j/\sqrt{2}))}$ and $H(z) = \frac{z}{z-1}$; **(b)** $h(n) = \sin(\frac{\pi}{4}n)u(n)$ and **(c)** $h(n) = u(n)$; **(d)** $H(z) = \frac{z}{z^2-2z+2} = \frac{0.5jz}{z-(1-j1)} - \frac{0.5jz}{z-(1+j1)}$ and $H(z) = \frac{z}{z+1}$; **(e)** $h(n) = (\sqrt{2})^n \sin(\frac{\pi}{4}n)u(n)$ and **(f)** $h(n) = (-1)^n u(n)$; **(g)** $H(z) = \frac{z(z-0.5)}{z^2-z+0.5} = \frac{0.5z}{z-(0.5-0.5j)} + \frac{0.5z}{z-(0.5+0.5j)}$ and $H(z) = \frac{z}{(z-1)^2}$; **(h)** $h(n) = (\frac{1}{\sqrt{2}})^n \cos(\frac{\pi}{4}n)u(n)$ and **(i)** $h(n) = nu(n)$; **(j)** $H(z) = \frac{z}{z-0.5}$ and $H(z) = \frac{z}{z-1.1}$.; **(k)** $h(n) = (0.5)^n u(n)$ and **(l)** $h(n) = (1.1)^n u(n)$; **(m)** $H(z) = \frac{z}{z+0.4}$ and $H(z) = \frac{z}{z+1.1}$; **(n)** $h(n) = (-0.4)^n u(n)$ (n) and **(o)** $h(n) = (-1.1)^n u(n)$

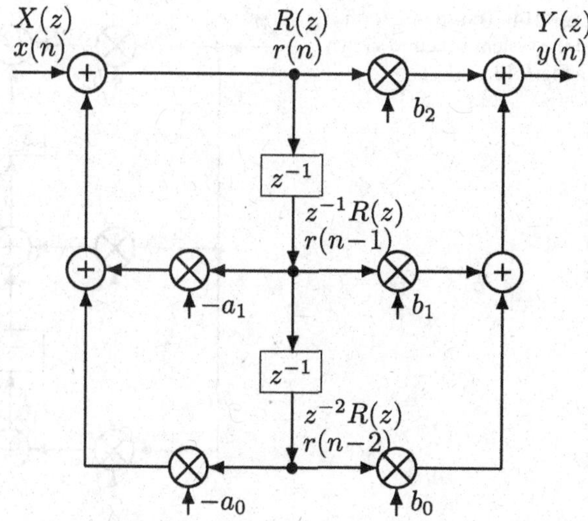

Fig. 9.13 Realization of a second-order system

points by adders and vice versa, and (iii) interchanging the input and output points. The transposed form of the system in Fig. 9.13 is shown in Fig. 9.14. This realization is known as the canonical form II realization. This form is derived as follows.

$$H(z) = \frac{Y(z)}{X(z)} = \frac{b_2 z^2 + b_1 z + b_0}{z^2 + a_1 z + a_0} = \frac{b_2 + b_1 z^{-1} + b_0 z^{-2}}{1 + a_1 z^{-1} + a_0 z^{-2}}$$

$$Y(z)(1 + a_1 z^{-1} + a_0 z^{-2}) = X(z)(b_2 + b_1 z^{-1} + b_0 z^{-2})$$

$$\begin{aligned}
Y(z) &= b_2 X(z) + z^{-1}(b_1 X(z) - a_1 Y(z)) + z^{-2}(b_0 X(z) - a_0 Y(z)) \\
&= b_2 X(z) + z^{-1}\{(b_1 X(z) - a_1 Y(z)) + z^{-1}(b_0 X(z) - a_0 Y(z))\} \\
&= b_2 X(z) + z^{-1}\{(b_1 X(z) - a_1 Y(z)) + z^{-1} r(2)(z)\} \\
&= b_2 X(z) + z^{-1} r(1)(z)
\end{aligned}$$

Therefore, the following difference equations characterize this system structure.

$$\begin{aligned}
y(n) &= b_2 x(n) + r(1)(n-1) \\
r(1)(n) &= b_1 x(n) - a_1 y(n) + r(2)(n-1) \\
r(2)(n) &= b_0 x(n) - a_0 y(n)
\end{aligned}$$

These realizations have the advantage of using the coefficients of the transfer function directly.

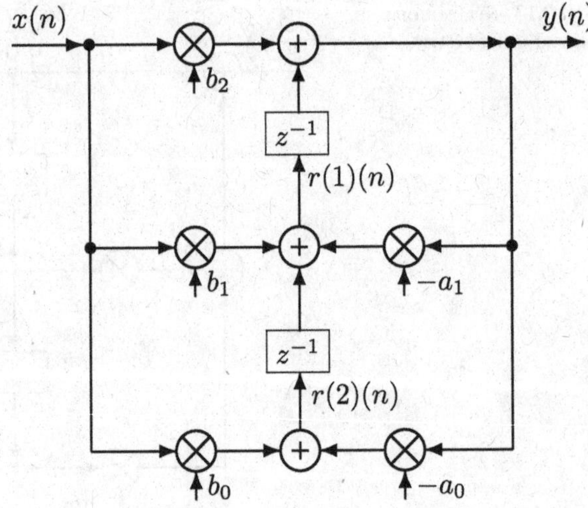

Fig. 9.14 Transposed form of the system structure shown in Fig. 9.13

While this type of realizations is applicable to system of any order, it becomes more sensitive to coefficient quantization due to the tendency of the poles and zeros to occur in clusters. Therefore, usually, a higher-order system is decomposed into first- and second-order sections connected in cascade or parallel. In the cascade form, the transfer function is decomposed into a product of first- and second-order transfer functions.

$$H(z) = H_1(z)H_2(z)\cdots H_m(z)$$

In the parallel form, the transfer function is decomposed into a sum of first- and second-order transfer functions.

$$H(z) = g + H_1(z) + H_2(z)+, \cdots, +H_m(z),$$

where g is a constant. Each section is independent, and clustering of poles and zeros is avoided as the maximum number of poles and zeros in each section is limited to 2. Each second-order section is realized as shown in Figs. 9.13 or 9.14.

9.5.8 Feedback Systems

In feedback systems, a fraction of the output signal is fed back and subtracted from the input signal to form the effective input signal. By using negative feedback, we can change the speed of response, reduce sensitivity, improve stability, and increase the range of operation of a system at the cost of reducing the open-loop gain. Consider the feedback system shown in Fig. 9.15. The feedback signal $R(z)$ can be expressed as

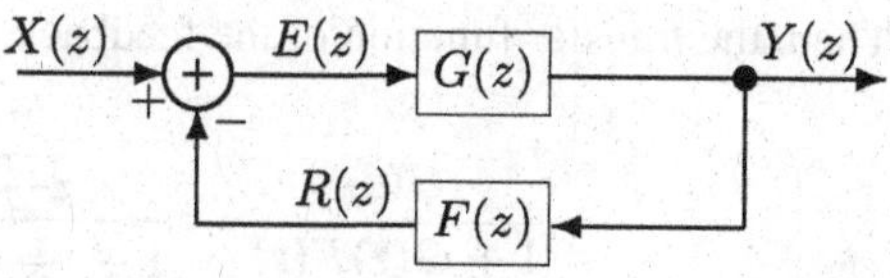

Fig. 9.15 Two systems connected in a feedback configuration

$$R(z) = F(z)Y(z),$$

where $F(z)$ is the feedback transfer function of the system and $Y(z)$ is the output. Now, the error signal $E(z)$ is

$$E(z) = X(z) - R(z) = X(z) - F(z)Y(z)$$

The output $Y(z)$ is expressed as

$$Y(z) = G(z)E(z) = G(z)(X(z) - F(z)Y(z)),$$

where $G(z)$ is the forward transfer function of the system. Therefore, the transfer function of the feedback system is given as

$$\frac{Y(z)}{X(z)} = \frac{G(z)}{1 + G(z)F(z)}$$

If $G(z)$ is very large, the transfer function of the feedback system approximates to the inverse of the feedback transfer function of the system.

$$\frac{Y(z)}{X(z)} \simeq \frac{1}{F(z)}$$

Consider the system with the transfer function

$$G(z) = \frac{z}{z - \frac{3}{2}}.$$

$G(z)$ has a pole at $z = \frac{3}{2}$, and, therefore, the system is unstable. We can make a stable feedback system, using this system in the forward path and another suitable system in the feedback path. Let the transfer function of the system in the feedback path be

$$F(z) = \frac{1}{z - \frac{1}{5}}$$

Then, the transfer function of the feedback system is

$$\frac{G(z)}{1+G(z)F(z)} = \frac{\frac{z}{z-\frac{3}{2}}}{1+\frac{z}{z-\frac{3}{2}}\frac{1}{z-\frac{1}{5}}} = \frac{z(z-\frac{1}{5})}{z^2-\frac{7}{10}z+\frac{3}{10}}$$

Now, both the poles of this system lie inside the unit circle, and, therefore, the system is stable.

9.6 Summary

- In this chapter, the theory of the one-sided z-transform and its properties and some applications have been described. As practical systems are causal, the one-sided z-transform is mostly used in practice.
- The z-transform is a generalized version of the Fourier analysis. The basis waveforms consist of sinusoids with varying amplitudes or exponentials with complex exponents. The larger set of basis waveforms makes this transform suitable for the analysis of a larger class of signals and systems.
- While the DTFT changes a sequence of numbers into a function of the purely imaginary complex variable $e^{j\omega}$, the z-transform changes a sequence of numbers into a function of the complex variable z with an expanded set of basis functions. That is, the DTFT of a signal is its z-transform with $z = e^{j\omega}$, if the DTFT exists.
- The z-transform corresponding to a one-dimensional sequence is two-dimensional (a surface), since it is a function of two variables (the real and imaginary parts of the complex frequency). In the frequency domain, a sequence is uniquely specified by its z-transform along with its ROC. The spectral values along any simply connected closed contour, encircling the origin, in the ROC can be used to reconstruct the corresponding time-domain sequence. There is no z-transform representation for a signal, which grows faster than an exponential.
- All practical signals satisfy the convergence condition and, therefore, have z-transform representation.
- The inverse z-transform is defined by a contour integral. However, for most practical purposes, the partial fraction method along with a short list of z-transform pairs is adequate to find the inverse z-transform.
- The z-transform is essential for the design and transient and stability analysis of discrete LTI systems. The z-transform of the impulse response of a system, the transfer function, is a frequency-domain model of the system.

Exercises

9.1 The nonzero values of a sequence $x(n)$ are specified as $\{x(-2) = 1, x(0) = 2, x(3) = -4\}$. Find the unilateral z-transform of

9.1.1 $x(n-3)$.
9.1.2 $x(n-1)$.
* 9.1.3 $x(n)$.
9.1.4 $x(n+1)$.
9.1.5 $x(n+2)$.
9.1.6 $x(n+4)$.

9.2 Find the nonzero values of the inverse z-transform of

9.2.1 $X(z) = 2 - 3z^{-2} + z^{-4}$.
9.2.2 $X(z) = z^{-2} - 2z^{-5}$.
9.2.3 $X(z) = -2 + 3z^{-1} - z^{-10}$.
* 9.2.4 $X(z) = 1 + z^{-1} - z^{-2}$.
9.2.5 $X(z) = z^{-2} + 2z^{-3}$.

9.3 Using the z-transform of $u(n)$ and $nu(n)$ and the shift property, find the z-transform of $x(n)$.

9.3.1 $x(n) = u(n-3) - u(n-5)$.
9.3.2 $x(n) = nu(n-3)$.
9.3.3 $x(n) = n,\ 0 \le n \le 4$ and $x(n) = 0$ otherwise.
* 9.3.4 $x(n) = (n-2)u(n)$.

9.4 The nonzero values of two sequences $x(n)$ and $h(n)$ are given. Using the z-transform, find the convolution of the sequences $y(n) = x(n) * h(n)$.

* 9.4.1 $\{x(0) = 2, x(2) = 3, x(4) = -2\}$ and $\{h(1) = 2, h(3) = -4\}$.
9.4.2 $\{x(1) = 3, x(4) = -4\}$ and $\{h(0) = -2, h(3) = 3\}$.
9.4.3 $\{x(2) = 3, x(4) = -2\}$ and $\{h(1) = 4, h(2) = 1\}$.
9.4.4 $\{x(0) = -4, x(3) = -1\}$ and $\{h(0) = 1, h(2) = -2\}$.
9.4.5 $\{x(2) = 3, x(4) = -1\}$ and $\{h(1) = 2, h(3) = 2\}$.

9.5 Using the multiplication by n property, find the z-transform of $x(n)$.

9.5.1 $x(n) = nu(n)$.
* 9.5.2 $x(n) = n2^n u(n)$.
9.5.3 $x(n) = nu(n-2)$.

9.6 Using the multiplication by a^n property, find the z-transform of $x(n)$.

9.6.1 $x(n) = 3^n u(n)$.
* 9.6.2 $x(n) = n4^n u(n)$.
9.6.3 $x(n) = 2^n \cos(n)u(n)$.

9.7 Using the summation property, find the sum $y(n) = \sum_{m=0}^{n} x(m)$.

9.7.1 $x(n) = \cos(\frac{2\pi}{4}n)u(n)$.
* 9.7.2 $x(n) = \sin(\frac{2\pi}{4}n)u(n)$.
9.7.3 $x(n) = e^{(j\frac{2\pi}{4}n)}u(n)$.
9.7.4 $x(n) = u(n)$.
9.7.5 $x(n) = (n)u(n)$.

9.8 Find the initial and final values of the sequence $x(n)$ corresponding to the transform $X(z)$, using the initial and final value properties.

9.8.1 $X(z) = \frac{z(3z+2)}{(z-\frac{1}{2})(z+\frac{1}{4})}$.
9.8.2 $X(z) = \frac{3z}{(z-2)(z+3)}$.
* 9.8.3 $X(z) = \frac{2z(z+3)}{(z-\frac{1}{2})(z-1)}$.
9.8.4 $X(z) = \frac{z}{(z-1)^2}$.
9.8.5 $X(z) = \frac{2z^2}{(z-1)(z+2)}$.

9.9 Given the sample values of the first period, find the z-transform of the semiperiodic function $x(n)u(n)$.

9.9.1 $\{1, 0, -1, 0\}$.
9.9.2 $\{0, 1, 0, -1\}$.
9.9.3 $\{1, j, -1, -j\}$.
* 9.9.4 $\{1, 1, -1, -1\}$.
9.9.5 $\{0, 1, 2, 1\}$.

9.10 Find the inverse z-transform of $X(z)$ using the inverse DTFT.

9.10.1 $X(z) = \frac{z}{z-5}$.
9.10.2 $X(z) = \frac{z}{z-0.8}$.
9.10.3 $X(z) = \frac{z}{(z-1)^2}$.

9.11 Find the inverse z-transform of

$$X(z) = \frac{z(2z+3)}{(z^2 - \frac{2}{15}z - \frac{1}{15})}$$

List the first four values of $x(n)$.

9.12 Find the inverse z-transform of

$$X(z) = \frac{(3z-1)}{(z^2 - \frac{11}{12}z + \frac{1}{6})}$$

List the first four values of $x(n)$.

* **9.13** Find the inverse z-transform of

$$X(z) = \frac{z(z+2)}{(z^2 + 2z + 2)}$$

List the first four values of $x(n)$.

9.14 Find the inverse z-transform of

$$X(z) = \frac{2z^2 + 1}{(z^2 - z - 6)}$$

List the first four values of $x(n)$.

9.15 Find the inverse z-transform of

$$X(z) = \frac{z}{(z^3 + \frac{3}{2}z^2 - \frac{1}{2})}$$

List the first four values of $x(n)$.

9.16 Find the inverse z-transform of

$$X(z) = \frac{z^2 - 1}{z^2(z + \frac{1}{3})}$$

List the first four values of $x(n)$.

9.17 Find the first four values of the inverse z-transform of $X(z)$ by the long division method.

9.17.1 $X(z) = \frac{2z^2+2z-3}{z^2-z+1}$.
9.17.2 $X(z) = \frac{z}{z^2+2z-2}$.
9.17.3 $X(z) = \frac{3z^2-z+2}{2z^2+z-3}$.

9.18 Using the z-transform, derive the closed-form expression of the impulse response $h(n)$ of the system governed by the difference equation

$$y(n) = x(n) + 2x(n-1) + x(n-2) + 3y(n-1) - 2y(n-2)$$

with input $x(n)$ and output $y(n)$. List the first four values of $h(n)$.

9.19 Given the difference equation of a system and the input signal $x(n)$, find the steady-state response of the system.

9.19.1 $y(n) = x(n) + 0.8y(n-1)$ and $x(n) = 2e^{j(\frac{2\pi}{4}n+\frac{\pi}{6})}u(n)$.
* 9.19.2 $y(n) = x(n) + 0.7y(n-1)$ and $x(n) = 3\cos(\frac{2\pi}{4}n - \frac{\pi}{3})u(n)$
9.19.3 $y(n) = x(n) + 0.5y(n-1)$ and $x(n) = 4\sin(\frac{2\pi}{4}n + \frac{\pi}{4})u(n)$.

9.20 Using the z-transform, derive the closed-form expression of the complete response of the system governed by the difference equation

$$y(n) = 2x(n) - x(n-1) + x(n-2) + \frac{7}{6}y(n-1) - \frac{1}{3}y(n-2)$$

with the initial conditions $y(-1) = 2$ and $y(-2) = -3$ and the input $x(n) = u(n)$, the unit-step function. List the first four values of $y(n)$. Deduce the expressions for the zero-input, zero-state, transient, and steady-state responses of the system.

* **9.21** Using the z-transform, derive the closed-form expression of the complete response of the system governed by the difference equation

$$y(n) = x(n) + 2x(n-1) - x(n-2) + \frac{5}{4}y(n-1) - \frac{3}{8}y(n-2)$$

with the initial conditions $y(-1) = 2$ and $y(-2) = 1$ and the input $x(n) = (-1)^n u(n)$. List the first four values of $y(n)$. Deduce the expressions for the zero-input, zero-state, transient, and steady-state responses of the system.

9.22 Using the z-transform, derive the closed-form expression of the complete response of the system governed by the difference equation

$$y(n) = 3x(n) - 3x(n-1) + x(n-2) + \frac{7}{12}y(n-1) - \frac{1}{12}y(n-2)$$

with the initial conditions $y(-1) = 1$ and $y(-2) = 2$ and the input $x(n) = nu(n)$, the unit-ramp function. List the first four values of $y(n)$. Deduce the expressions for the zero-input, zero-state, transient, and steady-state responses of the system.

9.23 Using the z-transform, derive the closed-form real-valued expression of the complete response of the system governed by the difference equation

$$y(n) = x(n) - 3x(n-1) + 2x(n-2) + y(n-1) - \frac{2}{9}y(n-2)$$

with the initial conditions $y(-1) = 3$ and $y(-2) = 2$ and the input $x(n) = (\frac{1}{4})^n u(n)$. List the first four values of $y(n)$. Deduce the expressions for the zero-input, zero-state, transient, and steady-state responses of the system.

9.24 Using the z-transform, derive the closed-form real-valued expression of the complete response of the system governed by the difference equation

$$y(n) = x(n) + x(n-1) - x(n-2) + \frac{3}{4}y(n-1) - \frac{1}{8}y(n-2)$$

with the initial conditions $y(-1) = 1$ and $y(-2) = 2$ and the input $x(n) = 2\cos(\frac{2\pi}{4}n - \frac{\pi}{6})u(n)$. List the first four values of $y(n)$. Deduce the expressions for the zero-input, zero-state, transient, and steady-state responses of the system.

9.25 Using the z-transform, derive the closed-form expression of the impulse response of the cascade system consisting of systems governed by the given difference equations with input $x(n)$ and output $y(n)$. List the first four values of the impulse response of the cascade system.

* 9.25.1 $y(n) = 2x(n) - x(n-1) + \frac{1}{4}y(n-1)$ and $y(n) = 3x(n) + x(n-1) - \frac{1}{3}y(n-1)$.
9.25.2 $y(n) = x(n) + x(n-1) - \frac{2}{3}y(n-1)$ and $y(n) = 2x(n) - x(n-1) - \frac{1}{5}y(n-1)$.
9.25.3 $y(n) = x(n) + 2x(n-1) + \frac{1}{3}y(n-1)$ and $y(n) = 3x(n) + 2x(n-1) + \frac{1}{2}y(n-1)$.

9.26 Using the z-transform, derive the closed-form expression of the impulse response of the combined system, connected in parallel, consisting of systems governed by the given difference equations with input $x(n)$ and output $y(n)$. List the first four values of the impulse response of the parallel system.

9.26.1 $y(n) = 2x(n) - x(n-1) + \frac{1}{4}y(n-1)$ and $y(n) = 3x(n) + x(n-1) - \frac{1}{3}y(n-1)$.
9.26.2 $y(n) = x(n) + x(n-1) - \frac{2}{3}y(n-1)$ and $y(n) = 2x(n) - x(n-1) - \frac{1}{5}y(n-1)$.
* 9.26.3 $y(n) = x(n) + 2x(n-1) + \frac{1}{3}y(n-1)$ and $y(n) = 3x(n) + 2x(n-1) + \frac{1}{2}y(n-1)$.

Chapter 10
The Laplace Transform

The generalization of the Fourier transform for continuous signals, by including sinusoids with exponentially varying amplitudes in the set of basis signals, is called the Laplace transform. This generalization makes the transform analysis applicable to a larger class of signals and systems. In Sect. 10.1, we develop the Laplace transform starting from the definition of the Fourier transform. In Sect. 10.2, the properties of the Laplace transform are described. In Sect. 10.3, the inverse Laplace transform is derived. Typical applications of the Laplace transform are presented in Sect. 10.4.

10.1 The Laplace Transform

We assume, in this chapter, that all the signals are causal, that is, $x(t) = 0,\ t < 0$, unless otherwise specified. This leads to the one-sided or unilateral version of the Laplace transform, which is mostly used for practical system analysis. If a signal $x(t)u(t)$ is not Fourier transformable, then its exponentially weighted version, $x(t)e^{-\sigma t}$, may be Fourier transformable for the positive real quantity $\sigma > 0$. If $x(t)u(t)$ is Fourier transformable, $x(t)e^{-\sigma t}$ may still be transformable for some values of $\sigma < 0$. The Fourier transform of this signal is

$$\int_0^{\infty} (x(t)e^{-\sigma t})e^{-j\omega t}\,dt$$

By combining the exponential factors, we get

$$X(\sigma + j\omega) = \int_0^{\infty} x(t)e^{-(\sigma+j\omega)t}\,dt \tag{10.1}$$

D. Sundararajan, *Signals and Systems*,
https://doi.org/10.1007/978-3-031-19377-4_10

This equation can be interpreted as the generalized Fourier transform of the signal $x(t)$ using exponentials with complex exponents or sinusoids with varying amplitudes as the basis signals. Therefore, a signal may be decomposed in terms of constant amplitude sinusoids or exponentially decaying sinusoids or exponentially growing sinusoids or an infinite combination of these types of sinusoids. By substituting $s = \sigma + j\omega$, we get the defining equation of the Laplace transform of $x(t)$ as

$$X(s) = \int_{0^-}^{\infty} x(t)e^{-st}\,dt$$

Note that the lower limit is assumed, in this book, to be 0^-, where $t = 0^-$ is the instant immediately before $t = 0$. This implies that a jump discontinuity or an impulse component of the function $x(t)$ at $t = 0$ is included in the integral. In addition, this lower limit enables the use of the initial conditions at $t = 0^-$ directly. In practical applications, we are more likely to know the initial conditions before the input signal is applied, rather than after.

While $X(j\omega)$ is the FT of $x(t)$, $X(s) = X(\sigma + j\omega)$ is the FT of $x(t)e^{-\sigma t}$ for all values of σ for which $\int_{0^-}^{\infty} |x(t)e^{-\sigma t}|dt < \infty$. If the value zero is included in these values of σ, then $X(j\omega)$ can be obtained from $X(s)$ by the substitution $s = j\omega$. The Laplace transform of $x(t)$, $X(s)$, exists for $\mathrm{Re}(s) > \sigma_0$ if $|x(t)| \leq Me^{\sigma_0 t}$ for some constants M and σ_0. For example, the signal e^{t^2} has no Laplace transform. In essence, the Laplace transform of a signal, whether it is converging or not, is the FT of all its versions, obtained by multiplying it with a real exponential of the form $e^{-\sigma t}$, so that the modified signal is guaranteed to converge.

The advantages of the Laplace transform include the pictorial description of the behavior of the system obtained by the use of the complex frequency; the ability to analyze unstable systems or systems with exponentially growing inputs; automatic inclusion of the initial conditions of the system in finding the output; and easier manipulation of the expressions involving the variable s rather than $j\omega$.

Example 10.1 Find the Laplace transform of the unit-impulse signal, $\delta(t)$.

Solution Using the Laplace transform definition, we get

$$X(s) = \int_{0^-}^{\infty} \delta(t)e^{-st}\,dt = 1, \text{ for all } s \qquad \text{and} \qquad \delta(t) \Longleftrightarrow 1, \text{ for all } s$$

This transform pair can also be obtained by applying a limit process to any function that degenerates into an impulse and its transform. ∎

The exponential signal, $e^{-at}u(t)$, is fundamental to the study of linear continuous systems, as it is more convenient to express the natural response of systems in that form.

Example 10.2 Find the Laplace transform of the real exponential signal, $x(t) = e^{-at}u(t)$. Deduce the Laplace transform of the unit step signal, $x(t) = u(t)$.

Solution Using the Laplace transform definition, we get

$$X(s) = \int_{0^-}^{\infty} e^{-at}u(t)e^{-st}dt = \int_{0^-}^{\infty} e^{-at}e^{-st}dt$$

$$= \int_{0^-}^{\infty} e^{-(s+a)t}dt = -\frac{e^{-(s+a)t}}{s+a}\bigg|_{0^-}^{\infty} = \frac{1}{s+a} - \frac{e^{-(s+a)t}}{s+a}\bigg|_{t=\infty}$$

For the integral to converge, $\lim_{t\to\infty} e^{-(s+a)t}$ must be equal to zero. This implies that the real part of $(s+a)$ is greater than zero and, hence, the convergence condition is $\mathrm{Re}(s) > -a$. This condition describes a region in the s-plane (a complex plane used for displaying the Laplace transform) that lies to the right of the vertical line characterized by the equation $\mathrm{Re}(s) = -a$. Note that the Fourier transform of $e^{-at}u(t)$ does not exist for negative values of a, whereas the Laplace transform exists for all values of a as long as $\mathrm{Re}(s) > -a$. Therefore, we get the Laplace transform pair

$$e^{-at}u(t) \Longleftrightarrow \frac{1}{s+a}, \quad \mathrm{Re}(s) > -a$$

This transform pair remains the same for complex-valued a with the convergence condition, $\mathrm{Re}(s) > \mathrm{Re}(-a)$.

With $a = 0$, we get the transform pair

$$u(t) \Longleftrightarrow \frac{1}{s}, \quad \mathrm{Re}(s) > 0$$

∎

The region, consisting the set of all values of s in the s-plane for which the defining integral of the Laplace transform converges, is called the region of convergence (ROC). For the signal in Example 10.2, the region to the right of the vertical line at $\mathrm{Re}(s) = \mathrm{Re}(-a)$ is the ROC.

The frequency content of a signal is usually displayed by the locations of zeros and poles and the magnitude of its Laplace transform. Figure 10.1a shows the pole-zero plot, and Fig. 10.1b shows the magnitude of the Laplace transform, $X(s) = \frac{1}{s+2}$, of the signal $e^{-2t}u(t)$. When $s = -2$, $|X(s)| = \infty$. This point marked by the symbol $\times$ in Fig. 10.1a is called a pole of $X(s)$ (the peak in Fig. 10.1b). Except for a constant factor, the Laplace transform of a signal can be reconstructed from its pole-zero plot. For all values of s in the ROC (the region to the right of the dotted vertical line at $\mathrm{Re}(s) = -2$ shown in Fig. 10.1a), $X(s)$ exists and is a valid representation of the signal. In general, the ROC of a Laplace transform is the region in the s-plane that is to the right of the vertical line passing through the rightmost pole location. If the ROC includes the imaginary axis, $\mathrm{Re}(s) = 0$, in the s-plane (as in Fig. 10.1a), then the FT can be obtained from the Laplace transform by replacing s with $j\omega$.

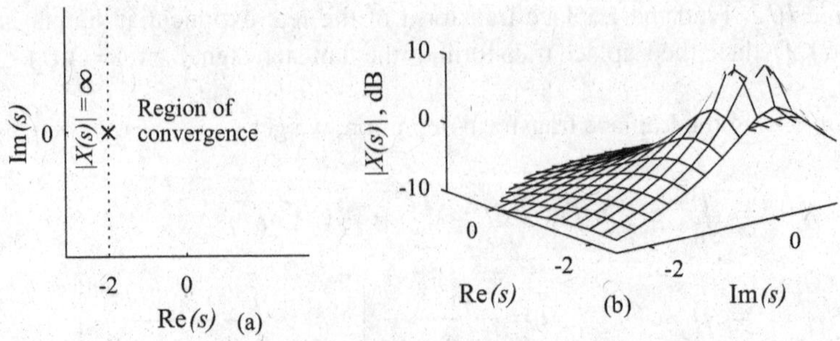

Fig. 10.1 (**a**) The pole-zero plot of the Laplace transform, $\frac{1}{s+2}$, of the signal, $e^{-2t}u(t)$; (**b**) the magnitude of the Laplace transform

Example 10.3 Find the Laplace transform of the signal $e^{j\omega_0 t}u(t)$. Deduce the Laplace transform of $\cos(\omega_0 t)u(t)$.

Solution Using the transform of $e^{-at}u(t)$ with $a = -j\omega_0$, we get

$$e^{j\omega_0 t}u(t) \Longleftrightarrow \frac{1}{s - j\omega_0}, \qquad \text{Re}(s - j\omega_0) = \text{Re}(s) > 0$$

Using the fact that $2\cos(\omega_0 t) = (e^{j\omega_0 t} + e^{-j\omega_0 t})$, we get

$$2X(s) = \frac{1}{s - j\omega_0} + \frac{1}{s + j\omega_0}, \qquad \text{Re}(s) > 0$$

$$\cos(\omega_0 t)u(t) \Longleftrightarrow \frac{s}{s^2 + \omega_0^2}, \qquad \text{Re}(s) > 0$$ ∎

Figure 10.2a shows the pole-zero plot, and Fig. 10.2b shows the magnitude of the Laplace transform, $\frac{s}{s^2+(\frac{\pi}{4})^2}$, of the signal $\cos(\frac{\pi}{4}t)u(t)$. When $s = \pm j\frac{\pi}{4}$, $|X(s)| = \infty$. These points marked by the symbol × in Fig. 10.2a are the poles of $X(s)$ (the peaks in Fig. 10.2b). When $s = 0$, $X(s) = 0$. This point marked by the symbol o in Fig. 10.2a is the zero of $X(s)$ (the valley in Fig. 10.2b).

Note that any periodic signal defined over the entire time domain, $-\infty < t < \infty$, has a FT but no Laplace transform. However, a causal periodic signal (identically zero for $t < 0$) has a Laplace transform exclusively of simple poles on the imaginary axis.

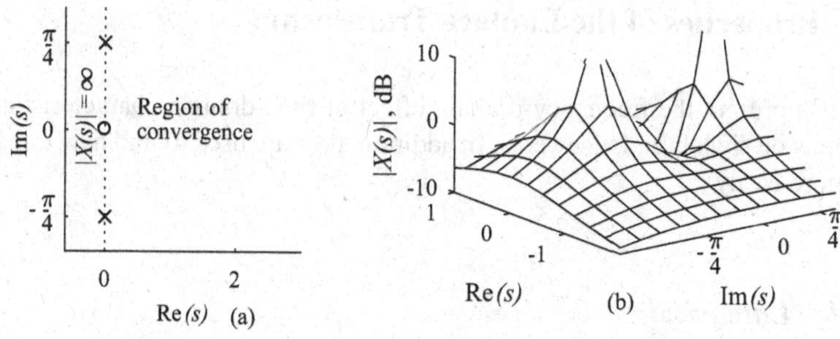

Fig. 10.2 (**a**) The pole-zero plot of the Laplace transform, $\frac{s}{s^2+(\frac{\pi}{4})^2}$, of the signal, $\cos(\frac{\pi}{4}t)u(t)$; (**b**) the magnitude of the Laplace transform

10.1.1 Relationship Between the Laplace Transform and the z-Transform

A relationship between the FT of a sampled signal and the DTFT of the corresponding discrete signal was derived in Chap. 8. Now, we derive a similar relationship between the Laplace transform and the z-transform. The sampled version of a signal $x(t)u(t)$ is $x_s(t) = \sum_{n=0}^{\infty} x(n)\delta(t-n)$, with a sampling interval of 1 second. As the Laplace transform of $\delta(t-n)$ is e^{-sn} and due to the linearity property of the Laplace transform, the Laplace transform of the sampled signal $x_s(t)$ is given by

$$X_s(s) = \sum_{n=0}^{\infty} x(n)e^{-sn}$$

With $z = e^s$, this equation becomes

$$X_s(s) = \sum_{n=0}^{\infty} x(n)z^{-n}$$

The right-hand side of this equation is the z-transform of $x(n)$.

For example, let $x(t) = e^{-2t}u(t)$. Then, the corresponding discrete signal is $x(n) = e^{-2n}u(n)$ with its z-transform $\frac{z}{z-e^{-2}}$. Now, the Laplace transform of the sampled version of $x(t)$, $x_s(t) = \sum_{n=0}^{\infty} e^{-2n}\delta(t-n)$, is $\frac{e^s}{e^s-e^{-2}}$, which is obtained from $\frac{z}{z-e^{-2}}$ by the substitution $z = e^s$.

10.2 Properties of the Laplace Transform

Properties present the frequency-domain effect of time-domain characteristics and operations on signals and vice versa. In addition, they are used to find new transform pairs more easily.

10.2.1 Linearity

The Laplace transform of a linear combination of signals is the same linear combination of their individual Laplace transforms. If $x(t) \iff X(s)$ and $y(t) \iff Y(s)$, then $ax(t)+by(t) \iff aX(s)+bY(s)$, where a and b are arbitrary constants. This property is due to the linearity of the defining integral of the Laplace transform. We use this property often to decompose a time-domain function in finding its Laplace transform (as in Example 10.3) and to decompose a transform in order to find its inverse.

10.2.2 Time Shifting

If $x(t)u(t) \iff X(s)$, then

$$x(t-t_0)u(t-t_0),\ t_0 \geq 0 \iff e^{-st_0}X(s)$$

Now, $e^{-st_0} = e^{-(\sigma+j\omega)t_0} = e^{-\sigma t_0}e^{-j\omega t_0}$. The term $e^{-j\omega t_0}$ is the linear shift of the phase of sinusoids, as in the case of the Fourier analysis. Due to the fact that the basis functions are sinusoids with varying amplitudes, we need another factor $e^{-\sigma t_0}$ to set the amplitude of the sinusoids appropriately so that the reconstructed waveform is the exact time-shifted version of $x(t)$.

Consider the waveform $x(t)u(t) = e^{-0.1t}u(t)$ and its shifted version $e^{-0.1(t-8)}u(t-8)$. The Laplace transforms of the two functions are, respectively, $\frac{1}{s+0.1}$ and $\frac{e^{-8s}}{s+0.1}$.

This property holds only for causal signals and for right shift only. Remember that the transform of the shifted signal is expressed in terms of that of the original signal, which is assumed to be zero for $t < 0$. For finding the transform of signals such as $x(t-t_0)u(t)$ and $x(t)u(t-t_0)$, express the signal so that the arguments of the signal and the unit-step signal are the same, and then apply the property. Of course, the transform can also be computed using the defining integral.

10.2.3 Frequency Shifting

If $x(t)u(t) \Longleftrightarrow X(s)$, then

$$e^{s_0 t}x(t)u(t) \Longleftrightarrow X(s - s_0)$$

Multiplying the signal $x(t)$ by the exponential $e^{s_0 t}$ amounts to changing the complex frequency of its spectral components by s_0. Therefore, the spectrum $X(s)$ is shifted in the s-plane by the amount s_0.

Consider finding the transform of the signal $e^{2t}u(t)$. This signal can be considered as the unit step, $u(t)$, multiplied by the exponential with $s_0 = 2$. Therefore, according to this property, the transform of $e^{2t}u(t)$ is the transform of $u(t)$, which is $\frac{1}{s}$, with the substitution $s = s - 2$, that is, $\frac{1}{s-2}$.

10.2.4 Time Differentiation

The time-differentiation property is used to express the transform of the derivative, $\frac{dx(t)}{dt}$, of a signal $x(t)$ in terms of its transform $X(s)$. If $x(t) \Longleftrightarrow X(s)$, then

$$\frac{dx(t)}{dt} \Longleftrightarrow sX(s) - x(0^-)$$

From the definition, the Laplace transform of the derivative of $x(t)$ is given by

$$\int_{0^-}^{\infty} \frac{dx(t)}{dt} e^{-st}\,dt$$

$$= x(t)e^{-st}\Big|_{0^-}^{\infty} + s\int_{0^-}^{\infty} x(t)e^{-st}\,dt$$

$$= -x(0^-) + sX(s)$$

We used the integration by parts property of integration. Note that, as $t \to \infty$, $x(t)e^{-st} \to 0$ in the ROC of $X(s)$ for all values of s.

As the signal, in the frequency domain, is expressed in terms of exponentials e^{st} and the derivative of the exponential is se^{st}, the differentiation of a signal in time domain corresponds to multiplication of its transform by the frequency variable s, in addition to a constant term due to the initial value of the signal at $t = 0^-$. The point is that two signals $x(t)$ and $x(t)u(t)$ have the same unilateral Laplace transform. However, the Laplace transforms of their derivatives will be different if $x(t)$ and $x(t)u(t)$ have different values of discontinuity at $t = 0$. The derivative of $x(t)$ with a different value of $x(0^-)$ differs, from that of $x(t)u(t)$, only at $t = 0$. $sX(s)$ is the derivative of $x(t)u(t)$, that is, the derivative of $x(t)$ with $x(0^-) = 0$.

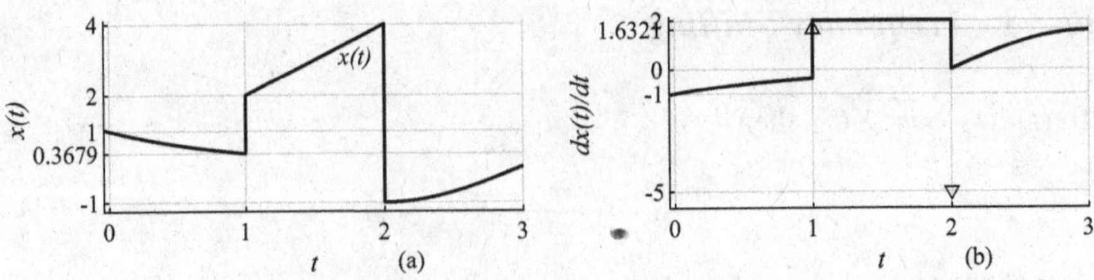

Fig. 10.3 (**a**) Signal $x(t)$ with step discontinuities and (**b**) its derivative

For example,

$$e^{-at}u(t) \Longleftrightarrow \frac{1}{s+a}, \ \mathrm{Re}(s) > -a$$

Then,

$$\frac{d\,(e^{-at}u(t))}{dt} = \delta(t) - ae^{-at}u(t) \Longleftrightarrow 1 - \left(\frac{a}{s+a}\right) = \frac{s}{s+a}$$

and

$$\frac{d\,(e^{-at})}{dt} = -a(e^{-at}) \Longleftrightarrow sX(s) - x(0^-) = \frac{s}{s+a} - 1 = -\frac{a}{s+a}$$

A signal $x(t)$, with step discontinuities, for example, at $t = 0$ of height $(x(0^+) - x(0^-))$ and at $t = t_1 > 0$ of height $(x(t_1^+) - x(t_1^-))$ can be expressed as

$$x(t) = x_c(t) + (x(0^+) - x(0^-))u(t) + (x(t_1^+) - x(t_1^-))u(t - t_1),$$

where $x_c(t)$ is $x(t)$ with the discontinuities removed and $x(t_1^+)$ and $x(t_1^-)$ are, respectively, the right- and left-hand limits of $x(t)$ at $t = t_1$. The derivative of $x(t)$ is given by the generalized function theory as

$$\frac{d\,x(t)}{dt} = \frac{d\,x_c(t)}{dt} + (x(0^+) - x(0^-))\delta(t) + (x(t_1^+) - x(t_1^-))\delta(t - t_1),$$

where $\frac{d\,x_c(t)}{dt}$ is the ordinary derivative of $x_c(t)$ at all t except at $t = 0$ and $t = t_1$. The Laplace transform of $\frac{d\,x(t)}{dt}$ is given by

$$sX_c(s) + (x(0^+) - x(0^-)) + (x(t_1^+) - x(t_1^-))e^{-st_1} = sX(s) - x(0^-)$$

Consider the signal, shown in Fig. 10.3a, from $t = 0^-$

$$\begin{aligned}x(t) &= e^{-t}(u(t) - u(t-1)) + 2t(u(t-1) - u(t-2)) \\ &\quad + \cos\left(\frac{\pi}{2}t\right)u(t-2) \\ &= x_c(t) + 1.6321u(t-1) - 5u(t-2),\end{aligned}$$

and its derivative

$$\begin{aligned}\frac{dx(t)}{dt} &= -e^{-t}(u(t) - u(t-1)) + 2(u(t-1) - u(t-2)) \\ &\quad - \left(\frac{\pi}{2}\right)\sin\left(\frac{\pi}{2}t\right)u(t-2) \\ &\quad + 1.6321\delta(t-1) - 5\delta(t-2),\end{aligned}$$

shown in Fig. 10.3b. The continuous part is shown by a line, and the impulse components are shown by triangles. The transform of $\frac{dx(t)}{dt}$ is

$$-\frac{1}{s+1} + \frac{e^{-1}e^{-s}}{s+1} + \frac{2e^{-s}}{s} - \frac{2e^{-2s}}{s} + \frac{(\frac{\pi}{2})^2 e^{-2s}}{s^2 + (\frac{\pi}{2})^2} + 1.6321e^{-s} - 5e^{-2s} = sX(s) - x(0^-)$$

The term $-\sin(\frac{\pi}{2}t)u(t-2)$ can be rewritten as $\sin(\frac{\pi}{2}(t-2))u(t-2)$, and then the time-shifting theorem can be applied to find its transform.

Note that $e^{-1} = 1/e \approx 0.3679$ and $tu(t-1) = u(t) + (t-1)u(t-1)$. The transform of $x(t)$ is

$$X(s) = \frac{1}{s+1} - \frac{e^{-1}e^{-s}}{s+1} + \frac{2e^{-s}}{s^2} - \frac{2e^{-2s}}{s^2} - \frac{se^{-2s}}{s^2 + (\frac{\pi}{2})^2} + 2\frac{e^{-s}}{s} - 4\frac{e^{-2s}}{s}$$

Remember that the value of $x(t)$ for $t < 0$ is ignored in computing the unilateral Laplace transform. The term $\cos(\frac{\pi}{2}t)u(t-2)$ can be rewritten as $-\cos(\frac{\pi}{2}(t-2))u(t-2)$, and then the time-shifting theorem can be applied to find its transform. The initial value is $x(0^-) = 1$. Now,

$$sX(s) - x(0^-) = \frac{s}{s+1} - \frac{se^{-1}e^{-s}}{s+1} + \frac{2e^{-s}}{s} - \frac{2e^{-2s}}{s} - \frac{s^2e^{-2s}}{s^2 + (\frac{\pi}{2})^2} + 2e^{-s} - 4e^{-2s} - 1.$$

This property can be extended, by repeated application, to find the transform of higher-order derivatives. For example,

$$\frac{d}{dt}\left(\frac{dx(t)}{dt}\right) = \frac{d^2x(t)}{dt^2} \Longleftrightarrow$$

$$s(sX(s) - x(0^-)) - \frac{dx(t)}{dt}\Big|_{t=0^-} = s^2X(s) - sx(0^-) - \frac{dx(t)}{dt}\Big|_{t=0^-}$$

One common application of this property is in the modeling of system components such as an inductor. The relationship between the current $i(t)$ through an inductor of value L henries and the voltage $v(t)$ across it is $v(t) = L\frac{di(t)}{dt}$. Assuming the initial value of current in the inductor is $i(0^-)$, using this property, we get the Laplace transform of the voltage across the inductor as $V(s) = L(sI(s) - i(0^-))$.

The application of time-differentiation and linearity properties reduces a differential equation into an algebraic equation, which can be easily solved. Consider solving the differential equation $\frac{dy(t)}{dt} + \frac{1}{2}y(t) = 0$, with the initial condition $y(0^-) = 3$. The solution using time-domain method (Chap. 4) is $y(t) = 3e^{-\frac{1}{2}t}u(t)$. Taking the transform of the differential equation, we get $sY(s) - 3 + \frac{1}{2}Y(s) = 0$. Solving for $Y(s)$, $Y(s) = \frac{3}{s+\frac{1}{2}}$. Finding the inverse transform, we get the same solution.

10.2.5 Integration

If $x(t) \Longleftrightarrow X(s)$, then

$$\int_{0-}^{t} x(\tau)\,d\tau \Longleftrightarrow \frac{1}{s}X(s)$$

As the signal, in the frequency domain, is expressed in terms of exponentials e^{st} and the integral of the exponential is $\frac{e^{st}}{s}$, the integration of a signal in time domain corresponds to a division of its transform by the frequency variable s. From another point of view, the product $\frac{1}{s}X(s)$ corresponds to the convolution of $x(t)$ and $u(t)$ in the time domain, which, of course, is equivalent to the integral of $x(t)$ from 0 to t. For example, the transform of the unit-step signal, which is the integral of the unit-impulse function with $X(s) = 1$, is $\frac{1}{s}$. Similarly, $tu(t) \Longleftrightarrow \frac{1}{s^2}$.

Consider the function $\sin(t)u(t)$ with its transform $\frac{1}{s^2+1}$. Using this property,

$$\int_{0-}^{t} \sin(\tau)d\tau \Longleftrightarrow \frac{1}{s(s^2+1)}$$

Finding the inverse transform, we get $(1 - \cos(t))u(t)$, which can be verified to be the time-domain integral of the sine function.

As the definite integral $\int_{-\infty}^{0-} x(\tau)\,d\tau$ is a constant,

$$\int_{-\infty}^{t} x(\tau)\,d\tau = \int_{-\infty}^{0-} x(\tau)\,d\tau + \int_{0-}^{t} x(\tau)\,d\tau \Longleftrightarrow \frac{1}{s}\int_{-\infty}^{0^-} x(\tau)\,d\tau + \frac{1}{s}X(s)$$

One common application of this property is in the modeling of system components such as a capacitor. The relationship between the current $i(t)$ through a capacitor of value C farads and the voltage $v(t)$ across it is $v(t) = \frac{1}{C}\int_{0^-}^{t} i(\tau)d\tau + v(0^-)$, where $v(0^-)$ is the initial voltage across the capacitor. Using this property, we get the Laplace transform of the voltage across the capacitor as $V(s) = \frac{I(s)}{sC} + \frac{v(0^-)}{s}$.

10.2.6 Time Scaling

If $x(t)u(t) \Longleftrightarrow X(s)$, then

$$x(at)u(at) \Longleftrightarrow \frac{1}{a}X\left(\frac{s}{a}\right),\ a > 0$$

The Laplace transform of $x(at)u(at)$, from the definition, is

$$\int_{0^-}^{\infty} x(at)u(at)e^{-st}dt$$

Substituting $at = \tau$, we get $t = \frac{\tau}{a}$ and $dt = \frac{d\tau}{a}$. Note that $u(at) = u(t), a > 0$. With these changes, the transform becomes

$$\frac{1}{a}\int_{0^-}^{\infty} x(\tau)e^{-\frac{s}{a}\tau}d\tau = \frac{1}{a}X\left(\frac{s}{a}\right)$$

Compression (expansion) of a signal in the time domain, by changing t to at, results in the expansion (compression) of its spectrum with the change s to $\frac{s}{a}$, in addition to scaling by $\frac{1}{a}$ (to take into account of the change in energy).

Consider the transform pair

$$e^{-2t}\sin(t)u(t) \Longleftrightarrow \frac{1}{s^2 + 4s + 5} = \frac{1}{(s + 2 - j)(s + 2 + j)}$$

The two poles are located at $-2 + j1$ and $-2 - j1$. With $a = 2$, we get

$$e^{-4t}\sin(2t)u(2t) \Longleftrightarrow \frac{1}{2}\frac{1}{\left(\frac{s}{2}\right)^2 + 4\left(\frac{s}{2}\right) + 5} = \frac{2}{(s + 4 - j2)(s + 4 + j2)}$$

The two poles are located at $-4 + j2$ and $-4 - j2$.

10.2.7 Convolution in Time

If $x(t)u(t) \Longleftrightarrow X(s)$ and $h(t)u(t) \Longleftrightarrow H(s)$, then

$$y(t) = x(t)u(t) * h(t)u(t) = \int_0^\infty x(\tau)h(t-\tau)d\tau \Longleftrightarrow X(s)H(s)$$

The FT of $x(t)e^{-\sigma t}$ is the Laplace transform $X(s)$ of $x(t)$. The convolution of $x(t)e^{-\sigma t}$ and $h(t)e^{-\sigma t}$ corresponds to $X(s)H(s)$ in the frequency domain. The inverse FT of $X(s)H(s)$, therefore, is the convolution of $x(t)e^{-\sigma t}$ and $h(t)e^{-\sigma t}$ given by

$$\int_0^\infty x(\tau)e^{-\sigma\tau}h(t-\tau)e^{-\sigma(t-\tau)}d\tau = e^{-\sigma t}\int_0^\infty x(\tau)h(t-\tau)d\tau = e^{-\sigma t}(x(t)*h(t))$$

As finding the inverse Laplace transform is the same as finding the inverse FT in addition to multiplying the signal by $e^{\sigma t}$, as will be seen later, we get the convolution of $x(t)$ and $h(t)$ by finding the inverse Laplace transform of $X(s)H(s)$.

Consider the convolution of $e^{2t}u(t)$ and $e^{-2t}u(t)$. The inverse of the product of their transforms,

$$\frac{1}{(s-2)(s+2)} = \frac{1}{4}\left(\frac{1}{(s-2)} - \frac{1}{(s+2)}\right),$$

is the convolution output $\frac{1}{4}(e^{2t} - e^{-2t})u(t)$.

10.2.8 Multiplication by t

If $x(t)u(t) \Longleftrightarrow X(s)$, then

$$tx(t)u(t) \Longleftrightarrow -\frac{dX(s)}{ds}$$

Differentiating the defining expression for $-X(s)$ with respect to s, we get

$$-\frac{dX(s)}{ds} = -\frac{d}{ds}\left(\int_{0^-}^\infty x(t)u(t)e^{-st}dt\right) = \int_{0^-}^\infty tx(t)e^{-st}dt$$

In general,

$$t^n x(t)u(t) \Longleftrightarrow (-1)^n\frac{d^n X(s)}{ds^n}, \ n = 0, 1, 2, \ldots$$

For example, $t\delta(t) = 0 \Longleftrightarrow -\frac{d(1)}{ds} = 0$. Another example is $tu(t) \Longleftrightarrow -\frac{d(\frac{1}{s})}{ds} = \frac{1}{s^2}$.

10.2.9 Initial Value

If only the initial and final values of a function $x(t)$ are required, these values can be found directly, from $X(s)$, using the following properties rather than finding the function $x(t)$ by inverting $X(s)$.

If $x(t) \Longleftrightarrow X(s)$ and the degree of the numerator polynomial of $X(s)$ is less than that of the denominator polynomial, then

$$x(0^+) = \lim_{s\to\infty} sX(s)$$

As $s \to \infty$, the value of any term with a higher-order denominator tends to zero, and

$$\lim_{s\to\infty} sX(s) = \lim_{s\to\infty} \left(\frac{sA_1}{s-s_1} + \frac{sA_2}{s-s_2} + \cdots + \frac{sA_N}{s-s_N} \right) = A_1 + A_2 + \cdots + A_N$$

The inverse transform of $X(s)$, as $t \to 0$, is

$$x(t) = A_1 e^{s_1 t} + A_2 e^{s_2 t} + \cdots + A_N e^{s_N t}$$

The right-hand limit of $x(t)$, as $t \to 0$, is

$$x(0^+) = A_1 + A_2 + \cdots + A_N = \lim_{s\to\infty} sX(s)$$

Similarly,

$$\frac{dx(t)}{dt}|_{t=0^+} = \lim_{s\to\infty} (s^2 X(s) - sx(0^-))$$

10.2.10 Final Value

If $x(t) \Longleftrightarrow X(s)$ and the ROC of sX(s) includes the $j\omega$ axis, then

$$\lim_{t\to\infty} x(t) = \lim_{s\to 0} sX(s)$$

As $t \to \infty$, the value $x(\infty)$, if it is nonzero, is solely due to the scaled unit-step component of $x(t)$. Multiplying $X(s)$ by s and setting $s = 0$ is just finding the partial fraction coefficient of the unit-step component of $x(t)$.

The initial and final values from the transform

$$\frac{1}{s(s+2)} = \frac{0.5}{s} - \frac{0.5}{s+1} \Longleftrightarrow (0.5 - 0.5e^{-t})u(t)$$

are

$$x(0^+) = \lim_{s\to\infty} \frac{1}{s+2} = 0 \quad \text{and} \quad \lim_{t\to\infty} x(t) = \lim_{s\to 0} \frac{1}{s+2} = \frac{1}{2}$$

The result is obvious from the time-domain response.

10.2.11 Transform of Semiperiodic Functions

Consider the function $x(t)u(t)$ that is periodic of period T for $t \geq 0$, that is, $x(t + T) = x(t),\ t \geq 0$. Let $x_1(t) = x(t)u(t) - x(t-T)u(t-T) \iff X_1(s)$. $x_1(t)$ is equal to $x(t)u(t)$ over its first period and is zero elsewhere. Then,

$$x(t)u(t) = x_1(t) + x_1(t-T) + \cdots + x_1(t-nT) + \cdots$$

Using the time-shifting property, the transform of $x(t)u(t)$ is

$$X(s) = X_1(s)(1 + e^{-sT} + \cdots + e^{-nsT} + \cdots) = \frac{X_1(s)}{1 - e^{-sT}}$$

Let us find the transform of a semiperiodic square wave, the first period of which is defined as

$$x_1(t) = \begin{cases} 1 \text{ for } 0 < t < 2 \\ 0 \text{ for } 2 < t < 4 \end{cases}$$

As $x_1(t) = (u(t) - u(t-2))$, $X_1(s) = \frac{(1-e^{-2s})}{s}$. From the property,

$$X(s) = \frac{1}{(1-e^{-4s})} \frac{(1-e^{-2s})}{s} = \frac{1}{s(1+e^{-2s})}$$

10.3 The Inverse Laplace Transform

Consider the transform pair $x(t)u(t) \iff \frac{1}{s-4}$, $\text{Re}(s) > 4$. Multiplying $x(t)u(t)$ by $e^{-5t}u(t)$ gives $x(t)e^{-5t}u(t) \iff \frac{1}{(s+5)-4} = \frac{1}{s+1}$, $\text{Re}(s) > -1$, due to the frequency-shifting property. Now, the ROC includes the $j\omega$ axis in the s-plane. Let us substitute $s = j\omega$ in $\frac{1}{s+1}$ to get $\frac{1}{j\omega+1}$. The inverse FT of this transform is the signal $e^{-t}u(t) = x(t)e^{-5t}u(t)$. Now, multiplying both sides by $e^{5t}u(t)$ gives the original time-domain signal $x(t)u(t) = e^{4t}u(t)$. This way of finding the inverse Laplace transform gives us a clear understanding of how the Laplace transform is the generalized version of the FT.

The inverse FT of $X(\sigma + j\omega)$, defined in (Eq. 10.1), is given as

$$x(t)e^{-\sigma t} = \frac{1}{2\pi}\int_{-\infty}^{\infty} X(\sigma + j\omega)e^{j\omega t}\,d\omega$$

Multiplying both sides by $e^{\sigma t}$, we get

$$x(t) = \frac{1}{2\pi}\int_{-\infty}^{\infty} X(\sigma + j\omega)e^{(\sigma + j\omega)t}\,d\omega$$

The complex frequency $(\sigma + j\omega)$ can be replaced by a complex variable $s = (\sigma + j\omega)$ with the limits of the integral changed to $\sigma - j\omega$ and $\sigma + j\omega$. As $ds = jd\omega$, we get the inverse Laplace transform of $X(s)$ as

$$x(t) = \frac{1}{2\pi j}\int_{\sigma - j\infty}^{\sigma + j\infty} X(s)e^{st}\,ds,$$

where σ is any real value that lies in the ROC of $X(s)$. Note that the integral converges to the value zero for $t < 0$ and to the mid-point value at any discontinuity of $x(t)$. This equation is not often used for finding the inverse transform, as it requires integration in the complex plane. The partial fraction method, which is essentially the same as that was described in Chap. 9, is commonly used. The difference is that the partial fraction terms are of the form $\frac{k}{s-p}$ in contrast to $\frac{kz}{z-p}$, as shown in the following examples.

10.3.1 Inverse Laplace Transform by Partial Fraction Expansion

In the transform method of system analysis, we find the forward transform of signals, do the required processing in the transform domain, and find the inverse transform to get the time-domain version of the processed signal. Most of the Laplace transforms of practical interest are rational functions (a ratio of two polynomials in s). The denominator polynomial can be factored into a product of first- or second-order terms. This type of Laplace transforms can be expressed as the sum of partial fractions with each denominator forming a factor. The inverse Laplace transforms of the individual fractions can be easily found from a short table of transform pairs, such as those of $\delta(t)$, $u(t)$, $tu(t)$, $t^2u(t)$, $e^{-at}u(t)$, and $te^{-at}u(t)$, shown in Table B.11. The sum of the individual inverses is the inverse of the given Laplace transform.

Two rational functions are added by converting them to a common denominator, add and then simplify. For example, the sum of the two rational functions is

$$X(s) = \frac{2}{(s+2)} + \frac{3}{(s+3)} = \frac{2(s+3)+3(s+2)}{(s+3)(s+2)} = \frac{(5s+12)}{(s^2+5s+6)}$$

Usually, we are given $X(s)$ to be inverted in the form on the right. The task is to find an equivalent expression like that on the left. The numerator polynomial of the rightmost expression is of order 1, whereas that of the denominator is of order 2. Partial fraction expansion of a rational function expresses it as a sum of appropriate fractions with the coefficient of each fraction to be found.

Example 10.4 Find the zero-state response of the system governed by the differential equation

$$\frac{dy}{dt} + 3y(t) = \frac{dx}{dt} + 2x(t)$$

with the input $x(t) = u(t)$, the unit-step function.

Solution The Laplace transforms of the terms of the differential equation are

$$x(t) \leftrightarrow \frac{1}{s}, \quad \frac{dx}{dt} \leftrightarrow 1 \quad y(t) \leftrightarrow Y(s), \quad \frac{dy}{dt} \leftrightarrow sY(s)$$

Substituting the corresponding transform for each term in the differential equation and solving for $Y(s)$, we get

$$Y(s) = \frac{s+2}{s(s+3)} = \frac{A}{s} + \frac{B}{(s+3)}$$

$$A = \left.\frac{(s+2)}{(s+3)}\right|_{s=0} = 2/3, \qquad B = \left.\frac{(s+2)}{s}\right|_{s=-3} = 1/3$$

$$Y(s) = \frac{2/3}{s} + \frac{1/3}{(s+3)}$$

Taking the inverse Laplace transform, we get the complete response.

$$y(t) = ((2/3) + (1/3)e^{-3t})u(t)$$

The steady-state response is $((2/3)u(t)$, and the transient response is $((1/3)e^{-3t})u(t)$. Letting $t = 0$, $y(0) = 1$. Letting $t \to \infty$, $y(\infty) = 2/3$. From the initial and final value properties also, we get

$$\lim_{s\to\infty} s\frac{2/3}{s} + s\frac{1/3}{(s+3)} = 1, \quad \lim_{s\to 0} s\frac{2/3}{s} + s\frac{1/3}{(s+3)} = 2/3$$

∎

Rational Function with the Same Order of Numerator and Denominator
Partial fraction expansion is applicable, only if the degree of the numerator polynomial is less than that of the denominator. In practical system analysis, we encounter only rational functions with the degree of the numerator polynomial less than or equal that of the denominator. In the case the degrees are equal, we divide the numerator polynomial by the denominator polynomial once to get a constant plus a proper function.

Example 10.5 Find the inverse of

$$X(s) = \frac{s+1}{s+3}$$

Solution

$$X(s) = \frac{s+1}{(s+3)} = 1 - \frac{2}{(s+3)}$$

Taking the inverse Laplace transform, we get the complete response.

$$x(t) = \delta(t) - 2e^{-3t}u(t)$$

Example 10.6 Find the zero-state response of the system governed by the differential equation

$$\frac{d^2y}{dt^2} + \frac{dy}{dt} + 2y = x,$$

using the Laplace transform, with the input $x(t) = e^{-t}u(t)$.

Solution The Laplace transforms of the terms of the differential equation are

$$x(t) \leftrightarrow \frac{1}{(s+1)}, \quad y(t) \leftrightarrow Y(s), \quad \frac{dy}{dt} \leftrightarrow sY(s), \quad \frac{d^2y}{dt^2} \leftrightarrow s^2Y(s)$$

Substituting the corresponding transform for each term in the differential equation and solving for $Y(s)$, we get

$$Y(s) = \frac{1}{(s^2+s+2)(s+1)} = \frac{0.5}{(s+1)} + \frac{-0.2500 + j0.0945}{(s+0.5000-j1.3229)}$$
$$+\frac{-0.2500 + j0.0945}{(s+0.5000+j1.3229)}$$

Suppose the two complex conjugate roots are

$$-a + jb \quad \text{and} \quad -a - jb$$

and the corresponding partial fraction coefficients are

$$Re^{j\theta} \quad \text{and} \quad Re^{-j\theta}$$

Then, the corresponding inverse transformation is

$$2Re^{-at}\cos(bt + \theta)$$

Taking the inverse Laplace transform, we get the complete response.

$$y(t) = (0.5e^{-t} + 0.5345e^{-0.5t}\cos(1.3229t - 2.7802))u(t)$$

■

Multiple-Order Poles
Each repeated linear factor $(s + a)^m$ contributes a sum of the form

$$\frac{A_m}{(s+a)^m} + \frac{A_{m-1}}{(s+a)^{m-1}} + \cdots + \frac{A_1}{(s+a)}$$

Example 10.7 Find the zero-state response of the system governed by the differential equation

$$\frac{d^3y}{dt^3} + 6\frac{d^2y}{dt^2} + 12\frac{dy}{dt} + 8y = \frac{d^2x}{dt^2} + \frac{dx}{dt} + 2x,$$

using the Laplace transform, with the input $x(t) = e^{-t}u(t)$.

Solution The Laplace transforms of the terms of the differential equation are

$$x(t) \leftrightarrow \frac{1}{(s+1)}, \quad \frac{dx}{dt} \leftrightarrow \frac{s}{(s+1)}, \quad \frac{d^2x}{dt^2} \leftrightarrow \frac{s^2}{(s+1)}$$

$$y(t) \leftrightarrow Y(s), \quad \frac{dy}{dt} \leftrightarrow sY(s), \quad \frac{d^2y}{dt^2} \leftrightarrow s^2Y(s)$$

Substituting the corresponding transform for each term in the differential equation and solving for $Y(s)$, we get

$$Y(s) = \frac{s^2 + s + 2}{(s^3 + 6s^2 + 12s + 8)(s+1)} = \frac{A}{(s+2)} + \frac{B}{(s+2)^2} + \frac{C}{(s+2)^3} + \frac{D}{(s+1)}$$

$$D = \left.\frac{s^2 + s + 2}{(s^3 + 6s^2 + 12s + 8)}\right|_{s=-1} = 2$$

$$C = \left.\frac{s^2 + s + 2}{(s + 1)}\right|_{s=-2} = -4$$

With C and D known, to find A, one method is to multiply both sides by s and let $s \to \infty$. That is,

$$\lim_{s\to\infty} s\frac{s^2 + s + 2}{(s^3 + 6s^2 + 12s + 8)(s + 1)} = \lim_{s\to\infty} \frac{As}{(s + 2)} + \frac{Bs}{(s + 2)^2} + \frac{Cs}{(s + 2)^3} + \frac{Ds}{(s + 1)}$$

We get $0 = 2 + A$ or $A = -2$. To find B, we replace s by a value other than the roots. Let $s = 0$, and we get

$$\frac{0 + 0 + 2}{(0 + 0 + 0 + 8)(0 + 1)} = \frac{-2}{(0 + 2)} + \frac{B}{(0 + 2)^2} + \frac{-4}{(0 + 2)^3} + \frac{2}{(0 + 1)} \text{ or } B = -1$$

Now,

$$Y(s) = \frac{s^2 + s + 2}{(s^3 + 6s^2 + 12s + 8)(s + 1)} = \frac{-2}{(s + 2)} + \frac{-1}{(s + 2)^2} + \frac{-4}{(s + 2)^3} + \frac{2}{(s + 1)}$$

∎

Taking the inverse Laplace transform, we get the complete response.

$$y(t) = (2e^{-t} - 2e^{-2t} - te^{-2t} - 2t^2e^{-2t})u(t)$$

Example 10.8 Find the inverse of

$$X(s) = \frac{1}{(s^2 + 4)^2}$$

Solution First, let us find the inverse using properties.

$$\sin(2t)u(t) \leftrightarrow \frac{2}{(s^2 + 4)}$$

Applying frequency-differentiating property, we get

$$t\sin(2t) \leftrightarrow \frac{4s}{(s^2 + 4)^2}$$

Now,

$$\frac{1}{(s^2+4)^2} = \left(\frac{4s}{(s^2+4)^2}\right)\left(\frac{1}{4s}\right)$$

Using the convolution property, the inverse is

$$\frac{1}{4} * t\sin(2t) = \frac{1}{4}\int_0^t \tau\sin(2\tau)d\tau = \frac{1}{16}(\sin(2t) - 2t\cos(2t))\,u(t)$$

By partial fraction, we get

$$X(s) = \frac{1}{(s^2+4)^2} = \frac{j0.0312}{s+j2} + \frac{-j0.0312}{s-j2} + \frac{-0.0625}{(s+j2)^2} + \frac{-0.0625}{(s-j2)^2}$$
$$= \frac{(1/8)}{(s^2+4)} - \frac{(1/8)(s^2-4)}{(s^2+4)^2}$$

Taking the inverse, we get the same result. ∎

10.4 Applications of the Laplace Transform

10.4.1 Transfer Function and the System Response

Consider the second-order differential equation of a causal LTI continuous system relating the input $x(t)$ and the output $y(t)$,

$$\frac{d^2y(t)}{dt^2} + a_1\frac{dy(t)}{dt} + a_0y(t) = b_2\frac{d^2x(t)}{dt^2} + b_1\frac{dx(t)}{dt} + b_0x(t).$$

Taking the Laplace transform of both sides, we get, assuming initial conditions are all zero,

$$(s^2 + a_1s + a_0)Y(s) = (b_2s^2 + b_1s + b_0)X(s)$$

The transfer function $H(s)$, which is the ratio of the transforms of the output and the input signals with the initial conditions zero, is obtained as

$$H(s) = \frac{Y(s)}{X(s)} = \frac{b_2s^2 + b_1s + b_0}{s^2 + a_1s + a_0} = \frac{\sum_{l=0}^{2} b_l s^l}{s^2 + \sum_{l=0}^{1} a_l s^l}$$

In general,

$$H(s) = \frac{Y(s)}{X(s)} = \frac{b_M s^M + b_{M-1}s^{M-1} + \cdots + b_1 s + b_0}{s^N + a_{N-1}s^{N-1} + \cdots + a_1 s + a_0}.$$

If the input to the system is the unit-impulse signal, then its transform is one, and $H(s) = Y(s)$. That is, the transform of the impulse response is the transfer function of the system. For stable systems, the frequency response $H(j\omega)$ is obtained from $H(s)$ by replacing s by $j\omega$.

10.4.2 Characterization of a System by Its Poles and Zeros

The numerator and denominator polynomials of the transfer function can be factored to get

$$H(s) = K\frac{(s - z_1)(s - z_2)\cdots(s - z_M)}{(s - p_1)(s - p_2)\cdots(s - p_N)} = K\frac{\prod_{l=1}^{M}(s - z_l)}{\prod_{l=1}^{N}(s - p_l)},$$

where K is a constant. As the coefficients of the polynomials of $H(s)$ are real for practical systems, the zeros and poles are real-valued, or they always occur as complex conjugate pairs.

The pole-zero plot of the transfer function $H(s)$ of a system is a pictorial description of its characteristics, such as speed of response, frequency selectivity, and stability. Poles located farther from the imaginary axis in the left half of the s-plane result in a fast-responding system with its transient response decaying rapidly. On the other hand, poles located close to the imaginary axis in the left half of the s-plane result in a sluggish system. Complex conjugate poles located in the left half of the s-plane result in an oscillatory transient response that decays with time. Complex conjugate poles located on the imaginary axis result in a steady oscillatory transient response. Poles located on the positive real axis in the left half of the s-plane result in exponentially decaying transient response. The frequency components of an input signal with frequencies close to a zero will be suppressed, while those close to a pole will be readily transmitted. Poles located symmetrically about the negative real axis in the left half of the s-plane and close to the imaginary axis in the passband result in a lowpass system that more readily transmits low-frequency signals than high-frequency signals. Zeros placed in the stopband further enhance the lowpass character of the frequency response. For example, pole-zero plots of some lowpass filters are shown in Figs. 10.1 and 10.20. The stability of a system can also be determined from its pole-zero plot, as presented later.

Example 10.9 Find the zero-input, zero-state, transient, steady-state, and complete responses of the system governed by the differential equation

$$\frac{d^2y(t)}{dt^2} + 4\frac{dy(t)}{dt} + 4y(t) = \frac{d^2x(t)}{dt^2} + \frac{dx(t)}{dt} + 2x(t)$$

with the initial conditions $y(0^-) = 2$ and $\frac{dy(t)}{dt}|_{t=0^-} = 3$ and the input $x(t) = u(t)$, the unit-step function.

Solution The Laplace transforms of the terms of the differential equation are

$$x(t) \Longleftrightarrow \frac{1}{s}, \qquad \frac{dx(t)}{dt} \Longleftrightarrow 1, \qquad \frac{d^2x(t)}{dt^2} \Longleftrightarrow s$$

$$y(t) \Longleftrightarrow Y(s), \qquad \frac{dy(t)}{dt} \Longleftrightarrow sY(s) - 2 \qquad \frac{d^2y(t)}{dt^2} \Longleftrightarrow s^2Y(s) - 2s - 3$$

Substituting the corresponding transform for each term in the differential equation and solving for $Y(s)$, we get

$$Y(s) = \frac{s^2 + s + 2}{s(s^2 + 4s + 4)} + \frac{2s + 11}{s^2 + 4s + 4}$$

The first term on the right-hand side is $H(s)X(s)$ and corresponds to the zero-state response. The second term is due to the initial conditions and corresponds to the zero-input response. Expanding into partial fractions, we get

$$Y(s) = \frac{0.5}{s} + \frac{0.5}{(s+2)} - \frac{2}{(s+2)^2} + \frac{2}{(s+2)} + \frac{7}{(s+2)^2}$$

Taking the inverse Laplace transform, we get the complete response.

$$\begin{aligned} y(t) &= (\overbrace{0.5 + 0.5e^{-2t} - 2te^{-2t}}^{\text{zero-state}} + \overbrace{2e^{-2t} + 7te^{-2t}}^{\text{zero-input}})u(t) \\ &= (0.5 + 2.5e^{-2t} + 5te^{-2t})u(t) \end{aligned}$$

The steady-state response is $0.5u(t)$, and the transient response is $(2.5e^{-2t} + 5te^{-2t})u(t)$. The initial and final values of $y(t)$ are 3 and 0.5, respectively. These values can be verified by applying the initial and final value properties to $Y(s)$. We can also verify that the initial conditions at $t = 0^-$ are satisfied by the zero-input component of the response. The zero-input, zero-state, and total responses are shown, respectively, in Fig. 10.4a,b, and c. Figure 10.5 shows the simulation diagram of the transfer function with initial conditions producing the total response. The top half of the diagram produces the zero-state response to the unit-step input. The zero-input response is produced by the bottom half. The summer unit adds the two responses to produce the total response. ∎

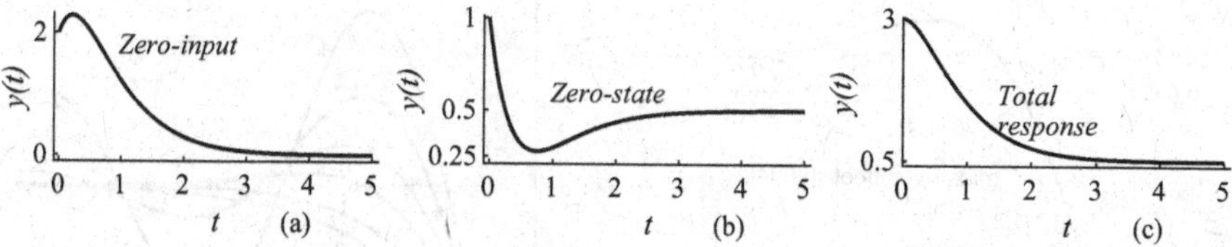

Fig. 10.4 The response of the system for unit-step signal: (**a**) zero-input response; (**b**) zero-state response; (**c**) total response

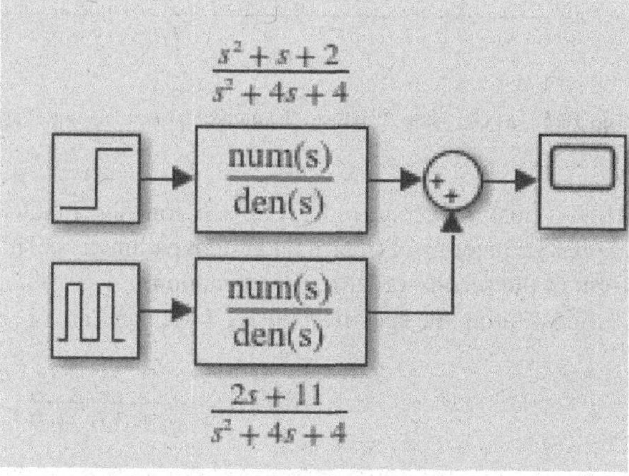

Fig. 10.5 The simulation diagram of the transfer function with initial conditions

10.4.3 *Unit-Step Response and Transient-Response Specifications*

In practice, the order of the control system is very high. But, for implementation advantages, they are usually decomposed into first- and second-order systems. Therefore, the analysis of first- and second-order systems is necessary for understanding the analysis and design of higher-order systems. The impulse response of stable systems does not have the steady-state component. Consequently, the unit-step signal is widely used as the test signal to analyze and design control systems.

Consider the open-loop transfer function of a second-order system with unity feedback in standard form.

$$G(s) = \frac{Y(s)}{E(s)} = \frac{\omega_n^2}{s^2 + 2\zeta\omega_n s},$$

where ζ is the damping ratio and ω_n is the undamped natural frequency in rad/sec. The corresponding closed-loop transfer function is

$$H(s) = \frac{Y(s)}{X(s)} = \frac{\omega_n^2}{s^2 + 2\zeta\omega_n s + \omega_n^2}$$

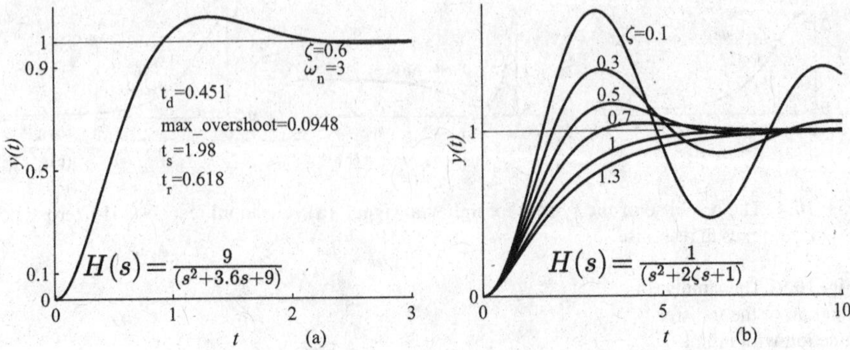

Fig. 10.6 (**a**) Unit-step response; (**b**) unit-step response with various damping ratios ζ

This is just a second-order Laplace transfer function, but the coefficients are expressed in terms of important system parameters. This form is called the standard form of the second-order transfer function.

Substituting the specific values $\zeta = 0.6$ and $\omega_n = 3$ rad/sec, we get

$$H(s) = \frac{9}{s^2 + 3.6s + 9}$$

Multiplying by the unit-step function $u(t) \leftrightarrow 1/s$, decomposing into partial fraction, and taking the inverse Laplace transform, we get the unit-step response as

$$y(t) = 1 + 1.25e^{-1.8t}\cos(2.4t + 2.4981)$$

The unit-impulse response is obtained by differentiating the unit-step response. Figure 10.6a shows the unit-step response of the second-order transfer function. The characteristic figures are shown in the figure. As both the transient and steady-state responses are critical for control systems, these specifications are quite important. In most systems, typically, the damping ratio is between 0.4 and 0.8 to avoid excessive overshoot and sluggish response. Note that maximum overshoot and rise time conflict each other.

Figure 10.6b shows the unit-step response of the second-order transfer function

$$H(s) = \frac{1}{(s^2 + 2\zeta s + 1)}$$

with $\omega_n = 1$ rad/sec and various values of ζ. The unit-step response is usually measured with zero initial conditions, which makes it easy to compare with those of other systems.

The response is of three types. If $\zeta = 1$, both the roots are real and the same (critically damped). If $\zeta < 1$, the roots are complex conjugates (underdamped) with negative real parts. If $\zeta > 1$, both the roots are real (overdamped). For the underdamped case $0 < \zeta < 1$, the unit-step response is given as

$$y(t) = \left(1 + \frac{e^{-\zeta\omega_n t}}{\sqrt{(1-\zeta^2)}} \cos(\omega_n\sqrt{(1-\zeta^2)}t + \left(\cos^{-1}(\zeta) + \frac{\pi}{2}\right))\right)u(t)$$

The **damped frequency** is $\omega_d = \omega_n\sqrt{1-\zeta^2}$ and usually less than ω_n. Substituting $\zeta = 0.6$ and $\omega_n = 3$ rad/sec in the general expression given above, we get the specific response for $y(t)$ shown in Fig. 6.3a. The damped frequency is 2.4 rad/sec, as can be seen from the expression for $y(t)$. The two roots of the underdamped ($\zeta < 1$) second-order transfer function in standard form, using the quadratic formula, are

$$-\zeta\omega_n \pm j\omega_n\sqrt{1-\zeta^2} = -\alpha \pm j\omega_d$$

with $\alpha = \zeta\omega_n$ and $\omega_d = \omega_n\sqrt{1-\zeta^2}$. The response becomes more oscillatory for lower values of ζ. For the overdamped case, the roots are

$$-\zeta\omega_n \pm \omega_n\sqrt{\zeta^2-1}$$

For the critically damped case $\zeta = 1$, the roots are $-\omega_n$. For $\zeta \geq 1$, the response never exceeds its final value. For the undamped case $\zeta = 0$, the roots are $\pm j\omega_n$. In this case, the response is a steady sinusoid. In the expression for unit-step response, α appears in the exponential term. Therefore, it controls the rise or decay of the response.

The maximum overshoot is

$$M_p = e^{-\frac{\pi\zeta}{\sqrt{1-\zeta^2}}}$$

and it occurs at

$$t_p = \frac{\pi}{\omega_n\sqrt{1-\zeta^2}}$$

The **maximum overshoot** of the response $y(t)$ is defined as the difference between the maximum value of $y(t)$ and the steady-state value $\lim_{t\to\infty} y(t)$. The **peak time** t_p is the time at which the first peak of the response occurs. For example, the value is 0.0948 at $t = 1.313$ sec. (or 9.48%) as shown in the figure.

The **settling time** t_s is defined as the time required for the response to be within certain percentage of its final value, typically 2%. The settling time t_s has to satisfy the condition

$$e^{-\zeta\omega_n t_s} < 0.02$$

That is,

$$e^{\zeta\omega_n t_s} = \frac{1}{0.02} = 50 \quad \text{or} \quad t_s = \frac{\log_e(50)}{\zeta\omega_n} = \frac{3.9120}{\zeta\omega_n} \cong \frac{4}{\zeta\omega_n} = 4\tau,$$

where τ is the time constant. Consider only the exponential part of the unit-step response

$$1.25e^{-1.8t}u(t) = 1.25e^{-\frac{t}{0.5556}}u(t)$$

At $t = 0.0.5556$, the response becomes

$$1.25e^{-1}u(t) = 1.25(0.3679) = 0.4599$$

That is, the value of the exponential decreases to 0.3679 of its initial value in one time constant $\tau = 0.5556$ s. In four time constants, the value decreases to $0.3679^4 = 0.0183$ which is less than 2% of its initial value. We approximated this to 2% to define the settling time. The system with a smaller time constant is faster that responds quickly to the input.

The **delay time** t_d is the time required for the response to reach 50% of its final value for the first time. The **rise time** is the time required for the response to rise from 10% to 90% (usually used for overdamped systems) or 0% to 100% (usually used for underdamped systems) of its final value.

Damping Ratio and Damping Factor

For the case of critical damping with $\zeta = 1$, the magnitude of the real part of the roots of the characteristics of the equation of the system is the same, ω_n, with the imaginary part zero. The factor $\zeta\omega_n$, called the damping factor, actually controls the damping of the system. Then, ζ can be considered as the damping ratio

$$\zeta = \frac{\zeta\omega_n}{\omega_n} = \frac{\text{actual damping factor}}{\text{damping factor at critical damping}}$$

10.4.4 System Stability

The zero-input response of a system depends solely on the locations of its poles. A system is considered stable if its zero-input response due to finite initial conditions converges, marginally stable if its zero-input response tends to a constant value or oscillates with a constant amplitude, and unstable if its zero-input response diverges. Commonly used marginally stable systems are oscillators, which produce a bounded zero-input response. The response corresponding to each pole p of a system is of the form e^{at}, where a is the location of the pole in the s-plane. If the real part of

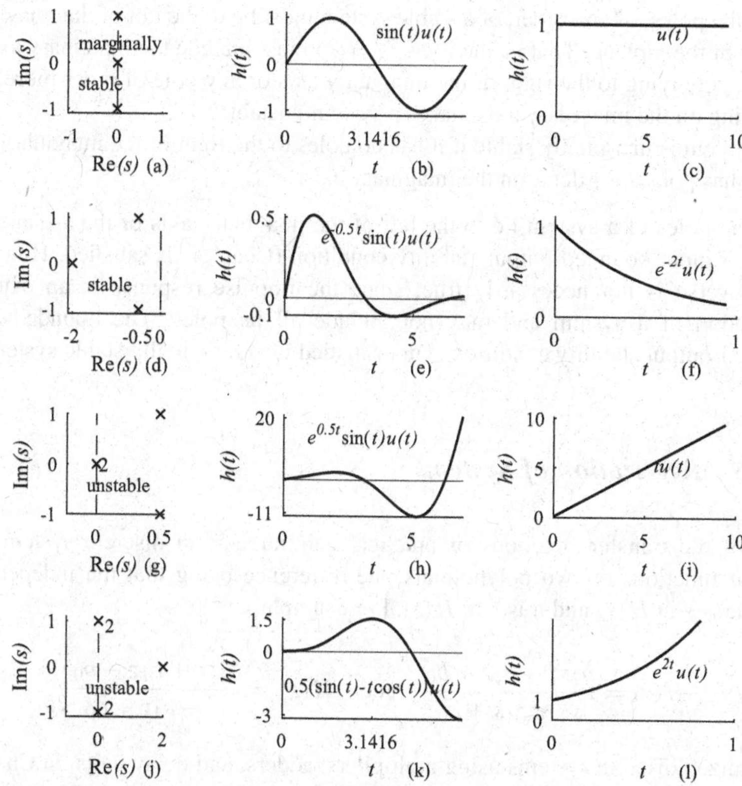

Fig. 10.7 The poles of some transfer functions $H(s)$ and the corresponding impulse responses $h(t)$. The imaginary axis is shown by a dashed line. **(a)** $H(s) = \frac{1}{s^2+1} = \frac{1}{(s+j)(s-j)}$ and $H(s) = \frac{1}{s}$; **(b)** $h(t) = \sin(t)u(t)$ and **(c)** $h(t) = u(t)$; **(d)** $H(s) = \frac{1}{(s+0.5)^2+1} = \frac{1}{(s+0.5+j)(s+0.5-j)}$ and $H(s) = \frac{1}{s+2}$; **(e)** $h(t) = e^{-0.5t}\sin(t)u(t)$ and **(f)** $h(t) = e^{-2t}u(t)$; **(g)** $H(s) = \frac{1}{(s-0.5)^2+1} = \frac{1}{(s-0.5+j)(s-0.5-j)}$ and $H(s) = \frac{1}{s^2}$; **(h)** $h(t) = e^{0.5t}\sin(t)u(t)$ and **(i)** $h(t) = tu(t)$; **(j)** $H(s) = \frac{1}{(s^2+1)^2} = \frac{1}{(s+j)^2(s-j)^2}$ and $H(s) = \frac{1}{s-2}$; **(k)** $h(t) = 0.5(\sin(t) - t\cos(t))u(t)$ and **(l)** $h(t) = e^{2t}u(t)$

a is less than zero, then e^{at} tends to zero as t tends to ∞. If the real part of a is greater than zero, then e^{at} tends to ∞ as t tends to ∞. If the real part of a is equal to zero, then e^{at} remains bounded as t tends to ∞. However, the response tends to infinity, for poles of order more than 1 lying on the imaginary axis of the s-plane, as the expression for the response includes a factor that is a function of t. Poles of any order lying to the left of the imaginary axis of the s-plane do not cause instability. Figure 10.7 shows pole locations of some transfer functions and the corresponding impulse responses. Therefore, we conclude that, in terms of the locations of the poles of a system:

- All the poles, of any order, of a stable system must lie to the left of the imaginary axis of the s-plane. That is, the ROC of $H(s)$ must include the imaginary axis.
- Any pole lying to the right of the imaginary axis or any pole of order more than 1 lying on the imaginary axis makes a system unstable.
- A system is marginally stable if it has no poles to the right of the imaginary axis and has poles of order 1 on the imaginary axis.

If all the poles of a system lie to the left of the imaginary axis of the s-plane, the bounded-input bounded-output stability condition (Chap. 4) is satisfied. However, the converse is not necessarily true, since the impulse response is an external description of a system and may not include all its poles. The bounded-input bounded-output stability condition is not satisfied by a marginally stable system.

10.4.5 Realization of Systems

Most of the transfer functions of practical continuous and discrete systems are rational functions of two polynomials, the difference being that the independent variable is s in $H(s)$ and it is z in $H(z)$. For example,

$$H(s) = \frac{b_2 s^2 + b_1 s + b_0}{s^2 + a_1 s + a_0} \quad \text{and} \quad H(z) = \frac{b_2 z^2 + b_1 z + b_0}{z^2 + a_1 z + a_0}$$

We realized discrete systems using multipliers, adders, and delay units, in Chap. 9. By comparison of the corresponding difference and differential equations, we find that the only difference is that integrators are required in realizing continuous systems instead of delay units. Therefore, the realization of continuous-time systems is the same as that for discrete systems, described in Chap. 9, except that delay units are replaced by integrators. Figure 10.8 shows the realization of a second-order continuous system. Integrators with feedback are used to simulate differential equations.

10.4.6 Frequency-Domain Representation of Circuits

By replacing each element in a circuit, along with their initial conditions, by the corresponding frequency-domain representation, we can analyze the circuit in a way similar to a resistor network. This procedure is quite effective for circuits with nonzero initial conditions compared with writing the differential equation and then finding the Laplace transform.

In time-domain representation, a capacitor with initial voltage $v(0^-)$ is modeled as an uncharged capacitor in series with a voltage source $v(0^-)u(t)$. The voltage-current relationship of a capacitor is

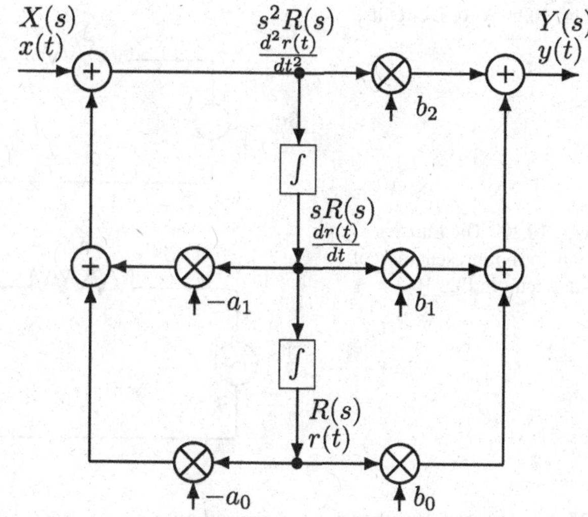

Fig. 10.8 The realization of a second-order continuous system

$$v(t) = \frac{1}{C}\int_{-\infty}^{t} i(\tau)d\tau = \frac{1}{C}\int_{-\infty}^{0^-} i(\tau)d\tau + \frac{1}{C}\int_{0^-}^{t} i(\tau)d\tau = v(0^-) + \frac{1}{C}\int_{0^-}^{t} i(\tau)d\tau$$

Taking the Laplace transform, the voltage across the capacitor is given as

$$V(s) = \frac{I(s)}{sC} + \frac{v(0^-)}{s}$$

The capacitor is modeled as an impedance $\frac{1}{sC}$ in series with an ideal voltage source $\frac{v(0^-)}{s}$. By taking the factor $\frac{1}{sC}$ out, an alternate representation is obtained as

$$V(s) = \frac{1}{sC}\left(I(s) + Cv(0^-)\right)$$

The voltage across the capacitor is due to the current $(I(s) + Cv(0^-))$ flowing through it. This representation, in the time domain, implies an uncharged capacitor in parallel with an impulsive current source $Cv(0^-)\delta(t)$.

In time-domain representation, an inductor with initial current $i(0^-)$ is modeled as an inductor, with no initial current, in series with an impulsive voltage source $Li(0^-)\delta(t)$. The voltage-current relationship of an inductor is

$$v(t) = L\frac{di(t)}{dt}$$

Taking the Laplace transform, the voltage across the inductor is given as

$$V(s) = L(sI(s) - i(0^-))$$

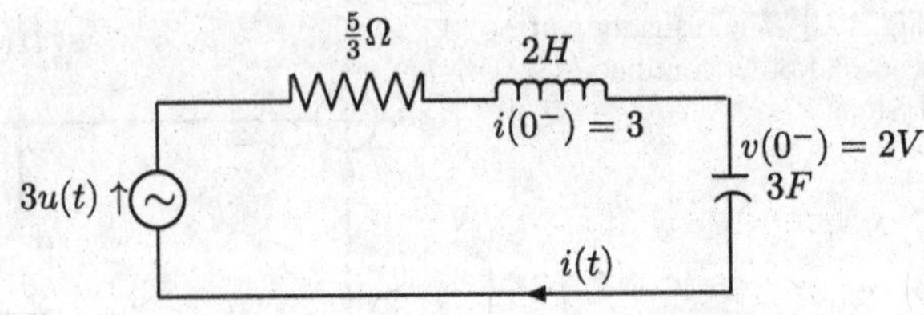

Fig. 10.9 A RCL circuit

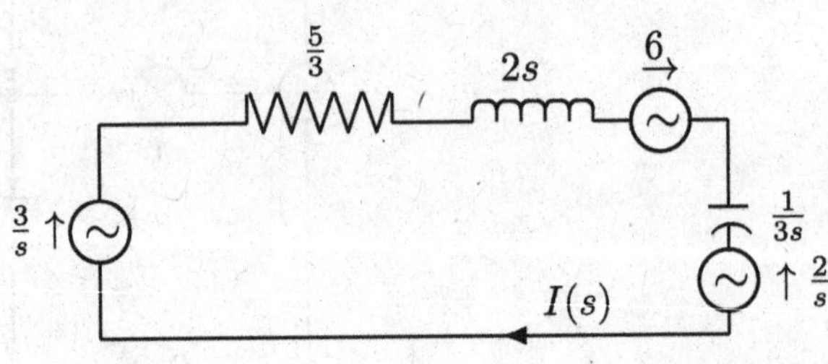

Fig. 10.10 The Laplace transform representation of the circuit in Fig. 10.5

The inductor is modeled as an impedance sL in series with an ideal voltage source $-Li(0^-)$. By taking the factor sL out, an alternate representation is obtained as

$$V(s) = sL\left(I(s) - \frac{i(0^-)}{s}\right)$$

The voltage across the inductor is due to the current $(I(s) - \frac{i(0^-)}{s})$ flowing through it.

Example 10.10 Find the current in the circuit, shown in Fig. 10.9, with the initial current through the inductor $i(0^-) = 3$ amperes and the initial voltage across capacitor $v(0^-) = 2$ volts and the input $x(t) = 3u(t)$ volts.

Solution The Laplace transform representation of the circuit in Fig. 10.5 is shown in Fig. 10.10.

The sum of the voltages in the circuit is

$$\frac{3}{s} + 6 - \frac{2}{s} = \frac{6s+1}{s}$$

The circuit impedance is

$$\frac{5}{3} + 2s + \frac{1}{3s} = \frac{6s^2+5s+1}{3s}$$

Dividing the voltage by the impedance, we get the current in the circuit as

$$I(s) = \frac{6s+1}{s}\,\frac{3s}{6s^2+5s+1} = \frac{(3s+\frac{1}{2})}{s^2+\frac{5}{6}s+\frac{1}{6}}$$

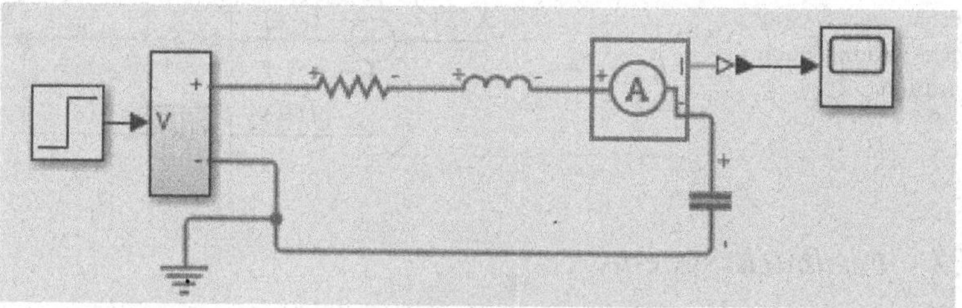

Fig. 10.11 The physical simulation diagram of the series circuit

Fig. 10.12 The current through the series circuit

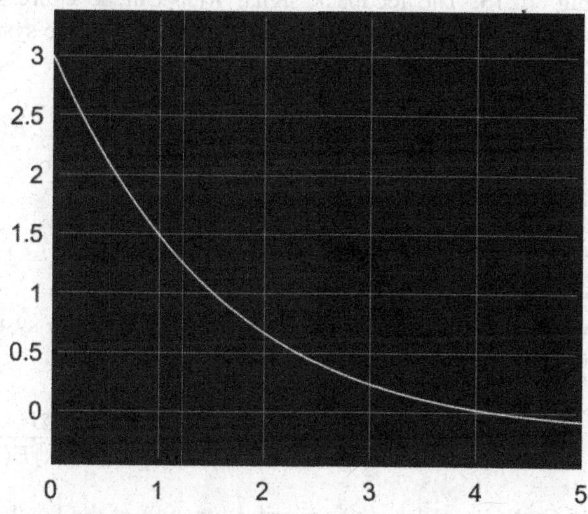

Expanding into partial fractions, we get

$$I(s) = \frac{6}{s+\frac{1}{2}} - \frac{3}{s+\frac{1}{3}}$$

Finding the inverse Laplace transform, we get the current in the circuit as

$$i(t) = (6e^{-\frac{1}{2}t} - 3e^{-\frac{1}{3}t})u(t)$$ ∎

The physical simulation diagram of the series circuit is shown in Fig. 10.11. The current through the series circuit is shown in Fig. 10.12. The horizontal axis is time and the vertical axis is current. The initial current is 3 ampere and the final current tends to zero.

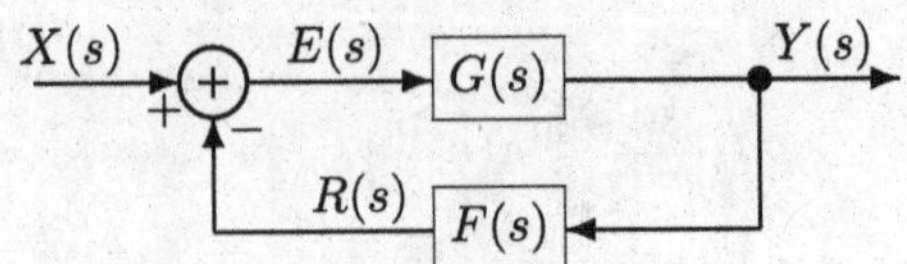

Fig. 10.13 Two systems connected in a feedback configuration

10.4.7 *Feedback Systems*

Consider the two systems connected in a feedback configuration, shown in Fig. 10.13. The feedback signal $R(s)$ can be expressed as $R(s) = F(s)Y(s)$, where $F(s)$ is the feedback transfer function of the system and $Y(s)$ is the output. Now, the error signal $E(s)$ is

$$E(s) = X(s) - R(s) = X(s) - F(s)Y(s)$$

The output $Y(s)$ is expressed as

$$Y(s) = G(s)E(s) = G(s)(X(s) - F(s)Y(s))$$

where $G(s)$ is the forward transfer function of the system. Therefore, the transfer function of the feedback system is given as

$$H(s) = \frac{Y(s)}{X(s)} = \frac{G(s)}{1 + G(s)F(s)}$$

If $G(s)$ is very large, the transfer function of the feedback system approximates to the inverse of the feedback transfer function of the system.

$$H(s) = \frac{Y(s)}{X(s)} \approx \frac{1}{F(s)}$$

10.4.8 *Bode Diagram*

The frequency response of systems, in two different formats, play an important in the analysis of feedback control systems. They are called Bode and Nyquist plots. The loop transfer function of the system, shown in Fig. 10.13, is $G(s)F(s)$. The frequency response is obtained as $G(j\omega)F(j\omega)$ by replacing s by $j\omega$. The frequency response can be easily obtained using an oscillator to provide sinusoidal input of frequencies of interest, applying the sinusoids to the system and measuring the change in the magnitude and phase of the input sinusoids. A large loop gain improves the performance of a system. However, it could also lead to instability of

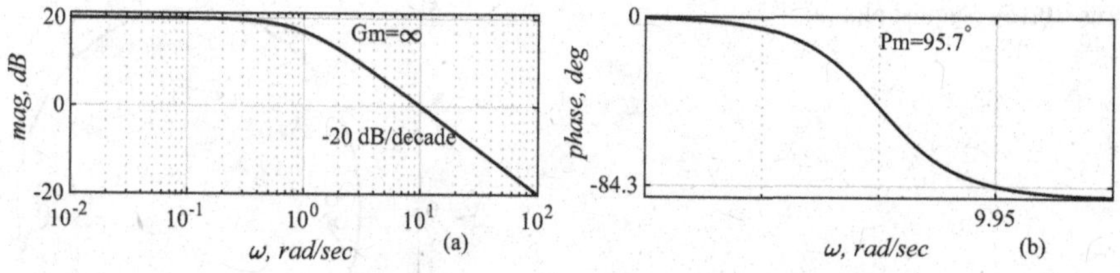

Fig. 10.14 Bode plot of $H(s) = \frac{10}{(s+1)}$ showing gain and phase margins

the system. Bode diagram is one of the methods to test the relative stability of a system.

Using decibels for the magnitude and logarithmic scales for the frequency, the plots of the magnitude and phase responses to sinusoidal inputs are known as Bode plots. These plots can be constructed easily using the asymptotic behavior of the responses and are used widely in practical feedback system analysis. Figure 10.14a and b shows, respectively, the magnitude and phase response of the Bode plot of the loop transfer function $H(s) = \frac{10}{(s+1)}$. It is a frequency response plot with the magnitude represented in decibels and phase represented in degrees with a logarithmic frequency scale. The frequency 1 rad/sec is called the corner frequency at which the slope of the plot changes from approximately 0 dB to −20 db/decade.

The phase crossover frequency is the frequency at which the phase of $H(j\omega)$ is -180 degrees. The gain margin is defined as $-20 \log 10|H(j\omega)|$ at the phase crossover frequency. When the phase never crosses -180 degrees, as in this example, the gain margin is defined as ∞. The gain crossover frequency is the frequency at which the gain of $H(j\omega)$ is 0 dB. The phase margin is defined as $180 + \angle|H(j\omega)|$ at the gain crossover frequency. For example, gain crossover frequency occurs at 9.95 rad/sec, and the phase margin is $180 - 84.3 = 95.7$ degrees. For minimum-phase systems with all the poles and zeros of the transfer function lying in the left half of the s-plane, both the measures have to be positive for a stable system.

10.4.9 The Nyquist Plot

The Nyquist or polar plot is a plot of the magnitude and phase response of the transfer function $H(j\omega)$ in polar coordinates, as the frequency ω varies from $-\infty$ to ∞. The result is that we get a single plot, rather than two plots as in the Bode plot. Figure 10.15 shows the Nyquist plot of the loop transfer function $\frac{1}{(s+1)}$. Each point on the Nyquist plot is the tip of a vector with some magnitude and phase value at a particular frequency ω.

Information about the stability from the Nyquist plot of $G(s)F(s)$ is obtained by observing its behavior about the critical point at $(-1 + j0)$. The plot never crosses

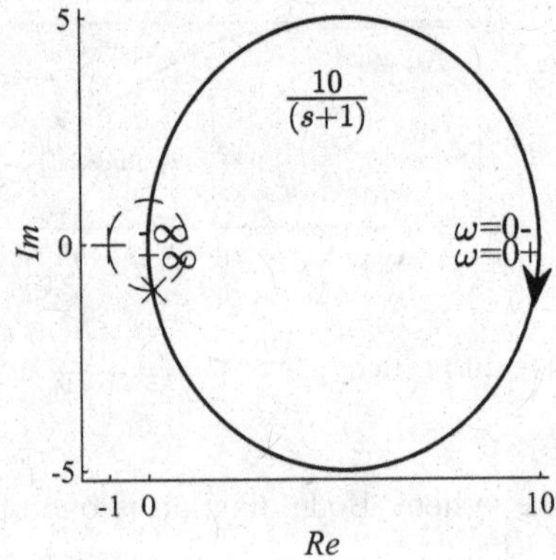

Fig. 10.15 Nyquist plot of
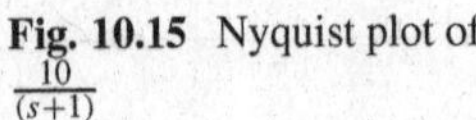
$\frac{10}{(s+1)}$

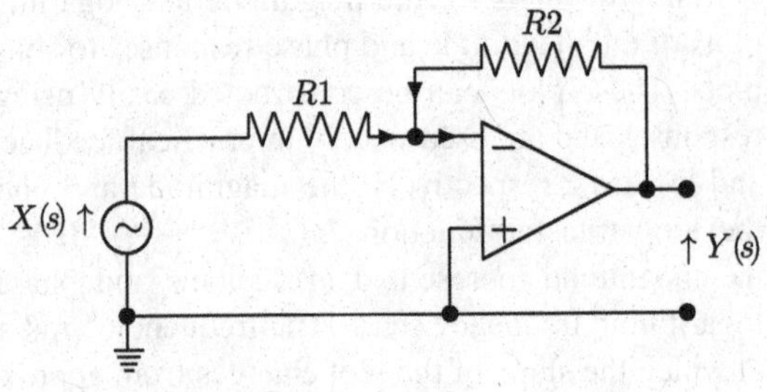

Fig. 10.16 Realization of a scalar multiplier unit using operational amplifier

the negative real axis, and, consequently, the gain margin is ∞. The unit circle is shown in dashed line. The unit circle intersects the plot at the point marked by a cross. The angle between the origin and this point is -84.3 degrees. Therefore, the phase margin is $180 - 84.3 = 95.7$ degrees, as obtained from the Bode plot.

10.4.9.1 Operational Amplifier Circuits

The frequency-domain representation of a scalar multiplier unit using operational amplifier is shown in Fig. 10.16. Operational amplifier circuits, shown in Fig. 10.16 with a triangular symbol, are very large gain (of the order of 10^6) amplifiers with almost infinite input impedance and zero output impedance. There are two input terminals, indicated by the symbols + and – (called, respectively, the noninverting and inverting input terminals), and one output terminal. The output voltage is specified as $v_0 = A(v_+ - v_-)$. As the gain A is very large, the voltage at the inverting terminal, in Fig. 10.16, is very small and can be considered as virtual ground. Further, the large input impedance makes the input terminal current negligible. Therefore, the currents in the forward and feedback paths must be almost equal and

$$\frac{X(s)}{R1} \approx -\frac{Y(s)}{R2}$$

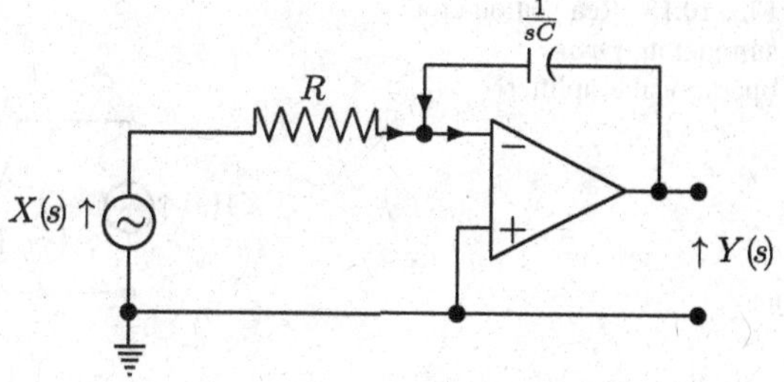

Fig. 10.17 Realization of an integrator unit using operational amplifier

The transfer function of the circuit is, therefore,

$$H(s) = \frac{Y(s)}{X(s)} \approx -\frac{R2}{R1}$$

In general, the elements in the circuit can be impedances, and the transfer function is expressed as

$$H(s) = \frac{Y(s)}{X(s)} \approx -\frac{Z2(s)}{Z1(s)}$$

The transfer function of the integrator circuit, shown in Fig. 10.17, is

$$H(s) = -\frac{Z2(s)}{Z1(s)} = -\frac{1}{sRC}$$

This is an ideal integrator with gain $-\frac{1}{RC}$. Let $x(t) = u(t)$, the unit-step signal. Then, $X(s) = \frac{1}{s}$ and $Y(s) = -\frac{1}{RCs}\frac{1}{s} = -\frac{1}{RCs^2}$. The inverse transform of $Y(s)$ is $y(t) = -\frac{1}{RC}tu(t)$, as the integral of unit step is the unit ramp. Compare this response with that of a passive RC network, $y(t) = (1 - e^{-\frac{t}{RC}})u(t) \approx \frac{1}{RC}tu(t)$. Due to the large gain of the amplifier and the feedback, we get an ideal response. In addition, the amplifier, due to its large input impedance, does not load the source of the input signal much and can feed several circuits at the output.

The output $Y(s)$ of the summer, shown in Fig. 10.18, is given as

$$Y(s) = -\left(\frac{R_f}{R1}X1(s) + \frac{R_f}{R2}X2(s)\right)$$

Remembering that the basic elements of a continuous system are scalar multipliers, integrators, and summers, we can build any system, however complex it may be, using the three operational amplifier circuits described.

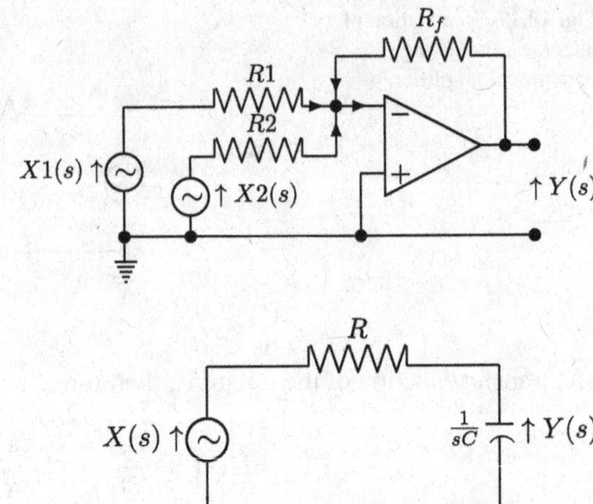

Fig. 10.18 Realization of a summer unit using operational amplifier

Fig. 10.19 The representation of a resistor-capacitor filter circuit in the frequency domain

10.4.10 Analog Filters

We present, in this subsection, an example of the design of lowpass filters. The rectangle, shown in Fig. 10.21 by dashed line, is the magnitude of the frequency response of an ideal analog lowpass filter. As the ideal filter is practically unrealizable, actual filters approximate the ideal filters to a desirable accuracy. While there are several types of filters with different characteristics, we describe the commonly used Butterworth filter.

10.4.10.1 Butterworth Filters

While active filters and digital filters are more commonly used, the word filter instantaneously reminds us the resistor-capacitor lowpass filter circuit shown in Fig. 10.19. The impedance, $\frac{1}{sC}$, of the capacitor is small at higher frequencies compared with that at lower frequencies. Therefore, the voltage across it is composed of high-frequency components with smaller amplitudes than low-frequency components compared with those of the input voltage. The reverse is the case for the voltage across the resistor. For example, there is no steady-state current with DC input (frequency = 0), and, therefore, all the input voltage appears across the capacitor.

In the Laplace transform model of the RC circuit, the input voltage is $X(s)$. The circuit impedance is $R + \frac{1}{sC}$. Therefore, the current in the circuit is $\frac{X(s)}{R+\frac{1}{sC}}$. The output voltage $Y(s)$ across the capacitor is

$$Y(s) = \left(\frac{X(s)}{R + \frac{1}{sC}}\right)\left(\frac{1}{sC}\right)$$

Therefore, the transfer function is

$$H(s) = \frac{Y(s)}{X(s)} = \frac{1}{1 + sRC}$$

Letting $s = j\omega$, we get the frequency response of the filter as

$$H(j\omega) = \frac{1}{1 + j\omega RC}$$

Let the cutoff frequency of the filter be $\omega_c = \frac{1}{RC} = 1$ radian/second. Then,

$$H(j\omega) = \frac{1}{1 + j\frac{\omega}{\omega_c}} = \frac{1}{1 + j\omega} \quad \text{and} \quad |H(j\omega)| = \frac{1}{\sqrt{1 + \omega^2}}$$

The filter circuit is a first-order system and a first-order lowpass Butterworth filter. For a Butterworth filter of order N, the magnitude of the frequency response, with $\omega_c = 1$, is

$$|H(j\omega)| = \frac{1}{\sqrt{1 + \omega^{(2N)}}}$$

The filter with $\omega_c = 1$ is called the normalized filter. From the transfer function of this filter, we can find the transfer function of other types of filters, such as highpass, with arbitrary cutoff frequencies using appropriate frequency transformations.

To find the transfer function of the normalized Butterworth filter, we substitute $\omega = \frac{s}{j}$ in the expression for the squared magnitude of the frequency response and get

$$|H(j\omega)|^2 = H(j\omega)H(-j\omega) = H(s)H(-s) = \frac{1}{1 + \omega^{(2N)}} = \frac{1}{1 + \left(\frac{s}{j}\right)^{(2N)}}$$

The poles of $H(s)H(-s)$ are obtained by solving the equation

$$s^{2N} = -(j)^{2N} = e^{j\pi(2n-1)}(e^{j\frac{\pi}{2}})^{(2N)} = e^{j\pi(2n-1+N)},$$

where n is an integer. Note that $e^{j\pi(2n-1)} = -1$ for an integral n and $e^{j\frac{\pi}{2}} = j$. As the transfer function $H(s)$ is to represent a stable system, all its poles must lie in the left half of the s-plane. Therefore, the poles p_n of $H(s)$, which are the N roots (of the $2N$th roots of $-(j)^{2N}$) with negative real parts, are specified as

$$p_n = e^{\frac{j\pi}{2N}(2n+N-1)}, \quad n = 1, 2, \ldots, N$$

The transfer function is given by

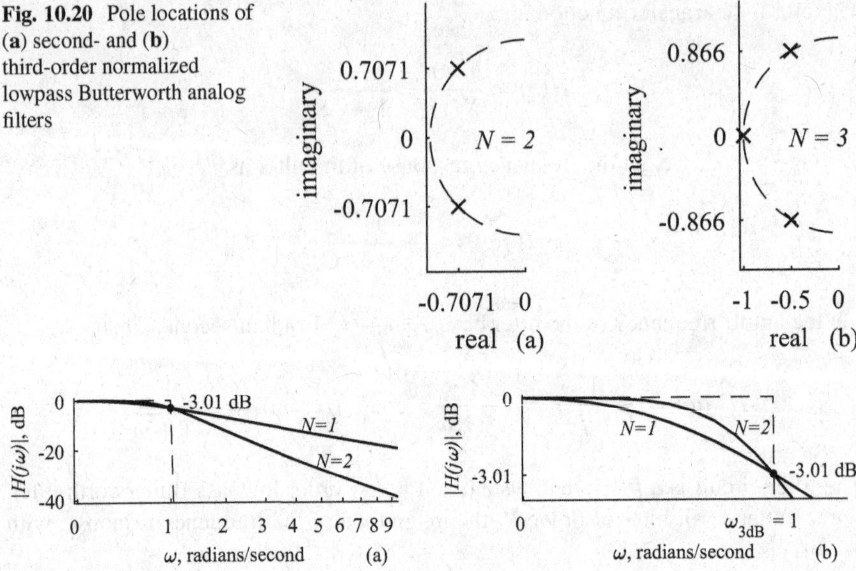

Fig. 10.20 Pole locations of (**a**) second- and (**b**) third-order normalized lowpass Butterworth analog filters

Fig. 10.21 (**a**) The magnitude of the frequency response of the first- and second-order normalized lowpass Butterworth analog filters; (**b**) the passbands are shown in an expanded scale

$$H(s) = \frac{1}{\prod_{n=1}^{N}(s - p_n)}$$

For $N = 1$, the pole is $p_1 = e^{\frac{j\pi}{2(1)}(2(1)+1-1)} = e^{j\pi} = -1$. The transfer function is specified as

$$H(s) = \frac{1}{(s + 1)}$$

The pole locations of the filter for $N = 2$ and $N = 3$ are shown in Fig. 10.20. The symmetrically located poles are equally spaced around the left half of the unit circle. There is a pole on the real axis for N odd.

Consider the magnitude of the frequency response of normalized Butterworth lowpass filters shown in Fig. 10.21. As the frequency response is an even function of ω, the figure shows the response for the positive half of the frequency range only. In both the passband and the stopband, the gain is monotonically decreasing. The asymptotic falloff rate, beyond the 3-dB frequency, is $-6N$ dB per octave (as the frequency is doubled) or $-20N$ dB per decade (as the frequency becomes ten times) approximately, where N is the order of the filter. Normalized filters of any order have the -3 dB ($-10\log_{10}(2)$ to be more precise) or $\frac{1}{\sqrt{2}}$ response point at the same frequency, $\omega_{3dB} = 1$ radian/second. A higher-order filter approximates the ideal response, shown by the dashed line, more closely compared with a lower-order filter.

10.5 Summary

- In this chapter, the theory of the one-sided Laplace transform, its properties, and some of its applications have been described. As practical systems are causal, the one-sided Laplace transform is mostly used in practice.
- The Laplace transform is a generalization of the Fourier transform. The basis waveforms include sinusoids with varying amplitudes or exponentials with complex exponents. The larger set of basis waveforms makes this transform suitable for the analysis of a larger class of signals and systems.
- The Laplace transform corresponding to a signal is a surface, since it is a function of two variables (the real and imaginary parts of the complex frequency). The Laplace transform of a signal along with its ROC uniquely represents the signal in the frequency domain. The spectral values along any straight line in the ROC can be used to reconstruct the corresponding time-domain signal.
- The inverse Laplace transform is defined by an integral in the complex plane. However, the partial fraction method, along with a short list of Laplace transform pairs, is adequate for most practical purposes to find the inverse Laplace transform.
- The Laplace transform is essential for the design and transient and stability analysis of continuous LTI systems. The Laplace transform of the impulse response of a system, the transfer function, is a frequency-domain model of the system.

Exercises

10.1 Find the Laplace transform of the unit-impulse signal, $\delta(t)$, by applying a limiting process to the rectangular pulse, defined as

$$x(t) = \begin{cases} \frac{1}{2a} & \text{for } -a < t < a \\ 0 & \text{otherwise} \end{cases}, \quad a > 0,$$

and its transform, as a tends to zero.

10.2 Find the Laplace transform of the function $x(t)$ using the time-shifting property and the transforms of $u(t)$, $tu(t)$, and $t^2u(t)$.

10.2.1 $x(t) = u(t-5)$.
10.2.2 $x(t) = 2,\ 0 \le t \le 4$, and $x(t) = 0$ otherwise.

10.2.3 $x(t) = 4,\ 1 \le t \le 3$, and $x(t) = 0$ otherwise.
*10.2.4 $x(t) = tu(t-2)$.
10.2.5 $x(t) = (t-3)u(t)$.
10.2.6 $x(t) = 2t^2u(t-2)$.

10.3 Find the Laplace transform of the function $x(t)$ using the frequency-shifting property.

10.3.1 $x(t) = e^{-2t}\cos(3t)u(t)$.
10.3.2 $x(t) = e^{-3t}\sin(2t)u(t)$.

10.4 Find the derivative $\frac{dx(t)}{dt}$ of $x(t)$. Verify that the Laplace transform of $\frac{dx(t)}{dt}$ is $sX(s) - x(0^-)$.

10.4.1 $x(t) = \cos(2t)$.
10.4.2 $x(t) = \cos(3t)u(t)$.
10.4.3 $x(t) = u(t) - u(t-1)$.
10.4.4 $x(t) = 3t(u(t-2) - u(t-4))$.
10.4.5

$$x(t) = \begin{cases} (t-1) & \text{for } t < 1 \\ 2 & \text{for } 1 < t < 3 \\ \cos(\frac{\pi}{3}t) & \text{for } t > 3 \end{cases}$$

10.4.6

$$x(t) = \begin{cases} 2e^t & \text{for } t < 0 \\ 3\sin(t) & \text{for } 0 < t < \frac{\pi}{2} \\ 4u(t-\frac{\pi}{2}) & \text{for } t > \frac{\pi}{2} \end{cases}$$

10.5 Given the Laplace transform $X(s)$ of $x(t)$, find the transform of $x(at)$ using the scaling property. Find the location of the poles and zeros of the two transforms. Find $x(t)$ and $x(at)$.

10.5.1 $X(s) = \frac{s+4}{s^2+5s+6}$ and $a = \frac{1}{2}$.
*10.5.2 $X(s) = \frac{s-1}{s^2+3s+2}$ and $a = 2$.
10.5.3 $X(s) = \frac{s-2}{s^2+1}$ and $a = 3$.

10.6 Using the Laplace transform, find the convolution, $y(t) = x(t) * h(t)$, of the functions $x(t)$ and $h(t)$.

10.6.1 $x(t) = e^{-2t}u(t)$ and $h(t) = u(t)$.
10.6.2 $x(t) = u(t)$ and $h(t) = u(t)$.
10.6.3 $x(t) = e^{3t}u(t)$ and $h(t) = e^{-4t}u(t)$.
10.6.4 $x(t) = e^{-2t}u(t)$ and $x(t) = e^{-2t}u(t)$.

*10.6.5 $x(t) = te^{-t}u(t)$ and $h(t) = e^{-t}u(t)$.
10.6.6 $x(t) = 2u(t-2)$ and $h(t) = 3u(t-3)$.
10.6.7 $x(t) = 2e^{-(t-2)}u(t-2)$ and $h(t) = 5u(t)$.

10.7 Find the Laplace transform of the function $x(t)$ using the multiplication by t property.

10.7.1 $x(t) = 4t\cos(2t)u(t)$.
10.7.2 $x(t) = 5t\sin(3t)u(t)$.

10.8 Find the initial and final values of the function $x(t)$ corresponding to the transform $X(s)$, using the initial and final value properties.

10.8.1 $X(s) = \frac{s+2}{(s+3)}$.
10.8.2 $X(s) = \frac{2}{s+3}$.
10.8.3 $X(s) = \frac{2}{s^2+1}$.
*10.8.4 $X(s) = \frac{3s^2+3s+2}{s(s^2+3s+2)}$.
10.8.5 $X(s) = \frac{s+2}{s(s-2)}$.
10.8.6 $X(s) = \frac{s+1}{(s-1)}$.

10.9 Find the Laplace transform of the semiperiodic signal $x(t)u(t)$, the first period of which is defined as follows.

10.9.1

$$x_1(t) = \begin{cases} 1 \text{ for } 0 < t < 2 \\ -1 \text{ for } 2 < t < 4 \end{cases}$$

10.9.2

$$x_1(t) = t \text{ for } 0 < t < 5$$

*10.9.3

$$x_1(t) = \begin{cases} t \text{ for } 0 < t < 2 \\ 4 - t \text{ for } 2 < t < 4 \end{cases}$$

10.9.4

$$x_1(t) = \sin(\omega t) \text{ for } 0 < t < \frac{\pi}{\omega}$$

10.10 Find the inverse Laplace transform of $X(s)$ using the inverse FT.

10.10.1 $x(t)u(t) \Longleftrightarrow X(s) = \frac{1}{s^2}$.
10.10.2 $x(t)u(t) \Longleftrightarrow X(s) = \frac{1}{s-2}$.
10.10.3 $x(t)u(t) \Longleftrightarrow X(s) = \frac{1}{s+2}$.

10.11 Find the inverse Laplace transform of

$$X(s) = \frac{s}{(s^2 + 3s + 2)}$$

10.12 Find the inverse Laplace transform of

$$X(s) = \frac{3s^2 + 2s + 3}{(s^2 + 5s + 6)}$$

10.13 Find the inverse Laplace transform of

$$X(s) = \frac{2s + 4}{(s^2 + 1)}$$

10.14 Find the inverse Laplace transform of

$$X(s) = \frac{s + 3}{(s^3 + 4s^2 + 5s + 2)}$$

* **10.15** Find the inverse Laplace transform of

$$X(s) = \frac{s + 2}{(s^3 + s^2)}$$

10.16 Find the inverse Laplace transform of

$$X(s) = \frac{s + 2e^{-3s}}{(s + 2)(s + 3)}$$

10.17 Find the inverse Laplace transform of

$$X(s) = \frac{se^{-s}}{(s + 1)(s + 3)}$$

10.18 Using the Laplace transform, derive the closed-form expression for the impulse response $h(t)$ of the system, with input $x(t)$ and output $y(t)$, governed by the given differential equation.

10.18.1 $\frac{d^2y(t)}{dt^2} + 6\frac{dy(t)}{dt} + 8y(t) = \frac{dx(t)}{dt} + x(t)$.
10.18.2 $\frac{d^2y(t)}{dt^2} + 3\frac{dy(t)}{dt} + 2y(t) = x(t)$.
*10.18.3 $\frac{d^2y(t)}{dt^2} - 4\frac{dy(t)}{dt} + 3y(t) = x(t)$.

10.19 Using the Laplace transform, find the zero-input, zero-state, transient, steady-state, and complete responses of the system governed by the differential equation

$$\frac{d^2y(t)}{dt^2} + 6\frac{dy(t)}{dt} + 8y(t) = 2\frac{dx(t)}{dt} + 3x(t)$$

with the initial conditions $y(0^-) = 2$ and $\frac{dy(t)}{dt}|_{t=0^-} = 3$ and the input $x(t) = u(t)$, the unit-step function. Find the initial and final values of the complete and zero-state responses.

* **10.20** Using the Laplace transform, find the zero-input, zero-state, transient, steady-state, and complete responses of the system governed by the differential equation

$$\frac{d^2y(t)}{dt^2} + 2\frac{dy(t)}{dt} + y(t) = x(t)$$

with the initial conditions $y(0^-) = 3$ and $\frac{dy(t)}{dt}|_{t=0^-} = -2$ and the input $x(t) = e^{-2t}u(t)$. Find the initial and final values of the complete and zero-state responses.

10.21 Using the Laplace transform, find the zero-input, zero-state, transient, steady-state, and complete responses of the system governed by the differential equation

$$\frac{d^2y(t)}{dt^2} + 5\frac{dy(t)}{dt} + 6y(t) = x(t)$$

with the initial conditions $y(0^-) = -1$ and $\frac{dy(t)}{dt}|_{t=0^-} = -2$ and the input $x(t) = tu(t)$. Find the initial and final values of the complete and zero-state responses.

10.22 Using the Laplace transform, find the zero-input, zero-state, transient, steady-state, and complete responses of the system governed by the differential equation

$$\frac{d^2y(t)}{dt^2} + 7\frac{dy(t)}{dt} + 12y(t) = x(t)$$

with the initial conditions $y(0^-) = 2$ and $\frac{dy(t)}{dt}|_{t=0^-} = -3$ and the input $x(t) = 2\cos(\frac{2\pi}{4}t - \frac{\pi}{6})u(t)$. Find the initial and final values of the complete and zero-state responses.

10.23 Given the differential equation of a system and the input signal $x(t)$, find the steady-state response of the system.

* 10.23.1 $\frac{dy(t)}{dt} + 0.5y(t) = x(t)$ and $x(t) = 3\cos(0.5t - \frac{\pi}{3})u(t)$.
10.23.2 $\frac{dy(t)}{dt} + y(t) = 2x(t)$ and $x(t) = 2\sin(t + \frac{\pi}{4})u(t)$.
10.23.3 $\frac{dy(t)}{dt} + y(t) = x(t)$ and $x(t) = 3e^{j(\sqrt{3}t - \frac{\pi}{6})}u(t)$.

10.24 Using the Laplace transform, derive the closed-form expression of the impulse response of the cascade system consisting of systems, with input $x(t)$ and output $y(t)$, governed by the given differential equations.

10.24.1 $\frac{dy(t)}{dt} + 2y(t) = \frac{dx(t)}{dt} + x(t)$ and $\frac{dy(t)}{dt} + 3y(t) = 2\frac{dx(t)}{dt} + 3x(t)$.
* 10.24.2 $\frac{dy(t)}{dt} - y(t) = x(t)$ and $\frac{dy(t)}{dt} = x(t)$.
10.24.3 $\frac{dy(t)}{dt} + 3y(t) = 2\frac{dx(t)}{dt} - x(t)$ and $\frac{dy(t)}{dt} + 2y(t) = 3\frac{dx(t)}{dt} + 2x(t)$.

10.25 Using the Laplace transform, derive the closed-form expression of the impulse response of the combined system, connected in parallel, consisting of systems, with input $x(t)$ and output $y(t)$, governed by the given differential equations.

10.25.1 $\frac{dy(t)}{dt} + 2y(t) = \frac{dx(t)}{dt} + x(t)$ and $\frac{dy(t)}{dt} + 3y(t) = 2\frac{dx(t)}{dt} + 3x(t)$.
10.25.2 $\frac{dy(t)}{dt} - y(t) = x(t)$ and $\frac{dy(t)}{dt} = x(t)$.
* 10.25.3 $\frac{dy(t)}{dt} + 3y(t) = 2\frac{dx(t)}{dt} - x(t)$ and $\frac{dy(t)}{dt} + 2y(t) = 3\frac{dx(t)}{dt} + 2x(t)$.

10.26 Using the Laplace transform representation of the circuit elements, find the current in the series resistor-inductor circuit, with $R = 2$ ohms, $L = 3$ henries, and the initial current through the inductor $i(0^-) = 4$ amperes, excited by the input voltage $x(t) = 10u(t)$ volts.

10.27 Using the Laplace transform representation of the circuit elements, find the current in the series resistor-inductor circuit, with $R = 3\Omega$ and $L = 4$ H. Assume zero initial current. The input voltage $x(t) = 10\delta(t)$ V.

10.28 Using the Laplace transform representation of the circuit elements, find the voltage across the capacitor in the series resistor-capacitor circuit, with $R = 2$ ohms,

$C = 1$ farad, and the initial voltage across capacitor $v(0^-) = 1$ volts, excited by the input voltage $x(t) = e^{-t}u(t)$ volts.

* **10.29** Using the Laplace transform representation of the circuit elements, find the voltage across the capacitor in the series resistor-capacitor circuit, with $R = 4\Omega$ and $C = 2$ F. Assume zero initial conditions. The input voltage is $x(t) = \delta(t)$ V.

10.30 Find the response of a differentiator to unit-step input signal:

(i) if the circuit is realized using resistor R and capacitor C and
(ii) if the circuit is realized using resistor R and capacitor C and an operational amplifier.

10.31 Find the transfer function $H(s)$ of fourth- and fifth-order Butterworth normalized lowpass filters.

Chapter 11
State-Space Analysis of Discrete Systems

So far, we studied three types of modeling of systems, the difference equation model, the convolution-summation model, and the transfer function model. Using these models, we concentrated on finding the output of a system corresponding to an input. However, in any system, there are internal variables. For example, the values of currents and voltages at various parts of a circuit are internal variables. The values of these variables are of interest in the analysis and design of systems. These values could indicate whether the components of a system work in their linear range and within their power ratings. Therefore, we need a model that also explicitly includes the internal description of systems. This type of model, which is a generalization of the other models of systems, is called the state-space model. In addition, it is easier to extend this model to the analysis of multi-input and multi-output, nonlinear, and time-varying systems.

In Sect. 11.1, we study the state-space model of some common realizations of systems. The time-domain and frequency-domain solutions of the state equations are presented, respectively, in Sects. 11.2 and 11.3. The linear transformation of state vector to obtain different realizations of systems is described in Sect. 11.4.

11.1 The State-Space Model

Consider the state-space model, shown in Fig. 11.1, of a second-order discrete system characterized by the difference equation

$$y(n) + a_1 y(n-1) + a_0 y(n-2) = b_2 x(n) + b_1 x(n-1) + b_0 x(n-2)$$

In addition to the input $x(n)$ and the output $y(n)$, we have shown two internal variables (called state variables), $q_1(n)$ and $q_2(n)$, of the system. State variables are a minimal set of variables (N for a Nth-order system) of a system so that a

D. Sundararajan, *Signals and Systems*,
https://doi.org/10.1007/978-3-031-19377-4_11

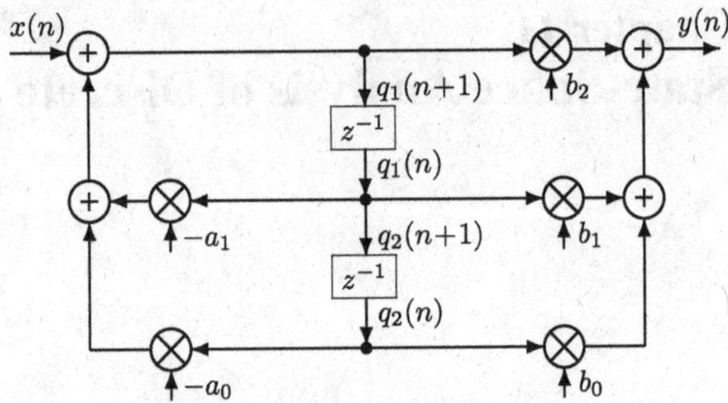

Fig. 11.1 A state-space model of the canonical form I of a second-order discrete system

knowledge of the values of these variables (the state of the system) at $n = k$ and those of the input for $n \geq k$ will enable the determination of the values of the state variables for all $n > k$ and the output for all $n \geq k$. An infinite number of different sets, each of N state variables, are possible for a particular Nth-order system.

From Fig. 11.1, we can write down the following state equations defining the state variables $q_1(n)$ and $q_2(n)$.

$$q_1(n+1) = -a_1q_1(n) - a_0q_2(n) + x(n)$$
$$q_2(n+1) = q_1(n)$$

The $(n+1)$th sample value of each state variable is expressed in terms of the nth sample value of all the state variables and the input. This form of the first-order difference equation is called the standard form. A second-order difference equation characterizing the system, shown in Fig. 11.1, has been decomposed into a set of two simultaneous first-order difference equations. These two equations may be combined into a first-order vector-matrix difference equation.

Selecting state variables as the output of the delay elements is a natural choice, since a delay element is characterized by a first-order difference equation. With that choice, we can write down a state equation at the input of each delay element. However, the state variables need not correspond to quantities that are physically observable in a system. In the state-space model of a system, in general, a Nth-order difference equation characterizing a system is decomposed into a set of N simultaneous first-order difference equations of a standard form. With a set of N simultaneous difference equations, we can solve for N unknowns. These are the N internal variables, called the state variables, of the system. The output is expressed as a linear combination of the state variables and the input. The concepts of impulse response, convolution, and transform analysis are all equally applicable to the state-space model. The difference is that, as the system is modeled using matrix and vector quantities, the system analysis involves matrix and vector quantities. One of the advantages of the state-space model is the easier modeling of systems with multiple inputs and outputs. For simplicity, we describe systems with single input and single output only. The output $y(n)$ of the system, shown in Fig. 11.1, is given by

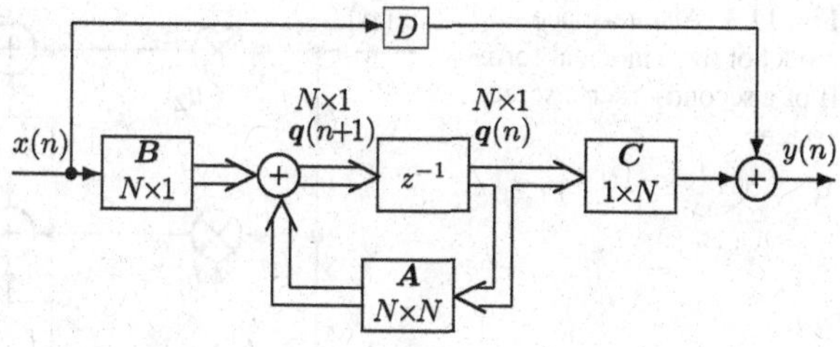

Fig. 11.2 Block diagram representation of the state-space model of a Nth-order system, with single input and single output

$$y(n) = -b_2a_1q_1(n) - b_2a_0q_2(n) + b_1q_1(n) + b_0q_2(n) + b_2x(n)$$

The output equation is an algebraic (not a difference) equation. We can write the state and output equations, using vectors and matrices, as

$$\begin{bmatrix} q_1(n+1) \\ q_2(n+1) \end{bmatrix} = \begin{bmatrix} -a_1 & -a_0 \\ 1 & 0 \end{bmatrix} \begin{bmatrix} q_1(n) \\ q_2(n) \end{bmatrix} + \begin{bmatrix} 1 \\ 0 \end{bmatrix} x(n)$$

$$y(n) = \begin{bmatrix} b_1 - b_2a_1 & b_0 - b_2a_0 \end{bmatrix} \begin{bmatrix} q_1(n) \\ q_2(n) \end{bmatrix} + b_2x(n)$$

Let us define the state vector $\boldsymbol{q}(n)$ as

$$\boldsymbol{q}(n) = \begin{bmatrix} q_1(n) \\ q_2(n) \end{bmatrix}$$

Then, with

$$\boldsymbol{A} = \begin{bmatrix} -a_1 & -a_0 \\ 1 & 0 \end{bmatrix}, \ \boldsymbol{B} = \begin{bmatrix} 1 \\ 0 \end{bmatrix}, \ \boldsymbol{C} = \begin{bmatrix} b_1 - b_2a_1 & b_0 - b_2a_0 \end{bmatrix}, \ D = b_2,$$

the general state-space model description is given as

$$\boldsymbol{q}(n+1) = \boldsymbol{A}\boldsymbol{q}(n) + \boldsymbol{B}x(n)$$
$$y(n) = \boldsymbol{C}\boldsymbol{q}(n) + Dx(n)$$

Block diagram representation of the state-space model of a Nth-order system, with single input and single output, is shown in Fig. 11.2.

Parallel lines terminating with an arrowhead indicate that the signal is a vector quantity.

Example 11.1 Derive the state-space model of the system governed by the difference equation

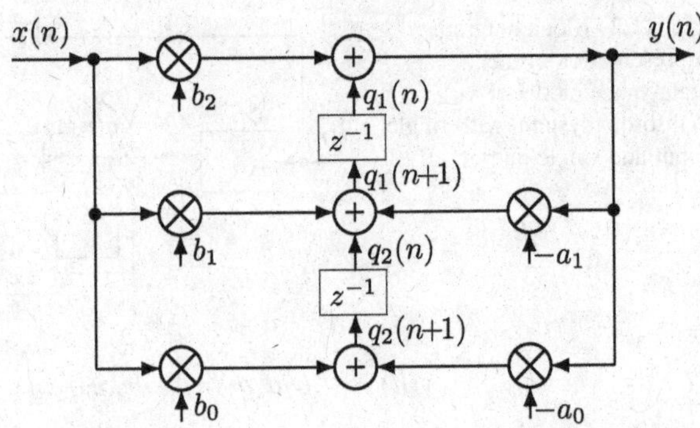

Fig. 11.3 A state-space model of the canonical form II of a second-order discrete system

$$y(n) - 2y(n-1) + 3y(n-2) = 2x(n) - 3x(n-1) + 4x(n-2)$$

Assign the state variables as shown in Fig. 11.1.

Solution With

$$\boldsymbol{A} = \begin{bmatrix} -a_1 & -a_0 \\ 1 & 0 \end{bmatrix} = \begin{bmatrix} 2 & -3 \\ 1 & 0 \end{bmatrix}, \quad \boldsymbol{B} = \begin{bmatrix} 1 \\ 0 \end{bmatrix},$$

$$\boldsymbol{C} = \begin{bmatrix} b_1 - b_2 a_1 & b_0 - b_2 a_0 \end{bmatrix} = \begin{bmatrix} 1 & -2 \end{bmatrix}, \quad D = b_2 = 2,$$

the state-space model of the system is

$$\boldsymbol{q}(n+1) = \begin{bmatrix} q_1(n+1) \\ q_2(n+1) \end{bmatrix} = \begin{bmatrix} 2 & -3 \\ 1 & 0 \end{bmatrix} \begin{bmatrix} q_1(n) \\ q_2(n) \end{bmatrix} + \begin{bmatrix} 1 \\ 0 \end{bmatrix} x(n)$$

$$y(n) = \begin{bmatrix} 1 & -2 \end{bmatrix} \begin{bmatrix} q_1(n) \\ q_2(n) \end{bmatrix} + 2x(n)$$

■

While there are several realizations of a system, some realizations are more commonly used. The realization, shown in Fig. 11.1, is called canonical form I. There is a dual realization that can be derived by using the transpose operation of a matrix. This realization, shown in Fig. 11.3, is called canonical form II and is characterized by the matrices defined, in terms of those of canonical form I, as

$$\overline{\boldsymbol{A}} = \boldsymbol{A}^T, \overline{\boldsymbol{B}} = \boldsymbol{C}^T, \overline{\boldsymbol{C}} = \boldsymbol{B}^T, \overline{D} = D,$$

The state-space model of the canonical form II of the system in Example 11.1 is

$$\boldsymbol{q}(n+1) = \begin{bmatrix} q_1(n+1) \\ q_2(n+1) \end{bmatrix} = \begin{bmatrix} 2 & 1 \\ -3 & 0 \end{bmatrix} \begin{bmatrix} q_1(n) \\ q_2(n) \end{bmatrix} + \begin{bmatrix} 1 \\ -2 \end{bmatrix} x(n)$$

$$y(n) = \begin{bmatrix} 1 & 0 \end{bmatrix} \begin{bmatrix} q_1(n) \\ q_2(n) \end{bmatrix} + 2x(n)$$

11.1.1 Parallel Realization

Consider a system characterized by the transfer function

$$H(z) = \frac{z^3 + z^2 - z + 1}{(z+1)(z^2 + 2z + 3)}$$

The transfer function can be expanded into partial fractions as

$$H(z) = 1 + \frac{1}{(z+1)} + \frac{-3z - 5}{(z^2 + 2z + 3)}$$

The state-space model, shown in Fig. 11.4 using canonical form I, is

$$\mathbf{q}(n+1) = \begin{bmatrix} q_1(n+1) \\ q_2(n+1) \\ q3(n+1) \end{bmatrix} = \begin{bmatrix} -1 & 0 & 0 \\ 0 & -2 & -3 \\ 0 & 1 & 0 \end{bmatrix} \begin{bmatrix} q_1(n) \\ q_2(n) \\ q3(n) \end{bmatrix} + \begin{bmatrix} 1 \\ 1 \\ 0 \end{bmatrix} x(n)$$

$$y(n) = \begin{bmatrix} 1 & -3 & -5 \end{bmatrix} \begin{bmatrix} q_1(n) \\ q_2(n) \\ q3(n) \end{bmatrix} + x(n)$$

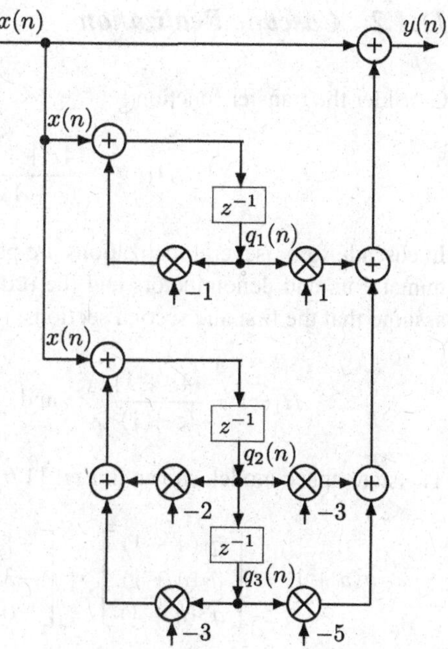

Fig. 11.4 A state-space model of the parallel realization of a third-order discrete system, using canonical form I

Consider the transfer function with a repeated pole

$$H(z) = \frac{2z^3 - z^2 + 3z - 1}{(z+1)(z+2)^2}$$

The transfer function can be expanded into partial fractions as

$$\begin{aligned} H(z) &= 2 + \frac{-7}{(z+1)} + \frac{-4}{(z+2)} + \frac{27}{(z+2)^2} \\ &= 2 + \frac{-7}{(z+1)} + \frac{1}{(z+2)}\left(-4 + \frac{27}{(z+2)}\right) \end{aligned}$$

The state-space model, shown in Fig. 11.5 using canonical form I, is

$$\boldsymbol{q}(n+1) = \begin{bmatrix} q_1(n+1) \\ q_2(n+1) \\ q3(n+1) \end{bmatrix} = \begin{bmatrix} -1 & 0 & 0 \\ 0 & -2 & 0 \\ 0 & 1 & -2 \end{bmatrix} \begin{bmatrix} q_1(n) \\ q_2(n) \\ q3(n) \end{bmatrix} + \begin{bmatrix} 1 \\ 1 \\ 0 \end{bmatrix} x(n)$$

$$y(n) = \begin{bmatrix} -7 & -4 & 27 \end{bmatrix} \begin{bmatrix} q_1(n) \\ q_2(n) \\ q3(n) \end{bmatrix} + 2x(n)$$

11.1.2 Cascade Realization

Consider the transfer function

$$H(z) = \frac{(4z+1)(z^2+3z+2)}{(z+1)(z^2+2z+3)}$$

In cascade form, several realizations are possible depending on the grouping of the numerators and denominators and the order of the sections in the cascade. Let us assume that the first and second sections, respectively, have the transfer functions

$$H_1(z) = \frac{(4z+1)}{(z+1)} \quad \text{and} \quad H_2(z) = \frac{(z^2+3z+2)}{(z^2+2z+3)}$$

The state-space model, shown in Fig. 11.6 using canonical form I, is

$$\boldsymbol{q}(n+1) = \begin{bmatrix} q_1(n+1) \\ q_2(n+1) \\ q3(n+1) \end{bmatrix} = \begin{bmatrix} -1 & 0 & 0 \\ -3 & -2 & -3 \\ 0 & 1 & 0 \end{bmatrix} \begin{bmatrix} q_1(n) \\ q_2(n) \\ q3(n) \end{bmatrix} + \begin{bmatrix} 1 \\ 4 \\ 0 \end{bmatrix} x(n)$$

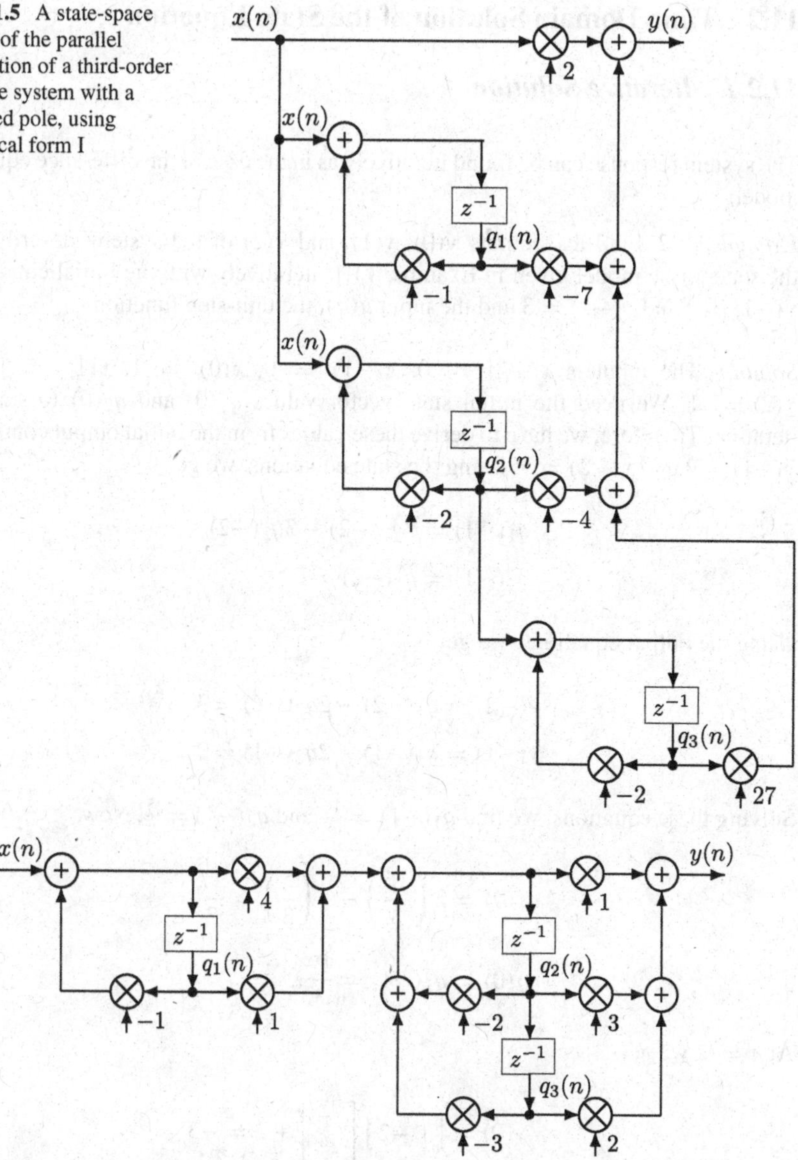

Fig. 11.5 A state-space model of the parallel realization of a third-order discrete system with a repeated pole, using canonical form I

Fig. 11.6 A state-space model of the cascade realization of a third-order discrete system, using canonical form I

$$y(n) = \begin{bmatrix} -3 \; 1 \; -1 \end{bmatrix} \begin{bmatrix} q_1(n) \\ q_2(n) \\ q3(n) \end{bmatrix} + 4x(n)$$

11.2 Time-Domain Solution of the State Equation

11.2.1 Iterative Solution

The system response can be found iteratively as in the case of the difference equation model.

Example 11.2 Find the outputs $y(0)$, $y(1)$, and $y(2)$ of the system, described by the state-space model given in Example 11.1, iteratively with the initial conditions $y(-1) = 2$ and $y(-2) = 3$ and the input $u(n)$, the unit-step function.

Solution The input is $x(-2) = 0$, $x(-1) = 0$, $x(0) = 1$, $x(1) = 1$, and $x(2) = 1$. We need the initial state vector values $q_1(0)$ and $q_2(0)$ to start the iteration. Therefore, we have to derive these values from the initial output conditions $y(-1) = 2$ and $y(-2) = 3$. Using the state equations, we get

$$q_1(-1) = 2q_1(-2) - 3q_2(-2)$$

$$q_2(-1) = q_1(-2)$$

Using the output equations, we get

$$y(-2) = q_1(-2) - 2q_2(-2) = 3$$

$$y(-1) = q_1(-1) - 2q_2(-1) = 2$$

Solving these equations, we find $q_1(-1) = \frac{16}{3}$ and $q_2(-1) = \frac{5}{3}$. Now,

$$q_1(0) = 2\left(\frac{16}{3}\right) - 3\left(\frac{5}{3}\right) = \frac{17}{3}$$

$$q_2(0) = q_1(-1) = \frac{16}{3}$$

At $n = 0$, we get

$$y(0) = \begin{bmatrix} 1 & -2 \end{bmatrix} \begin{bmatrix} \frac{17}{3} \\ \frac{16}{3} \end{bmatrix} + 2 = -3$$

$$\boldsymbol{q}(1) = \begin{bmatrix} q_1(1) \\ q_2(1) \end{bmatrix} = \begin{bmatrix} 2 & -3 \\ 1 & 0 \end{bmatrix} \begin{bmatrix} \frac{17}{3} \\ \frac{16}{3} \end{bmatrix} + \begin{bmatrix} 1 \\ 0 \end{bmatrix} 1 = \begin{bmatrix} -\frac{11}{3} \\ \frac{17}{3} \end{bmatrix}$$

At $n = 1$, we get

$$y(1) = \begin{bmatrix} 1 & -2 \end{bmatrix} \begin{bmatrix} -\frac{11}{3} \\ \frac{17}{3} \end{bmatrix} + 2 = -13$$

$$\boldsymbol{q}(2) = \begin{bmatrix} q_1(2) \\ q_2(2) \end{bmatrix} = \begin{bmatrix} 2 & -3 \\ 1 & 0 \end{bmatrix} \begin{bmatrix} -\frac{11}{3} \\ \frac{17}{3} \end{bmatrix} + \begin{bmatrix} 1 \\ 0 \end{bmatrix} 1 = \begin{bmatrix} -\frac{70}{3} \\ -\frac{11}{3} \end{bmatrix}$$

At $n = 2$, we get

$$y(2) = \begin{bmatrix} 1 & -2 \end{bmatrix} \begin{bmatrix} -\frac{70}{3} \\ -\frac{11}{3} \end{bmatrix} + 2 = -14$$

■

11.2.2 Closed-Form Solution

In the state-space model also, the convolution-summation gives the zero-state response of a system in the time domain. Substituting $n = 0$ in the state equation, we get

$$\boldsymbol{q}(1) = \boldsymbol{A}\boldsymbol{q}(0) + \boldsymbol{B}x(0)$$

Similarly, for $n = 1$ and $n = 2$, we get

$$\begin{aligned} \boldsymbol{q}(2) &= \boldsymbol{A}\boldsymbol{q}(1) + \boldsymbol{B}x(1) \\ &= \boldsymbol{A}(\boldsymbol{A}\boldsymbol{q}(0) + \boldsymbol{B}x(0)) + \boldsymbol{B}x(1) \\ &= \boldsymbol{A}^2\boldsymbol{q}(0) + \boldsymbol{A}\boldsymbol{B}x(0) + \boldsymbol{B}x(1) \\ \boldsymbol{q}(3) &= \boldsymbol{A}\boldsymbol{q}(2) + \boldsymbol{B}x(2) \\ &= \boldsymbol{A}(\boldsymbol{A}^2\boldsymbol{q}(0) + \boldsymbol{A}\boldsymbol{B}x(0) + \boldsymbol{B}x(1)) + \boldsymbol{B}x(2) \\ &= \boldsymbol{A}^3\boldsymbol{q}(0) + \boldsymbol{A}^2\boldsymbol{B}x(0) + \boldsymbol{A}\boldsymbol{B}x(1) + \boldsymbol{B}x(2)\,. \end{aligned}$$

Proceeding in this way, we get the general expression for the state vector as

$$\begin{aligned} \boldsymbol{q}(n) &= \boldsymbol{A}^n\boldsymbol{q}(0) + \boldsymbol{A}^{n-1}\boldsymbol{B}x(0) + \boldsymbol{A}^{n-2}\boldsymbol{B}x(1) + \cdots + \boldsymbol{B}x(n-1) \\ &= \overbrace{\boldsymbol{A}^n\boldsymbol{q}(0)}^{q_{zi}(n)} + \overbrace{\sum_{m=0}^{n-1} \boldsymbol{A}^{n-1-m}\boldsymbol{B}x(m)}^{q_{zs}(n)}, \quad n = 1, 2, 3, \ldots \end{aligned}$$

The first and the second expressions on the right-hand side are, respectively, the zero-input and zero-state components of the state vector $\boldsymbol{q}(n)$. The second expression is the convolution-summation $\boldsymbol{A}^{n-1}u(n-1) * \boldsymbol{B}x(n)$. Convolution of two matrices is similar to multiplication operation of two matrices with the multiplication of the elements replaced by the convolution of the elements. Once we know the state vector, we get the output of the system using the output equation as

$$y(n) = \boldsymbol{C}\boldsymbol{q}(n) + \boldsymbol{D}x(n)$$

$$= \overbrace{\boldsymbol{C}\boldsymbol{A}^n\boldsymbol{q}(0)}^{y_{zi}(n)} + \overbrace{\sum_{m=0}^{n-1} \boldsymbol{C}\boldsymbol{A}^{n-1-m}\boldsymbol{B}x(m) + \boldsymbol{D}x(n)}^{y_{zs}(n)}, \ n = 1, 2, 3, \ldots$$

The term $\boldsymbol{C}\boldsymbol{A}^n\boldsymbol{q}(0)$ is the zero-input component, and the other two terms constitute the zero-state component of the system response $y(n)$. The zero-input response of the system depends solely on the matrix $\boldsymbol{A}^n$. This matrix is called the state transition or fundamental matrix of the system. This matrix, for a Nth-order system, is evaluated, using the Cayley-Hamilton theorem, as

$$\boldsymbol{A}^n = c_0\boldsymbol{I} + c_1\boldsymbol{A} + c_2\boldsymbol{A}^2 + \cdots + c_{N-1}\boldsymbol{A}^{(N-1)}$$

where

$$\begin{bmatrix} c_0 \\ c_1 \\ \cdots \\ c_{N-1} \end{bmatrix} = \begin{bmatrix} 1 & \lambda_1 & \lambda_1^2 & \cdots & \lambda_1^{N-1} \\ 1 & \lambda_2 & \lambda_2^2 & \cdots & \lambda_2^{N-1} \\ & & \cdots & & \\ 1 & \lambda_N & \lambda_N^2 & \cdots & \lambda_N^{N-1} \end{bmatrix}^{-1} \begin{bmatrix} \lambda_1^n \\ \lambda_2^n \\ \cdots \\ \lambda_N^n \end{bmatrix}$$

and $\lambda_1, \lambda_2, \ldots, \lambda_N$ are the N distinct characteristic values of $\boldsymbol{A}$. The characteristic equation of the matrix $\boldsymbol{A}$ is $\det(z\boldsymbol{I} - \boldsymbol{A}) = 0$, where the abbreviation det stands for determinant and $\boldsymbol{I}$ is the identity matrix of the same size of that of $\boldsymbol{A}$. The expanded form of $\det(z\boldsymbol{I} - \boldsymbol{A})$ is a polynomial in z called the characteristic polynomial of $\boldsymbol{A}$. The roots, which are the solutions of the characteristic equation, of this polynomial are the characteristic values of $\boldsymbol{A}$.

For a value λ_r repeated m times, the first row corresponding to that value will remain the same as for a distinct value and the $m - 1$ successive rows will be successive derivatives of the first row with respect to λ_r. For example, with the first value of a fourth-order system repeating two times, we get

$$\begin{bmatrix} c_0 \\ c_1 \\ c_2 \\ c_3 \end{bmatrix} = \begin{bmatrix} 1 & \lambda_1 & \lambda_1^2 & \lambda_1^3 \\ 0 & 1 & 2\lambda_1 & 3\lambda_1^2 \\ 1 & \lambda_2 & \lambda_2^2 & \lambda_2^3 \\ 1 & \lambda_3 & \lambda_3^2 & \lambda_3^3 \end{bmatrix}^{-1} \begin{bmatrix} \lambda_1^n \\ n\lambda_1^{n-1} \\ \lambda_2^n \\ \lambda_3^n \end{bmatrix}$$

Example 11.3 Derive the characteristic polynomial, and determine the characteristic roots of the system with the state-space model as given in Example 11.1.

Solution

$$A = \begin{bmatrix} 2 & -3 \\ 1 & 0 \end{bmatrix}, \quad (zI - A) = z\begin{bmatrix} 1 & 0 \\ 0 & 1 \end{bmatrix} - \begin{bmatrix} 2 & -3 \\ 1 & 0 \end{bmatrix} = \begin{bmatrix} z-2 & 3 \\ -1 & z \end{bmatrix}$$

The characteristic polynomial of the system, given by the determinant of this matrix, is

$$z^2 - 2z + 3$$

The characteristic roots, which are the roots of this polynomial, are

$$\lambda_1 = 1 + j\sqrt{2} \quad \text{and} \quad \lambda_2 = 1 - j\sqrt{2}$$

■

Example 11.4 Find a closed-form expression for the output $y(n)$ of the system, described by the state-space model given in Example 11.1, using the time-domain method, with the initial conditions $y(-1) = 2$ and $y(-2) = 3$ and the input $u(n)$, the unit-step function.

Solution The initial state vector was determined, from the given initial output conditions, in Example 11.2 as

$$q_1(0) = \frac{17}{3}, \quad q_2(0) = \frac{16}{3}$$

The characteristic values, as determined in Example 11.3, are

$$\lambda_1 = 1 + j\sqrt{2} \quad \text{and} \quad \lambda_2 = 1 - j\sqrt{2}$$

The transition matrix is given by

$$\begin{aligned} A^n &= c_0 I + c_1 A \\ &= c_0\begin{bmatrix} 1 & 0 \\ 0 & 1 \end{bmatrix} + c_1\begin{bmatrix} 2 & -3 \\ 1 & 0 \end{bmatrix} = \begin{bmatrix} c_0 + 2c_1 & -3c_1 \\ c_1 & c_0 \end{bmatrix} \end{aligned}$$

where

$$\begin{aligned} \begin{bmatrix} c_0 \\ c_1 \end{bmatrix} &= \begin{bmatrix} 1 & \lambda_1 \\ 1 & \lambda_2 \end{bmatrix}^{-1} \begin{bmatrix} \lambda_1^n \\ \lambda_2^n \end{bmatrix} \\ &= \frac{j}{2\sqrt{2}}\begin{bmatrix} 1 - j\sqrt{2} & -1 - j\sqrt{2} \\ -1 & 1 \end{bmatrix}\begin{bmatrix} (1 + j\sqrt{2})^n \\ (1 - j\sqrt{2})^n \end{bmatrix} \\ &= \frac{j}{2\sqrt{2}}\begin{bmatrix} (1 - j\sqrt{2})(1 + j\sqrt{2})^n + (-1 - j\sqrt{2})(1 - j\sqrt{2})^n \\ -(1 + j\sqrt{2})^n + (1 - j\sqrt{2})^n \end{bmatrix} \end{aligned}$$

$$A^n = \frac{j}{2\sqrt{2}}\begin{bmatrix} -(1+j\sqrt{2})^{(n+1)} + (1-j\sqrt{2})^{(n+1)} & 3(1+j\sqrt{2})^n - 3(1-j\sqrt{2})^n \\ -(1+j\sqrt{2})^n + (1-j\sqrt{2})^n & 3(1+j\sqrt{2})^{(n-1)} - 3(1-j\sqrt{2})^{(n-1)} \end{bmatrix}$$

As a check on $\boldsymbol{A}^n$, verify that $\boldsymbol{A}^n = \boldsymbol{I}$ with $n = 0$ and $\boldsymbol{A}^n = \boldsymbol{A}$ with $n = 1$.

The zero-input component of the state vector is

$$\boldsymbol{q}_{zi}(n) = \boldsymbol{A}^n \boldsymbol{q}(0) = \frac{j}{6\sqrt{2}}\begin{bmatrix} (31 - j17\sqrt{2})(1+j\sqrt{2})^n - (31+j17\sqrt{2})(1-j\sqrt{2})^n \\ (-1-j16\sqrt{2})(1+j\sqrt{2})^n + (1-j16\sqrt{2})(1-j\sqrt{2})^n \end{bmatrix}$$

Using the fact that the sum a complex number and its conjugate is twice the real part of either of the numbers, we get

$$\boldsymbol{q}_{zi}(n) = \begin{bmatrix} \frac{17}{3}(\sqrt{3})^n \cos(\tan^{-1}(\sqrt{2})n) - \frac{31}{3\sqrt{2}}(\sqrt{3})^n \sin(\tan^{-1}(\sqrt{2})n)) \\ \frac{16}{3}(\sqrt{3})^n \cos(\tan^{-1}(\sqrt{2})n) + \frac{1}{3\sqrt{2}}(\sqrt{3})^n \sin(\tan^{-1}(\sqrt{2})n)) \end{bmatrix}$$

The zero-input response $y_{zi}(n)$ is given by

$$\boldsymbol{C}\boldsymbol{A}^n\boldsymbol{q}(0) = \begin{bmatrix} 1 & -2 \end{bmatrix}\begin{bmatrix} \frac{17}{3}(\sqrt{3})^n \cos(\tan^{-1}(\sqrt{2})n) - \frac{31}{3\sqrt{2}}(\sqrt{3})^n \sin(\tan^{-1}(\sqrt{2})n)) \\ \frac{16}{3}(\sqrt{3})^n \cos(\tan^{-1}(\sqrt{2})n) + \frac{1}{3\sqrt{2}}(\sqrt{3})^n \sin(\tan^{-1}(\sqrt{2})n)) \end{bmatrix}$$

$$= (-5(\sqrt{3})^n \cos(\tan^{-1}(\sqrt{2})n) - \frac{11}{\sqrt{2}}(\sqrt{3})^n \sin(\tan^{-1}(\sqrt{2})n))u(n)$$

The first four values of the zero-input response $y_{zi}(n)$ are

$$y_{zi}(0) = -5, \quad y_{zi}(1) = -16, \quad y_{zi}(2) = -17, \quad y_{zi}(3) = 14$$

The zero-state component of the state vector is

$$q_{zs}(n) = \sum_{m=0}^{n-1} \boldsymbol{A}^{n-1-m}\boldsymbol{B}x(m)$$

The convolution-summation, $\boldsymbol{A}^{n-1}u(n-1) * \boldsymbol{B}x(n)$, can be evaluated, using the shift theorem of convolution (Chap. 4), by evaluating $\boldsymbol{A}^n u(n) * \boldsymbol{B}x(n)$ first and then replacing n by $n-1$.

$$\boldsymbol{B}x(n) = \begin{bmatrix} 1 \\ 0 \end{bmatrix} u(n) = \begin{bmatrix} u(n) \\ 0 \end{bmatrix}$$

$$\boldsymbol{A}^n * \boldsymbol{B}x(n) = \frac{j}{2\sqrt{2}}\begin{bmatrix} (-(1+j\sqrt{2})^{(n+1)} + (1-j\sqrt{2})^{(n+1)}) * u(n) \\ (-(1+j\sqrt{2})^n + (1-j\sqrt{2})^n) * u(n) \end{bmatrix}$$

Since the first operand of the convolutions is the sum of two complex conjugate expressions and the convolution of $p(n)$ and $u(n)$ is equivalent to the sum of the first $n+1$ values of $p(n)$, we get

$$\boldsymbol{A}^n * \boldsymbol{B}x(n)$$

$$= \begin{bmatrix} 2\,\mathrm{Re}\left\{\left(-\frac{1}{2} - \frac{j}{2\sqrt{2}}\right)\sum_{m=0}^{n}(1+j\sqrt{2})^m\right\} \\ 2\,\mathrm{Re}\left\{\left(-\frac{j}{2\sqrt{2}}\right)\sum_{m=0}^{n}(1+j\sqrt{2})^m\right\} \end{bmatrix}$$

$$= \begin{bmatrix} 2\,\mathrm{Re}\left\{\left(\frac{1}{2} - \frac{j}{2\sqrt{2}}\right)\left(\frac{1-(1+j\sqrt{2})^{n+1}}{1-(1+j\sqrt{2})}\right)\right\} \\ 2\,\mathrm{Re}\left\{\left(-\frac{j}{2\sqrt{2}}\right)\left(\frac{1-(1+j\sqrt{2})^{n+1}}{1-(1+j\sqrt{2})}\right)\right\} \end{bmatrix}$$

$$= \begin{bmatrix} \frac{1}{2} - \frac{1}{2}(\sqrt{3})^{(n+1)}\cos(\tan^{-1}(\sqrt{2})(n+1)) + \frac{1}{\sqrt{2}}(\sqrt{3})^{(n+1)}\sin(\tan^{-1}(\sqrt{2})(n+1)) \\ \frac{1}{2} - \frac{1}{2}(\sqrt{3})^{(n+1)}\cos(\tan^{-1}(\sqrt{2})(n+1)) \end{bmatrix}$$

Replacing $n = n-1$, we get

$$\boldsymbol{q}_{zs}(n) = \boldsymbol{A}^{n-1} * \boldsymbol{B}x(n)$$

$$= \begin{bmatrix} \frac{1}{2} - \frac{1}{2}(\sqrt{3})^n\cos(\tan^{-1}(\sqrt{2})n) + \frac{1}{\sqrt{2}}(\sqrt{3})^n\sin(\tan^{-1}(\sqrt{2})n) \\ \frac{1}{2} - \frac{1}{2}(\sqrt{3})^n\cos(\tan^{-1}(\sqrt{2})) \end{bmatrix}$$

The zero-state response is given by multiplying the state vector with the $\boldsymbol{C}$ vector and adding the input signal as

$$y_{zs}(n) =$$

$$\begin{bmatrix} 1 & -2 \end{bmatrix}\begin{bmatrix} \frac{1}{2} - \frac{1}{2}(\sqrt{3})^n\cos(\tan^{-1}(\sqrt{2})n) + \frac{1}{\sqrt{2}}(\sqrt{3})^n\sin(\tan^{-1}(\sqrt{2})n) \\ \frac{1}{2} - \frac{1}{2}(\sqrt{3})^n\cos(\tan^{-1}(\sqrt{2})) \end{bmatrix} u(n-1)$$

$$+\,2u(n)$$

$$= \left(-\frac{1}{2} + \frac{1}{2}(\sqrt{3})^n\cos(\tan^{-1}(\sqrt{2})n) + \frac{1}{\sqrt{2}}(\sqrt{3})^n\sin(\tan^{-1}(\sqrt{2})n)\right)u(n-1) + 2u(n)$$

$$= \left(1.5 + \frac{1}{2}(\sqrt{3})^n\cos(\tan^{-1}(\sqrt{2})n) + \frac{1}{\sqrt{2}}(\sqrt{3})^n\sin(\tan^{-1}(\sqrt{2})n)\right)u(n)$$

The first four values of the zero-state response $y_{zs}(n)$ are

$$y_{zs}(0) = 2, \quad y_{zs}(1) = 3, \quad y_{zs}(2) = 3, \quad y_{zs}(3) = 0$$

Adding the zero-input and the zero-state components, we get the total response of the system as

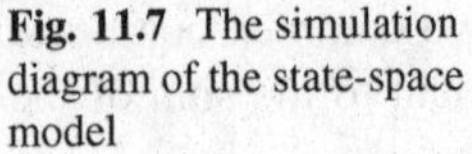

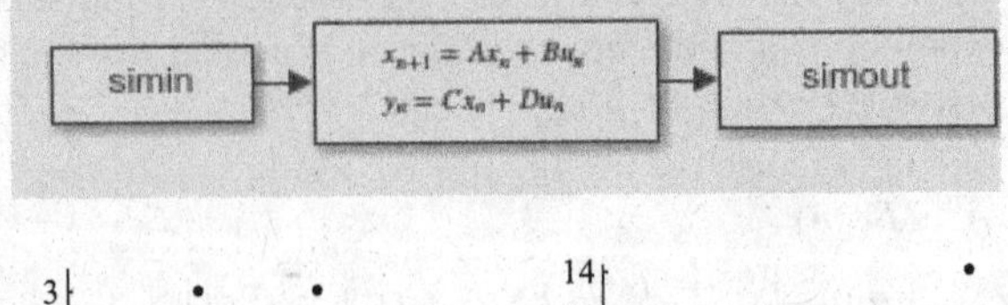

Fig. 11.7 The simulation diagram of the state-space model

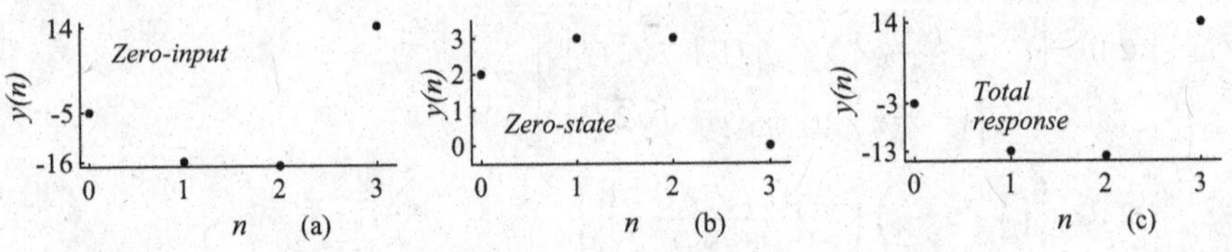

Fig. 11.8 Various components of the response of the system in Example 11.4

$$y(n) = 1.5 - 4.5(\sqrt{3})^n \cos(\tan^{-1}(\sqrt{2})n) - \frac{10}{\sqrt{2}}(\sqrt{3})^n \sin(\tan^{-1}(\sqrt{2})n), \quad n = 0, 1, 2, \ldots$$

The first four values of the total response $y(n)$ are

$$y(0) = -3, \quad y(1) = -13, \quad y(2) = -14, \quad y(3) = 14$$ ■

Figure 11.7 shows the simulation diagram of the state-space model, with initial conditions, producing the total response. The input is the samples of the discrete unit-step signal, which has to be loaded into the simin block by executing the given input program. The initial values of the state variables are set in the simulation block. Various components of the response of the system in Example 11.4 are shown in Fig. 11.8.

11.2.3 The Impulse Response

The impulse response, $h(n)$, is the output of an initially relaxed system with the input $x(n) = \delta(n)$ and is given by

$$h(n) = \sum_{m=0}^{n-1} \boldsymbol{C}\boldsymbol{A}^{n-1-m}\boldsymbol{B}x(m) + Dx(n) = \boldsymbol{C}\boldsymbol{A}^{n-1}\boldsymbol{B}u(n-1) + D\delta(n)$$

Example 11.5 Find the closed-form expression for the impulse response of the system, described by the state-space model given in Example 11.1, using the time-domain method.

Solution The impulse response is given by

$$\begin{aligned}
h(n) &= \boldsymbol{C}\boldsymbol{A}^{n-1}\boldsymbol{B}u(n-1) + D\delta(n)\\
&= \begin{bmatrix}1 & -2\end{bmatrix}\frac{j}{2\sqrt{2}}\begin{bmatrix}(-(1+j\sqrt{2})^n + (1-j\sqrt{2})^n)\\(-(1+j\sqrt{2})^{(n-1)} + (1-j\sqrt{2})^{(n-1)})\end{bmatrix} + 2\delta(n)\\
&= \frac{1}{-j2\sqrt{2}}\left[(1-j\sqrt{2})(1+j\sqrt{2})^{n-1} + (-1-j\sqrt{2})(1-j\sqrt{2})^{n-1}\right] + 2\delta(n)\\
&= 2\delta(n) + \Bigg((\sqrt{3})^{n-1}\cos((\tan^{-1}(\sqrt{2}))(n-1))\\
&\quad - \frac{1}{\sqrt{2}}(\sqrt{3})^{n-1}\sin((\tan^{-1}(\sqrt{2}))(n-1))\Bigg)u(n-1),\ n = 0, 1, 2, \ldots
\end{aligned}$$

The first four values of the impulse response $h(n)$ are

$$h(0) = 2, \quad h(1) = 1, \quad h(2) = 0, \quad h(3) = -3 \quad \blacksquare$$

To get the impulse response by simulation, we set the initial conditions zero and the discrete impulse input.

11.3 Frequency-Domain Solution of the State Equation

The z-transform of a vector function, such as $\boldsymbol{q}(n)$, is defined to be the vector function $\boldsymbol{Q}(z)$, where the elements are the transforms of the corresponding elements of $\boldsymbol{q}(n)$. Taking the z-transform of the state equation, we get

$$z\boldsymbol{Q}(z) - z\boldsymbol{q}(0) = \boldsymbol{A}\boldsymbol{Q}(z) + \boldsymbol{B}X(z)$$

We have used the left shift property of the z-transform, and $\boldsymbol{q}(0)$ is the initial state vector. Solving for $\boldsymbol{Q}(z)$, we get

$$\boldsymbol{Q}(z) = \overbrace{(z\boldsymbol{I} - \boldsymbol{A})^{-1}z\boldsymbol{q}(0)}^{\boldsymbol{Q}_{zi}(z)} + \overbrace{(z\boldsymbol{I} - \boldsymbol{A})^{-1}\boldsymbol{B}X(z)}^{\boldsymbol{Q}_{zs}(z)}$$

The inverse z-transforms of the first and the second expressions on the right-hand side yield, respectively, the zero-input and zero-state components of the state vector $\boldsymbol{q}(n)$. Taking the z-transform of the output equation, we get

$$Y(z) = \boldsymbol{C}\boldsymbol{Q}(z) + DX(z)$$

Now, substituting for $\boldsymbol{Q}(z)$, we get

$$Y(z) = \overbrace{\boldsymbol{C}z(z\boldsymbol{I} - \boldsymbol{A})^{-1}\boldsymbol{q}(0)}^{Y_{zi}(z)} + \overbrace{(\boldsymbol{C}(z\boldsymbol{I} - \boldsymbol{A})^{-1}\boldsymbol{B} + D)X(z)}^{Y_{zs}(z)}$$

The inverse z-transforms of the first and the second expressions on the right-hand side yield, respectively, the zero-input and zero-state components of the system response $y(n)$. The inverse z-transform of $(z(z\boldsymbol{I} - \boldsymbol{A})^{-1})$, by correspondence with the equation for state vector in time domain, is $\boldsymbol{A}^n$, the transition or fundamental matrix of the system. With the system initial conditions zero, the transfer function is given by

$$H(z) = \frac{Y(z)}{X(z)} = (\boldsymbol{C}(z\boldsymbol{I} - \boldsymbol{A})^{-1}\boldsymbol{B} + D)$$

Example 11.6 Find a closed-form expression for the output $y(n)$ of the system, described by the state-space model given in Example 11.1, using the frequency-domain method, with the initial conditions $y(-1) = 2$ and $y(-2) = 3$ and the input $u(n)$, the unit-step function.

Solution The initial state vector

$$\boldsymbol{q}(0) = \begin{bmatrix} \frac{17}{3} \\ \frac{16}{3} \end{bmatrix}$$

is derived in Example 11.2 from the given initial output conditions.

$$(z\boldsymbol{I} - \boldsymbol{A}) = \begin{bmatrix} z-2 & 3 \\ -1 & z \end{bmatrix} \quad \text{and} \quad (z\boldsymbol{I} - \boldsymbol{A})^{-1} = \begin{bmatrix} \frac{z}{z^2-2z+3} & -\frac{3}{z^2-2z+3} \\ \frac{1}{z^2-2z+3} & \frac{z-2}{z^2-2z+3} \end{bmatrix}$$

As a check on $(z\boldsymbol{I} - \boldsymbol{A})^{-1}$, we use the initial value theorem of the z-transform to verify that

$$\lim_{z\to\infty} z(z\boldsymbol{I} - \boldsymbol{A})^{-1} = \boldsymbol{I} = \boldsymbol{A}^0$$

The transform of the zero-input component of the state vector is

$$\begin{aligned} Q_{zi}(z) &= z(z\boldsymbol{I} - \boldsymbol{A})^{-1}\boldsymbol{q}(0) \\ &= z\begin{bmatrix} \frac{z}{z^2-2z+3} & -\frac{3}{z^2-2z+3} \\ \frac{1}{z^2-2z+3} & \frac{z-2}{z^2-2z+3} \end{bmatrix}\begin{bmatrix} \frac{17}{3} \\ \frac{16}{3} \end{bmatrix} = \frac{z}{3}\begin{bmatrix} \frac{17z-48}{z^2-2z+3} \\ \frac{16z-15}{z^2-2z+3} \end{bmatrix} \\ &= \begin{bmatrix} \frac{(\frac{17}{6}+j\frac{31}{6\sqrt{2}})z}{z-1-j\sqrt{2}} + \frac{(\frac{17}{6}-j\frac{31}{6\sqrt{2}})z}{z-1+j\sqrt{2}} \\ \frac{(\frac{8}{3}-j\frac{1}{6\sqrt{2}})z}{z-1-j\sqrt{2}} + \frac{(\frac{8}{3}+j\frac{1}{6\sqrt{2}})z}{z-1+j\sqrt{2}} \end{bmatrix} \end{aligned}$$

Finding the inverse z-transform and simplifying, we get the zero-input component of the state vector as

$$\boldsymbol{q}_{zi}(n) = \begin{bmatrix} \frac{17}{3}(\sqrt{3})^n \cos(\tan^{-1}(\sqrt{2})n) - \frac{31}{3\sqrt{2}}(\sqrt{3})^n \sin(\tan^{-1}(\sqrt{2})n)) \\ \frac{16}{3}(\sqrt{3})^n \cos(\tan^{-1}(\sqrt{2})n) + \frac{1}{3\sqrt{2}}(\sqrt{3})^n \sin(\tan^{-1}(\sqrt{2})n)) \end{bmatrix} u(n)$$

The transform of the zero-state component of the state vector is

$$\begin{aligned} Q_{zs}(z) &= z(z\boldsymbol{I} - \boldsymbol{A})^{-1}\boldsymbol{B}X(z) \\ &= \begin{bmatrix} \frac{z}{z^2-2z+3} & -\frac{3}{z^2-2z+3} \\ \frac{1}{z^2-2z+3} & \frac{z-2}{z^2-2z+3} \end{bmatrix} \begin{bmatrix} \frac{z}{z-1} \\ 0 \end{bmatrix} = z \begin{bmatrix} \frac{z}{(z-1)(z^2-2z+3)} \\ \frac{1}{(z-1)(z^2-2z+3)} \end{bmatrix} \\ &= \begin{bmatrix} \frac{(\frac{1}{2})z}{z-1} - \frac{\frac{1}{4}(1+j\sqrt{2})z}{z-1-j\sqrt{2}} - \frac{\frac{1}{4}(1-j\sqrt{2})z}{z-1+j\sqrt{2}} \\ \frac{(\frac{1}{2})z}{z-1} - \frac{\frac{1}{4}z}{z-1-j\sqrt{2}} - \frac{\frac{1}{4}z}{z-1+j\sqrt{2}} \end{bmatrix} \end{aligned}$$

Finding the inverse z-transform and simplifying, we get the zero-state component of the state vector as

$$\boldsymbol{q}_{zs}(n) = \begin{bmatrix} \frac{1}{2} - \frac{1}{2}(\sqrt{3})^n \cos(\tan^{-1}(\sqrt{2})n) + \frac{1}{\sqrt{2}}(\sqrt{3})^n \sin(\tan^{-1}(\sqrt{2})n) \\ \frac{1}{2} - \frac{1}{2}(\sqrt{3})^n \cos(\tan^{-1}(\sqrt{2})) \end{bmatrix} u(n)$$

Using the output equation, the output can be computed as given in Example 11.4. ∎

Example 11.7 Find a closed-form expression for the impulse response of the system, described by the state-space model given in Example 11.1, using the frequency-domain method.

Solution The transfer function of a system is given by

$$H(z) = (\boldsymbol{C}(z\boldsymbol{I} - \boldsymbol{A})^{-1}\boldsymbol{B} + D)$$

$$H(z) = \begin{bmatrix} 1 & -2 \end{bmatrix} \begin{bmatrix} \frac{z}{z^2-2z+3} & -\frac{3}{z^2-2z+3} \\ \frac{1}{z^2-2z+3} & \frac{z-2}{z^2-2z+3} \end{bmatrix} \begin{bmatrix} 1 \\ 0 \end{bmatrix} + 2 = \frac{z-2}{(z^2-2z+3)} + 2$$

Expanding into partial fractions, we get

$$H(z) = 2 + \frac{0.5 + j\frac{1}{2\sqrt{2}}}{z - 1 - j\sqrt{2}} + \frac{0.5 - j\frac{1}{2\sqrt{2}}}{z - 1 + j\sqrt{2}}$$

Finding the inverse z-transform and simplifying, we get

$$h(n) = 2\delta(n) + ((\sqrt{3})^{n-1}\cos((\tan^{-1}(\sqrt{2}))(n-1))$$

$$-\frac{1}{\sqrt{2}}(\sqrt{3})^{n-1}\sin((\tan^{-1}(\sqrt{2}))(n-1)))u(n-1),\ n = 0, 1, 2, \ldots$$ ■

Example 11.8 Find the zero-input, zero-state, transient, steady-state, and complete responses of the system governed by the difference equation

$$y(n) = 2x(n) - x(n-1) + 3x(n-2) + \frac{9}{20}y(n-1) - \frac{1}{20}y(n-2)$$

with the initial conditions $y(-1) = 3$ and $y(-2) = 2$ and the input $x(n) = u(n)$, the unit-step function.

Solution The corresponding state-space model is

$$\boldsymbol{A} = \begin{bmatrix} 0.4500 & -0.0500 \\ 1.0000 & 0 \end{bmatrix}, \quad \boldsymbol{B} = \begin{bmatrix} 1 \\ 0 \end{bmatrix},$$

$$\boldsymbol{C} = \begin{bmatrix} -0.1000 & 2.9000 \end{bmatrix}, \quad D = 2$$

We need the initial state vector values $q_1(0)$ and $q_2(0)$ to start the iteration. Therefore, we have to derive these values from the initial output conditions $y(-1) = 3$ and $y(-2) = 2$. Using the state equations, we get

$$q_1(-1) = 0.45q_1(-2) - 0.05q_2(-2)$$

$$q_2(-1) = q_1(-2)$$

Using the output equations, we get

$$y(-2) = -0.1q_1(-2) + 2.9q_2(-2) = 2$$

$$y(-1) = -0.1q_1(-1) + 2.9q_2(-1) = 3$$

Solving these equations, we find $q_1(-1) = 0.4360$ and $q_2(-1) = 1.0495$. Now,

$$q_1(0) = 0.45(0.4360) - 1.0495(0.05) = 0.1437$$

$$q_2(0) = q_1(-1) = 0.4360$$

The initial state vector is

$$q(0) = \begin{bmatrix} 0.1437 \\ 0.4360 \end{bmatrix}$$

$$(zI - A) = \begin{bmatrix} z - 0.45 & 0.05 \\ -1 & z \end{bmatrix} \quad \text{and} \quad (zI - A)^{-1} = \begin{bmatrix} \frac{z}{z^2-0.45z+0.05} & -\frac{0.05}{z^2-0.45z+0.05} \\ \frac{1}{z^2-0.45z+0.05} & \frac{z-0.45}{z^2-0.45z+0.05} \end{bmatrix}$$

As a check on $(zI - A)^{-1}$, we use the initial value theorem of the z-transform to verify that

$$\lim_{z\to\infty} z(zI - A)^{-1} = I = A^0$$

The transform of the zero-input component of the state vector is

$$Q_{zi}(z) = z(zI - A)^{-1}q(0)$$

$$= z\begin{bmatrix} \frac{z}{z^2-0.45z+0.05} & -\frac{0.05}{z^2-0.45z+0.05} \\ \frac{1}{z^2-0.45z+0.05} & \frac{z-0.45}{z^2-0.45z+0.05} \end{bmatrix}\begin{bmatrix} 0.1437 \\ 0.4360 \end{bmatrix}$$

$$= z\begin{bmatrix} \frac{0.1437z-0.0218}{z^2-0.45z+0.05} \\ \frac{0.4360z-0.0525}{z^2-0.45z+0.05} \end{bmatrix}$$

$$Y_{zi}(z) = \begin{bmatrix} -0.1 & 2.9 \end{bmatrix} z \begin{bmatrix} \frac{0.1437z-0.0218}{z^2-0.45z+0.05} \\ \frac{0.4360z-0.0525}{z^2-0.45z+0.05} \end{bmatrix} = z\frac{1.25z - 0.1500}{z^2 - 0.45z + 0.05},$$

which is the same as that obtained in Chap. 9 Example 9.11.

The transform of the zero-state component of the state vector is

$$Q_{zs}(z) = (zI - A)^{-1}BX(z)$$

$$= \begin{bmatrix} \frac{z}{z^2-0.45z+0.05} & -\frac{0.05}{z^2-0.45z+0.05} \\ \frac{1}{z^2-0.45z+0.05} & \frac{z-0.45}{z^2-0.45z+0.05} \end{bmatrix}\begin{bmatrix} \frac{z}{z-1} \\ 0 \end{bmatrix} = z\begin{bmatrix} \frac{z}{(z-1)(z^2-0.45z+0.05)} \\ \frac{1}{(z-1)(z^2-0.45z+0.05)} \end{bmatrix}$$

$$Y_{zs}(z) = C(zI - A)^{-1}BX(z) = z\begin{bmatrix} -0.1 & 2.9 \end{bmatrix}\begin{bmatrix} \frac{z}{(z-1)(z^2-0.45z+0.05)} \\ \frac{1}{(z-1)(z^2-0.45z+0.05)} \end{bmatrix}$$

$$= z\frac{-0.1z + 2.9}{(z - 1)(z^2 - 0.45z + 0.05)}$$

Adding the direct input component, we get

$$Y_{zs}(z) = z\left(\frac{-0.1z + 2.9}{z^3 - 1.45z^2 + 0.5z - 0.05} + \frac{2}{z-1}\right) = z\frac{2z^2 - z + 3}{z^3 - 1.45z^2 + 0.5z - 0.05},$$

which is the same as that obtained in Chap. 9 Example 9.11. ■

11.4 Linear Transformation of State Vectors

For a specific input-output relationship of a system, the system can have different internal structures. By a linear transformation of a state vector, we can obtain another vector implying different internal structure of the system. Let us find the state-space model of a system with state vector $\boldsymbol{q}$ using another state vector $\overline{\boldsymbol{q}}$ such that $\overline{\boldsymbol{q}} = \boldsymbol{P}\boldsymbol{q}$ and $\boldsymbol{q} = \boldsymbol{P}^{-1}\overline{\boldsymbol{q}}$, where $\boldsymbol{P}$ is the $N \times N$ transformation matrix and $\boldsymbol{P}^{-1}$ exists. With the new state vector, the state equation can be written as

$$\boldsymbol{P}^{-1}\overline{\boldsymbol{q}}(n+1) = \boldsymbol{A}\boldsymbol{P}^{-1}\overline{\boldsymbol{q}}(n) + \boldsymbol{B}x(n)$$

Premultiplying by $\boldsymbol{P}$, we get

$$\overline{\boldsymbol{q}}(n+1) = \boldsymbol{P}\boldsymbol{A}\boldsymbol{P}^{-1}\overline{\boldsymbol{q}}(n) + \boldsymbol{P}\boldsymbol{B}x(n)$$

With $\overline{\boldsymbol{A}} = \boldsymbol{P}\boldsymbol{A}\boldsymbol{P}^{-1}$ and $\overline{\boldsymbol{B}} = \boldsymbol{P}\boldsymbol{B}$, the state equation can be written as

$$\overline{\boldsymbol{q}}(n+1) = \overline{\boldsymbol{A}}\overline{\boldsymbol{q}}(n) + \overline{\boldsymbol{B}}x(n)$$

With $\overline{\boldsymbol{C}} = \boldsymbol{C}\boldsymbol{P}^{-1}$, the output equation can be written as

$$y(n) = \overline{\boldsymbol{C}}\overline{\boldsymbol{q}}(n) + Dx(n)$$

Some properties of $\boldsymbol{A}$ and $\overline{\boldsymbol{A}}$ matrices can be used to check the computation of $\overline{\boldsymbol{A}}$. The determinants of $\boldsymbol{A}$ and $\overline{\boldsymbol{A}}$ are equal. The determinants of $(z\boldsymbol{I} - \boldsymbol{A})$ and $(z\boldsymbol{I} - \overline{\boldsymbol{A}})$ are the same. The traces (sum of the diagonal elements) of $\boldsymbol{A}$ and $\overline{\boldsymbol{A}}$ are equal.

Example 11.9 Derive the state-space model of the system in Example 11.1 with the new state vector that is related to old state vector as

$$\overline{q}_1(n) = 2q_1(n) + q_2(n)$$

$$\overline{q}_2(n) = q_1(n) - q_2(n)$$

Verify that the transfer function remains the same using either state-space model.

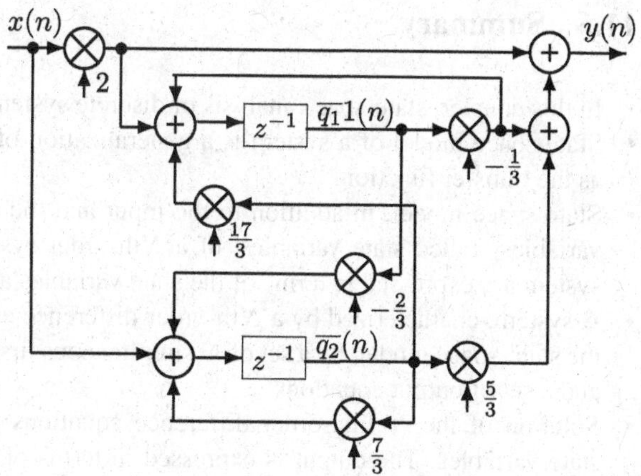

Fig. 11.9 The state-space model of a second-order discrete system with the new state vector

Solution

$$\boldsymbol{P} = \begin{bmatrix} 2 & 1 \\ 1 & -1 \end{bmatrix}, \quad \boldsymbol{P}^{-1} = \begin{bmatrix} \frac{1}{3} & \frac{1}{3} \\ \frac{1}{3} & -\frac{2}{3} \end{bmatrix}$$

$$\overline{\boldsymbol{A}} = \boldsymbol{P}\boldsymbol{A}\boldsymbol{P}^{-1} = \begin{bmatrix} 2 & 1 \\ 1 & -1 \end{bmatrix} \begin{bmatrix} 2 & -3 \\ 1 & 0 \end{bmatrix} \begin{bmatrix} \frac{1}{3} & \frac{1}{3} \\ \frac{1}{3} & -\frac{2}{3} \end{bmatrix} = \begin{bmatrix} -\frac{1}{3} & \frac{17}{3} \\ -\frac{2}{3} & \frac{7}{3} \end{bmatrix}$$

$$\overline{\boldsymbol{B}} = \boldsymbol{P}\boldsymbol{B} = \begin{bmatrix} 2 & 1 \\ 1 & -1 \end{bmatrix} \begin{bmatrix} 1 \\ 0 \end{bmatrix} = \begin{bmatrix} 2 \\ 1 \end{bmatrix}$$

$$\overline{\boldsymbol{C}} = \boldsymbol{C}\boldsymbol{P}^{-1} = \begin{bmatrix} 1 & -2 \end{bmatrix} \begin{bmatrix} \frac{1}{3} & \frac{1}{3} \\ \frac{1}{3} & -\frac{2}{3} \end{bmatrix} = \begin{bmatrix} -\frac{1}{3} & \frac{5}{3} \end{bmatrix}$$

The state-space model of a second-order discrete system with the new state vector is shown in Fig. 11.9. The transfer function, computed using the new state-space model, is

$$H(z) = \begin{bmatrix} -\frac{1}{3} & \frac{5}{3} \end{bmatrix} \begin{bmatrix} z + \frac{1}{3} & -\frac{17}{3} \\ \frac{2}{3} & z - \frac{7}{3} \end{bmatrix}^{-1} \begin{bmatrix} 2 \\ 1 \end{bmatrix} + 2 = \frac{z-2}{(z^2 - 2z + 3)} + 2,$$

which is the same as that obtained in Example 11.7. ■

Diagonalization, controllability, and observability are similar to continuous state-space systems as presented in the next chapter.

11.5 Summary

- In this chapter, state-space analysis of discrete systems has been presented.
- State-space model of a system is a generalization of input-output models, such as the transfer function.
- State-space model, in addition to the input and the output, includes N internal variables, called state variables, of a Nth-order system. All the outputs of the system are expressed in terms of the state variables and the input.
- A system, characterized by a Nth-order difference equation, is characterized, in the state-space model, by a set of N simultaneous first-order difference equations and a set of output equations.
- Solution of the N first-order difference equations yields the values of the N state variables. The output is expressed in terms of these values and the input. Solution of the state equations can be obtained by time-domain or frequency-domain methods.
- The state-space model of a system can be derived from its difference equation, transfer function, or realization diagram.
- The state-space model is not unique, since there are infinite realizations of a system with the same input-output relationship.
- Since it is an internal description of the system, by using linear transformation of the state vector, we can obtain another realization of the system, although of the same input-output relationship, with different characteristics, such as amount of quantization noise, number of components required, sensitivity to parameter variations, etc.
- State-space models can be easily extended to the analysis of time-varying and nonlinear systems and systems with multiple inputs and multiple outputs.

Exercises

11.1 Given the difference equation governing a second-order system, with input $x(n)$ and output $y(n)$, (a) find the state-space model of the system realized as shown in Fig. 11.1, and (b) find the state-space model of the system realized as shown in Fig. 11.3. Find the first four values of the impulse response of the system, iteratively, using both the state-space models, and verify that they are equal.

11.1.1 $y(n) - 5y(n-1) + 3y(n-2) = -6x(n) + 4x(n-1) - 2x(n-2)$
11.1.2 $y(n) + 5y(n-1) + 4y(n-2) = 5x(n) - 2x(n-1) - 6x(n-2)$
11.1.3 $y(n) + 3y(n-1) + 2y(n-2) = 4x(n) - 5x(n-1) + 6x(n-2)$

11.2 Given the difference equation governing a second-order system, (a) find the state-space model of the system realized as shown in Fig. 11.1, and (b) find the state-space model of the system realized as shown in Fig. 11.3. Find the outputs $y(0)$, $y(1)$, and $y(2)$ of the system for the input $x(n)$, iteratively, using both the state-space models, and verify that they are equal. The initial conditions of the system are $y(-1) = 1$ and $y(-2) = 2$.

11.2.1 $y(n) - \frac{5}{4}y(n-1) + \frac{3}{8}y(n-2) = 3x(n) - 4x(n-1) - 2x(n-2)$, $x(n) = \left(\frac{1}{2}\right)^n u(n)$.
11.2.2 $y(n)+2y(n-1)+4y(n-2) = 4x(n)-2x(n-1)-6x(n-2)$, $x(n) = u(n)$.
11.2.3 $y(n) - y(n-1) + 2y(n-2) = 2x(n) - 3x(n-1) + 2x(n-2)$, $x(n) = (-1)^n u(n)$.

11.3 Given the difference equation governing a second-order system, with input $x(n)$ and output $y(n)$, find the state-space model of the system realized as shown in Fig. 11.1. Derive the closed-form expression of the impulse response of the system using the time-domain state-space method. Give the first four values of the impulse response.

* 11.3.1 $y(n) + y(n-1) + \frac{2}{9}y(n-2) = x(n) - 3x(n-1) + 2x(n-2)$.
11.3.2 $y(n) - y(n-1) + y(n-2) = 2x(n) + 3x(n-1) + 4x(n-2)$.
11.3.3 $y(n) + 3y(n-1) + 2y(n-2) = 3x(n) - 4x(n-1) + 2x(n-2)$.

11.4 Given the difference equation governing a second-order system, find the state-space model of the system realized as shown in Fig. 11.1. Derive the closed-form expression of the zero-input and zero-state components of the state vector, the zero-input and zero-state components of the response, and the total response of the system, using the time-domain state-space method, for the input $x(n)$. Give the first four values of the zero-input, zero-state, and total responses. The initial conditions of the system are $y(-1) = -1$ and $y(-2) = 2$.

11.4.1 $y(n) + \frac{5}{6}y(n-1) + \frac{1}{6}y(n-2) = 4x(n) + 2x(n-1) - x(n-2)$, $x(n) = \left(\frac{1}{2}\right)^n u(n)$.
* 11.4.2 $y(n)+y(n-1)+\frac{1}{4}y(n-2) = 2x(n)-x(n-1)+x(n-2)$, $x(n) = u(n)$.
11.4.3 $y(n)+3y(n-1)+2y(n-2) = -2x(n)-x(n-1)+3x(n-2)$, $x(n) = \cos(\frac{2\pi}{4}n)u(n)$.

11.5 Given the difference equation governing a second-order system, with input $x(n)$ and output $y(n)$, find the state-space model of the system realized as shown in Fig. 11.1. Derive the closed-form expression of the impulse response of the system using the frequency-domain state-space method. Give the first four values of the impulse response.

11.5.1 $y(n) + y(n-1) + \frac{2}{9}y(n-2) = x(n) - 2x(n-1) - 2x(n-2)$.
11.5.2 $y(n) - \frac{3}{4}y(n-1) + \frac{1}{8}y(n-2) = 3x(n) - 2x(n-1) + x(n-2)$.
* 11.5.3 $y(n) + \frac{2}{3}y(n-1) + \frac{1}{9}y(n-2) = 2x(n) + x(n-1) + x(n-2)$.
11.5.4 $y(n) + \sqrt{2}y(n-1) + y(n-2) = x(n-1)$.

11.6 Given the difference equation governing a second-order system, find the state-space model of the system realized as shown in Fig. 11.1. Derive the closed-form expression of the zero-input and zero-state components of the state vector, the zero-input and zero-state components of the response, and the total response of the system, using the frequency-domain state-space method, for the given input $x(n)$ and the initial conditions $y(-1)$ and $y(-2)$. Give the first four values of the zero-input, zero-state, and total responses.

11.6.1 $y(n) - \frac{5}{6}y(n-1) + \frac{1}{6}y(n-2) = x(n-1), \quad x(n) = \sin(\frac{2\pi}{4}n)u(n)$, $y(-1) = 0$ and $y(-2) = 0$.

* 11.6.2 $y(n) + y(n-1) + \frac{1}{4}y(n-2) = x(n) + x(n-1) + x(n-2), \quad x(n) = (\frac{1}{3})^n u(n), \quad y(-1) = 1$ and $y(-2) = 1$.

11.6.3 $y(n) + y(n-1) + y(n-2) = x(n) - 2x(n-1) + x(n-2), \quad x(n) = u(n), \quad y(-1) = 2$ and $y(-2) = 1$.

11.7 The state-space model of a system is given. Derive another state-space model of the system using the given transformation matrix $\boldsymbol{P}$. Verify that the transfer function remains the same using either state-space model. Further verify that (i) the traces and determinants of matrices $\boldsymbol{A}$ and $\overline{\boldsymbol{A}}$ are equal and (ii) the determinants of $(z\boldsymbol{I} - \boldsymbol{A})$ and $(z\boldsymbol{I} - \overline{\boldsymbol{A}})$ are the same.

11.7.1

$$\boldsymbol{A} = \begin{bmatrix} 1 & 2 \\ 1 & 3 \end{bmatrix}, \quad \boldsymbol{B} = \begin{bmatrix} 1 \\ 2 \end{bmatrix}, \quad \boldsymbol{C} = \begin{bmatrix} 2 & 2 \end{bmatrix}, \quad D = 1, \quad \boldsymbol{P} = \begin{bmatrix} 1 & 1 \\ 1 & -1 \end{bmatrix}$$

11.7.2

$$\boldsymbol{A} = \begin{bmatrix} 3 & -1 \\ 2 & 3 \end{bmatrix}, \quad \boldsymbol{B} = \begin{bmatrix} 1 \\ 2 \end{bmatrix}, \quad \boldsymbol{C} = \begin{bmatrix} -2 & 1 \end{bmatrix}, \quad D = 3, \quad \boldsymbol{P} = \begin{bmatrix} 2 & 3 \\ 1 & 1 \end{bmatrix}$$

11.7.3

$$\boldsymbol{A} = \begin{bmatrix} 2 & -1 \\ 2 & 1 \end{bmatrix}, \quad \boldsymbol{B} = \begin{bmatrix} 1 \\ -1 \end{bmatrix}, \quad \boldsymbol{C} = \begin{bmatrix} 2 & 3 \end{bmatrix}, \quad D = 3, \quad \boldsymbol{P} = \begin{bmatrix} 0 & 1 \\ 1 & 0 \end{bmatrix}$$

Chapter 12
State-Space Analysis of Continuous Systems

The state-space analysis of continuous systems is similar to that of discrete systems. The realization diagrams are the same with the delay elements replaced by integrators. Therefore, we concentrate, in this chapter, on time-domain and frequency-domain solutions of the state equation. Further, diagonalization, controllability, and observability, which are similar to that of the discrete systems, are presented in detail. The state-space model is presented in Sect. 12.1. Time-domain and frequency-domain solutions of the state equation are presented, respectively, in Sects. 12.2 and 12.3. The linear transformation of state vectors to obtain different realizations of systems is described in Sect. 12.4. The topics of diagonalization, similarity transformation, controllability, and observability are addressed in Sects. 12.5–12.8.

12.1 The State-Space Model

Consider the state-space model, shown in Fig. 12.1, of a second-order continuous system, characterized by the differential equation

$$\ddot{y}(t) + a_1\dot{y}(t) + a_0 y(t) = b_2\ddot{x}(t) + b_1\dot{x}(t) + b_0 x(t)$$

(In this chapter, a dot over a variable indicates its first derivative and two dots indicates its second derivative. For example, $\dot{y}(t) = \frac{dy(t)}{dt}$ and $\ddot{y}(t) = \frac{d^2y(t)}{dt^2}$.) In addition to the input $x(t)$ and the output $y(t)$, we have shown two internal variables (called the state variables), $q_1(t)$ and $q_2(t)$, of the system. State variables are a minimal set of variables (N for a Nth-order system) of a system so that a knowledge of the values of these variables (the state of the system) at $t = t_0$ and those of the input for $t \geq t_0$ will enable the determination of the values of the state variables for

D. Sundararajan, *Signals and Systems*,
https://doi.org/10.1007/978-3-031-19377-4_12

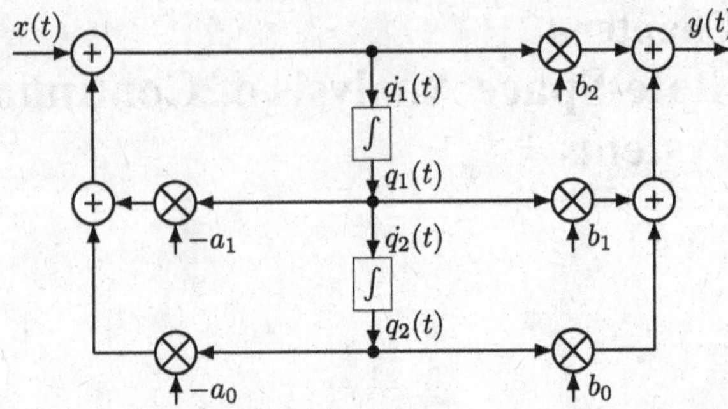

Fig. 12.1 A state-space model of the canonical form I of a second-order continuous system

$t > t_0$ and the output for $t \geq t_0$. An infinite number of different sets, each of N state variables, are possible for a particular Nth-order system.

From Fig. 12.1, we can write down the following state equations defining the state variables $q_1(t)$ and $q_2(t)$.

$$\dot{q}_1(t) = -a_1 q_1(t) - a_0 q_2(t) + x(t)$$
$$\dot{q}_2(t) = q_1(t)$$

The first derivative of each state variable is expressed in terms of all the state variables and the input. No derivatives of either the state variables or the input is permitted, on the right-hand side, to have the equation in a standard form. A second-order differential equation characterizing the system, shown in Fig. 12.1, has been decomposed into a set of two simultaneous first-order differential equations. These two equations may be combined into a first-order vector-matrix differential equation.

Selecting state variables as the output of the integrators is a natural choice, since an integrator is characterized by a first-order differential equation. With that choice, we can write down a state equation at the input of each integrator. However, the state variables need not correspond to quantities that are physically observable in a system. In the state-space model of a system, in general, a Nth-order differential equation characterizing a system is decomposed into a set of N simultaneous first-order differential equations of a standard form. With a set of N simultaneous differential equations, we can solve for N unknowns. These are the N internal variables, called the state variables, of the system. The output is expressed as a linear combination of the state variables and the input. The concepts of impulse response, convolution, and transform analysis are all equally applicable to the state-space model. The difference is that, as the system is modeled using matrix and vector quantities, the system analysis involves matrix and vector quantities. One of the advantages of the state-space model is the easier extension to multiple inputs and outputs. For simplicity, we describe systems with single input and single output only. The output $y(t)$ of the system, shown in Fig. 12.1, is given by

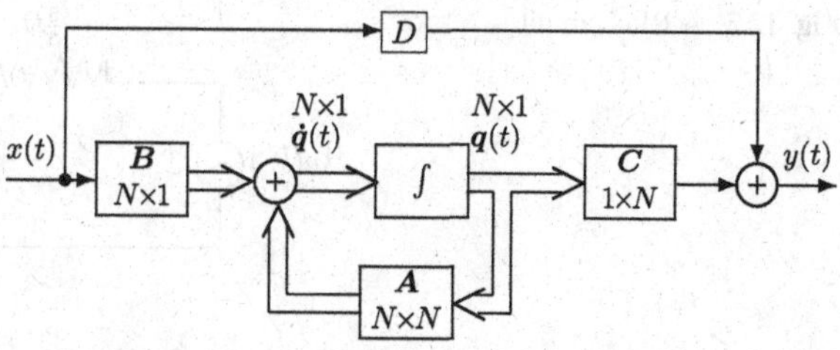

Fig. 12.2 Block diagram representation of the state-space model of a Nth-order continuous system, with single input and single output

$$y(t) = -b_2 a_1 q_1(t) - b_2 a_0 q_2(t) + b_1 q_1(t) + b_0 q_2(t) + b_2 x(t)$$

The output equation is an algebraic (not a differential) equation. We can write the state and output equations, using vectors and matrices, as

$$\begin{bmatrix} \dot{q}_1(t) \\ \dot{q}_2(t) \end{bmatrix} = \begin{bmatrix} -a_1 & -a_0 \\ 1 & 0 \end{bmatrix} \begin{bmatrix} q_1(t) \\ q_2(t) \end{bmatrix} + \begin{bmatrix} 1 \\ 0 \end{bmatrix} x(t)$$

$$y(t) = \begin{bmatrix} b_1 - b_2 a_1 & b_0 - b_2 a_0 \end{bmatrix} \begin{bmatrix} q_1(t) \\ q_2(t) \end{bmatrix} + b_2 x(t)$$

Let us define the state vector $\boldsymbol{q}(t)$ as

$$\boldsymbol{q}(t) = \begin{bmatrix} q_1(t) \\ q_2(t) \end{bmatrix}$$

Then, with

$$\boldsymbol{A} = \begin{bmatrix} -a_1 & -a_0 \\ 1 & 0 \end{bmatrix}, \quad \boldsymbol{B} = \begin{bmatrix} 1 \\ 0 \end{bmatrix},$$

$$\boldsymbol{C} = \begin{bmatrix} b_1 - b_2 a_1 & b_0 - b_2 a_0 \end{bmatrix}, \quad D = b_2,$$

the general state-space model description for continuous systems is given as

$$\dot{\boldsymbol{q}}(t) = \boldsymbol{A}\boldsymbol{q}(t) + \boldsymbol{B}x(t)$$

$$y(t) = \boldsymbol{C}\boldsymbol{q}(t) + \boldsymbol{D}x(t)$$

The block diagram representation of the state-space model of a Nth-order continuous system, with single input and single output, is shown in Fig. 12.2. Parallel lines terminating with an arrowhead indicate that the signal is a vector quantity.

Fig. 12.3 A RLC circuit

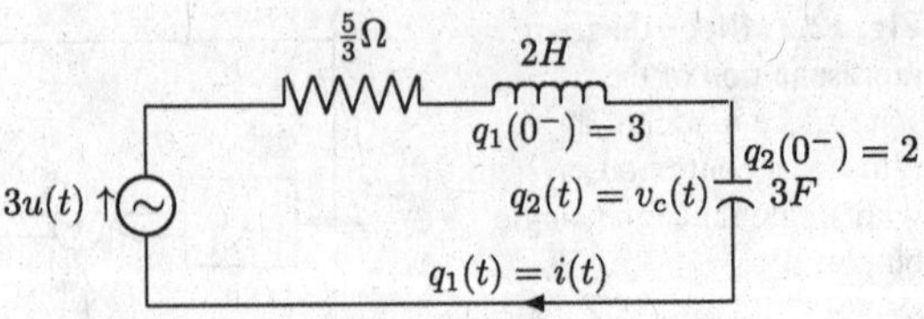

Example 12.1 Consider the RLC circuit analyzed in Example 10.5, shown in Fig. 12.3. It is a series circuit with a resistor of $\frac{5}{3}$ ohms, an inductance of 2 henries, and a capacitor of 3 farads. The initial current through the inductor is 3 amperes, and the initial voltage across the capacitor is 2 volts. This circuit is excited with a voltage source of $3u(t)$ volts. Assuming the capacitor voltage as the output of the circuit, find a state-space model of the circuit.

Solution The first step in finding the model is to write down the equations governing the circuit using the circuit theorems and the input-output behavior of the components. Let the current through the circuit be $i(t)$, the voltage across the capacitor be $v_c(t)$, the input voltage be $x(t)$, and the output voltage be $y(t)$. The sum of the voltages across the components of the circuit must be equal to the input voltage. Therefore, we get

$$2\dot{i}(t) + \frac{5}{3}i(t) + v_c(t) = x(t)$$

The current in the circuit is given by $i(t) = 3\dot{v}_c(t)$. The next step is to select the minimum set of state variables required. Let the current through the inductor, $i(t)$, be the first state variable $q_1(t)$. Let the capacitor voltage, $v_c(t)$, be the second state variable $q_2(t)$. The next step is to substitute the state variables for the variables in the circuit differential equations. After substituting, these equations are rearranged such that only the first derivatives of the state variables appear on the left side and no derivatives appear on the right side. For this example, we get

$$\dot{q}_1(t) = -\frac{5}{6}q_1(t) - \frac{1}{2}q_2(t) + \frac{1}{2}x(t)$$

$$\dot{q}_2(t) = \frac{1}{3}q_1(t)$$

These are the state equations of the circuit. The output equation of the circuit is $y(t) = v_c(t) = q_2(t)$. Using matrices, we get the state-space model as

$$\dot{\mathbf{q}}(t) = \begin{bmatrix} \dot{q}_1(t) \\ \dot{q}_2(t) \end{bmatrix} = \begin{bmatrix} -\frac{5}{6} & -\frac{1}{2} \\ \frac{1}{3} & 0 \end{bmatrix} \begin{bmatrix} q_1(t) \\ q_2(t) \end{bmatrix} + \begin{bmatrix} \frac{1}{2} \\ 0 \end{bmatrix} x(t)$$

$$y(t) = \begin{bmatrix} 0 & 1 \end{bmatrix} \begin{bmatrix} q_1(t) \\ q_2(t) \end{bmatrix}$$

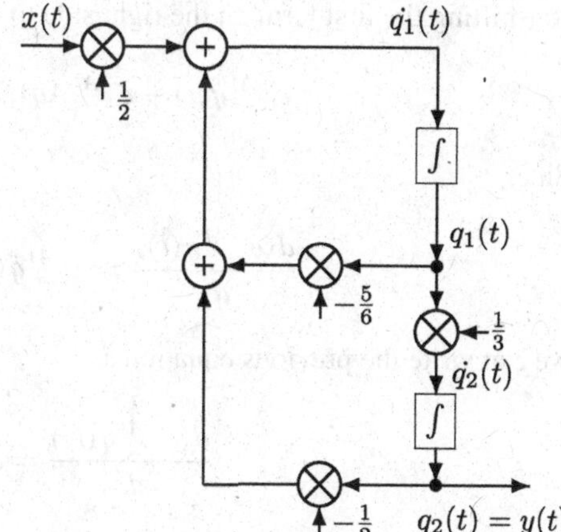

Fig. 12.4 State-space model of the RCL circuit shown in Fig. 12.3

Therefore,

$$\boldsymbol{A} = \begin{bmatrix} -\frac{5}{6} & -\frac{1}{2} \\ \frac{1}{3} & 0 \end{bmatrix}, \quad \boldsymbol{B} = \begin{bmatrix} \frac{1}{2} \\ 0 \end{bmatrix}, \boldsymbol{C} = \begin{bmatrix} 0\ 1 \end{bmatrix}, \quad D = 0$$

The state-space model of the RCL circuit is shown in Fig. 12.4. It is similar to that shown in Fig. 12.1, except that there is a multiplier with coefficient $\frac{1}{3}$ between the two integrators. ■

12.2 Time-Domain Solution of the State Equation

We have to find the solution to the state equation. For this purpose, we need the exponential of a matrix $e^{\boldsymbol{A}t}$ and its derivative. Similar to the infinite series defining an exponential of a scalar,

$$e^{\boldsymbol{A}t} = \boldsymbol{I} + \boldsymbol{A}t + \boldsymbol{A}^2\frac{t^2}{2!} + \boldsymbol{A}^3\frac{t^3}{3!} + \cdots$$

This series is absolutely and uniformly convergent for all values of t. Therefore, it can be differentiated or integrated term by term.

$$\frac{d(e^{\boldsymbol{A}t})}{dt} = \boldsymbol{A} + \boldsymbol{A}^2t + \boldsymbol{A}^3\frac{t^2}{2!} + \boldsymbol{A}^4\frac{t^3}{3!} + \cdots = \boldsymbol{A}e^{\boldsymbol{A}t} = e^{\boldsymbol{A}t}\boldsymbol{A}$$

By premultiplying both sides of state equation by $e^{-\boldsymbol{A}t}$, we get

$$e^{-\boldsymbol{A}t}\dot{\boldsymbol{q}}(t) = e^{-\boldsymbol{A}t}\boldsymbol{A}\boldsymbol{q}(t) + e^{-\boldsymbol{A}t}\boldsymbol{B}x(t)$$

By shifting the first term on the right side to the left, we get

$$e^{-\mathbf{A}t}\dot{\boldsymbol{q}}(t) - e^{-\mathbf{A}t}\boldsymbol{A}\boldsymbol{q}(t) = e^{-\mathbf{A}t}\boldsymbol{B}x(t)$$

Since

$$\frac{d(e^{-\mathbf{A}t}\boldsymbol{q}(t))}{dt} = e^{-\mathbf{A}t}\dot{\boldsymbol{q}}(t) - e^{-\mathbf{A}t}\boldsymbol{A}\boldsymbol{q}(t),$$

we can write the previous equation as

$$\frac{d(e^{-\mathbf{A}t}\boldsymbol{q}(t))}{dt} = e^{-\mathbf{A}t}\boldsymbol{B}x(t)$$

Integrating both sides of this equation from 0^- to t, we get

$$e^{-\mathbf{A}t}\boldsymbol{q}(t)|_{0^-}^{t} = \int_{0^-}^{t} e^{-\mathbf{A}\tau}\boldsymbol{B}x(\tau)d\tau$$

Applying the limit and then premultiplying both sides by $e^{\mathbf{A}t}$, we get

$$\boldsymbol{q}(t) = \overbrace{e^{\mathbf{A}t}\boldsymbol{q}(0^-)}^{q_{zi}(t)} + \overbrace{\int_{0^-}^{t} e^{\mathbf{A}(t-\tau)}\boldsymbol{B}x(\tau)d\tau}^{q_{zs}(t)}$$

The first and the second expressions on the right-hand side are, respectively, the zero-input and zero-state components of the state vector $\boldsymbol{q}(t)$. Note that the part of the expression

$$\int_{0^-}^{t} e^{\mathbf{A}(t-\tau)}\boldsymbol{B}x(\tau)d\tau$$

is the convolution of the matrices $e^{\mathbf{A}t}$ and $\boldsymbol{B}x(t)$, $e^{\mathbf{A}t} * \boldsymbol{B}x(t)$. Convolution of matrices is the same as the multiplication of two matrices, except that the product of two elements is replaced by their convolution. If the initial state vector values are given at $t = t_0^-$, rather than at $t = 0^-$, the state equation is modified as

$$\boldsymbol{q}(t) = e^{\mathbf{A}(t-t_0)}\boldsymbol{q}(t_0^-) + \int_{t_0^-}^{t} e^{\mathbf{A}(t-\tau)}\boldsymbol{B}x(\tau)d\tau$$

The matrix $e^{\mathbf{A}t}$ is called the state-transition matrix or the fundamental matrix of the system.

Once we know the state vector, we get the output of the system using the output equation as

$$y(t) = C\left(e^{At}q(0^-) + \int_{0^-}^{t} e^{A(t-\tau)}Bx(\tau)d\tau\right) + Dx(t)$$

$$= \overbrace{Ce^{At}q(0^-)}^{y_{zi}(t)} + \overbrace{C\int_{0^-}^{t} e^{A(t-\tau)}Bx(\tau)d\tau + Dx(t)}^{y_{zs}(t)}$$

The first expression on the right-hand side is the zero-input component of the system response $y(t)$, and the other two expressions yield the zero-state component. The zero-input response of the system depends solely on the state-transition matrix e^{At}. This matrix, for a Nth-order system, is evaluated, using the Cayley-Hamilton theorem, as

$$e^{At} = c_0 I + c_1 A + c_2 A^2 + \cdots + c_{N-1} A^{(N-1)}$$

where

$$\begin{bmatrix} c_0 \\ c_1 \\ \cdots \\ c_{N-1} \end{bmatrix} = \begin{bmatrix} 1 & \lambda_1 & \lambda_1^2 & \cdots & \lambda_1^{N-1} \\ 1 & \lambda_2 & \lambda_2^2 & \cdots & \lambda_2^{N-1} \\ & & \cdots & & \\ 1 & \lambda_N & \lambda_N^2 & \cdots & \lambda_N^{N-1} \end{bmatrix}^{-1} \begin{bmatrix} e^{\lambda_1 t} \\ e^{\lambda_2 t} \\ \cdots \\ e^{\lambda_N t} \end{bmatrix}$$

and $\lambda_1, \lambda_2, \ldots, \lambda_N$ are the N distinct characteristic roots of A. For a root λ_r repeated m times, the first row corresponding to that root will remain the same as for a distinct root, and the $m - 1$ successive rows will be successive derivatives of the first row with respect to λ_r. For example, with the first root of a fourth-order system repeating two times, we get

$$\begin{bmatrix} c_0 \\ c_1 \\ c_2 \\ c_3 \end{bmatrix} = \begin{bmatrix} 1 & \lambda_1 & \lambda_1^2 & \lambda_1^3 \\ 0 & 1 & 2\lambda_1 & 3\lambda_1^2 \\ 1 & \lambda_2 & \lambda_2^2 & \lambda_2^3 \\ 1 & \lambda_3 & \lambda_3^2 & \lambda_3^3 \end{bmatrix}^{-1} \begin{bmatrix} e^{\lambda_1 t} \\ te^{\lambda_1 t} \\ e^{\lambda_2 t} \\ e^{\lambda_3 t} \end{bmatrix}$$

Example 12.2 Find a closed-form expression for the output $y(t)$ of the system, described by the differential equation

$$\ddot{y}(t) + 4\dot{y}(t) + 4y(t) = \ddot{x}(t) + \dot{x}(t) + 2x(t)$$

using the time-domain method, with the initial conditions $y(0^-) = 2$ and $\dot{y}(0^-) = 3$ and the input $u(t)$, the unit-step function. Assume canonical form I realization of the system as shown in Fig. 12.1.

Solution

$$\boldsymbol{A} = \begin{bmatrix} -4 & -4 \\ 1 & 0 \end{bmatrix}, \quad \boldsymbol{B} = \begin{bmatrix} 1 \\ 0 \end{bmatrix}, \boldsymbol{C} = \begin{bmatrix} -3 & -2 \end{bmatrix}, \quad D = 1$$

The initial state vector has to be found from the given initial output conditions using the state and output equations. From the state equation, we get

$$\dot{q}_1(0^-) = -4q_1(0^-) - 4q_2(0^-)$$

$$\dot{q}_2(0^-) = q_1(0^-)$$

Note that the input $x(t)$ is zero at $t = 0^-$. From the output equation, we get

$$-3\dot{q}_1(0^-) - 2\dot{q}_2(0^-) = 3$$

$$-3q_1(0^-) - 2q_2(0^-) = 2$$

Solving these equations, we get the initial state vector as

$$q_1(0^-) = -\frac{15}{8}, \qquad q_2(0^-) = \frac{29}{16}$$

The characteristic polynomial of a system is given by the determinant of the matrix $(s\boldsymbol{I} - \boldsymbol{A})$, where $\boldsymbol{I}$ is the identity matrix of the same size as $\boldsymbol{A}$. While we can write down the characteristic polynomial from the differential equation, we just show how it can be found using the matrix $\boldsymbol{A}$. For this example,

$$(s\boldsymbol{I} - \boldsymbol{A}) = s\begin{bmatrix} 1 & 0 \\ 0 & 1 \end{bmatrix} - \begin{bmatrix} -4 & -4 \\ 1 & 0 \end{bmatrix} = \begin{bmatrix} s+4 & 4 \\ -1 & s \end{bmatrix}$$

The characteristic polynomial of the system, given by the determinant of this matrix, is

$$s^2 + 4s + 4$$

With each of the infinite different realizations of a system, we get the $\boldsymbol{A}$ matrix with different values. However, as the system is the same, its characteristic polynomial, given by the determinant of $(s\boldsymbol{I} - \boldsymbol{A})$, will be the same for any valid $\boldsymbol{A}$. The characteristic roots, which are the roots of this polynomial, are $\lambda_1 = -2$ and $\lambda_2 = -2$. The transition matrix is given by

$$e^{\boldsymbol{A}t} = c_0\boldsymbol{I} + c_1\boldsymbol{A}$$

where

$$\begin{bmatrix} c_0 \\ c_1 \end{bmatrix} = \begin{bmatrix} 1 & \lambda_1 \\ 0 & 1 \end{bmatrix}^{-1} \begin{bmatrix} e^{\lambda_1 t} \\ te^{\lambda_1 t} \end{bmatrix} = \begin{bmatrix} 1 & 2 \\ 0 & 1 \end{bmatrix} \begin{bmatrix} e^{-2t} \\ te^{-2t} \end{bmatrix}$$
$$= \begin{bmatrix} e^{-2t} + 2te^{-2t} \\ te^{-2t} \end{bmatrix}$$

$$e^{At} = c_0 \begin{bmatrix} 1 & 0 \\ 0 & 1 \end{bmatrix} + c_1 \begin{bmatrix} -4 & -4 \\ 1 & 0 \end{bmatrix} = \begin{bmatrix} c_0 - 4c_1 & -4c_1 \\ c_1 & c_0 \end{bmatrix}$$
$$= \begin{bmatrix} e^{-2t} - 2te^{-2t} & -4te^{-2t} \\ te^{-2t} & e^{-2t} + 2te^{-2t} \end{bmatrix}$$

Since $\boldsymbol{q}(t) = e^{\boldsymbol{A}t}\boldsymbol{q}(0)$, with $t = 0$, we get $\boldsymbol{q}(0) = e^{\boldsymbol{A}0}\boldsymbol{q}(0)$. That is, $\boldsymbol{I} = e^{\boldsymbol{A}0}$. This result, which can be used to check the state-transition matrix, is also obvious from the infinite series for $e^{\boldsymbol{A}t}$.

The state vector $\boldsymbol{q}(t)$ can be computed as follows.

$$\begin{aligned}
\boldsymbol{q}(t) &= \begin{bmatrix} e^{-2t} - 2te^{-2t} & -4te^{-2t} \\ te^{-2t} & e^{-2t} + 2te^{-2t} \end{bmatrix} \begin{bmatrix} -\frac{15}{8} \\ \frac{29}{16} \end{bmatrix} \\
&+ \begin{bmatrix} e^{-2t} - 2te^{-2t} & -4te^{-2t} \\ te^{-2t} & e^{-2t} + 2te^{-2t} \end{bmatrix} * \left(\begin{bmatrix} 1 \\ 0 \end{bmatrix} u(t) \right) \\
&= \begin{bmatrix} e^{-2t} - 2te^{-2t} & -4te^{-2t} \\ te^{-2t} & e^{-2t} + 2te^{-2t} \end{bmatrix} \begin{bmatrix} -\frac{15}{8} \\ \frac{29}{16} \end{bmatrix} \\
&+ \begin{bmatrix} \int_0^t (e^{-2\tau} - 2\tau e^{-2\tau}) d\tau \\ \int_0^t (\tau e^{-2\tau}) d\tau \end{bmatrix} \\
&= \begin{bmatrix} -\frac{15}{8}e^{-2t} - \frac{7}{2}te^{-2t} \\ \frac{29}{16}e^{-2t} + \frac{7}{4}te^{-2t} \end{bmatrix} + \begin{bmatrix} te^{-2t} \\ -\frac{1}{4}e^{-2t} - \frac{1}{2}te^{-2t} + \frac{1}{4} \end{bmatrix} \\
&= \begin{bmatrix} -\frac{15}{8}e^{-2t} - \frac{5}{2}te^{-2t} \\ \frac{1}{4} + \frac{25}{16}e^{-2t} + \frac{5}{4}te^{-2t} \end{bmatrix}
\end{aligned}$$

The output $y(t)$ can be computed using the output equation. The zero-input component of the output is given by

$$y_{zi}(t) = \begin{bmatrix} -3 & -2 \end{bmatrix} \begin{bmatrix} -\frac{15}{8}e^{-2t} - \frac{7}{2}te^{-2t} \\ \frac{29}{16}e^{-2t} + \frac{7}{4}te^{-2t} \end{bmatrix}$$
$$= 2e^{-2t} + 7te^{-2t}$$

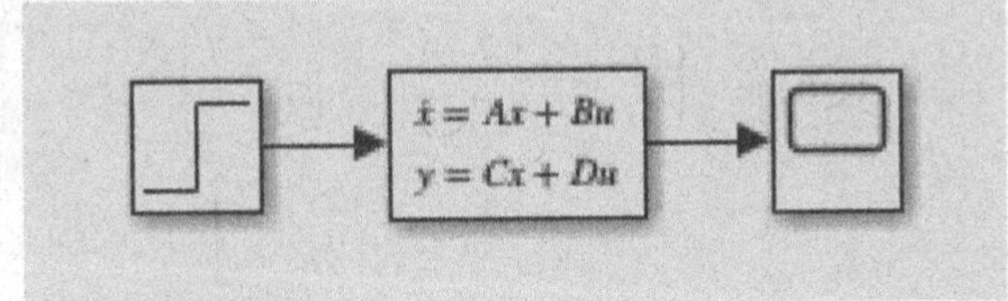

Fig. 12.5 The simulation diagram of the transfer function

The zero-state component of the output is given by

$$y_{zs}(t) = \begin{bmatrix} -3 & -2 \end{bmatrix} \begin{bmatrix} te^{-2t} \\ -\frac{1}{4}e^{-2t} - \frac{1}{2}te^{-2t} + \frac{1}{4} \end{bmatrix} + 1$$
$$= 0.5 + 0.5e^{-2t} - 2te^{-2t}$$

The total response of the system is the sum of the zero-input and zero-state components of the response and is given as

$$y(t) = (0.5 + 2.5e^{-2t} + 5te^{-2t})u(t)$$ ■

The simulation diagram of the state-space model, shown in Fig. 12.5, produces the total output. The input is the unit-step signal. The initial values of the state variables are set in the simulation block.

Example 12.3 Find the closed-form expression for the impulse response of the system, described by the state-space model given in Example 12.2, using the time-domain method.

Solution

$$\boldsymbol{C}e^{\boldsymbol{A}t} = \begin{bmatrix} -3 & -2 \end{bmatrix} \begin{bmatrix} c_0 - 4c_1 & -4c1 \\ c_1 & c_0 \end{bmatrix}$$
$$= \begin{bmatrix} -3c_0 + 10c_1 & -2c_0 + 12c_1 \end{bmatrix}$$

Since the convolution output of a function with the unit impulse is itself and the vector $\boldsymbol{B}$ is a constant, the impulse response is given by

$$h(t) = \boldsymbol{C}e^{\boldsymbol{A}t}\boldsymbol{B} + D\delta(t) = -3c_0 + 10c_1 + \delta(t)$$

$$h(t) = (\delta(t) - 3e^{-2t} + 4te^{-2t})u(t)$$ ■

This result can also be obtained by differentiating the zero-state response of the previous example, as the unit-impulse response is the derivative of the unit-step response.

12.3 Frequency-Domain Solution of the State Equation

The Laplace transform of a vector function, such as $\boldsymbol{q}(t)$, is defined to be the vector function $\boldsymbol{Q}(s)$, where the elements are the transforms of the corresponding elements of $\boldsymbol{q}(t)$. Taking the Laplace transform of the state equation, we get

$$s\boldsymbol{Q}(s) - \boldsymbol{q}(0^-) = \boldsymbol{A}\boldsymbol{Q}(s) + \boldsymbol{B}X(s)$$

We have used the time-differentiation property of the Laplace transform, and $\boldsymbol{q}(0^-)$ is the initial state vector. Since $\boldsymbol{I}\boldsymbol{Q}(s) = \boldsymbol{Q}(s)$, where $\boldsymbol{I}$ is the identity matrix of the same size as the matrix $\boldsymbol{A}$, and collecting the terms involving $\boldsymbol{Q}(s)$, we get

$$(s\boldsymbol{I} - \boldsymbol{A})\boldsymbol{Q}(s) = \boldsymbol{q}(0^-) + \boldsymbol{B}X(s)$$

The inclusion of the identity matrix is necessary to combine the terms involving $\boldsymbol{Q}(s)$. Premultiplying both sides by $(s\boldsymbol{I} - \boldsymbol{A})^{-1}$, which is the inverse of $(s\boldsymbol{I} - \boldsymbol{A})$, we get

$$\boldsymbol{Q}(s) = \overbrace{(s\boldsymbol{I} - \boldsymbol{A})^{-1}\boldsymbol{q}(0^-)}^{q_{zi}(s)} + \overbrace{(s\boldsymbol{I} - \boldsymbol{A})^{-1}\boldsymbol{B}X(s)}^{q_{zs}(s)}$$

The inverse Laplace transforms of the first and the second expressions on the right-hand side are, respectively, the zero-input and zero-state components of the state vector $\boldsymbol{q}(t)$. Taking the Laplace transform of the output equation, we get

$$Y(s) = \boldsymbol{C}\boldsymbol{Q}(s) + \boldsymbol{D}X(s)$$

Substituting for $\boldsymbol{Q}(s)$, we get

$$Y(s) = \overbrace{\boldsymbol{C}(s\boldsymbol{I} - \boldsymbol{A})^{-1}\boldsymbol{q}(0^-)}^{y_{zi}(s)} + \overbrace{(\boldsymbol{C}(s\boldsymbol{I} - \boldsymbol{A})^{-1}\boldsymbol{B} + \boldsymbol{D})X(s)}^{y_{zs}(s)}$$

The inverse Laplace transforms of the first and the second expressions on the right-hand side are, respectively, the zero-input and zero-state components of the system response $y(t)$. Comparing with the expression for $\boldsymbol{Q}(t)$, we find that the inverse Laplace transform of $((s\boldsymbol{I} - \boldsymbol{A})^{-1})$ is $e^{\boldsymbol{A}t}$, the transition or fundamental matrix of the system. With the system initial conditions zero, the transfer function is given by

$$H(s) = \frac{Y(s)}{X(s)} = (\boldsymbol{C}(s\boldsymbol{I} - \boldsymbol{A})^{-1}\boldsymbol{B} + \boldsymbol{D})$$

Example 12.4 Solve the problem of Example 12.2 using the frequency-domain method.

Solution The initial state vector is

$$\boldsymbol{q}(0^-) = \begin{bmatrix} -\frac{15}{8} \\ \frac{29}{16} \end{bmatrix}$$

as derived in Example 12.2 from the given initial output conditions.

$$(s\boldsymbol{I} - \boldsymbol{A}) = s\begin{bmatrix} 1 & 0 \\ 0 & 1 \end{bmatrix} - \begin{bmatrix} -4 & -4 \\ 1 & 0 \end{bmatrix} = \begin{bmatrix} s+4 & 4 \\ -1 & s \end{bmatrix}$$

$$(s\boldsymbol{I} - \boldsymbol{A})^{-1} = \frac{1}{s^2+4s+4}\begin{bmatrix} s & -4 \\ 1 & s+4 \end{bmatrix} = \begin{bmatrix} \frac{s}{s^2+4s+4} & \frac{-4}{s^2+4s+4} \\ \frac{1}{s^2+4s+4} & \frac{s+4}{s^2+4s+4} \end{bmatrix}$$

We used the fact that $\boldsymbol{I} = e^{\boldsymbol{A}0}$ to check the computation of $e^{\boldsymbol{A}t}$. In the frequency domain, the corresponding check, using the initial value theorem of the Laplace transform, is $\lim_{s\to\infty} s(s\boldsymbol{I} - \boldsymbol{A})^{-1} = \boldsymbol{I}$.

The zero-input component of the state vector is

$$\boldsymbol{q}_{zi}(s) = (s\boldsymbol{I} - \boldsymbol{A})^{-1}\boldsymbol{q}(0^-) = \begin{bmatrix} \frac{s}{s^2+4s+4} & \frac{-4}{s^2+4s+4} \\ \frac{1}{s^2+4s+4} & \frac{s+4}{s^2+4s+4} \end{bmatrix}\begin{bmatrix} -\frac{15}{8} \\ \frac{29}{16} \end{bmatrix}$$

$$= \begin{bmatrix} \frac{-\frac{15}{8}s - \frac{29}{4}}{s^2+4s+4} \\ \frac{\frac{29}{16}s + \frac{43}{8}}{s^2+4s+4} \end{bmatrix} = \begin{bmatrix} -\frac{\frac{7}{2}}{(s+2)^2} - \frac{\frac{15}{8}}{s+2} \\ \frac{\frac{7}{4}}{(s+2)^2} + \frac{\frac{29}{16}}{s+2} \end{bmatrix}$$

Taking the inverse Laplace transform, we get

$$\boldsymbol{q}_{zi}(t) = \begin{bmatrix} -\frac{15}{8}e^{-2t} - \frac{7}{2}te^{-2t} \\ \frac{29}{16}e^{-2t} + \frac{7}{4}te^{-2t} \end{bmatrix}$$

The zero-state component of the state vector is

$$\boldsymbol{q}_{zs}(s) = (s\boldsymbol{I} - \boldsymbol{A})^{-1}\boldsymbol{B}X(s) = \begin{bmatrix} \frac{s}{s^2+4s+4} & \frac{-4}{s^2+4s+4} \\ \frac{1}{s^2+4s+4} & \frac{s+4}{s^2+4s+4} \end{bmatrix}\begin{bmatrix} 1 \\ 0 \end{bmatrix}\frac{1}{s}$$

$$= \begin{bmatrix} \frac{1}{s^2+4s+4} \\ \frac{1}{s(s^2+4s+4)} \end{bmatrix} = \begin{bmatrix} \frac{1}{(s+2)^2} \\ \frac{\frac{1}{4}}{s} - \frac{\frac{1}{2}}{(s+2)^2} - \frac{\frac{1}{4}}{s+2} \end{bmatrix}$$

Taking the inverse Laplace transform, we get

$$\boldsymbol{q}_{zs}(t) = \begin{bmatrix} te^{-2t} \\ \frac{1}{4} - \frac{1}{4}e^{-2t} - \frac{1}{2}te^{-2t} \end{bmatrix}$$

Using the output equation, the output can be computed as given in Example 12.2. ■

Example 12.5 Find the closed-form expression for the impulse response of the system, described by the state-space model given in Example 12.2, using the frequency-domain method.

Solution The transfer function is given by

$$H(s) = \frac{Y(s)}{X(s)} = (\boldsymbol{C}(s\boldsymbol{I} - \boldsymbol{A})^{-1}\boldsymbol{B} + \boldsymbol{D})$$

$$H(s) = \begin{bmatrix} -3 & -2 \end{bmatrix} \begin{bmatrix} \frac{s}{s^2+4s+4} & \frac{-4}{s^2+4s+4} \\ \frac{1}{s^2+4s+4} & \frac{s+4}{s^2+4s+4} \end{bmatrix} \begin{bmatrix} 1 \\ 0 \end{bmatrix} + 1$$

$$= \frac{-3s - 2}{(s^2 + 4s + 4)} + 1 = 1 - \frac{3}{s+2} + \frac{4}{(s+2)^2}$$

Finding the inverse Laplace transform, we get

$$h(t) = (\delta(t) - 3e^{-2t} + 4te^{-2t})u(t)$$

■

12.4 Linear Transformation of State Vectors

In common with discrete systems, for a specific input-output relationship of a continuous system, the system can have different internal structures. By a linear transformation of a state vector, we can obtain another vector implying different internal structure of the system. Let us find the state-space model of a system with state vector $\boldsymbol{q}$ using another state vector $\overline{\boldsymbol{q}}$ such that $\overline{\boldsymbol{q}} = \boldsymbol{P}\boldsymbol{q}$ and $\boldsymbol{q} = \boldsymbol{P}^{-1}\overline{\boldsymbol{q}}$, where $\boldsymbol{P}$ is the $N \times N$ transformation matrix and $\boldsymbol{P}^{-1}$ exists. With the new state vector, the state equation can be written as

$$\boldsymbol{P}^{-1}\dot{\overline{\boldsymbol{q}}}(t) = \boldsymbol{A}\boldsymbol{P}^{-1}\overline{\boldsymbol{q}}(t) + \boldsymbol{B}x(t)$$

Premultiplying by $\boldsymbol{P}$, we get

$$\dot{\overline{\boldsymbol{q}}}(t) = \boldsymbol{P}\boldsymbol{A}\boldsymbol{P}^{-1}\overline{\boldsymbol{q}}(t) + \boldsymbol{P}\boldsymbol{B}x(t)$$

With $\overline{\boldsymbol{A}} = \boldsymbol{P}\boldsymbol{A}\boldsymbol{P}^{-1}$ and $\overline{\boldsymbol{B}} = \boldsymbol{P}\boldsymbol{B}$, the state equation can be written as

$$\dot{\overline{\boldsymbol{q}}}(t) = \overline{\boldsymbol{A}}\overline{\boldsymbol{q}}(t) + \overline{\boldsymbol{B}}x(t)$$

With $\overline{C} = CP^{-1}$, the output equation can be written as

$$y(t) = \overline{C}\overline{q}(t) + Dx(t)$$

Some properties of A and $\overline{A}$ matrices can be used to check the computation of $\overline{A}$. The determinants of A and $\overline{A}$ are equal. The determinants of $(sI - A)$ and $(sI - \overline{A})$ are the same. The traces (sum of the diagonal elements) of A and $\overline{A}$ are equal.

Example 12.6 Derive the state-space model of the system in Example 12.1 with the new state vector that is related to old state vector as

$$\overline{q}_1(t) = q_1(t) + 2q_2(t)$$
$$\overline{q}_2(t) = -3q_1(t) + 4q_2(t)$$

Verify that the transfer function remains the same using either state-space model.

Solution

$$P = \begin{bmatrix} 1 & 2 \\ -3 & 4 \end{bmatrix}, \quad P^{-1} = \begin{bmatrix} \frac{4}{10} & -\frac{2}{10} \\ \frac{3}{10} & \frac{1}{10} \end{bmatrix}$$

$$\overline{A} = PAP^{-1} = \begin{bmatrix} 1 & 2 \\ -3 & 4 \end{bmatrix} \begin{bmatrix} -\frac{5}{6} & -\frac{1}{2} \\ \frac{1}{3} & 0 \end{bmatrix} \begin{bmatrix} \frac{4}{10} & -\frac{2}{10} \\ \frac{3}{10} & \frac{1}{10} \end{bmatrix} = \begin{bmatrix} -\frac{13}{60} & -\frac{1}{60} \\ \frac{119}{60} & -\frac{37}{60} \end{bmatrix}$$

$$\overline{B} = PB = \begin{bmatrix} 1 & 2 \\ -3 & 4 \end{bmatrix} \begin{bmatrix} \frac{1}{2} \\ 0 \end{bmatrix} = \begin{bmatrix} \frac{1}{2} \\ -\frac{3}{2} \end{bmatrix}$$

$$\overline{C} = CP^{-1} = \begin{bmatrix} 0 & 1 \end{bmatrix} \begin{bmatrix} \frac{4}{10} & -\frac{2}{10} \\ \frac{3}{10} & \frac{1}{10} \end{bmatrix} = \begin{bmatrix} \frac{3}{10} & \frac{1}{10} \end{bmatrix}$$

The state-space model of a second-order continuous system with the new state vector is shown in Fig. 12.6. This realization requires more components than that shown in Fig. 12.4. However, it must be noted that, while the minimum number of components is of great importance, there are other criteria, such as less coefficient sensitivity, that could decide which of the realizations of a system is suitable for a particular application.

The transfer function, using the new state-space model, is computed as follows.

$$(sI - \overline{A}) = s\begin{bmatrix} 1 & 0 \\ 0 & 1 \end{bmatrix} - \begin{bmatrix} -\frac{13}{60} & -\frac{1}{60} \\ \frac{119}{60} & -\frac{37}{60} \end{bmatrix} = \begin{bmatrix} s + \frac{13}{60} & \frac{1}{60} \\ -\frac{119}{60} & s + \frac{37}{60} \end{bmatrix}$$

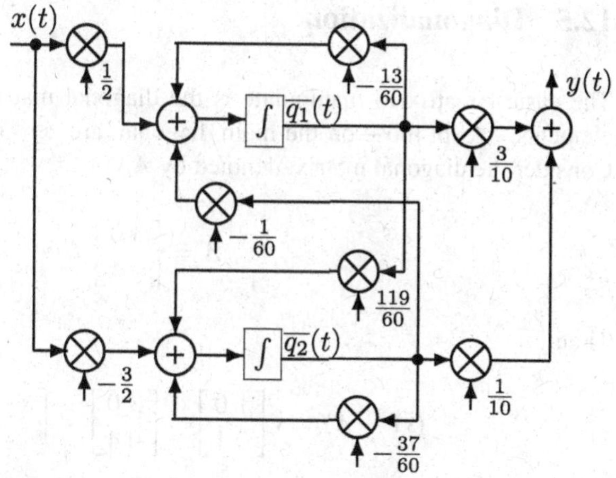

Fig. 12.6 The state-space model of a second-order continuous system with the new state vector

$$(s\boldsymbol{I}-\overline{\boldsymbol{A}})^{-1}=\frac{1}{s^2+\frac{5}{6}s+\frac{1}{6}}\begin{bmatrix} s+\frac{37}{60} & -\frac{1}{60} \\ \frac{119}{60} & s+\frac{13}{60}\end{bmatrix}=\begin{bmatrix} \frac{s+\frac{37}{60}}{s^2+\frac{5}{6}s+\frac{1}{6}} & \frac{-\frac{1}{60}}{s^2+\frac{5}{6}s+\frac{1}{6}} \\ \frac{\frac{119}{60}}{s^2+\frac{5}{6}s+\frac{1}{6}} & \frac{s+\frac{13}{60}}{s^2+\frac{5}{6}s+\frac{1}{6}}\end{bmatrix}$$

$$H(s)=\begin{bmatrix}\frac{3}{10} & \frac{1}{10}\end{bmatrix}\begin{bmatrix} \frac{s+\frac{37}{60}}{s^2+\frac{5}{6}s+\frac{1}{6}} & \frac{-\frac{1}{60}}{s^2+\frac{5}{6}s+\frac{1}{6}} \\ \frac{\frac{119}{60}}{s^2+\frac{5}{6}s+\frac{1}{6}} & \frac{s+\frac{13}{60}}{s^2+\frac{5}{6}s+\frac{1}{6}}\end{bmatrix}\begin{bmatrix}\frac{1}{2} \\ -\frac{3}{2}\end{bmatrix}=\frac{\frac{1}{6}}{(s^2+\frac{5}{6}s+\frac{1}{6})}$$

The transfer function, using the old state-space model, is computed as follows.

$$(s\boldsymbol{I}-\boldsymbol{A})=s\begin{bmatrix}1 & 0 \\ 0 & 1\end{bmatrix}-\begin{bmatrix}-\frac{5}{6} & -\frac{1}{2} \\ \frac{1}{3} & 0\end{bmatrix}=\begin{bmatrix}s+\frac{5}{6} & \frac{1}{2} \\ -\frac{1}{3} & s\end{bmatrix}$$

$$(s\boldsymbol{I}-\boldsymbol{A})^{-1}=\frac{1}{s^2+\frac{5}{6}s+\frac{1}{6}}\begin{bmatrix} s & -\frac{1}{2} \\ \frac{1}{3} & s+\frac{5}{6}\end{bmatrix}=\begin{bmatrix} \frac{s}{s^2+\frac{5}{6}s+\frac{1}{6}} & \frac{-\frac{1}{2}}{s^2+\frac{5}{6}s+\frac{1}{6}} \\ \frac{\frac{1}{3}}{s^2+\frac{5}{6}s+\frac{1}{6}} & \frac{s+\frac{5}{6}}{s^2+\frac{5}{6}s+\frac{1}{6}}\end{bmatrix}$$

$$H(s)=[0 \quad 1]\begin{bmatrix} \frac{s}{s^2+\frac{5}{6}s+\frac{1}{6}} & \frac{-\frac{1}{2}}{s^2+\frac{5}{6}s+\frac{1}{6}} \\ \frac{\frac{1}{3}}{s^2+\frac{5}{6}s+\frac{1}{6}} & \frac{s+\frac{5}{6}}{s^2+\frac{5}{6}s+\frac{1}{6}}\end{bmatrix}\begin{bmatrix}\frac{1}{2} \\ 0\end{bmatrix}=\frac{\frac{1}{6}}{(s^2+\frac{5}{6}s+\frac{1}{6})},$$

which is the same as that obtained above. ■

12.5 Diagonalization

The easiest matrix to manipulate is the diagonal matrix. A square matrix whose elements, except those on the main diagonal, are zero is called a diagonal matrix. Consider the diagonal matrix, denoted by $\boldsymbol{\Lambda}$,

$$\boldsymbol{\Lambda} = \begin{bmatrix} 3 & 0 \\ 0 & 4 \end{bmatrix}$$

Then,

$$(s\boldsymbol{I} - \boldsymbol{\Lambda}) = s\begin{bmatrix} 1 & 0 \\ 0 & 1 \end{bmatrix} - \begin{bmatrix} 3 & 0 \\ 0 & 4 \end{bmatrix} = \begin{bmatrix} s-3 & 0 \\ 0 & s-4 \end{bmatrix}$$

The characteristic polynomial of the system, given by the determinant of this matrix, is

$$s^2 - 7s + 12 = (s-3)(s-4)$$

Therefore, the eigenvalues of a diagonal matrix are its diagonal elements. A square matrix possess equivalent diagonal form. Let us find the transformation of a matrix $\boldsymbol{A}$ so that the resultant matrix is a diagonal one.

A nonzero vector $\boldsymbol{v}$ is an eigenvector of a $N \times N$ matrix $\boldsymbol{A}$ with eigenvalue λ if

$$\boldsymbol{A}\boldsymbol{v} = \lambda\boldsymbol{v}$$

A $N \times N$ matrix $\boldsymbol{A}$ is diagonalizable if and only if $\boldsymbol{A}$ has N linearly independent eigenvectors. Assume that $\boldsymbol{A}$ has N independent column eigenvectors

$$\boldsymbol{v}_1, \boldsymbol{v}_1, \ldots, \boldsymbol{v}_N$$

corresponding to N eigenvalues

$$\lambda_1, \lambda_2, \ldots, \lambda_N$$

Let

$$\boldsymbol{P} = [\boldsymbol{v}_1 : \boldsymbol{v}_1 :, \ldots, \boldsymbol{v}_N]$$

and $\boldsymbol{\Lambda}$ be a diagonal matrix with the ith element on the diagonal λ_i. Now,

$$\begin{aligned} \boldsymbol{A}\boldsymbol{P} &= \boldsymbol{A}[\boldsymbol{v}_1 : \boldsymbol{v}_1 :, \ldots, : \boldsymbol{v}_N] \\ &= [\boldsymbol{A}\boldsymbol{v}_1 : \boldsymbol{A}\boldsymbol{v}_1 :, \ldots, : \boldsymbol{A}\boldsymbol{v}_N] \end{aligned}$$

$$= [\lambda_1 \boldsymbol{v_1} : \lambda_2 \boldsymbol{v_1} :, \ldots, : \lambda_N \boldsymbol{v_N}]$$

$$= [\boldsymbol{v_1} : \boldsymbol{v_1} :, \ldots, : \boldsymbol{v_N}]\boldsymbol{\Lambda} = \boldsymbol{P\Lambda}$$

Since $\boldsymbol{AP} = \boldsymbol{P\Lambda}$,

$$\boldsymbol{P}^{-1}\boldsymbol{AP} = \boldsymbol{\Lambda}$$

Consider the 2×2 matrix $\boldsymbol{A}$,

$$\boldsymbol{A} = \begin{bmatrix} 2 & 3 \\ 2 & 1 \end{bmatrix}$$

Then,

$$(s\boldsymbol{I} - \boldsymbol{A}) = s \begin{bmatrix} 1 & 0 \\ 0 & 1 \end{bmatrix} - \begin{bmatrix} 2 & 3 \\ 2 & 1 \end{bmatrix} = \begin{bmatrix} s-2 & -3 \\ -2 & s-1 \end{bmatrix}$$

The characteristic polynomial of the system, given by the determinant of this matrix, is

$$s^2 - 3s - 4 = (s+1)(s-4)$$

The roots of this equation are the two eigenvalues $\{\lambda_1 = -1, \lambda_2 = 4\}$. For finding the eigenvectors, we use the equation

$$(\lambda \boldsymbol{I} - \boldsymbol{A})\boldsymbol{v} = 0$$

For $\lambda = -1$, we get

$$\begin{bmatrix} -1-2 & -3 \\ -2 & -1-1 \end{bmatrix} \begin{bmatrix} v(0) \\ v(1) \end{bmatrix} = 0$$

The resulting equations are

$$-3v(0) - 3v(1) = 0$$
$$-2v(0) - 2v(1) = 0$$

As these equations are dependent, we take one of them and give any nontrivial solution. For example, $\{v(0) = 1, v(1) = -1\}$. For $\lambda = 4$, we get

$$\begin{bmatrix} 4-2 & -3 \\ -2 & 4-1 \end{bmatrix} \begin{bmatrix} v(0) \\ v(1) \end{bmatrix} = 0$$

Similarly, $\{v(0) = 2, v(1) = 4/3\}$. Therefore,

$$\boldsymbol{P} = \begin{bmatrix} 1 & 2 \\ -1 & 4/3 \end{bmatrix}$$

The first and second columns are, respectively, the eigenvectors corresponding to eigenvalues -1 and 4.

$$\boldsymbol{P}^{-1}\boldsymbol{A}\boldsymbol{P} = \begin{bmatrix} 0.4 & -0.6 \\ 0.3 & 0.3 \end{bmatrix} \begin{bmatrix} 2 & 3 \\ 2 & 1 \end{bmatrix} \begin{bmatrix} 1 & 2 \\ -1 & 4/3 \end{bmatrix} = \boldsymbol{\Lambda} = \begin{bmatrix} -1 & 0 \\ 0 & 4 \end{bmatrix}$$

Consider the 3×3 matrix $\boldsymbol{A}$,

$$\boldsymbol{A} = \begin{bmatrix} 3 & 1 & 2 \\ 1 & 2 & 3 \\ 3 & 2 & 1 \end{bmatrix}$$

Then,

$$(s\boldsymbol{I} - \boldsymbol{A}) = s \begin{bmatrix} 1 & 0 & 0 \\ 0 & 1 & 0 \\ 0 & 0 & 1 \end{bmatrix} - \begin{bmatrix} 3 & 1 & 2 \\ 1 & 2 & 3 \\ 3 & 2 & 1 \end{bmatrix} = \begin{bmatrix} s-3 & -1 & -2 \\ -1 & s-2 & -3 \\ -3 & -2 & s-1 \end{bmatrix}$$

The characteristic polynomial of the system, given by the determinant of this matrix, is

$$s^3 - 6s^2 - 2s + 12 = (s - 6)(s - \sqrt{2})(s + \sqrt{2})$$

The roots of this equation are the three eigenvalues $\{\lambda_1 = 6, \lambda_2 = \sqrt{2}, \lambda_2 = -\sqrt{2}\}$. For finding the eigenvectors, we use the equation

$$(\lambda \boldsymbol{I} - \boldsymbol{A})\boldsymbol{v} = 0$$

For $\lambda = 6$, we get

$$\begin{bmatrix} 6-3 & -1 & -2 \\ -1 & 6-2 & -3 \\ -3 & -2 & 6-1 \end{bmatrix} \begin{bmatrix} v(0) \\ v(1) \\ v(2) \end{bmatrix} = 0$$

Solving, we get $\{v(0) = 1, v(1) = 1, v(2) = 1\}$. For $\lambda = \sqrt{2}$, we get

$$\begin{bmatrix} \sqrt{2}-3 & -1 & -2 \\ -1 & \sqrt{2}-2 & -3 \\ -3 & -2 & \sqrt{2}-1 \end{bmatrix} \begin{bmatrix} v(0) \\ v(1) \\ v(2) \end{bmatrix} = 0$$

Solving, we get $\{v(0) = -0.6631, v(1) = 1, v(2) = 0.0258\}$. For $\lambda = -\sqrt{2}$, we get

$$\begin{bmatrix} -\sqrt{2}-3 & -1 & -2 \\ -1 & -\sqrt{2}-2 & -3 \\ -3 & -2 & -\sqrt{2}-1 \end{bmatrix} \begin{bmatrix} v(0) \\ v(1) \\ v(2) \end{bmatrix} = 0$$

Solving, we get $\{v(0) = -0.2720, v(1) = -0.7990, v(2) = 1\}$. Therefore,

$$\boldsymbol{P} = \begin{bmatrix} 1 & -0.6631 & -0.2720 \\ 1 & 1 & -0.7990 \\ 1 & 0.0258 & 1 \end{bmatrix}$$

The first, second, and third columns are, respectively, the eigenvectors corresponding to eigenvalues 1, 2, and 3.

$$\boldsymbol{P}^{-1}\boldsymbol{A}\boldsymbol{P} = \begin{bmatrix} 0.4118 & 0.2647 & 0.3235 \\ -0.7258 & 0.5132 & 0.2126 \\ -0.3931 & -0.2779 & 0.6710 \end{bmatrix} \begin{bmatrix} 3 & 1 & 2 \\ 1 & 2 & 3 \\ 3 & 2 & 1 \end{bmatrix} \begin{bmatrix} 1 & -0.6631 & -0.2720 \\ 1 & 1 & -0.7990 \\ 1 & 0.0258 & 1 \end{bmatrix}$$

$$= \boldsymbol{\Lambda} = \begin{bmatrix} 6 & 0 & 0 \\ 0 & \sqrt{2} & 0 \\ 0 & 0 & -\sqrt{2} \end{bmatrix}$$

12.6 Similarity Transformation

Transformations are used to simplify the solution of a problem. For example, Fourier representation reduces convolution to multiplication. Using logarithms, multiplication is reduced to addition. There are two versions of transformation commonly used. In Sect. 12.4, the new state variables is related to the original state variables by the relation $\overline{\boldsymbol{q}} = \boldsymbol{P}\boldsymbol{q}$. In the other version, the relation becomes $\overline{\boldsymbol{q}} = \boldsymbol{P}^{-1}\boldsymbol{q}$. Therefore, by replacing $\boldsymbol{P}$ by $\boldsymbol{P}^{-1}$, we get the other version from a given version. Either version yields the same results, if consistently is used.

Let $\boldsymbol{A}$ be a nonsingular square matrix. Then, $\boldsymbol{A}$ and $\boldsymbol{P}^{-1}\boldsymbol{A}\boldsymbol{P}$ are said to be similar, where $\boldsymbol{P}$ is a nonsingular matrix of the same size as $\boldsymbol{A}$. The transformation from $\boldsymbol{A}$ to $\boldsymbol{P}^{-1}\boldsymbol{A}\boldsymbol{P}$ is called the similarity transformation. Two similar matrices have the same eigenvalues and, hence, the same characteristic polynomial. This transformation produces a similar diagonal matrix to that of $\boldsymbol{A}$.

Example 12.7 Find a closed-form expression for the state output $q(t)$ of the system, described by

$$\boldsymbol{A} = \begin{bmatrix} 2 & 1 \\ 3 & 4 \end{bmatrix}, \quad \boldsymbol{B} = \begin{bmatrix} 1 \\ 0 \end{bmatrix},$$

using the frequency-domain method, with zero initial conditions and the input $u(t)$, the unit-step function.

Solution

$$(s\boldsymbol{I} - \boldsymbol{A}) = s\begin{bmatrix} 1 & 0 \\ 0 & 1 \end{bmatrix} - \begin{bmatrix} 2 & 1 \\ 3 & 4 \end{bmatrix} = \begin{bmatrix} s-2 & -1 \\ -3 & s-4 \end{bmatrix}$$

$$(s\boldsymbol{I} - \boldsymbol{A})^{-1} = \frac{1}{s^2 - 6s + 5}\begin{bmatrix} s-4 & 1 \\ 3 & s-2 \end{bmatrix} = \begin{bmatrix} \frac{s-4}{s^2-6s+5} & \frac{1}{s^2-6s+5} \\ \frac{3}{s^2-6s+5} & \frac{s-2}{s^2-6s+5} \end{bmatrix}$$

The eigenvalues are $\{1, 5\}$.

The zero-state component of the state vector is

$$\boldsymbol{q}_{zs}(s) = (s\boldsymbol{I} - \boldsymbol{A})^{-1}\boldsymbol{B}X(s) = \begin{bmatrix} \frac{s-4}{s^2-6s+5} & \frac{1}{s^2-6s+5} \\ \frac{3}{s^2-6s+5} & \frac{s-2}{s^2-6s+5} \end{bmatrix}\begin{bmatrix} 1 \\ 0 \end{bmatrix}\frac{1}{s}$$

$$= \begin{bmatrix} \frac{s-4}{s(s^2-6s+5)} \\ \frac{3}{s(s^2-6s+5)} \end{bmatrix} = \begin{bmatrix} \frac{-0.8}{s} + \frac{0.75}{s-1} + \frac{0.05}{s-5} \\ \frac{0.6}{s} - \frac{0.75}{s-1} + \frac{0.15}{s-5} \end{bmatrix}$$

Taking the inverse Laplace transform, we get

$$\boldsymbol{q}_{zs}(t) = \begin{bmatrix} -0.8 + 0.75e^{t} + 0.05e^{5t} \\ 0.6 - 0.75e^{t} + 0.15e^{5t} \end{bmatrix} u(t)$$

Using the similarity transformation method also, the state output can be computed. The state transition matrix associated with the diagonal system matrix $\boldsymbol{A}$ is

$$\begin{bmatrix} e^{t} & 0 \\ 0 & e^{5t} \end{bmatrix} u(t)$$

$$\boldsymbol{P}e^{\boldsymbol{A}t}\boldsymbol{P}^{-1} = \begin{bmatrix} 1 & 1 \\ -1 & 3 \end{bmatrix}\begin{bmatrix} e^{t} & 0 \\ 0 & e^{5t} \end{bmatrix}\begin{bmatrix} 0.75 & -0.25 \\ 0.25 & 0.25 \end{bmatrix}$$

$$= e^{\boldsymbol{A}t} = \begin{bmatrix} 0.75e^{t} + 0.25e^{5t} & -0.25e^{t} + 0.25e^{5t} \\ -0.75e^{t} + 0.75e^{5t} & 0.25e^{t} + 0.75e^{5t} \end{bmatrix} u(t)$$

which is the same as the inverse Laplace transform of

$$(s\boldsymbol{I} - \boldsymbol{A})^{-1} = \begin{bmatrix} \frac{s-4}{s^2-6s+5} & \frac{1}{s^2-6s+5} \\ \frac{3}{s^2-6s+5} & \frac{s-2}{s^2-6s+5} \end{bmatrix}$$

■

12.7 Controllability

A system is said to be controllable, if it is possible to control the state of a system from its initial value to any other value in a finite interval of time. One of the necessary and sufficient conditions for complete controllability of a system is that the rank of the matrix

$$[\boldsymbol{B} \mid \boldsymbol{AB} \mid \cdots \mid \boldsymbol{A}^{N-1}\boldsymbol{B}]$$

be N, where $\boldsymbol{A}$ and $\boldsymbol{B}$ are the system and input matrices, respectively.

Another test for controllability is that $\boldsymbol{P}^{-1}\boldsymbol{B}$ has all nonzero elements, after the system has been transformed to diagonal form.

Example 12.8 Determine the controllability of the system, described by

$$\boldsymbol{A} = \begin{bmatrix} 0 & -1 \\ 2 & 3 \end{bmatrix}, \quad \boldsymbol{B} = \begin{bmatrix} 0 \\ 1 \end{bmatrix}$$

Solution

$$(s\boldsymbol{I} - \boldsymbol{A}) = s\begin{bmatrix} 1 & 0 \\ 0 & 1 \end{bmatrix} - \begin{bmatrix} 0 & -1 \\ 2 & 3 \end{bmatrix} = \begin{bmatrix} s & 1 \\ -2 & s-3 \end{bmatrix}$$

The characteristic polynomial of the system, given by the determinant of this matrix, is

$$s^2 - 3s + 2 = (s-1)(s-2)$$

The roots of this equation are the two eigenvalues $\{\lambda_1 = 1, \lambda_2 = 2\}$. For finding the eigenvectors, we use the equation

$$(\lambda\boldsymbol{I} - \boldsymbol{A})\boldsymbol{v} = 0$$

For $\lambda = 1$, we get

$$\begin{bmatrix} 1 & 1 \\ -2 & 1-3 \end{bmatrix}\begin{bmatrix} v(0) \\ v(1) \end{bmatrix} = 0$$

We get

$$v(0) + v(1) = 0$$
$$-2v(0) - 2v(1) = 0$$

As these equations are dependent, we take one of them and give any nontrivial solution. For example, $\{v(0) = 1, v(1) = -1\}$. For $\lambda = 2$, we get

$$\begin{bmatrix} 2 & 1 \\ -2 & 2-3 \end{bmatrix}\begin{bmatrix} v(0) \\ v(1) \end{bmatrix} = 0$$

Similarly, $\{v(0) = 1, v(1) = -2\}$. Therefore,

$$\boldsymbol{P} = \begin{bmatrix} 1 & 1 \\ -1 & -2 \end{bmatrix}.$$

The first and second columns are, respectively, the eigenvectors corresponding to eigenvalues 1 and 2.

$$\boldsymbol{P}^{-1}\boldsymbol{A}\boldsymbol{P} = \begin{bmatrix} 2 & 1 \\ -1 & -1 \end{bmatrix}\begin{bmatrix} 0 & -1 \\ 2 & 3 \end{bmatrix}\begin{bmatrix} 1 & 1 \\ -1 & -2 \end{bmatrix} = \boldsymbol{\Lambda} = \begin{bmatrix} 1 & 0 \\ 0 & 2 \end{bmatrix}$$

The rank of the matrix

$$[\boldsymbol{B} \mid \boldsymbol{A}\boldsymbol{B}] = \begin{bmatrix} 1 & 1 \\ 0 & -1 \end{bmatrix}$$

is 2. Therefore, the system is controllable.

$$\boldsymbol{P}^{-1}\boldsymbol{B} = \begin{bmatrix} 1 \\ -1 \end{bmatrix}$$

has no zero entries, and, therefore, the system is controllable. ■

With

$$\boldsymbol{B} = \begin{bmatrix} 1 \\ -1 \end{bmatrix}$$

the rank of the matrix

$$[\boldsymbol{B} \mid \boldsymbol{A}\boldsymbol{B}] = \begin{bmatrix} 1 & 1 \\ -1 & -1 \end{bmatrix}$$

is 1. Therefore, the system is uncontrollable.

$$P^{-1}B = \begin{bmatrix} 1 \\ 0 \end{bmatrix}$$

has one zero entry, and, therefore, the system is uncontrollable.

12.8 Observability

A system is said to be observable, if it is possible to determine the state of a system from its outputs in a finite interval of time. One of the necessary and sufficient conditions for complete observability of a system is that the rank of the matrix

$$[C^T \mid A^T C^T \mid \cdots \mid (A^*)^{N-1} C^T]$$

is N, where A^T and C^T are the transposes of the system and output matrices, respectively. Another test for observability is that CP^{-1} has all nonzero elements, after the system has been transformed to diagonal form.

Example 12.9 Determine the observability of the system, described by

$$A = \begin{bmatrix} 0 & -1 \\ 2 & 3 \end{bmatrix}, \quad C = \begin{bmatrix} 1 & -1 \end{bmatrix}$$

Solution

$$(sI - A) = s\begin{bmatrix} 1 & 0 \\ 0 & 1 \end{bmatrix} - \begin{bmatrix} 0 & -1 \\ 2 & 3 \end{bmatrix} = \begin{bmatrix} s & 1 \\ -2 & s-3 \end{bmatrix}$$

The characteristic polynomial of the system, given by the determinant of this matrix, is

$$s^2 - 3s + 2 = (s-1)(s-2)$$

The roots of this equation are the two eigenvalues $\{\lambda_1 = 1, \lambda_2 = 2\}$. Therefore, the transformation matrix from the last example is

$$P = \begin{bmatrix} 2 & 1 \\ -1 & 1 \end{bmatrix}$$

The first and second columns are, respectively, the eigenvectors corresponding to eigenvalues 2 and 5.

$$P^{-1}AP = \begin{bmatrix} \frac{1}{3} & -\frac{1}{3} \\ \frac{1}{3} & \frac{2}{3} \end{bmatrix} \begin{bmatrix} 3 & 2 \\ 1 & 4 \end{bmatrix} \begin{bmatrix} 2 & 1 \\ -1 & 1 \end{bmatrix} = \Lambda = \begin{bmatrix} 2 & 0 \\ 0 & 5 \end{bmatrix}$$

The rank of the matrix

$$[C^T \mid A^T C^T] = \begin{bmatrix} 1 & -2 \\ -1 & -4 \end{bmatrix}$$

is 2. Therefore, the system is observable.

$$CP = \begin{bmatrix} 2 & 3 \end{bmatrix}$$

has no zero entries, and, therefore, the system is observable. ■

With

$$C = \begin{bmatrix} 1 & 1 \end{bmatrix},$$

the rank of the matrix

$$[C^T \mid A^T C^T] = \begin{bmatrix} 1 & 2 \\ 1 & 2 \end{bmatrix}$$

is 1. Therefore, the system is unobservable.

$$CP = \begin{bmatrix} 0 \\ -1 \end{bmatrix}$$

has a zero entry, and, therefore, the system is unobservable.

12.9 Summary

- In this chapter, state-space analysis of continuous systems has been presented.
- State-space model of a system is a generalization of input-output models, such as the transfer function.
- State-space model, in addition to the input and the output, includes N internal variables of the system, called state variables, for a Nth-order system. All the outputs of the system are expressed in terms of the state variables and the input.
- A system, characterized by a Nth-order differential equation, is characterized, in the state-space model, by a set of N simultaneous first-order differential equations and a set of output equations.

- Solution of the N first-order differential equations yields the values of the state variables. The output is expressed in terms of these values and the input. Solution of the state equations can be obtained by time-domain or frequency-domain methods.
- The state-space model of a system can be derived from its differential equation, transfer function, or realization diagram.
- The state-space model is not unique, since there are infinite realizations of a system with the same input-output relationship.
- Since it is an internal description of the system, by using linear transformation of the state vector, we can obtain another realization of the system, although of the same input-output relationship, with different characteristics, such as sensitivity to parameter variations, number of components required, etc.
- State-space models can be easily extended to the analysis of time-varying and nonlinear systems and systems with multiple inputs and multiple outputs.

Exercises

12.1 Find the zero-input and zero-state components of the output of the circuit, described in Example 12.1, using the time-domain state-space method. Find the total output also.

12.2 Consider the series RLC circuit with a resistor of 9 ohms, an inductance of 3 henries, and a capacitor of $\frac{1}{6}$ farads. The initial current through the inductor is 2 amperes, and the initial voltage across the capacitor is 3 volts. This circuit is excited with a voltage source $x(t) = 2e^{-3t}u(t)$ volts. Assuming the current in the circuit as the output and the current through the inductor, $q1$, and the voltage across the capacitor, $q2$, as the state variables, find the state-space model of the circuit. Find the zero-input and zero-state components of the output of the circuit using the frequency-domain state-space method. Find the total output also.

12.3 Consider the series RLC circuit with a resistor of 8 ohms, an inductance of 2 henries, and a capacitor of $\frac{1}{6}$ farads. The initial current through the inductor is 4 amperes, and the initial voltage across the capacitor is 3 volts. This circuit is excited with a voltage source $x(t) = 3u(t)$ volts. Assuming the inductor voltage as the output and the current through the inductor, $q1$, and the voltage across the capacitor, $q2$, as the state variables, find the state-space model of the circuit. Find the zero-input and zero-state components of the output of the circuit using the time-domain state-space method. Find the total output also.

* **12.4** Consider the series RLC circuit with a resistor of 2 ohms, an inductance of 1 henry, and a capacitor of 1 farad. The initial current through the inductor is 0 ampere, and the initial voltage across the capacitor is 0 volt. This circuit is excited with a voltage source $x(t) = 4e^{-t}u(t)$ volts. Assuming the voltage across the resistor

as the output and the current through the inductor, $q1$, and the voltage across the capacitor, $q2$, as the state variables, find the state-space model of the circuit. Find the output of the circuit using the frequency-domain state-space method.

12.5 Consider the system described by the differential equation

$$\ddot{y}(t) + 5\dot{y}(t) + 6y(t) = 2\ddot{x}(t) - 3\dot{x}(t) + 4x(t)$$

with the initial conditions $y(0^-) = 2$ and $\dot{y}(0^-) = 1$ and the input $x(t) = 2u(t)$. Assign two state variables to the output of each integrator, and assume canonical form I realization of the system as shown in Fig. 12.1. Find the zero-input and zero-state components of the output of the system using the time-domain state-space method. Find the total output also.

* **12.6** Consider the system described by the differential equation

$$\ddot{y}(t) + 4\dot{y}(t) + 3y(t) = \ddot{x}(t) - 2\dot{x}(t) + 3x(t)$$

with the initial conditions $y(0^-) = 3$ and $\dot{y}(0^-) = 1$ and the input $x(t) = 3e^{-2t}u(t)$. Assign two state variables to the output of each integrator, and assume canonical form I realization of the system as shown in Fig. 12.1. Find the zero-input and zero-state components of the output of the system using the time-domain state-space method. Find the total output also.

12.7 Consider the system described by the differential equation

$$\ddot{y}(t) + 5\dot{y}(t) + 4y(t) = x(t)$$

with the input $x(t) = \sin(t + \frac{\pi}{3})u(t)$. Assign two state variables to the output of each integrator, and assume canonical form I realization of the system as shown in Fig. 12.1. Find the zero-state output of the system using the time-domain state-space method.

12.8 Find the impulse response of the system characterized by the differential equation, with input $x(t)$ and output $y(t)$,

$$\ddot{y}(t) + 2\dot{y}(t) + y(t) = \ddot{x}(t) + \dot{x}(t) + 2x(t)$$

using the time-domain state-space method. Assign two state variables to the output of each integrator, and assume canonical form I realization of the system as shown in Fig. 12.1.

* **12.9** Find the impulse response of the system characterized by the differential equation, with input $x(t)$ and output $y(t)$,

$$\ddot{y}(t) + \frac{5}{6}\dot{y}(t) + \frac{1}{6}y(t) = \dot{x}(t) + x(t)$$

using the time-domain state-space method. Assign two state variables to the output of each integrator, and assume canonical form I realization of the system as shown in Fig. 12.1.

12.10 Find the impulse response of the system characterized by the differential equation, with input $x(t)$ and output $y(t)$,

$$\ddot{y}(t) + 6\dot{y}(t) + 5y(t) = 2\ddot{x}(t)$$

using the time-domain state-space method. Assign two state variables to the output of each integrator, and assume canonical form I realization of the system as shown in Fig. 12.1.

* **12.11** Consider the system described by the differential equation

$$\ddot{y}(t) + 3\dot{y}(t) + 2y(t) = 3\ddot{x}(t) - \dot{x}(t) + 4x(t)$$

with the initial conditions $y(0^-) = 2$ and $\dot{y}(0^-) = 3$ and the input $x(t) = 3u(t)$. Assign two state variables to the output of each integrator, and assume canonical form I realization of the system as shown in Fig. 12.1. Find the zero-input and zero-state components of the output of the system using the frequency-domain state-space method. Find the total output also.

12.12 Consider the system described by the differential equation

$$\ddot{y}(t) + 6\dot{y}(t) + 9y(t) = -2\ddot{x}(t) + \dot{x}(t) - 3x(t)$$

with the initial conditions $y(0^-) = -2$ and $\dot{y}(0^-) = -3$ and the input $x(t) = 2e^{-4t}u(t)$. Assign two state variables to the output of each integrator, and assume canonical form I realization of the system as shown in Fig. 12.1. Find the zero-input and zero-state components of the output of the system using the frequency-domain state-space method. Find the total output also.

12.13 Consider the system described by the differential equation

$$\ddot{y}(t) + 6\dot{y}(t) + 8y(t) = x(t)$$

with the input $x(t) = \cos(2t - \frac{\pi}{6})u(t)$. Assign two state variables to the output of each integrator, and assume canonical form I realization of the system as shown in Fig. 12.1. Find the zero-state output of the system using the frequency-domain state-space method.

12.14 Find the impulse response of the system characterized by the differential equation, with input $x(t)$ and output $y(t)$,

$$\ddot{y}(t) + 4\dot{y}(t) + 3y(t) = 3\ddot{x}(t) - 2\dot{x}(t) + x(t)$$

using the frequency-domain state-space method. Assign two state variables to the output of each integrator, and assume canonical form I realization of the system as shown in Fig. 12.1.

* **12.15** Find the impulse response of the system characterized by the differential equation, with input $x(t)$ and output $y(t)$,

$$\ddot{y}(t) + 2\dot{y}(t) + y(t) = -2\ddot{x}(t) + 3\dot{x}(t) - 4x(t)$$

using the frequency-domain state-space method. Assign two state variables to the output of each integrator, and assume canonical form I realization of the system as shown in Fig. 12.1.

12.16 Find the impulse response of the system characterized by the differential equation, with input $x(t)$ and output $y(t)$,

$$\ddot{y}(t) + 7\dot{y}(t) + 12y(t) = -3\dot{x}(t) + 2x(t)$$

using the frequency-domain state-space method. Assign two state variables to the output of each integrator, and assume canonical form I realization of the system as shown in Fig. 12.1.

12.17 Derive the state-space model of the system in Example 12.1 with the new state vector $\overline{\boldsymbol{q}}$ that is related to old state vector $\boldsymbol{q}$ as

$$\overline{q}_1(t) = q_2(t)$$
$$\overline{q}_2(t) = q_1(t)$$

Verify that the transfer function remains the same using either state-space model. Further verify that (i) the traces and determinants of matrices $\boldsymbol{A}$ and $\overline{\boldsymbol{A}}$ are equal and (ii) the determinants of $(s\boldsymbol{I} - \boldsymbol{A})$ and $(s\boldsymbol{I} - \overline{\boldsymbol{A}})$ are the same.

12.18 Derive the state-space model of the system in Example 12.1 with the new state vector $\overline{\boldsymbol{q}}$ that is related to old state vector $\boldsymbol{q}$ as

$$\overline{q}_1(t) = q_1(t) + q_2(t)$$
$$\overline{q}_2(t) = q_1(t) - q_2(t)$$

Verify that the transfer function remains the same using either state-space model. Further verify that (i) the traces and determinants of matrices $\boldsymbol{A}$ and $\overline{\boldsymbol{A}}$ are equal and (ii) the determinants of $(s\boldsymbol{I} - \boldsymbol{A})$ and $(s\boldsymbol{I} - \overline{\boldsymbol{A}})$ are the same.

12.19 Derive the state-space model of the system in Example 12.1 with the new state vector $\overline{\boldsymbol{q}}$ that is related to old state vector $\boldsymbol{q}$ as

$$\overline{q}_1(t) = q_1(t)$$

$$\overline{q}_2(t) = q_1(t) + q_2(t)$$

Verify that the transfer function remains the same using either state-space model. Further verify that (i) the traces and determinants of matrices $\boldsymbol{A}$ and $\overline{\boldsymbol{A}}$ are equal and (ii) the determinants of $(s\boldsymbol{I} - \boldsymbol{A})$ and $(s\boldsymbol{I} - \overline{\boldsymbol{A}})$ are the same.

Appendix A
Complex Numbers

The complex number system is an extension of the real number system. A complex number is an ordered pair of real numbers, a two-element vector. The complex number $z = 2 + j1$, called its rectangular form , is shown in Fig. A.1. The two real numbers a and b are called, respectively, the real and imaginary parts of the complex number z, and $j = \sqrt{-1}$ is the imaginary unit. The necessity for complex numbers is that it is more efficient to represent related entities in the vector form. For example, at a given frequency, a sinusoid is defined by its amplitude and phase. In signal analysis, the complex form of representing the amplitude and phase of a sinusoid is more convenient than by two scalars. In a Cartesian coordinate system, a point is represented by its distance from a set of perpendicular lines that intersect at the origin of the system. A Cartesian coordinate system in which the horizontal and vertical axes represent, respectively, the real and imaginary parts of a complex number is called a complex plane. Complex numbers $z = a + jb$ and $p = c + jd$ are equal, if and only if $a = c$ and $b = d$.

A complex number $z = a + jb$ can be written in its polar or exponential form $Ae^{j\theta}$. The representation of the complex number $z = 2 + j1$ is $\sqrt{5}e^{j26.5651}$ using degree measure for the angle, as shown in Fig. A.1. The magnitude A and phase θ are, respectively,

$$A = \sqrt{a^2 + b^2} \qquad \text{and} \qquad \theta = \tan^{-1}\frac{b}{a}$$

The inverse relations are

$$a = A\cos(\theta) \qquad \text{and} \qquad b = A\sin(\theta)$$

The real number system is a subset of the complex number system. Therefore, all the operations, if the imaginary parts are zero, reduce to real arithmetic operations.

Addition and Subtraction

D. Sundararajan, *Signals and Systems*,
https://doi.org/10.1007/978-3-031-19377-4

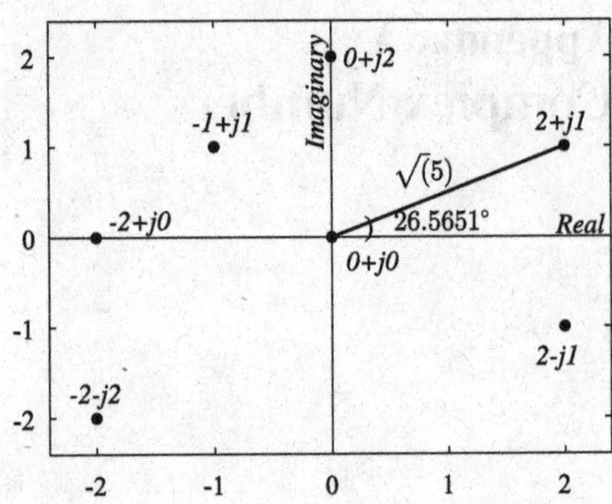

Fig. A.1 The complex plane with some complex numbers

Let the two numbers be $z = a + jb$ and $p = c + jd$. Then,

$$q = z \pm p = (a \pm c) + j(b \pm d)$$

With $z = 2 + j3$ and $p = 1 - j4$, $q = z + p = 3 - j1$ and $q = z - p = 1 + j7$.

Multiplication
Let the two numbers be $z = a + jb$ and $p = c + jd$. Then,

$$q = (z)(p) = (a + jb)(c + jd) = (ac - bd) + j(ad + bc),$$

where $j^2 = -1$. With $z = 2 + j3$ and $p = 1 - j4$, $q = (z)(p) = 14 - j5$.

In polar form,

$$q = (z)(p) = (a + jb)(c + jd) = Ae^{j\theta}Ce^{j\phi} = ACe^{j(\theta+\phi)}$$

$$(2 + j3)(1 - j4) = 3.6056e^{j0.9828}4.1231e^{-j1.3258} = 14.8661e^{-j0.3430} = 14 - j5$$

using radian measure.

Complex Conjugate
The conjugate of a complex number $z = a + jb$ is $z^* = a - jb$, obtained by replacing j by $-j$. z^* is the mirror image of z about the real axis in the complex plane. In polar form, the conjugate of $Ae^{j\theta}$ is $Ae^{-j\theta}$. Obviously, the product of a complex number with its conjugate is its magnitude squared, A^2. That is,

$$(z)z^* = (a + jb)(a - jb) = a^2 + b^2$$

$$z + z^* = 2a \quad \text{and} \quad z - z^* = j2b$$

Division

With $z = a + jb$ and $p = c + jd$,

$$q = \frac{z}{p} = \frac{zp^*}{pp^*} = \frac{zp^*}{|p|^2} = \frac{ac + bd}{c^2 + d^2} + j\frac{bc - ad}{c^2 + d^2}$$

In polar form,

$$q = \frac{z}{p} = \frac{Ae^{j\theta}}{Ce^{j\phi}} = \frac{A}{C}e^{j(\theta - \phi)}$$

For example,

$$\frac{(14 - j5)}{(2 + j3)} = \frac{14.8661e^{-j0.3430}}{3.6056e^{j0.9828}} = 4.1231e^{-j1.3258} = (1 - j4)$$

Powers and Roots of Complex Numbers

Since $x^2 \geq 0$ for all real numbers, the quadratic equation $x^2 = -1$ has no solution in the real number system. In the complex number system, the two roots are j and $-j$, and, in fact, every polynomial equation does have a solution.

$$z^N = (Ae^{j\theta})^N = A^n e^{jN\theta}$$

Replacing N by $1/N$ and adding $2k\pi$ to θ, we get

$$z^{\frac{1}{N}} = +\sqrt[N]{A}e^{\frac{j(\theta + 2k\pi)}{N}} = +\sqrt[N]{A}\left(\cos\left(\frac{(\theta + 2k\pi)}{N}\right) + j\sin\left(\frac{(\theta + 2k\pi)}{N}\right)\right),$$
$$k = 0, 1, 2, \ldots, N-1$$

With $A = 1$ and $\theta = 0$, we get the Nth roots of unity, which form the DFT basis functions.

$$1^{\frac{1}{N}} = \cos\left(\frac{2k\pi}{N}\right) + j\sin\left(\frac{2k\pi}{N}\right), \quad k = 0, 1, 2, \ldots, N - 1$$

For example, with $N = 4$, we get the roots as $\{1, j, -1, -j\}$. Each root raised to the power of 4 will yield 1. Since the magnitude of the roots is 1, their angles add to $\{0, 2\pi, 4\pi, -2\pi\}$. The complex number with these arguments is 1.

Appendix B
Transform Pairs and Properties

See Tables B.1, B.2, B.3, B.4, B.5, B.6, B.7, B.8, B.9, B.10, B.11, B.12.

Table B.1 DFT pairs

$x(n)$, period $= N$	$X(k)$, period $= N$
$\delta(n)$	1
1	$N\delta(k)$
$e^{j(\frac{2\pi}{N}mn)}$	$N\delta(k-m)$
$\cos(\frac{2\pi}{N}mn)$	$\frac{N}{2}(\delta(k-m)+\delta(k-(N-m)))$
$\sin(\frac{2\pi}{N}mn)$	$\frac{N}{2}(-j\delta(k-m)+j\delta(k-(N-m)))$
$x(n) = \begin{cases} 1 \text{ for } n = 0, 1, \ldots, L-1 \\ 0 \text{ for } n = L, L+1, \ldots, N-1 \end{cases}$	$e^{(-j\frac{\pi}{N}(L-1)k)} \frac{\sin(\frac{\pi}{N}kL)}{\sin(\frac{\pi}{N}k)}$

D. Sundararajan, *Signals and Systems*,
https://doi.org/10.1007/978-3-031-19377-4

Table B.2 DFT properties

Property	$x(n), h(n)$, period $= N$	$X(k), H(k)$, period $= N$
Linearity	$ax(n) + bh(n)$	$aX(k) + bH(k)$
Duality	$\frac{1}{N}X(N \mp n)$	$x(N \pm k)$
Time shifting	$x(n \pm m)$	$e^{\pm j\frac{2\pi}{N}mk}X(k)$
Frequency shifting	$e^{\mp j\frac{2\pi}{N}mn}x(n)$	$X(k \pm m)$
Time convolution	$\sum_{m=0}^{N-1} x(m)h(n-m)$	$X(k)H(k)$
Frequency convolution	$x(n)h(n)$	$\frac{1}{N}\sum_{m=0}^{N-1} X(m)H(k-m)$
Time expansion	$h(mn) = \begin{cases} x(n) & \text{for } n = 0, 1, \ldots, N-1 \\ 0 & \text{otherwise} \end{cases}$ where m is any positive integer	$H(k) = X(k \bmod N)$, $k = 0, 1, \ldots, mN-1$
Time reversal	$x(N-n)$	$X(N-k)$
Conjugation	$x^*(N \pm n)$	$X^*(N \mp k)$
Parseval's theorem	$\sum_{n=0}^{N-1} \lvert x(n)\rvert^2$	$\frac{1}{N}\sum_{k=0}^{N-1} \lvert X(k)\rvert^2$

Table B.3 FS pairs

$x(t)$, period $= T$	$X_{cs}(k),\ \omega_0 = \frac{2\pi}{T}$
$\begin{cases} 1 & \text{for } \lvert t\rvert < a \\ 0 & \text{for } a < \lvert t\rvert \le \frac{T}{2} \end{cases}$	$\frac{\sin(k\omega_0 a)}{k\pi}$
$\sum_{n=-\infty}^{\infty} \delta(t-nT)$	$\frac{1}{T}$
$e^{jk_0\omega_0 t}$	$\delta(k-k_0)$
$\cos(k_0\omega_0 t)$	$0.5(\delta(k+k_0) + \delta(k-k_0))$
$\sin(k_0\omega_0 t)$	$0.5j(\delta(k+k_0) - \delta(k-k_0))$

Table B.4 FS properties

Property	$x(t), h(t)$, period $= T$	$X_{cs}(k), H_{cs}(k),\ \omega_0 = \frac{2\pi}{T}$
Linearity	$ax(t) + bh(t)$	$aX_{cs}(k) + bH_{cs}(k)$
Time shifting	$x(t \pm t_0)$	$e^{\pm jk\omega_0 t_0}X_{cs}(k)$
Frequency shifting	$x(t)e^{\pm jk_0\omega_0 t}$	$X_{cs}(k \mp k_0)$
Time convolution	$\int_0^T x(\tau)h(t-\tau)d\tau$	$TX_{cs}(k)H_{cs}(k)$
Frequency convolution	$x(t)h(t)$	$\sum_{l=-\infty}^{\infty} X_{cs}(l)H_{cs}(k-l)$
Time scaling	$x(at), a > 0$, Period $= \frac{T}{a}$	$X_{cs}(k),\ \omega_0 = a\frac{2\pi}{T}$
Time reversal	$x(-t)$	$X_{cs}(-k)$
Time differentiation	$\frac{d^n x(t)}{dt^n}$	$(jk\omega_0)^n X_{cs}(k)$
Time integration	$\int_{-\infty}^{t} x(\tau)d\tau$	$\frac{X_{cs}(k)}{jk\omega_0}$, if $(X_{cs}(0) = 0)$
Parseval's theorem	$\frac{1}{T}\int_0^T \lvert x(t)\rvert^2 dt$	$\sum_{k=-\infty}^{\infty} \lvert X_{cs}(k)\rvert^2$
Conjugate symmetry	$x(t)$ real	$X_{cs}(k) = X_{cs}^*(-k)$
Even symmetry	$x(t)$ real and even	$X_{cs}(k)$ real and even
Odd symmetry	$x(t)$ real and odd	$X_{cs}(k)$ imaginary and odd

Table B.5 DTFT pairs

$x(n)$	$X(e^{j\omega})$
$\begin{cases} 1 \text{ for } -N \le n \le N \\ 0 \text{ otherwise} \end{cases}$	$\frac{\sin(\omega \frac{(2N+1)}{2})}{\sin(\frac{\omega}{2})}$
$\frac{\sin(an)}{\pi n}$, $0 < a \le \pi$	$\begin{cases} 1 \text{ for } \lvert\omega\rvert < a \\ 0 \text{ for } a < \lvert\omega\rvert \le \pi \end{cases}$
$a^n u(n)$, $\lvert a\rvert < 1$	$\frac{1}{1-ae^{-j\omega}}$
$(n+1)a^n u(n)$, $\lvert a\rvert < 1$	$\frac{1}{(1-ae^{-j\omega})^2}$
$a^{\lvert n\rvert}$, $\lvert a\rvert < 1$	$\frac{1-a^2}{1-2a\cos(\omega)+a^2}$
$a^n \sin(\omega_0 n)u(n)$, $\lvert a\rvert < 1$	$\frac{(a)e^{-j\omega}\sin(\omega_0)}{1-2(a)e^{-j\omega}\cos(\omega_0)+(a)^2e^{-j2\omega}}$
$a^n \cos(\omega_0 n)u(n)$, $\lvert a\rvert < 1$	$\frac{1-(a)e^{-j\omega}\cos(\omega_0)}{1-2(a)e^{-j\omega}\cos(\omega_0)+(a)^2e^{-j2\omega}}$
$\delta(n)$	1
$\sum_{k=-\infty}^{\infty} \delta(n-kN)$	$\frac{2\pi}{N}\sum_{k=-\infty}^{\infty} \delta(\omega - \frac{2\pi}{N}k)$
$u(n)$	$\pi\delta(\omega) + \frac{1}{1-e^{-j\omega}}$
1	$2\pi\delta(\omega)$
$e^{j\omega_0 n}$	$2\pi\delta(\omega-\omega_0)$
$\cos(\omega_0 n)$	$\pi(\delta(\omega+\omega_0)+\delta(\omega-\omega_0))$
$\sin(\omega_0 n)$	$j\pi(\delta(\omega+\omega_0)-\delta(\omega-\omega_0))$

Table B.6 DTFT properties

Property	$x(n), h(n)$	$X(e^{j\omega}), H(e^{j\omega})$
Linearity	$ax(n)+bh(n)$	$aX(e^{j\omega})+bH(e^{j\omega})$
Time shifting	$x(n \pm n_0)$	$e^{\pm j\omega n_0}X(e^{j\omega})$
Frequency shifting	$x(n)e^{\pm j\omega_0 n}$	$X(e^{j(\omega\mp\omega_0)})$
Time convolution	$\sum_{m=-\infty}^{\infty} x(m)h(n-m)$	$X(e^{j\omega})H(e^{j\omega})$
Frequency convolution	$x(n)h(n)$	$\frac{1}{2\pi}\int_0^{2\pi} X(e^{jv})H(e^{j(\omega-v)})dv$
Time expansion	$h(n)$ $h(an)=x(n)$, $a>0$ is a positive integer and $h(n)=0$ zero otherwise	$H(e^{j\omega})=X(e^{ja\omega})$
Time reversal	$x(-n)$	$X(e^{-j\omega})$
Conjugation	$x^*(\pm n)$	$X^*(e^{\mp j\omega})$
Difference	$x(n)-x(n-1)$	$(1-e^{-j\omega})X(e^{j\omega})$
Summation	$\sum_{l=-\infty}^{n} x(l)$	$\frac{X(e^{j\omega})}{(1-e^{-j\omega})}+\pi X(e^{j0})\delta(\omega)$
Frequency differentiation	$(n)^m x(n)$	$(j)^m \frac{d^m X(e^{j\omega})}{d\omega^m}$
Parseval's theorem	$\sum_{n=-\infty}^{\infty} \lvert x(n)\rvert^2$	$\frac{1}{2\pi}\int_0^{2\pi} \lvert X(e^{j\omega})\rvert^2 d\omega$
Conjugate symmetry	$x(n)$ real	$X(e^{j\omega})=X^*(e^{-j\omega})$
Even symmetry	$x(n)$ real and even	$X(e^{j\omega})$ real and even
Odd symmetry	$x(n)$ real and odd	$X(e^{j\omega})$ imaginary and odd

Table B.7 FT pairs

$x(t)$	$X(j\omega)$
$u(t+a)-u(t-a)$	$2\frac{\sin(\omega a)}{\omega}$
$\frac{\sin(\omega_0 t)}{\pi t}$	$u(\omega+\omega_0)-u(\omega-\omega_0)$
$e^{-at}u(t),\ \text{Re}(a)>0$	$\frac{1}{a+j\omega}$
$te^{-at}u(t),\ \text{Re}(a)>0$	$\frac{1}{(a+j\omega)^2}$
$e^{-a\|t\|},\ \text{Re}(a)>0$	$\frac{2a}{a^2+\omega^2}$
$\frac{1}{a}((t+a)u(t+a)-2tu(t)+(t-a)u(t-a))$	$a\left(\frac{\sin(\omega\frac{a}{2})}{\omega\frac{a}{2}}\right)^2$
$e^{-at}\sin(\omega_0 t)u(t),\ \text{Re}(a)>0$	$\frac{\omega_0}{(a+j\omega)^2+\omega_0^2}$
$e^{-at}\cos(\omega_0 t)u(t),\ \text{Re}(a)>0$	$\frac{a+j\omega}{(a+j\omega)^2+\omega_0^2}$
$\delta(t)$	1
$\sum_{n=-\infty}^{\infty}\delta(t-nT)$	$\frac{2\pi}{T}\sum_{k=-\infty}^{\infty}\delta(\omega-k\frac{2\pi}{T})$
$u(t)$	$\pi\delta(\omega)+\frac{1}{j\omega}$
1	$2\pi\delta(\omega)$
$e^{j\omega_0 t}$	$2\pi\delta(\omega-\omega_0)$
$\cos(\omega_0 t)$	$\pi(\delta(\omega+\omega_0)+\delta(\omega-\omega_0))$
$\sin(\omega_0 t)$	$j\pi(\delta(\omega+\omega_0)-\delta(\omega-\omega_0))$

Table B.8 FT properties

Property	$x(t)$, $h(t)$	$X(j\omega)$, $H(j\omega)$
Linearity	$ax(t)+bh(t)$	$aX(j\omega)+bH(j\omega)$
Duality	$X(\pm t)$	$2\pi x(\mp j\omega)$
Time shifting	$x(t\pm t_0)$	$X(j\omega)e^{\pm j\omega t_0}$
Frequency shifting	$x(t)e^{\pm j\omega_0 t}$	$X(j(\omega\mp\omega_0))$
Time convolution	$x(t)*h(t)$	$X(j\omega)H(j\omega)$
Frequency convolution	$x(t)h(t)$	$\frac{1}{2\pi}(X(j\omega)*H(j\omega))$
Time scaling	$x(at)$, $a\neq 0$ and real	$\frac{1}{\|a\|}X(j\frac{\omega}{a})$
Time reversal	$x(-t)$	$X(-j\omega)$
Conjugation	$x^*(\pm t)$	$X^*(\mp j\omega)$
Time differentiation	$\frac{d^n x(t)}{dt^n}$	$(j\omega)^n X(j\omega)$
Time integration	$\int_{-\infty}^{t}x(\tau)d\tau$	$\frac{X(j\omega)}{j\omega}+\pi X(j0)\delta(\omega)$
Frequency differentiation	$t^n x(t)$	$(j)^n\frac{d^n X(j\omega)}{d\omega^n}$
Parseval's theorem	$\int_{-\infty}^{\infty}\|x(t)\|^2dt$	$\frac{1}{2\pi}\int_{-\infty}^{\infty}\|X(j\omega)\|^2d\omega$
Autocorrelation	$x(t)*x(-t)=\int_{-\infty}^{\infty}x(\tau)x(\tau-t)d\tau$	$\|X(j\omega)\|^2$
Conjugate symmetry	$x(t)$ real	$X(j\omega)=X^*(-j\omega)$
Even symmetry	$x(t)$ real and even	$X(j\omega)$ real and even
Odd symmetry	$x(t)$ real and odd	$X(j\omega)$ imaginary and odd

Table B.9 z-transform pairs

$x(n)$	$X(z)$	ROC
$\delta(n)$	1	$\|z\| \geq 0$
$\delta(n-p),\ p>0$	z^{-p}	$\|z\| > 0$
$u(n)$	$\frac{z}{z-1}$	$\|z\| > 1$
$a^n u(n)$	$\frac{z}{z-a}$	$\|z\| > \|a\|$
$na^n u(n)$	$\frac{az}{(z-a)^2}$	$\|z\| > \|a\|$
$nu(n)$	$\frac{z}{(z-1)^2}$	$\|z\| > \|1\|$
$\cos(\omega_0 n)u(n)$	$\frac{z(z-\cos(\omega_0))}{z^2-2z\cos(\omega_0)+1}$	$\|z\| > 1$
$\sin(\omega_0 n)u(n)$	$\frac{z\sin(\omega_0)}{z^2-2z\cos(\omega_0)+1}$	$\|z\| > 1$
$a^n\cos(\omega_0 n)u(n)$	$\frac{z(z-a\cos(\omega_0))}{z^2-2az\cos(\omega_0)+a^2}$	$\|z\| > \|a\|$
$a^n\sin(\omega_0 n)u(n)$	$\frac{az\sin(\omega_0)}{z^2-2az\cos(\omega_0)+a^2}$	$\|z\| > \|a\|$

Table B.10 z-transform properties

Property	$x(n)u(n),\ h(n)u(n)$	$X(z),\ H(z)$
Linearity	$ax(n)u(n)+bh(n)u(n)$	$aX(z)+bH(z)$
Left shift	$x(n+m)u(n),\ m>0$	$z^m X(z) - z^m \sum_{n=0}^{m-1} x(n)z^{-n}$
Right shift	$x(n-m)u(n),\ m>0$	$z^{-m}X(z) + z^{-m}\sum_{n=1}^{m} x(-n)z^n$
Multiplication by a^n	$a^n x(n)u(n)$	$X(\frac{z}{a})$
Time convolution	$x(n)u(n) * h(n)u(n)$	$X(z)H(z)$
Summation	$\sum_{m=0}^{n} x(m)$	$\frac{z}{z-1}X(z)$
Multiplication by n	$nx(n)u(n)$	$-z\frac{dX(z)}{dz}$
Initial value	$x(0)$	$\lim_{z\to\infty} X(z)$
Final value	$\lim_{n\to\infty} x(n)$	$\lim_{z\to 1}((z-1)X(z))$ ROC of $(z-1)X(z)$ includes the unit circle

Table B.11 Laplace transform pairs

$x(t)$	$X(s)$	ROC
$\delta(t)$	1	All s
$u(t)$	$\frac{1}{s}$	$\mathrm{Re}(s) > 0$
$t^n u(t),\ n = 0, 1, 2, \ldots$	$\frac{n!}{s^{n+1}}$	$\mathrm{Re}(s) > 0$
$e^{-at} u(t)$	$\frac{1}{s+a}$	$\mathrm{Re}(s) > -a$
$t^n e^{-at} u(t),\ n = 0, 1, 2, \ldots$	$\frac{n!}{(s+a)^{n+1}}$	$\mathrm{Re}(s) > -a$
$\cos(\omega_0 t)\, u(t)$	$\frac{s}{s^2+\omega_0^2}$	$\mathrm{Re}(s) > 0$
$\sin(\omega_0 t)\, u(t)$	$\frac{\omega_0}{s^2+\omega_0^2}$	$\mathrm{Re}(s) > 0$
$e^{-at} \cos(\omega_0 t)\, u(t)$	$\frac{s+a}{(s+a)^2+\omega_0^2}$	$\mathrm{Re}(s) > -a$
$e^{-at} \sin(\omega_0 t)\, u(t)$	$\frac{\omega_0}{(s+a)^2+\omega_0^2}$	$\mathrm{Re}(s) > -a$
$t \cos(\omega_0 t)\, u(t)$	$\frac{s^2-\omega_0^2}{(s^2+\omega_0^2)^2}$	$\mathrm{Re}(s) > 0$
$t \sin(\omega_0 t)\, u(t)$	$\frac{2\omega_0 s}{(s^2+\omega_0^2)^2}$	$\mathrm{Re}(s) > 0$

Table B.12 Laplace transform properties

Property	$x(t)u(t), h(t)u(t)$	$X(s), H(s)$
Linearity	$ax(t) + bh(t)$	$aX(s) + bH(s)$
Time shifting	$x(t - t_0)u(t - t_0),\ t_0 \geq 0$	$X(s)e^{-st_0}$
Frequency shifting	$x(t)u(t)e^{s_0 t}$	$X(s - s_0)$
Time convolution	$x(t) * h(t)$	$X(s)H(s)$
Time scaling	$x(at), a > 0$ and real	$\frac{1}{a}X(\frac{s}{a})$
Time differentiation	$\frac{dx(t)}{dt}$	$sX(s) - x(0^-)$
Time differentiation	$\frac{d^2x(t)}{dt^2}$	$s^2X(s) - sx(0^-) - \frac{dx(t)}{dt}\|_{t=0^-}$
Time integration	$\int_{0^-}^{t} x(\tau)d\tau$	$\frac{X(s)}{s}$
Time integration	$\int_{-\infty}^{t} x(\tau)d\tau$	$\frac{X(s)}{s} + \frac{1}{s}\int_{-\infty}^{0^-} x(\tau)d\tau$
Frequency differentiation	$tx(t)u(t)$	$-\frac{dX(s)}{ds}$
Frequency differentiation	$t^n x(t)u(t),\ n = 0, 1, 2, \ldots$	$(-1)^n \frac{d^n X(s)}{ds^n}$
Initial value	$x(0^+)$	$\lim_{s\to\infty} sX(s)$, if $X(s)$ is strictly proper
Final value	$\lim_{t\to\infty} x(t)$	$\lim_{s\to 0} sX(s)$, (ROC of $sX(s)$ includes the $j\omega$ axis)

Appendix C
Useful Mathematical Formulas

C.1 Trigonometric Identities

Pythagorean Identity

$$\sin^2 x + \cos^2 x = 1$$

Addition and Subtraction Formulas

$$\sin(x \pm y) = \sin x \cos y \pm \cos x \sin y$$

$$\cos(x \pm y) = \cos x \cos y \mp \sin x \sin y$$

Double-Angle Formulas

$$\cos 2x = \cos^2 x - \sin^2 x = 2\cos^2 x - 1 = 1 - 2\sin^2 x$$

$$\sin 2x = 2 \sin x \cos x$$

Product Formulas

$$2 \sin x \cos y = \sin(x - y) + \sin(x + y)$$

$$2 \cos x \sin y = -\sin(x - y) + \sin(x + y)$$

$$2 \sin x \sin y = \cos(x - y) - \cos(x + y)$$

D. Sundararajan, *Signals and Systems*,
https://doi.org/10.1007/978-3-031-19377-4

$$2\cos x\cos y = \cos(x - y) + \cos(x + y)$$

Sum and Difference Formulas

$$\sin x \pm \sin y = 2\sin\frac{x \pm y}{2}\cos\frac{x \mp y}{2}$$

$$\cos x + \cos y = 2\cos\frac{x + y}{2}\cos\frac{x - y}{2}$$

$$\cos x - \cos y = -2\sin\frac{x + y}{2}\sin\frac{x - y}{2}$$

Other Formulas

$$\sin(-x) = \sin(2\pi - x) = -\sin x$$

$$\cos(-x) = \cos(2\pi - x) = \cos x$$

$$\sin(\pi \pm x) = \mp\sin x$$

$$\cos(\pi \pm x) = -\cos x$$

$$\cos\left(\frac{\pi}{2} \pm x\right) = \mp\sin x$$

$$\sin\left(\frac{\pi}{2} \pm x\right) = \cos x$$

$$\cos\left(\frac{3\pi}{2} \pm x\right) = \pm\sin x$$

$$\sin\left(\frac{3\pi}{2} \pm x\right) = -\cos x$$

$$e^{\pm jx} = \cos x \pm j\sin x$$

$$\cos x = \frac{e^{jx} + e^{-jx}}{2}$$

$$\sin x = \frac{e^{jx} - e^{-jx}}{2j}$$

C.2 Series Expansions

$$e^{jx} = 1 + (jx) + \frac{(jx)^2}{2!} + \frac{(jx)^3}{3!} + \frac{(jx)^4}{4!} + \cdots + \frac{(jx)^r}{(r)!} + \cdots$$

$$\cos(x) = 1 - \frac{x^2}{2!} + \frac{x^4}{4!} - \cdots + (-1)^r \frac{x^{2r}}{(2r)!} - \cdots$$

$$\sin(x) = x - \frac{x^3}{3!} + \frac{x^5}{5!} - \cdots + (-1)^r \frac{x^{2r+1}}{(2r+1)!} - \cdots$$

$$\sin^{-1} x = x + \frac{1}{2}\frac{x^3}{3} + \frac{(1)(3)}{(2)(4)}\frac{x^5}{5} + \frac{(1)(3)(5)}{(2)(4)(6)}\frac{x^7}{7} + \cdots, \quad |x| < 1$$

$$\cos^{-1} x = \frac{\pi}{2} - \sin^{-1} x, \quad |x| < 1$$

C.3 Summation Formulas

$$\sum_{k=0}^{N-1} (a + kd) = \frac{N(2a + (N-1)d)}{2}$$

$$\sum_{k=0}^{N-1} ar^k = \frac{a(1 - r^N)}{1 - r}, \quad r \neq 1$$

$$\sum_{k=0}^{\infty} r^k = \frac{1}{1 - r}, \quad |r| < 1$$

$$\sum_{k=0}^{\infty} kr^k = \frac{r}{(1-r)^2}, \quad |r| < 1$$

$$1 + \cos(t) + \cos(2t) + \cdots + \cos(Nt) = \frac{1}{2} + \frac{\sin(0.5(2N+1)t)}{2\sin(0.5t)}$$

C.4 Indefinite Integrals

$$\int u dv = uv - \int v du$$

$$\int e^{at} dt = \frac{e^{at}}{a}$$

$$\int te^{at} dt = \frac{e^{at}}{a^2}(at - 1)$$

$$\int e^{bt} \sin(at) dt = \frac{e^{bt}}{a^2 + b^2}(b \sin(at) - a \cos(at))$$

$$\int e^{bt} \cos(at) dt = \frac{e^{bt}}{a^2 + b^2}(b \cos(at) + a \sin(at))$$

$$\int \sin(at) dt = -\frac{1}{a} \cos(at)$$

$$\int \cos(at) dt = \frac{1}{a} \sin(at)$$

$$\int t \sin(at) dt = \frac{1}{a^2}(\sin(at) - at \cos(at))$$

$$\int t \cos(at) dt = \frac{1}{a^2}(\cos(at) + at \sin(at))$$

$$\int \sin^2(at) dt = \frac{t}{2} - \frac{1}{4a} \sin(2at)$$

$$\int \cos^2(at)dt = \frac{t}{2} + \frac{1}{4a}\sin(2at)$$

C.5 Differentiation Formulas

$$\frac{d(uv)}{dt} = u\frac{dv}{dt} + v\frac{du}{dt}$$

$$\frac{d(\frac{u}{v})}{dt} = \frac{v\frac{du}{dt} - u\frac{dv}{dt}}{v^2}$$

$$\frac{d(x^n)}{dt} = nx^{n-1}$$

$$\frac{d(e^{at})}{dt} = ae^{at}$$

$$\frac{d(\sin(at))}{dt} = a\cos(at)$$

$$\frac{d(\cos(at))}{dt} = -a\sin(at)$$

C.6 L'Hôpital's Rule

$$\text{If } \lim_{x\to a} f(x) = 0 \text{ and } \lim_{x\to a} g(x) = 0, \text{ or}$$

$$\text{If } \lim_{x\to a} f(x) = \infty \text{ and } \lim_{x\to a} g(x) = \infty, \text{ then}$$

$$\lim_{x\to a}\frac{f(x)}{g(x)} = \lim_{x\to a}\frac{\frac{df(x)}{dx}}{\frac{dg(x)}{dx}}$$

The rule can be applied as many times as necessary.

C.7 Matrix Inversion

A rectangular array of numbers is called a matrix. For example, a $N \times N$ matrix $\boldsymbol{A}$ is given by

$$\mathbf{A} = \begin{bmatrix} a_{11} & a_{21} & \cdots & a_{N1} \\ a_{12} & a_{22} & \cdots & a_{N2} \\ . & . & \cdots & . \\ a_{1N} & a_{2N} & \cdots & a_{NN} \end{bmatrix},$$

where a_{ij} are the elements of matrix. The elements may be constants, variables, or functions. The horizontal lines are the row vectors, and the vertical lines are the column vectors. A matrix with M rows and N columns is referred to as a $M \times N$ matrix. The subscripts ij in a_{ij} refer to the element in the ith row and jth column. The minimum of the maximum number of linearly independent rows and columns in a matrix is its rank.

A cofactor M_{ij} of a_{ij} is $(-1)^{i+j}$ multiplied by the determinant of $(N-1) \times (N-1)$ matrix obtained by deleting the ith row and the jth column of the $N \times N$ matrix.

The determinant of a 2×2 matrix is given by

$$\begin{vmatrix} a_{11} & a_{12} \\ a_{21} & a_{22} \end{vmatrix} = a_{11}a_{22} - a_{12}a_{21}$$

The determinant D of a 3×3 matrix is given by

$$D = \begin{vmatrix} a_{11} & a_{12} & a_{13} \\ a_{21} & a_{22} & a_{23} \\ a_{31} & a_{32} & a_{33} \end{vmatrix} = a_{11} \begin{vmatrix} a_{22} & a_{23} \\ a_{32} & a_{33} \end{vmatrix} - a_{21} \begin{vmatrix} a_{12} & a_{13} \\ a_{32} & a_{33} \end{vmatrix} + a_{31} \begin{vmatrix} a_{12} & a_{13} \\ a_{22} & a_{23} \end{vmatrix}$$

or

$$D = a_{11}M_{11} - a_{21}M_{21} + a_{31}M_{31}$$

cofactor Mij check.
The inverse, $\mathbf{A}^{-1}$, of a nonsingular $N \times N$ matrix $\mathbf{A}$ is defined as

$$\mathbf{A}^{-1} = \frac{1}{\det \mathbf{A}} \begin{bmatrix} M_{11} & M_{21} & \cdots & M_{N1} \\ M_{12} & M_{22} & \cdots & M_{N2} \\ . & . & \cdots & . \\ M_{1N} & M_{2N} & \cdots & M_{NN} \end{bmatrix}$$

where M_{ij} is the cofactor of a_{ji} in $\mathbf{A}$.

The inverse, $\mathbf{A}^{-1}$, of a 2×2 matrix

$$\mathbf{A} = \begin{bmatrix} a & b \\ c & d \end{bmatrix} \text{ is defined as } \mathbf{A}^{-1} = \frac{1}{ad - bc} \begin{bmatrix} d & -b \\ -c & a \end{bmatrix}$$

provided $ad - bc \neq 0$.

Answers to Selected Exercises

Chapter 1

1.1.2 Energy $\frac{100}{9}$.
1.3.3

$$x_e(n) = \begin{cases} \frac{(0.4)^n}{2} \text{ for } n > 0 \\ 1 \text{ for } n = 0 \\ \frac{(0.4)^{-n}}{2} \text{ for } n < 0 \end{cases}$$

$$x_e(-3) = 0.032, x_e(-2) = 0.08, x_e(-1) = 0.20, x_e(0) = 1,$$

$$x_e(1) = 0.20, x_e(2) = 0.08, x_e(3) = 0.032$$

$$x_0(n) = \begin{cases} \frac{(0.4)^n}{2} \text{ for } n > 0 \\ 0 \text{ for } n = 0 \\ -\frac{(0.4)^{-n}}{2} \text{ for } n < 0 \end{cases}$$

$$x_o(-3) = -0.032, x_o(-2) = -0.08, x_o(-1) = -0.20, x_o(0) = 0,$$

$$x_o(1) = 0.20, x_o(2) = 0.08, x_o(3) = 0.032$$

$$x(n) = x_e(n) + x_0(n)$$

$$x(-3) = 0, x(-2) = 0, x(-1) = 0, x(0) = 1.0000, x(1) = 0.4000,$$
$$x(2) = 0.16, x(3) = 0.064$$

D. Sundararajan, *Signals and Systems*,
https://doi.org/10.1007/978-3-031-19377-4

The sum of the values of the even component is 1.624 and that of the signal is also 1.624.

1.4.2 0

1.6.5 Periodic with period 9.

1.7.6

$$x(n) = 2\sqrt{3}\cos\left(\frac{\pi}{6}n\right) + 2\sin\left(\frac{\pi}{6}n\right)$$

$$3.4641, 4, 3.4641, 2, 0, -2, -3.4641, -4, -3.4641, -2, 0, 2$$

1.8.3

$$x(n) = 2\sqrt{3}\cos\left(\frac{\pi}{6}n - \frac{\pi}{6}\right)$$

$$3, 3.4641, 3, 1.7321, 0, -1.7321, -3, -3.4641, -3, -1.7321, 0, 1.7321$$

1.10.5 $x(n) = 5.9544e^{j(\frac{\pi}{3}n+0.6984)}$.

$$1.7321 + j1, j2, -1.7321 + j1, -1.7321 - j1, -j2, 1.7321 - j1$$

$$2.8284 + j2.8284, -1.0353 + j3.8637, -3.8637 + j1.0353, -2.8284 - j2.8284,$$

$$1.0353 - j3.8637, 3.8637 - j1.0353$$

$$4.5605 + j3.8284, -1.0353 + j5.8637, -5.5958 + j2.0353, -4.5605 - j3.8284,$$

$$1.0353 - j5.8637, 5.5958 - j2.0353$$

1.11.3 $x(n) = (0.5)^n$.

$$x(0) = 1, x(1) = 0.5, x(2) = 0.25, x(3) = 0.125, x(4) = 0.0625, x(5) = 0.0313$$

1.13.4 $-3\sin(5\frac{2\pi}{8}n + \frac{\pi}{3}), 3\sin(11\frac{2\pi}{8}n - \frac{\pi}{3}), -3\sin(13\frac{2\pi}{8}n + \frac{\pi}{3})$.

1.14.3 11 samples per second.

1.15.5

$$0, -\frac{\sqrt{3}}{2}, -\frac{\sqrt{3}}{2}, 0, \frac{\sqrt{3}}{2}, \frac{\sqrt{3}}{2}$$

$$x(n-7) = \cos\left(\frac{2\pi}{6}n + \frac{\pi}{6}\right)$$

$$\frac{\sqrt{3}}{2}, 0, -\frac{\sqrt{3}}{2}, -\frac{\sqrt{3}}{2}, 0, \frac{\sqrt{3}}{2}$$

1.16.3

$$\frac{\sqrt{3}}{2}, \frac{\sqrt{3}}{2}, 0, -\frac{\sqrt{3}}{2}, -\frac{\sqrt{3}}{2}, 0$$

$x(-n+1) = \cos(\frac{2\pi}{6}n - \frac{\pi}{6}) = x(n)$

$$\frac{\sqrt{3}}{2}, \frac{\sqrt{3}}{2}, 0, -\frac{\sqrt{3}}{2}, -\frac{\sqrt{3}}{2}, 0$$

1.17.3

$$0.5, 1, 0.5, -0.5, -1, -0.5$$

$x(-n+1) = \cos(\frac{2\pi}{6}n)$

$$1, 0.5, -0.5, -1, -0.5, 0.5$$

1.18.8

$$x(-3) = 0, x(-2) = 0, x(-1) = 0, x(0) = -1,$$
$$x(1) = -2, x(2) = -1, x(3) = 1$$

Shifted and scaled waveform samples

$$x(-3) = -1, x(-2) = -1, x(-1) = 2, x(0) = -1,$$
$$x(1) = -1, x(2) = 0, x(3) = 0$$

Chapter 2

2.1.9 Energy 4.
2.3.4

$$x_e(t) = \begin{cases} \frac{3}{2}|t|, & |t| < 1 \\ 0 & \text{otherwise} \end{cases}$$

$$x_o(t) = \begin{cases} \frac{3}{2}t, & -1 < t < 1 \\ 0 & \text{otherwise} \end{cases}$$

The integral of the odd component is zero. The integral of the even component is 1.5 and that of the signal is also 1.5.

2.4.4 3.

2.5.3

$$-1.2622, -1.4975, -1.5000, -1.5000, \text{ and } -1.5000$$

2.6.3 7.3891.

2.7.2

$$x(t) \approx \sum_{n=0}^{3} \cos(\frac{\pi}{6}(n)(1))\delta_q(t-(n)(1))(1)$$

$$x(t) \approx \sum_{n=0}^{7} \cos(\frac{\pi}{6}(n)(0.5))\delta_q(t-(n)(0.5))(0.5)$$

2.8.3

$$-6e^{-3t}u(t) + 2\delta(t)$$

2.9.4 0.

2.10.3

$$x(t) = -\frac{5}{\sqrt{2}}\cos(2\pi t) - \frac{5}{\sqrt{2}}\sin(2\pi t)$$

$$\frac{5}{8}, \frac{9}{8}, \frac{13}{8}$$

2.11.4

$$x(t) = \sqrt{2}\cos(\frac{2\pi}{6}t - \frac{\pi}{4})$$

$$\frac{9}{4}, \frac{21}{4}, \frac{33}{4}$$

2.12.3 $x(t) = 1.3483\cos(\frac{2\pi}{6}t - 2.9699)$.

$$1.5, -1.5, -3$$

$$2.8284, -1.0353, -3.8637$$

$$-1.3284, -0.4647, 0.8637$$

2.14.2

$$1, 2$$

The shift of $x(t)$ to the right by one second makes the positive peak of the shifted waveform, $\sin(\frac{2\pi}{6}t - \frac{\pi}{6})$, occurs after one second of the occurrence of that of the given sinusoid.

2.15.2

$$\frac{3}{4}, \frac{5}{4}$$

2.16.4

$$\frac{10}{3}, 10$$

2.17.5

$$x(-3) = 0, x(-2) = 0, x(-1) = 0, x(0) = -1.7321,$$
$$x(1) = -1.7321, x(2) = 0, x(3) = 1.7321$$

Shifted waveform

$$x(-3) = -1.7321, x(-2) = -1.7321, x(-1) = 0, x(0) = 1.7321,$$

$$x(1) = 1.7321, x(2) = 0, x(3) = -1.7321$$

Scaled and shifted waveform

$$x(-3) = 0, x(-2) = 0, x(-1) = -1.7321, x(0) = 1.7321,$$

$$x(1) = 0, x(2) = -1.7321, x(3) = 1.7321$$

Chapter 3

3.1.3

$$h(n) = 3\left(-\frac{1}{3}\right)^n u(n)$$

$$h(0) = 3, h(1) = -1, h(2) = \frac{1}{3}, h(3) = -\frac{1}{9}, h(4) = \frac{1}{27}, h(5) = -\frac{1}{81}$$

3.2.4

$$h(n) = 3\delta(n) + (-1)^n u(n),\ n = 0, 1, 2, \ldots$$

$$h(0) = 4, h(1) = -1, h(2) = 1, h(3) = -1, h(4) = 1, h(5) = -1$$

3.3.3 Linear.
3.4.3 Time-invariant.
3.5.3 $\{y(n), n = -1, 0, 1, 2, 3, 4, 5\} = \{6, 10, 13, 28, 19, 16, 16\}$.
3.6.5

$$y(n) = 0.9(1 - (0.6)^{n-2})u(n-3)$$

$$y(0) = 0, y(1) = 0, y(2) = 0, y(3) = 0.36, y(4) = 0.576, y(5) = 0.7056$$

3.10 $8.4276\cos(\frac{2\pi}{5}n + \frac{\pi}{4} - 0.9964)$.
$4.2138e^{j(\frac{2\pi}{5}n - 0.9964)}$.
3.13 The zero-state response is

$$y(n) = \frac{20}{9} + \frac{4}{3}n - \frac{20}{9}\left(\frac{1}{4}\right)^n$$

The zero-input response is

$$\left(\frac{1}{4}\right)^{(n+1)}$$

The complete response is

$$y(n) = \frac{20}{9} + \frac{4}{3}n - \frac{71}{36}\left(\frac{1}{4}\right)^n,\quad n = 0, 1, 2, \ldots$$

$$y(0) = 0.2500,\ y(1) = 3.0625,\ y(2) = 4.7656,$$
$$y(3) = 6.1914,\ y(4) = 7.5479,\ y(5) = 8.8870$$

The transient response is

$$-\frac{71}{36}\left(\frac{1}{4}\right)^n,\ n = 0, 1, 2, \ldots$$

The steady-state response is $(\frac{20}{9} + \frac{4}{3}n)u(n)$.

3.17.2 (i)

$$h(n) = \left(\frac{25}{3}\right)\delta(n) - 4(\frac{1}{5})^n - \left(\frac{7}{3}\right)\left(-\frac{3}{5}\right)^n$$

The first four values of $h(n)$ are

$$\{2, 0.6, -1, 0.472\}$$

(ii)

$$h(n) = \left(\frac{50}{3}\right)\delta(n) - \left(\frac{14}{3}\right)\left(-\frac{3}{5}\right)^n - 11\left(\frac{1}{5}\right)^n,\ n = 0, 1, 2, \ldots$$

The first four values of $h(n)$ are

$$\{1, 0.6, -2.12, 0.92\}$$

Chapter 4

4.1.9 Nonlinear.
4.2.4 Time-invariant.
4.3.4

$$y(t) = tu(t) - 2(t-3)u(t-3) + (t-6)u(t-6)$$

$$y(0) = 0,\ y(1) = 1,\ y(2) = 2,\ y(3) = 3,\ y(4) = 2,\ y(5) = 1$$

4.7.2

$$h(t) = 2\delta(t) + 5e^{t}u(t)$$

$$y(t) = (-3 + 5e^{t})u(t)$$

4.11 The zero-input response is $3e^{-t}u(t)$. The zero-state response is $(2\sin(t) - 2e^{-t})u(t)$. The complete response is

$$y(t) = (2\sin(t) + e^{-t})u(t)$$

The transient response is $e^{-t}u(t)$.
The steady-state response is $(2\sin(t))u(t)$.

4.14 $y(t) = 4\sin(\frac{2\pi}{6}t - \frac{\pi}{6} + 1.5247)$.
$y(t) = 2e^{j(\frac{2\pi}{6}t+1.5247)}$.

Chapter 5

5.1.3

$$x(n) = \frac{1}{4}\left(1 + 8\cos\left(\frac{2\pi}{4}n - \frac{\pi}{3}\right) - 3\cos(\pi n)\right)$$

5.2.2

$$\{x(0) = 0, x(1) = -3 + \sqrt{3}, x(2) = 2, x(3) = -3 - \sqrt{3}\}$$

$$\{X(0) = -4, X(1) = -2 - j2\sqrt{3}, X(2) = 8, X(3) = -2 + j2\sqrt{3}\}$$

5.3.4

$$\left\{x(0) = -2 - \frac{3}{\sqrt{2}}, x(1) = -6 + \frac{3}{\sqrt{2}}, x(2) = -2 + \frac{3}{\sqrt{2}}, x(3) = -6 - \frac{3}{\sqrt{2}}\right\}$$

5.5.1

$$\{x(0) = 2.25, x(1) = 0.25, x(2) = 0.25, x(3) = -1.75 - j1\}$$

5.6.2

$$X(k) = \{6 - j3, -j1, j11, -2 + j1\}$$

$$x(-14) = 2 + j2,\ x(43) = 1 - j4$$

$$X(12) = 6 - j3,\ X(-7) = -j1$$

5.9.2

$$\{208, 224, 208, 224\}$$

5.11.3

$$\{-4, -13, 24, -9\}$$

Chapter 6

6.2.4 $\omega_0 = 2$.

$$X_c(0) = \frac{3}{8}, X_c(1) = \frac{1}{2}, X_c(2) = \frac{1}{8}$$

$$X_p(0) = \frac{3}{8}, \quad X_p(1) = \frac{1}{2}, \theta(1) = 0, \quad X_p(2) = \frac{1}{8}, \theta(2) = 0$$

$$X_{cs}(0) = \frac{3}{8}, X_{cs}(\pm 1) = \frac{1}{4}, X_{cs}(\pm 2) = \frac{1}{16}$$

6.3.5 $\omega_0 = 2\pi$.

$$X_c(0) = 1, X_c(1) = \frac{\sqrt{3}}{2}, X_s(1) = \frac{1}{2}, X_c(3) = \sqrt{3}, X_s(3) = 1$$

$$X_p(0) = 1, X_p(1) = 1, \theta(1) = -\frac{\pi}{6}, X_p(3) = 2, \theta(3) = -\frac{\pi}{6}$$

$$X_{cs}(0) = 1, X_{cs}(1) = \frac{1}{4}(\sqrt{3} - j1), X_{cs}(3) = \frac{1}{2}(\sqrt{3} - j1),$$

$$X_{cs}(-1) = \frac{1}{4}(\sqrt{3} + j1), X_{cs}(-3) = \frac{1}{2}(\sqrt{3} + j1)$$

6.4.2 $\omega_0 = \frac{1}{63}$.

$$X_c(0) = 2, X_s(14) = -2, X_c(27) = -5$$

6.9.6 $X_{cs}(0) = \frac{2}{\pi}$ and $X_{cs}(k) = \frac{2}{\pi(1-4k^2)}$, $k \neq 0$.

6.11

$$x(t) = \frac{1}{2} - \frac{4}{\pi^2}\left(\cos\frac{\pi}{2}t + \frac{1}{9}\cos 3\frac{\pi}{2}t + \frac{1}{25}\cos 5\frac{\pi}{2}t \cdots\right)$$

$$x(t+2) = \frac{1}{2} + \frac{4}{\pi^2}\left(\cos\frac{\pi}{2}t + \frac{1}{9}\cos 3\frac{\pi}{2}t + \frac{1}{25}\cos 5\frac{\pi}{2}t \cdots\right)$$

$$3x(t) - 2 = -\frac{1}{2} - \frac{12}{\pi^2}\left(\cos\frac{\pi}{2}t + \frac{1}{9}\cos 3\frac{\pi}{2}t + \frac{1}{25}\cos 5\frac{\pi}{2}t \cdots\right)$$

$$\frac{\pi^2}{8} = 1 + \frac{1}{9} + \frac{1}{25} + \frac{1}{49} + \cdots$$

The DFT approximation of the trigonometric FS coefficients are $\{X_c(0) = \frac{1}{2}, X_c(1) = -\frac{1}{2}, X_c(2) = 0\}$.
The power of the signal is $\frac{1}{3}$.
The power of the signal, up to the third harmonic, is 0.3331.
The power of the signal, up to the fifth harmonic, is 0.3333.

6.15

$$y(t) = \frac{1}{\sqrt{5}}e^{j(2t+\tan^{-1}(\frac{-2}{1}))} + \frac{1}{\sqrt{10}}e^{j(3t+\tan^{-1}(\frac{-3}{1}))}$$

Chapter 7

7.2

$$X(e^{j\omega}) = \frac{\sin(\frac{2N+1}{2}\omega)}{\sin(\frac{\omega}{2})}$$

$X(e^{j0}) = 11$ and $X(e^{j\pi}) = -1$.

7.4

$$X(e^{j\omega}) = \frac{(a)e^{-j\omega}\sin(\omega_0)}{1 - 2(a)e^{-j\omega}\cos(\omega_0) + (a)^2 e^{-j2\omega}}$$

$X(e^{j0}) = 0.7/1.49$ and $X(e^{j\pi}) = -0.7/1.49$.

7.9.5

$$X(e^{j\omega}) = \frac{0.36e^{-j2\omega}}{1 - 0.6e^{-j\omega}}$$

7.10.2 $\{X(e^{j0}) = 10(\frac{2\pi}{4})\delta(\omega), X(e^{j\frac{2\pi}{4}}) = (2 + j2)(\frac{2\pi}{4})\delta(\omega - \frac{2\pi}{4}), X(e^{j2\frac{2\pi}{4}}) = 2(\frac{2\pi}{4})\delta(\omega - 2\frac{2\pi}{4}), X(e^{j3\frac{2\pi}{4}}) = (2 - j2)(\frac{2\pi}{4})\delta(\omega - 3\frac{2\pi}{4})\}$.

7.16.2 $2(0.5)^n u(n) - (0.25)^n u(n)$.

7.21.1 $y(n) = \{x(n) * h(n),\ n = 1, 2, \ldots, 6\} = \{-2, 1, -8, -4, -5, -12\}$.

7.24.2 $\frac{1}{1-(0.6)e^{j4\omega}}, \quad -\pi < \omega \leq \pi$.

7.26.4 $Y(e^{j\omega}) = \frac{1}{(1-0.6e^{-j\omega})(1-e^{-j\omega})} + 2.5\pi\delta(\omega)$.

7.29 The IDFT values are {1.0082, 0.3024, 0.0907, 0.0272}.
The exact values of $x(n)$ are $\{x(0) = 1, x(1) = 0.3, x(2) = 0.09, x(3) = 0.027\}$.

7.31

$$h(n) = \left(\frac{56}{5}\left(-\frac{2}{3}\right)^n - \frac{51}{5}\left(-\frac{1}{4}\right)^n\right)u(n)$$

The first four values of the impulse response are 1, −4.9167, 4.3403, −3.1591.

7.35

$$y(n) = 55\left(\frac{1}{2}\right)^n - 90\left(\frac{1}{3}\right)^n + 38\left(\frac{1}{5}\right)^n, \quad n = 0, 1, \ldots$$

7.36.2 $x_H(n) = -0.5\sin(2\pi n/6)$.

Chapter 8

8.7

$$X(j\omega) = \frac{1}{(2 + j\omega)^2}, \quad X(j0) = \frac{1}{4}$$

8.14.2

$$X(j\omega) = \begin{cases} 2\pi & \text{for } |\omega| < 3 \\ 0 & \text{for } |\omega| > 3 \end{cases}$$

8.15.3

$$X(j\omega) = \frac{\sin((\omega-\omega_0)a)}{(\omega-\omega_0)} + \frac{\sin((\omega+\omega_0)a)}{(\omega+\omega_0)}$$

8.16

$$x(t) = \begin{cases} -2 \text{ for } t < 0 \\ (t-2) \text{ for } 0 < t < 4 \\ 2 \text{ for } t > 4 \end{cases}$$

8.18.2

$$\frac{-j\pi}{2}(\delta(\omega-\omega_0) - \delta(\omega+\omega_0)) - \frac{\omega_0}{(\omega^2-\omega_0^2)}$$

8.20.9

$$\pi\delta(\omega) - \frac{1}{j\omega}$$

8.22.4

$$X(j\omega) = \frac{-1 + e^{-j\omega} + e^{-j2\omega} - e^{-j3\omega}}{\omega^2}$$

8.23.6

$$Y(j\omega) = \pi\delta(\omega) + \frac{1}{j\omega} - \frac{1}{1+j\omega}$$

8.24.4

$$\frac{(1+j\omega)e^{-j\omega} - 1}{\omega^2}$$

8.28.3

$$\left\{X(0) = -1, X(3) = \frac{j3}{2}, X(-3) = -\frac{j3}{2}, X(5) = 1, X(-5) = 1, X(-7) = 6\right\}$$

$$X(j\omega) = \pi(-2\delta(\omega) + j3(\delta(\omega-3) - \delta(\omega+3)) + 2(\delta(\omega-5) + \delta(\omega+5)) + 12\delta(\omega+7))$$

8.29.4

$$X_{cs}(k) = \frac{j}{k\pi},\ k \neq 0 \quad \text{and} \quad X_{cs}(0) = 1$$

8.31.2

$$X(j\omega) = \frac{2}{1+\omega^2} \quad \text{and} \quad X_s(j\omega) = \frac{1}{T_s}\sum_{k=-\infty}^{\infty} \frac{2}{1+(\omega - k\omega_s)^2}, \; \omega_s = \frac{2\pi}{T_s}$$

$$X(j0) = 2, \quad X_s(j0) = 200.0017, \; T_s = 0.01, \quad X_s(j0) = 20.0167, \; T_s = 0.1,$$

$$X_s(j0) = 2.1640, \; T_s = 1, \quad X_s(j0) = 1.0001, \; T_s = 10$$

8.33 The exact values of the FT are $X(j0) = 1$ and $X(j\pi) = \frac{4}{\pi^2} = 0.4053$. The four samples of the signal are $\{x(0) = 1, x(1) = 0.5, x(2) = 0, x(3) = 0.5\}$ and the DFT is $\{X(0) = 2, X(1) = 1, X(2) = 0, X(3) = 1\}$. As the sampling interval is 0.5 second, the first two samples of the spectrum obtained by the DFT are $0.5\{2, 1\} = \{1, 0.5\}$.

Chapter 9

9.1.3 $X(z) = 2 - 4z^{-3}$.
9.2.4 $\{x(0) = 1, x(1) = 1, x(2) = -1\}$.
9.3.4

$$X(z) = \left(\frac{-2z^2 + 3z}{(z-1)^2}\right)$$

9.4.1 The nonzero values of $y(n)$ are $\{y(1) = 4, y(3) = -2, y(5) = -16, y(7) = 8\}$.
9.5.2

$$X(z) = \frac{2z}{(z-2)^2}$$

9.6.2

$$X(z) = \frac{4z}{(z-4)^2}$$

9.7.2 $y(n) = (0.5 + 0.5\sin(\frac{2\pi}{4}n) - 0.5\cos(\frac{2\pi}{4}n))u(n)$.
9.8.3 $x(0) = 2$. $x(\infty) = 16$.
9.9.4

$$X(z) = \frac{z(z+1)}{(z^2+1)}$$

9.13

$$x(n) = (\sqrt{2})^{n+1} \cos\left(\frac{3\pi}{4}n - \frac{\pi}{4}\right) u(n)$$

$$x(0) = 1, \quad x(1) = 0, \quad x(2) = -2, \quad x(3) = 4$$

9.19.2 $y(n) = (0.8192)(3)\cos(\frac{2\pi}{4}n - \frac{\pi}{3} - 0.6107)u(n)$.

9.21

$$y(n) = -\frac{16}{21}(-1)^n + \frac{325}{56}\left(\frac{3}{4}\right)^n - \frac{23}{12}\left(\frac{1}{2}\right)^n, \ n = 0, 1, 2, \ldots$$

The first four values of $y(n)$ are

$$\{3.1250, 4.1563, 2.0234, 2.9707\}$$

The zero-input response is

$$\frac{27}{8}\left(\frac{3}{4}\right)^n - \frac{5}{4}\left(\frac{1}{2}\right)^n$$

The zero-state response is

$$-\frac{16}{21}(-1)^n + \frac{17}{7}\left(\frac{3}{4}\right)^n - \frac{2}{3}\left(\frac{1}{2}\right)^n$$

The transient response is

$$\frac{325}{56}\left(\frac{3}{4}\right)^n - \frac{23}{12}\left(\frac{1}{2}\right)^n$$

The steady-state response is $-\frac{16}{21}(-1)^n u(n)$.

9.25.1

$$h(n) = 12\delta(n) - \left(6\left(\frac{1}{4}\right)^n\right)u(n)$$

The first four values of the impulse response are

$$\{6, -1.5, -0.3750, -0.0938\}$$

9.26.3

$$h(n) = -10\delta(n) + \left(7\left(\frac{1}{3}\right)^n + 7\left(\frac{1}{2}\right)^n\right)u(n)$$

The first four values of the impulse response are

$$\{4, 5.8333, 2.5278, 1.1343\}$$

Chapter 10

10.2.4

$$X(s) = \left(\frac{e^{-2s}}{s^2} + \frac{2e^{-2s}}{s}\right)$$

10.5.2 The poles of $X(s)$ are located at $s = -1$ and $s = -2$. The zero is located at $s = 1$. $x(t) = (-2e^{-t} + 3e^{-2t})u(t)$.
The transform of the scaled signal is

$$\frac{s-2}{(s+2)(s+4)}$$

The poles are located at $s = -2$ and $s = -4$. The zero is located at $s = 2$. $x(at) = (-2e^{-2t} + 3e^{-4t})u(t)$.

10.6.5

$$y(t) = \frac{1}{2}\left(t^2e^{-t}\right)u(t)$$

10.8.4 $x(0^+) = 3$. $x(\infty) = 1$.

10.9.3

$$X(s) = \frac{(1 - e^{-2s})}{s^2(1 + e^{-2s})}$$

10.15

$$x(t) = \left(2t - 1 + e^{-t}\right)u(t)$$

10.18.3

$$h(t) = (-0.5e^t + 0.5e^{3t})u(t)$$

10.20

$$y(t) = (\overbrace{e^{-2t} + te^{-t} - e^{-t}}^{\text{zero-state}} + \overbrace{te^{-t} + 3e^{-t}}^{\text{zero-input}})u(t)$$
$$= (e^{-2t} + 2te^{-t} + 2e^{-t})u(t)$$

The steady-state response is $e^{-2t}u(t)$ and the transient response is $(2te^{-t} + 2e^{-t})u(t)$. The initial and final values of $y(t)$ are 3 and 0 , respectively. The initial and final values of the zero-state response are 0 and 0 , respectively.

10.23.1 $y(t) = 3\sqrt{2}\cos(0.5t - \frac{\pi}{3} - \frac{\pi}{4})u(t)$.

10.24.2

$$h(t) = \left(e^{t} - 1\right)u(t)$$

10.25.3

$$h(t) = 5\delta(t) - \left(7e^{-3t} + 4e^{-2t}\right)u(t)$$

10.29 $v(t) = \frac{1}{8}e^{-\frac{1}{8}t}u(t)$.

Chapter 11

11.3.1

$$h(n) = \delta(n) + \left(\frac{28}{3}\left(-\frac{1}{3}\right)^{n-1} - \frac{40}{3}\left(-\frac{2}{3}\right)^{n-1}\right)u(n-1),\ n = 0, 1, 2, \ldots$$

$$h(0) = 1, \quad h(1) = -4, \quad h(2) = \frac{52}{9}, \quad h(3) = -\frac{44}{9}$$

11.4.2 The zero-input component of the state-vector is given by

$$\begin{bmatrix} -\frac{1}{8}(-\frac{1}{2})^n \\ \frac{1}{4}(-\frac{1}{2})^n \end{bmatrix}$$

The zero-input response is given by

$$\left(\frac{1}{2}\left(-\frac{1}{2}\right)^n\right)u(n)$$

The first four values of the zero-input response $y(n)$ are

$$y(0) = \frac{1}{2}, \quad y(1) = -\frac{1}{4}, \quad y(2) = \frac{1}{8}, \quad y(3) = -\frac{1}{16}$$

The zero-state component of the state-vector is given by

$$\begin{bmatrix} \frac{2}{3}\left(1-(-\frac{1}{2})^n\right)+\left(-\frac{2}{9}+\frac{2}{9}(-\frac{1}{2})^n-\frac{2}{3}n(-\frac{1}{2})^n\right) \\ -2\left(-\frac{2}{9}+\frac{2}{9}(-\frac{1}{2})^n-\frac{2}{3}n(-\frac{1}{2})^n\right) \end{bmatrix} u(n)$$

The zero-state response is given by

$$\left(\frac{8}{9}+\frac{10}{9}\left(-\frac{1}{2}\right)^n+\frac{8}{3}n\left(-\frac{1}{2}\right)^n\right)u(n)$$

The first four values of the zero-state response $y(n)$ are

$$y(0) = 2, \quad y(1) = -1, \quad y(2) = \frac{5}{2}, \quad y(3) = -\frac{1}{4}$$

The total response is

$$y(n) = \left(\frac{8}{9}+\frac{29}{18}\left(-\frac{1}{2}\right)^n+\frac{8}{3}n\left(-\frac{1}{2}\right)^n\right)u(n), \quad n = 0, 1, 2, \ldots$$

The first four values of the total response $y(n)$ are

$$y(0) = \frac{5}{2}, \quad y(1) = -\frac{5}{4}, \quad y(2) = \frac{21}{8}, \quad y(3) = -\frac{5}{16}$$

11.5.3

$$h(n) = \left(9\delta(n) - 7\left(-\frac{1}{3}\right)^n + 8n\left(-\frac{1}{3}\right)^n\right)u(n), \ n = 0, 1, 2, \ldots$$

The first four values of the sequence $h(n)$ are

$$h(0) = 2, \quad h(1) = -\frac{1}{3}, \quad h(2) = 1, \quad h(3) = -\frac{17}{27}$$

11.6.2 The zero-input component of the state vector is

$$\boldsymbol{q}(n) = \begin{bmatrix} -\frac{1}{4}n\left(-\frac{1}{2}\right)^{n-1} + \frac{4}{3}\left(-\frac{1}{2}\right)^{n} \\ \frac{1}{2}n\left(-\frac{1}{2}\right)^{n-1} - \frac{5}{3}\left(-\frac{1}{2}\right)^{n} \end{bmatrix} u(n)$$

The zero-input response is given by

$$\left(-\frac{3}{4}n\left(-\frac{1}{2}\right)^n-\frac{5}{4}\left(-\frac{1}{2}\right)^n\right)u(n)$$

The first four values of the zero-input response $y(n)$ are

$$y(0)=-\frac{5}{4},\quad y(1)=1,\quad y(2)=-\frac{11}{16},\quad y(3)=\frac{7}{16}$$

The zero-state component of the state vector is

$$\boldsymbol{q}(n)=\begin{bmatrix}\frac{3}{5}n(-\frac{1}{2})^{n-1}-\frac{12}{25}\left(-\frac{1}{2}\right)^n+\frac{12}{25}\left(\frac{1}{3}\right)^n\\ -\frac{6}{5}n\left(-\frac{1}{2}\right)^{n-1}-\frac{36}{25}\left(-\frac{1}{2}\right)^n+\frac{36}{25}\left(\frac{1}{3}\right)^n\end{bmatrix}u(n)$$

The zero-state response is given by

$$\left(\frac{9}{5}n\left(-\frac{1}{2}\right)^n-\frac{27}{25}\left(-\frac{1}{2}\right)^n+\frac{52}{25}\left(\frac{1}{3}\right)^n\right)u(n)$$

The first four values of the zero-state response $y(n)$ are

$$y(0)=1,\quad y(1)=\frac{1}{3},\quad y(2)=\frac{31}{36},\quad y(3)=-\frac{25}{54}$$

The total response is

$$y(n)=(\frac{21}{20}n(-\frac{1}{2})^n-\frac{233}{100}(-\frac{1}{2})^n+\frac{52}{25}(\frac{1}{3})^n)u(n),\quad n=0,1,2,\ldots$$

The first four values of the total response $y(n)$ are

$$y(0)=-\frac{1}{4},\quad y(1)=\frac{4}{3},\quad y(2)=\frac{25}{144},\quad y(3)=-\frac{11}{432}$$

Chapter 12

12.4

$$\boldsymbol{A}=\begin{bmatrix}-2&-1\\1&0\end{bmatrix},\quad \boldsymbol{B}=\begin{bmatrix}1\\0\end{bmatrix},\boldsymbol{C}=\begin{bmatrix}2\ 0\end{bmatrix},\quad D=0$$

$$(-4t^2e^{-t} + 8te^{-t})u(t)$$

12.6 The zero-input component of the output is given by

$$5e^{-t} - 2e^{-3t}$$

The zero-state component of the output is given by

$$9e^{-t} - 33e^{-2t} + 27e^{-3t}$$

The total response of the system is

$$y(t) = (14e^{-t} - 33e^{-2t} + 25e^{-3t})u(t)$$

12.9

$$h(t) = (-3e^{-\frac{1}{2}t} + 4e^{-\frac{1}{3}t})u(t)$$

12.11 The zero-input component of the output is given by

$$7e^{-t} - 5e^{-2t}$$

The zero-state component of the output is given by

$$6 - 24e^{-t} + 27e^{-2t}$$

The total response of the system is

$$y(t) = (6 - 17e^{-t} + 22e^{-2t})u(t)$$

12.15

$$h(t) = (-2\delta(t) + 7e^{-t} - 9te^{-t})u(t)$$

Bibliography

E.A. Guillemin, *The Mathematics of Circuit Analysis* (Wiley, New York, 1959)
E.A. Guillemin, *Theory of Linear Physical Systems* (Wiley, New York, 1963)
S. Haykin, B. Van Veen, *Signals and Systems* (Wiley, New York, 2003)
B.P. Lathi, *Linear Systems and Signals* (Oxford University Press, New York, 2005)
D. Sundararajan, *Signals and Systems – A Practical Approach* (Wiley, Singapore, 2008)
D. Sundararajan, *Discrete Wavelet Transform, A Signal Processing Approach* (Wiley, Singapore, 2015)
D. Sundararajan, *Control Systems - An Introduction* (Springer, Switzerland, 2022)
The Mathworks, *Matlab Signal Processing Tool Box User's Guide* (The Mathworks, Inc., USA, 2021)

D. Sundararajan, *Signals and Systems*,
https://doi.org/10.1007/978-3-031-19377-4

Index

D. Sundararajan, *Signals and Systems*,
https://doi.org/10.1007/978-3-031-19377-4

Printed in the United States
by Baker & Taylor Publisher Services